# INNOVATIONS 2004

*World Innovations in Engineering Education and Research*

# INNOVATIONS 2004

## *World Innovations in Engineering Education and Research*

*Edited by*


**Win Aung**
*International Network for Engineering Education and Research, Potomac, MD, USA*
*National Science Foundation, Arlington, VA, USA*

**Robert Altenkirch**
*New Jersey Institute of Technology*
*Newark, New Jersey, USA*

**Tomas Cermak**
*VSB – Technical University of Ostrava*
*Ostrava, Czech Republic*

**Robin W. King**
*University of South Australia*
*Mawson Lakes, Australia*

**Luis Manuel Sanchez Ruiz**
*Universidad Politécnica de Valencia*
*Valencia, Spain*



International Network for Engineering Education & Research
iNEER

BEGELL HOUSE PUBLISHING

**INNOVATIONS 2004: World Innovations in Engineering Education and Research**



**Library of Congress Cataloging-in-Publication Data**

Innovations 2004 : world innovations in engineering education and research / edited by Win Aung ... [et al.].
p. cm.
Includes bibliographical references and index.
ISBN 0-9741252-1-0
1. Engineering--Study and teaching. I. Aung, W.
T65.3.I48 2004
620'.0071'1--dc22

2004009784

# CONTENTS

# PREFACE

Covering topics ranging from curriculum changes to new, more effective approaches for teaching and learning, the forty chapters in this book showcase recent and ongoing developments in engineering education in many parts of the world. These articles show the trend towards adoption of learning-based pedagogical approaches. There is a broad recognition about the need to improve the quality of engineering education. New developments are being reported that deal with team-based as well as self-directed learning. Attention to outcomes assessment of education innovations is now common in the world's leading institutions.

This volume starts with a guest commentary by Bordogna (Chapter 1), a leader of science and engineering education and research in the U.S., who writes from the U.S. perspective. He states that science and engineering is inherently international, but that he also believes in the need to recruit into engineering more of the native talents. To make this happen, he says, we must render engineering careers more appealing to the young generation and this will require a new model of engineering education, in which educators must understand how to produce engineers who are innovators, master integrators, holistic designers, change agents, and more.

Writing in the context of the nation of Colombia, Caro (2) outlines the challenges and opportunities facing the civil engineering profession in that country. She presents the current situation and the needs, including the actions needed to produce a modern civil engineering program in Colombia.

Focusing on a new interactive knowledge model, Bourgeois and Badillo (3) of France examine the implications of information technology in knowledge transmission, and conclude that the role of the instructor – the "pedagogue" – is crucial. In Israel, Waks and Merdler (4) explore aspects related to student creativity and conclude that both originality and innovation depend on the integration of disciplinary and interdisciplinary knowledge.

Several chapters in the book discuss new education programs and activities at specific institutions, and reflect the manner in which these institutions have responded to emerging opportunities in engineering education arising from new technological development. New education and research programs include that in NeuroEngineering at the University of California, Los Angeles (5), and in communications engineering education in Taiwan (6)(7). In Australia, four universities in the inner Melbourne area have collaborated to develop a new Master's degree program in microelectronics engineering (8).

To improve learning, student retention of prior course materials is important, and this can be facilitated through web-based instruction, as shown by Schmahl (9) at Miami University in Ohio, USA. A remotely controlled heat exchanger experiment has been implemented by Colton (10) at the Massachusetts Institute of Technology and student responses have been assessed. From the University of Auckland in New Zealand comes a report by Tedford et al. (11) concerning a multimedia-based learning system that delivers immersive projects via a virtual company environment. Online delivery can be used also to improve distance communication and virtual teamwork skills of students, as demonstrated by Bohemia (12) of the University of Western Sydney in Australia, who discusses the online delivery of a course on design management.

Often, large lecture classes result when two or more classes are combined into one, and the large number of students in a large room often inhibits students from participating actively in the lectures. A solution adopted in Norway, as described by Sandnes and Talberg (13), is to have students in a class use mobile telephones to dial into a server and keying in questions and comments, which are then shown to the class and addressed by the instructor by projecting on screens located around the room. In Singapore, to encourage student attendance in large classes, instructors are presenting well-balanced course materials tailored to learners' performance that is monitored during and after a class by adaptive feedback (14). The problem of scheduling hands-on laboratory experiments for large classes is eased by using online access to remote, physical laboratory equipment, as is done in the U.S. (15) and in Norway (16).

Some of the recent education innovations are aimed at encouraging interactive learning, creative problem-solving and hands-on manipulation. In Europe, nine educational institutions in eight countries have collaborated to develop eMerge (17), an advance educational network infrastructure that permits dissemination of online laboratory experiments. In addition to remote manipulation of physical experiments, the use of technology in education includes the virtual laboratory concept. Here, Hutchinson and Kuester (18) at the University of California Irvine uses visualization and display technologies to implement a framework that facilitates 3-dimensional spatial learning in a specially designed lecture room. In addition, the LabVIEW platform has been applied by Holbert et al. (19) to provide flexible scheduling and remote access for experimentation in an electrical engineering laboratory, allowing students in the U.S. and Romania to share equipment such as a spectrum analyzer. Bachiller et al. (20) of the Universidad Politécnica de Valencia in Spain describe an interactive multimedia course in which students are provided with a self-learning tool for communications laboratory equipment.

In another course at the same university, Cerdá et al. (21) deals with the integration of theory and laboratory classes in circuit and electronic system design.

Gerecke et al. (22) at the University of Hannover, Germany, have developed a modular educational robotic toolbox that incorporates ease of experimentation with the flexibility to work on multiple robot platforms. At the Technical University of Catalonia in Spain, Sánchez Robert (23) has applied the cooperative learning method in a digital electronics course that involves the learning of theory, problem solution, project design, portfolio preparation and the use of examinations to assess the progress of students.

The focus on ABET Engineering Criteria 2000, as exemplified by the work of Sharma et al. (24), has led to a widening emphasis on learner-centered pedagogy and acceptance of learning assessment as an integral part of new education development. This is illustrated by the recent work of Saliceti-Piazza et al. in Puerto Rico (25), Aziz in Australia (26), Berg et al. in the U.S. (27), Gillet in Switzerland (28), and Böhne et al. in Germany (29).

Convinced that laboratory classes can be used to inculcate student learning of scientific concept as well as organizational skills, Montero et al. in Spain (30) have identified eight characteristics of learning that can be applied in class. A similar emphasis on learning strategies is featured in the work by Duque et al. (31) at the Universidad de los Andes in Colombia, in which problem-based learning, teaching-for-understanding, virtual peer evaluation, and hands-on laboratory classes have been employed in automatic control courses. Concerning interaction with industry, Farrell et al. (32) report from Rowan University in the U.S. that all undergraduate engineering students at their institution are required to take a sequence known as Engineering Clinics in which students work in multi-disciplinary teams on semester-long or year-long projects sponsored by industry.

With a view toward developing more effective teaching tools, a team comprising of educators from the U.S. and Iran, led by Eydgahi (33) of the University of Maryland Eastern Shore, has designed and implemented a MATLAB-based toolbox with a graphical user interface applicable to computing reduced order models of a large system. A collaborative effort among educators in LICEF Research Center and Télé-université in Canada, and the Research Unit PRINCE, Faculty of Science of Tunis in Tunisia (34), has led to development of a graphic system for use in electrical engineering and computer science education and research that simplifies the tasks of verification and execution of Grafcet, the French/European standard graphical tool, without the need of a real programmable logic controller or an operative part. In addressing the difficulties associated with adopting technical language and terminology in non-English speaking countries, Jian and Sandnes (35) conclude that students are a valuable and appropriate resource for discovering suitable terminology. In a chapter dealing with an improved pedagogical approach in dynamic programming, Iqbal and Alvi (36) use typical optimization problems to demonstrate a method that enables students to build graphical templates that take advantage of the power of visual processing.

Racial and gender diversity and inclusiveness, and outreach to elementary, middle and high school have been a hallmark of much of the recent education developments in the U.S. From Florida Atlantic University, Hsu et al. (37) highlight a summer enrichment program that has been used in part to encourage gifted and talented high school students to pursue careers in engineering. At the University of Puerto Rico at Humacao, the attention of Ramos and co-workers (38) is directed at the roles of undergraduate research and international cooperation in promoting the participation of women in research and education, while at Oslo University College in Norway, the issue of student diversity, relating to WIB students, that is, students With Immigrant Background, is addressed by Torvatn and Sandnes (39).

To conclude this volume, Grüter et al. (40) address the high failure rates and lack of student motivation and capabilities in applying mathematics to real world problems. Working from Germany's Hochschule Bremen, University of Allied Sciences, they present teleVISE, a web-based tutoring system for mathematical modeling and reasoning.

In selecting the above chapters, we reviewed 68 articles authored by 214 individuals. A total of 249 reviewers from 58 countries helped with peer review. We appreciate the contributions made by all of the authors, including those whose articles we could not accept for one reason or another, including page limitation imposed on this volume. Through their in-depth comments, reviewers have contributed immensely to this publication. We have enjoyed and benefited from our interaction with authors and reviewers alike. To all of them, we do herewith once again extend our heart-felt thanks.

We also thank Robert Aung and Linnea Hasegawa without whose professional help the publication of this volume would not have been possible.

Last but not least, we thank our sponsors, including our industrial sponsors: Wayne Johnson of Hewlett Packard Company; Randy Hinrichs of Microsoft Corporation; and William Begell of Begell House Publishers, for their friendship, advice and continued support.

**iNEER Board of Editors**

**Win Aung**
iNEER, Arlington, VA, USA
**Robert Altenkirch**
New Jersey Institute of Technology, Newark, New Jersey, USA
**Tomas Cermak**
VSB – Technical University of Ostrava, Ostrava, Czech Republic
**Robin King**
University of South Australia, Adelaide, Australia
**Luis Manuel Sanchez Ruiz**
Polytechnic University of Valencia, Valencia, Spain

April 16, 2004

# BOARD OF EDITORS

**WIN AUNG**
In addition to being Secretary-General of iNEER, Win Aung is affiliated with the U.S. National Science Foundation (NSF) where he holds the position of Senior Staff Associate – Engineering Education. Prior to joining NSF, he carried out research at Bell Telephone Laboratories in Whippany, NJ. A member of the U.S. Senior Executive Service and the recipient of the first NSF Federal Engineer of the Year Award, he is a Fellow of American Society of Mechanical Engineers (ASME). He has written extensively on research and education and is a frequent speaker on campuses around the world and at international conferences. He has over 120 technical publications, and has edited or co-edited 9 books. In 1999, he received the ASME Classic Paper Award, and the *Doctorem Honoris Causa* (honorary doctorate degree) from VSB – Technical University of Ostrava (VSB), Czech Republic. He is the founder of the International Conference on Engineering Education series and is a co-founder of the International Network for Engineering Education and Research (iNEER). He is a member of the Scientific Board of VSB. He is a member of the ASME Board on Education, and was an editor of the Transactions of ASME, Journal of Heat Transfer.

**ROBERT ALTENKIRCH**
Robert Altenkirch is President of New Jersey Institute of Technology, Newark, New Jersey, USA. Prior to that he was Vice President for Research at Mississippi State University (MSU), and served as Dean of College of Engineering and Architecture at Washington State University, and Dean of Engineering at MSU. A mechanical engineer, Altenkirch received the Ph.D. from Purdue, and has hundreds of publications and presentations to his credit. Altenkirch is a fellow of ASME, a member of Phi Eta Sigma, Tau Beta Pi, Sigma Xi, ASEE and the iNEER Board.

**TOMAS CERMAK**
Professor Tomas Cermak was elected as Rector of VSB – Technical University of Ostrava, Ostrava, Czech Republic in elections held in November 2002. He assumed the position on February 1, 2003 succeeding Vaclav Roubicek. Cermak first came to VSB in 1964 as a senior assistant, rising to become Head of the Department of Electrical Machines and Drives at VSB in 1985. In 1990 he became Rector, which position he held until 1997. In 1997, he was appointed as Vice Rector for R&D and Foreign Affairs. His service record includes membership of the Scientific Board of VSB and of the State Committee for Scientific Degrees. He chairs the Engineering Division of the Grant Agency of the Czech Republic. He was past member of the Scientific Board of the Technical University of Kosice, Slovak Republic; the Czech Engineering Academy; Vice President of Czech Universities Council; and General Assembly of Czech Academy of

Science. An early believer of the value of international cooperation in engineering education and research, he was actively involved with the organization of ICEE since ICEE-1997. He is the current Chair of the iNEER Board, and is a member of the ICEE and iCEER International Steering Committees. He is Chairman of the Board of Vesuvius (formerly Hinckley) Slavia Group. He is also Chairman and Vice Chairman of the Supervisory Boards of OZOBEL and Vitkovice Steel Company, respectively. He teaches courses in control of electrical drives and machines; power energy systems, among others. He has written 9 textbooks, and published 50 papers in journals and conference proceedings, with another 35 research papers based on industrial collaboration. He holds 4 patents.

**ROBIN W. KING**

Robin King is Pro Vice Chancellor, Division of Information Technology, Engineering & the Environment at the University of South Australia, Mawson Lakes, Australia. He received his Ph.D. in Electronic Engineering from Imperial College, University of London in 1972. He was a research engineer with the British Broadcasting Corporation from 1971 – 1974, and has held several academic positions in Papua New Guinea, UK and Australia. As an Associate Professor at the University of Sydney from 1989 – 1996 he led the speech technology research group, and was Associate Dean (undergraduate). He has more than 80 publications, and has been involved with a number of progressive engineering education initiatives, and with the Australasian Association for Engineering Education. In his current position he is responsible for the management and strategic development of a wide range of academic and research programs that span information technology and engineering, the built and natural environment and mathematics. He is a member of the Australian Council of Engineering Deans, and continues to be involved with accreditation of Australian engineering degrees through his work with the Institution of Engineers, Australia.

**LUIS MANUEL SÁNCHEZ RUIZ**

Luis Manuel Sánchez Ruiz has been affiliated with Universidad Politécnica de Valencia (UPV), Spain since 1980 and has been Professor of Mathematics since 2000. He graduated and received his Ph.D. from Universidad de Valencia in 1980 and 1988, respectively, and was a Visiting Professor at the University of Florida, Gainesville, FL, USA on several occasions during 1992-99. His current research interests include functional analysis from both theoretical and applied points of view. He has published over 80 papers in scientific journals and conference proceedings and more than 10 textbooks on mathematics for engineers, and is co-author of the research monograph *Metrizable Barrelled Spaces* published by Longman. The former Academic Coordinator of Mediterranean University of Science and Technology, he is responsible for several research projects granted by the Spanish Ministry of Education. He is currently a reviewer for several international journals, publishing companies and research projects, as well as a member of the Editorial Board of Scientiae Mathematicae Japonicae and of the online Scientiae Mathematicae. He is a member of the International Steering Committee of International Conference on Engineering Education and was General Chair of ICEE-2003 held in Valencia.

# REVIEWERS

A.A. Burluka, *United Kingdom*
A. Betâmio de Almeida, *Portugal*
A. Haji-Sheikh, *USA*
A.Y. Montgomery, *Australia*
A.A. Kassim, *Singapore*
Abbas El-Zein, *Lebanon*
Adam Crawford, *United Kingdom*
Adam J.-L. Hou, *Taiwan, ROC*
Adam Moore, *United Kingdom*
Adnan A. Basma, *U. Arab Emirates*
Adnan Akay, *USA*
Adrian Bastiani, *Australia*
Alan McPhail, *Australia*
Alan V. Sahakian, *USA*
Alcínia Zita Sampaio, *Portugal*
Alessandro La Neve, *Brazil*
Alex Hariz, *Australia*
Alexander Mamishev, *USA*
Amine Berqia, *Portugal*
Amit K. Roy Chowdhury, *USA*
Anastasia S. Safigianni, *Greece*
Andrea Samolejova, *Czech Republic*
Andreas Schlachetzki, *Germany*
Andrew T. Metters, *USA*
Andy Koronios, *Australia*
Angelo J. Perna, *USA*
Antonio D. de Figueiredo, *Portugal*
Antonios Zavaliangos, *USA*
Arne Gjengedal, *Norway*
Ata Kaban, *United Kingdom*
Avinash K. Sinha, *India*
Azizan Zainal Abidin, *Malaysia*
Barry J. Farbrother, *USA*
Bart Massey, *USA*
Belén García Mora, *Spain*
Bernabé Marí, *Spain*
Betâmio de Almeida, *Portugal*
Beverly J. Davis, *USA*
Brandon Muramatsu, *USA*
Brenton Dansie, *Australia*
Brian Lithgow, *Australia*
C.-K. Lee, *Taiwan,, ROC*
Carlo Cattani, *Italy*
Carlos Gonzalia, *Sweden*
Charles Nippert, *USA*
Charles R. Glagola, *USA*
Charles-Antoine Brunet, *Canada*
Chin-Hsiang Cheng, *Taiwan, ROC*
Chris Smaill, *New Zealand*
Clara Amelia de Oliveira, *Brazil*
Corrado Poli, *USA*
Craig Mcdonald, *Australia*
D.C. Levy, *Australia*
D.D. Sworder, *USA*
Damián Ginestar Peiró, *Spain*
David Airey, *Australia*
David Dowling, *Australia*
David J. Ahlgren, *USA*
David Lowe, *Australia*
David M. Panton, *Australia*
David M. Holifield, *United Kingdom*
Denis Gracanin, *USA*
Des Jenkinson, *United Kingdom*
Dominique Bourgeois, *France*
Donald R. Falkenburg, *USA*
Donald Senich, *USA*

Doug Hargreaves, *Australia*
Drago Hercog, *Slovenia*
Edward E. Anderson, *USA*
Eli Fromm, *USA*
Elijah Kannatey-Asibu, *USA*
Elizabeth M. Laws, *United Kingdom*
Ernst L. Leiss, *USA*
Fiona Underwood, *Australia*
Frances L. Van Scoy, *USA*
Gary E. Ford, *USA*
Gavin Melles, *Australia*
George M. Papadourakis, *Greece*
Geraldine B. Milano, *USA*
Gilles Y. Delisle, *Canada*
Gloria M. Rogers, *USA*
Guy Gauthier, *Canada*
H.E. Hanrahan, *South Africa*
Harald Haugen, *Norway*
Henry N. Jerez, *USA*
Herman Mann, *Czech Republic*
Hong Wu, *Norway*
Howard Kimmel, *USA*
Hsu Chun-liang, *Taiwan, ROC*
Hugo Fuks, *Brazil*
Ian Benest, *United Kingdom*
Ian F.C. Smith, *Switzerland*
Idalia Ramos, *Puerto Rico*
Igor M. Verner, *Israel*
Ingvar Gustavsson, *Sweden*
Ishiang Shih, *Canada*
J.A. Gilles Doiron, *Singapore*
J. Keith Nisbett, *USA*
J.N. Bandyopadhyay, *India*
Jaime Riera, *Spain*
James A Newell, *USA*
James P Trevelyan, *Australia*
Janis Swan, *New Zealand*
Jaromír Hrad, *Czech Republic*
Jaromir Polak, *Czech Republic*
Javid Atai, *Australia*
Jean Armstrong, *Australia*
Jeffrey E. Froyd, *USA*
Jerzy Mościński, *Poland*
Jim Henry, *USA*
Jiun-Jih Miau, *Taiwan, ROC*
Jocelyn Armarego, *Australia*
Joe Tront, *USA*
Johannes Görke, *Germany*
John C. Lusth, *USA*
John Carpinelli, *USA*
John Hoddinott, *Canada*
John Steffen, *USA*
John W. Prados, *USA*
John Webster, *New Zealand*
Jonathan Lopez-Morell, *USA*
Jong H. Kim, *South Korea*
Jordan Berg, *USA*
Jorma Kyyrä, *Finland*
José A. R. Parise, *Brazil*
Jose Antonio Moraño, *Spain*
José Galindo, *Spain*
José R. Ferrer Villanueva, *Spain*
Juan-Miguel Martínez-Rubio, *Spain*
Juh Wah Chen, *USA*
K. Krishnamurthy, *USA*
K.L. DeVries, *USA*
Karen E. Schmahl, *USA*
Karmela Aleksic-Maslac, *Croatia*
Keith Holbert, *USA*
Keith Stuart, *Spain*
Kenneth J. Hill, *United Kingdom*
Kevin D. Taylor, *USA*
Kevin F. Sweeney, *USA*
Kurt Gramoll, *USA*
Lakshmi V. Munukutla, *USA*
Larry E. Erickson, *USA*
Lawrence Jenkins, *India*
Liang-Rui Chen, *Taiwan, ROC*
Liu, Shu-Fen, *Taiwan, ROC*
Loren Schwiebert, *USA*

Low Sew Ming, *Malaysia*
Lucy Morse, *USA*
Lueny Morell, *USA*
Luiz Carlos Scavarda, *Brazil*
Luiz Gargione, *Brzail*
M. Brumsen, *Netherland*
M.M. Mwinyiwiwa, *Tanzania*
Marcela BORECKÁ, *Czech Republic*
Marcela Mora Vargas, *Colombia*
Maria H.L. Baptista Braz, *Portugal*
María Isabel Morales, *Spain*
Maria Joao M. Martins, *Portugal*
Mariana Ruiz Morales, *Mexico*
Marion Hagler, *USA*
Marios Kassinopoulos, *Cyprus*
Mary P. Harper, *USA*
Matilde P. Legua, *Spain*
Mehter Mohammed Allam, *India*
Melvyn Dodridge, *United Kingdon*
Michael H. Cole, *USA*
Michael Khonsari, *USA*
Michel Robert, *France*
Miguel Castro, *Puerto Rico*
Morteza Biglari-Abhari, *New Zealand*
Muhammad Sohail Ahmed, *USA*
Myra Kogen, *USA*
N.K. Kishore, *India*
Nariman Sepehri, *Canada*
Nicholas Bilalis, *Greece*
Norbert Gruenwald, *Germany*
Ovid A. "Gus" Plumb, *USA*
P.N. Balaguru, *USA*
Pardo Mustillo, *Canada*
Patrick Badillo, *France*
Patrio Chiu, *Canada*
Pedro M. Gameiro Henriques, *Portugal*
Petr Filip, *USA*
Peter H. Meckl, *USA*
Peter Kazanzides, *USA*
Peter Moson, *Hungary*
Petr Polak, *Czech Republic*
R.S. Gutierrez, *USA*
Rabin Raut, *Canada*
Rabindranath Raut, *Canada*
Reidar Kvadsheim, *Norway*
Ren-Hung Hwang, *Taiwan, ROC*
Richard Gale, *USA*
Richard M. Felder, *USA*
Richard S. Cowan, *USA*
Richard S. Figliola, *USA*
Richard Zollars, *USA*
Robert English, *USA*
Robin Ford, *Australia*
Rogelio Palomera-Garcia, *Puerto Rico*
Rohit Dua, *USA*
Roman Taraban, *USA*
Ron J. Kovac, *USA*
Ronald Sterkenburg, *USA*
Rui Miguel Lage Ferreira, *Portugal*
S. Kakac, *USA*
S. M. Aziz, *Australia*
Salvador Coll, *Spain*
Sam Hsu, *USA*
Sameh Issa, *Jordan*
Sami Kara, *Australia*
Sandnes Frode, *Norway*
Sarah Williamson, *United Kingdom*
Serge Abrate, *USA*
Sheikh A. Akbar, *USA*
Shlomo Waks, *Israel*
Sohail Anwar, *USA*
Sonia Bartolomei_Suárez, *Puerto Rico*
Soumen Das, *India*
Stephanie Farrell, *USA*
Stephen B. Seidman, *USA*
Stephen Hansen, *Australia*
Stephen Johnston, *Australia*
Stuart H. Munson-McGee, *USA*
Surendra Shrestha, *Australia*
Surita R. Bhatia, *USA*

Sylvain Coulombe, *Canada*
T.J. Brailsford, *United Kingdom*
Tatsuya Kikuchi, *Japan*
Thomas J. Kim, *USA*
Thomas K. Jewell, *USA*
Thomas R. Fischer, *USA*
Thomas W. Healy, *Australia*
Tim Dallas, *USA*
Tim Wilkinson, *Australia*
Timothy L.J. Ferris, *Australia*
Tor A. Fjeldly, *Norway*
Tracie Barber, *Australia*
Tsun-Wei Chang, *Taiwan, ROC*
Vaclav Havlicek, *Czech Republic*
Vicent Rodrigo Peñarrocha, *Spain*
Vikram M. Gadre, *India*
Vojislav Ilic, *Australia*
Warren Batchelor, *Australia*
Warren Yates, *Australia*
Wasim Saman, *Australia*
Withit Chatlatanagulchai, *USA*
Xiren Cao, *China*

# Chapter 1

# Guest Commentary: Engineering Education – Enabling a Nation's Capacity to Perform

JOSEPH BORDOGNA
*Deputy Director and Chief Operating Officer, National Science Foundation, Arlington, VA 22230, USA. E-mail: jbordogn@nsf.gov*

## EDUCATING ENGINEERS

Today, many in the U.S. consider the pursuit of educating engineers to be *decisive* to the nation's future – and I am one of them. Educating engineers has occupied the greater part of my life – as an engineering student, as a naval officer, as an engineer in industry, as a professor and dean of engineering, and now in my role with the National Science Foundation (NSF). As a steward of the health of our nation's science and engineering enterprise, NSF works at the frontier of research and education, where risks and rewards are high and where potential benefits to society are most promising. More than most, engineers have opportunities to ride the crest of the wave and arrive first at new shores.

Engineers are the ones responsible for getting things done and out the door in our society. We make things, and we make them right. At our best, we also make the right things. That has always been the role and responsibility of engineers in the continuing progress of society.

One particular application of those skills is the *making of engineers,* meaning *educating* engineers. Our design of the educational process manifests our intent for society. Thus, we must be very conscious of what we do and the path we take. Viewed in this light, the question of central importance becomes this: How do we make the *right engineers* for our times and for the future?

The amazing transformations that new knowledge and technology continuously trigger in our contemporary world bring education squarely into the limelight. In our knowledge-based economy and society, education is an investment in our nation's future. Engineering education *in particular* is an investment in our nation's capacity to perform. This is an awesome responsibility, ranging from an everyday improvement of society's operative underpinnings to responding smartly to sudden change.

## ENGINEERING EDUCATION IN THE U.S.

Society's appetite for high performance in every sector is expanding. Standing at the fulcrum of scientific and technological change, engineers are expected to foster progress toward a daunting array of ends – creating new knowledge, artifacts, and systems; stimulating economic development; creating wealth and jobs; sharpening the nation's competitive edge; raising our prospects for more productive and satisfying lives; caring for the environment; and strengthening our national security. Engineering education is at the very heart of these issues. Demands are increasing for a holistic breed of engineers – graduates with the skill to work across intellectual, social, and cultural boundaries. This integrative capability is key to successful performance in an increasingly diverse and complex international work environment.

Although some might quibble, most would agree that U.S. engineering education is the best in the world. The frontier research, cutting-edge tools, and skill sets that characterize U.S. engineering schools are the modern equivalent of Alexandria in the days of the great library. Our schools of engineering are intellectual magnets, drawing students from every nation of the world. Today's global engineers are contemporary versions of the mobile scholars of classical times – diffusing new knowledge and technology and thus contributing to our common future on the planet. Science and engineering have always been international, democratic, and collegial in this sense.

But this raises an obvious and sobering question. If U.S. engineering education is the best in the world, why aren't U.S. students flocking to the engineering fold? For some years, this question has been a serious concern and focus for the National Science Foundation, as well as for engineers. More recently, it has become a hot topic well beyond the engineering and scientific community. Industry, academe, government at all levels, and parents want to know if we are on the right track.

Will our youngsters have the skills needed to thrive in a competitive global economy driven by innovation and rapid technological change? Will there be a robustly capable supply of knowledge workers to meet increasing demands for them in our society and workplace? Will foreign nationals trained in U.S. engineering schools increasingly return to their home countries, depleting the ranks of engineers in the U.S.? Will an exodus of international talent, combined with growing numbers of engineers trained in other nations throughout the world, dull the competitive edge we enjoy in the United States? Academe worries about losing half of its engineering research capacity if international talent disappears from our graduate programs.

There is a growing clamor to recruit more of our native talent to the engineering fold, and NSF is focused on this issue. Unequivocally, more U.S. women and under-represented minorities are needed in our engineering workforce. This is a strategic as well as equitable manifestation of our intent in the design of engineering education. We need their talent and their perspectives. Without them, we will one day awaken, like the fictional Rip Van Winkle, to a world that has passed us by.

In 1822, James Madison wrote [1]:

> *"A popular Government, without popular information, or the means of acquiring it, is but a prologue to a farce or a tragedy; or, perhaps, both. Knowledge will forever govern ignorance; and a people who mean to be their own governors must arm themselves with the power which knowledge gives."*

Madison's words are the most fundamental rationale for an open door policy to an engineering education, an open-door policy that educates and enables our own citizens to be contributing participants in our great democratic system, as well as continuing the successful policy of embracing those from abroad.

This new perspective on an old policy will make us a genuine welcoming nation to both talent from abroad and from the nation's women and underrepresented minorities. It can never be one or the other. The Statue of Liberty's torch must light the way for those within our borders as well as those from outside. If we engineers aspire to be leaders, this is the challenge we must accept and meet.

This challenge requires a transformation – a continuing revolution, if you will – in engineering education. Our engineering education must *go beyond best* to confront the new global realities. We need a model of engineering education suitable to a new world in which change and complexity are the rule, a world continuously transformed by new knowledge and the technology it makes possible, a world linked globally, where differences and divisions that have not been integrated can have immediate and large-scale consequences. We must understand how to make the *right* engineers for our times and our future. Only then can engineering careers become a promising and appealing choice for more of our young people.

## THE HOLISTIC ENGINEER

Fast-paced technological change lifts society's expectation that engineers be holistic designers, astute makers, trusted innovators, harm avoiders, change agents, master integrators, enterprise enablers, knowledge handlers, and technology stewards. The best engineers in all eras have lived up to this description, and more; however, in our own times, we have moved to a whole new threshold of capabilities that breach with the past and that will catapult us beyond today's horizons. They are nanoscope, terascope, cognition, complexity, and holism.

Each is characterized by a fertile research frontier. They also overlap and interweave in subtle and intricate ways that add tremendous power to an engineer's tool kit. This newly minted synergy is what enables us to vastly enrich the scope and span of our knowledge and design the world of the future.

Ubiquitously available terascale computing - terascope for short - is computing technology that takes us three orders of magnitude beyond prevailing computing capabilities. Crossing that boundary of $10^{12}$ operations per second launches us to new universes.

Terascale is only one aspect of a much broader transformation. Our vision here is to reach *terascope* competency and catapult capability into a whole new era of science and engineering. In essence, we want to create a *tera universe or era* for science and engineering and a freshly robust national cyberinfrastructure [2]. Within this infrastructure, we'll enjoy tera-ops power, terabyte storage, terabit connectivity, and tera-instrument interfaces. What an addition to our engineering tool kit!

The new *nano* capability brings together many disciplines of science and engineering to work in collaboration. Nano-*scale* is three orders of magnitude smaller than most of today's human-made devices. Here, too, our vision is *nanoscope*: to create an overarching, enabling field, not unlike the role of information technologies today. Nanotechnology gives engineers the ability to manipulate matter one atom or molecule at a time. We will be able to make a wish list of qualities and characteristics to build into a new material or machine. Nanostructures are at the confluence of the smallest human-made devices and the large molecules of living systems. We are now at the point of being able to connect machines to individual living cells.

Our tera and nano capabilities open vistas that were barely visible only 20 years ago. They vastly expand our options for designing our future. For that reason, they have raised new questions about the choices and directions we should take as we move forward.

Enter cognition, the third arrow in our quiver. The dictionary defines cognition as the mental process by which knowledge is acquired. Most of us would simply say this is learning. But learning is anything but simple. Cognition encompasses a panoply of individual human competencies – acquiring knowledge, solving problems, making decisions, communicating, and creating, among others. A rich knowledge base in cognitive science is being developed jointly by linguists, psychologists, philosophers, computer scientists, engineers, mathematicians, neuroscientists, and others.

Like tera and nano, the scope of this capability takes us to new territories. Cognition illuminates the full gamut of human and social dynamics that characterize our institutions at all levels of complexity. We already speak of *learning organizations* in this context. We are now beginning to address the complexities of large-scale change; for example, democratization, globalization, and the causes of war – as well as how we design engineering education. For these reasons, NSF has recently announced a competition to create *Science of Learning Centers*.

This brings us to the fourth capability, complexity. Mitch Waldrop, in his book *Complexity*, writes about a point we often refer to as "the edge of chaos". That is, "where the components of a system never quite lock into place, and yet never quite dissolve into turbulence either . . . . The edge of chaos is where new ideas and innovative genotypes are forever nibbling away at the edges of the status quo. . . ." This territory of complexity is *a space of opportunity*, a place to make a marriage of unlike partners or disparate ideas. It allows us to explore interdependencies on vastly broader scales and across traditional boundaries. It provides a window on the heart of change in its myriad forms.

Holism, the fifth of our new capabilities, is closely linked to complexity. If complexity enables us to understand broader connections, holism teaches us that combinations of things have a power and capability greater than the sum of their separate parts. It spurs the search for new synergies and enables us to investigate and anticipate emerging structures and systems.

The hallmark of these new capabilities is their potential to transform. They help to connect, recompose, and expand core science and engineering disciplines. Each is nothing short of revolutionary, but in combination they are truly breathtaking in scope. They enable us to find solutions in unlikely places. One of the super chargers of our current age of cross-boundary discovery is the ability it gives us to shift from one context to another with agility, borrowing concepts and models along the way.

That's only for starters. As we integrate these new capabilities into engineering education, they will transform the very nature of learning environments. These new capabilities have the power to raise engineering education to a new dimension, to go beyond best to a newly transformed excellence.

## Do not Jettison what Works

I don't want to give the impression that we must abandon what we already know how to do so well in preparing engineers. We can't afford to jettison the tried and true. I envision the skills we routinely teach our students as paired with newly emerging ones. Both are components of a holistic engineering education.

We can already see some of what is ahead for academe in the not too distant future. We are making progress in integrating research and education. Industry-campus partnerships have taken root in the fertile soil at the frontier of knowledge. Other changes are slower in coming. We are struggling to find a fit between the global reach of our new information and communications technologies and our campus-centric systems. As we set out to revolutionize engineering education, there will be disconcerting moments when our vision of the end is not crystal clear, but the need to move forward is compelling. That is one of the characteristics of working at the frontier and of innovation in its many manifestations.

Forty years ago, in times of great national purpose similar to our own, President John F. Kennedy described the core of the educational enterprise in these words:

> *"Let us think of education as the means of developing our greatest abilities, because in each of us there is a private hope and dream which, fulfilled, can be translated into benefit for everyone and greater strength for our nation."*

We are closer than ever before to making this a reality in engineering. If we want to be leaders, we should welcome transformation and renewal, with all the risks and rewards they promise. If we do so, we will be well on our way to making the *right* engineers for our times and for the future.

To help us keep our focus on the inflection points in our progress – that is, to anticipate changes in our learning curve – I end with a quote from that great guru of human and social dynamics, William Shakespeare [3]:

> *"The day shall not be up so soon as I, to try the fair adventure of tomorrow."*

## REFERENCES

1. James Madison, Letter to W. T. Barry, August 4, 1822.
2. Daniel E. Atkins, et al., "Revolutionizing Science and Engineering Through Cyberinfrastructure", Report of the National Science Foundation Blue-Ribbon Advisory Panel on Cyberinfrastructure, January 2003. http://www.cise.nsf.gov/sci/reports/atkins.pdf.
3. William Shakespeare, King John, Act V; Scene 5, Bishop's Palace; speaker, Lewis (The Dauphin).

**Joseph Bordogna** is Deputy Director and Chief Operating Officer of the U.S. National Science Foundation. He served previously as head of NSF's Directorate for Engineering. He currently serves on the U.S. President's Management Council, has chaired committees on manufacturing, environmental technologies, and automotive technologies within the U.S. President's National Science and Technology Council, and was a member of the U.S.-Japan Joint Optoelectronics Project. He previously served at the University of Pennsylvania as Alfred Fitler Moore Professor of Engineering, Director of the Moore School of Electrical Engineering, Dean of the School of Engineering and Applied Science, and faculty master of Stouffer College House, a living learning student residence at the university. He received the B.S.E.E. and Ph.D. degrees from the University of Pennsylvania and the S.M. degree from the Massachusetts Institute of Technology. His career includes experience as a line officer in the U.S. Navy and an engineer in industry. He has made contributions to the engineering profession in a variety of areas, including early laser communications systems, electro-optic recording materials, holographic television playback systems, and early space-capsule recovery. He was a founder of PRIME (Philadelphia Regional Introduction for Minorities to Engineering) and served on the board of the Philadelphia Partnership for Education, community coalitions providing, respectively, supportive academic programs for K-12 students and teachers. He is a fellow of the AAAS, the ASEE, the International Engineering Consortium, and is a Fellow and Past President of the IEEE.

*This commentary is adapted from a speech presented by the author at the 2003 annual conference of the American Society for Engineering Education, and an article published by the author in the Fall 2003 issue of The Bent of Tau Beta Pi.*

# Chapter 2

# The Paradigm of Civil Engineering Education within the Colombian Context

SILVIA CARO SPINEL
*University of Los Andes, Bogotá, Colombia. E-mail: scaro@uniandes.edu.co*

*Colombia needs to develop a reliable, efficient and optimal civil infrastructure system for facing the socio-economic challenges of this century. The civil engineering profession has a significant role to play in this process, and current civil engineering undergraduate programs are not training professionals in this direction. The pedagogic concepts and teaching and learning methodologies are very traditional and do not provide the tools needed to ensure sustainable development. There is evidence indicating that civil engineering programs should be modified substantially. The new Colombian civil engineer must have not only a basic and technical understanding of design systems, but also the ability to identify and solve complex problems, to communicate in an effective and efficient way, to work as a part of multi-disciplinary groups and to understand the responsibility of his/her professional performance on society. This paper presents the current situation for civil engineering programs, the requirements for a civil engineering program that responds to the Colombian context, and some concrete actions to be considered for the construction of modern programs.*

## INTRODUCTION

Statistical data show that the level of poverty in most South American countries is growing quickly, and that economic development is seriously delayed [1]. Within this context, infrastructure development plays a key role since it provides the basic support for economic improvement and social welfare. Some examples of the required infrastructure are: water supply and sewage systems, housing and transportation. Therefore, it is clear that civil engineering has a very important role to play in the development of a country. Civil engineers should be outstanding professionals who are able to face the challenges imposed by both global trends and local conditions. They must have the skills to identify complex problems, which usually involve other engineering fields, and solve them in an efficient and effective way by using the resources available.

The two main difficulties related to the civil engineering profession in Colombia are:

- The number of freshman students has decreased dramatically during the last decade (Figure 1).
- Academic civil engineering programs have significant deficiencies.

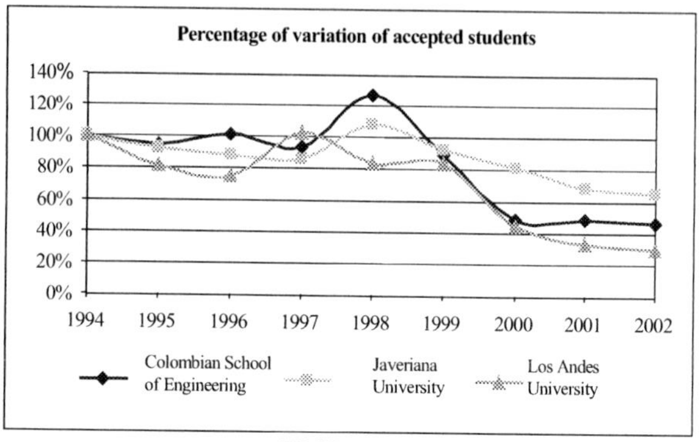


FIGURE 1

PERCENTAGE OF VARIATION OF ACCEPTED STUDENTS FOR THREE CIVIL ENGINEERING UNDERGRADUATE PROGRAMS

Regarding the first point, some universities have reported a reduction of 50% in the number of newly accepted students in two consecutive years. This situation is even more dramatic in light of the fact that only 50-60% of accepted students in undergraduate civil engineering programs make the decision to register. The situation is caused mainly by a severe economic crisis that has been occurring in Colombia, as in other Latin-American countries, since the early 1990s. The unemployment rate has increased from 9% in 1994 to 14% in 2004, affecting mainly the construction industry. Since the employment opportunities for civil engineering professionals are limited, most students who had considered studying civil engineering are moving to other engineering programs. For instance, the current distribution of the students in the School of Engineering at the University of Los Andes (Bogotá, Colombia) shows that civil engineering has poor participation (8%).

Regarding the educational deficiencies, most academic programs were designed to respond to the country's needs during the first part of the 20th century. At the time, most universities did not have consolidated graduate programs, and students were prepared to cover all possible subjects during the undergraduate period. This policy led to over-design of programs and professionals who were more technicians than engineers. Nowadays, the main universities offer master and doctoral programs, which meet international standards. Nevertheless, in the last 50 years there have not been significant changes in undergraduate programs; furthermore, the paradigms about knowledge, teaching and learning in the country are still out of date and the design of more flexible, interdisciplinary and humanistic programs has not been considered relevant.

World and regional changes have forced some Colombian universities to re-think the way civil engineers are educated, especially under the new global principles of teaching and learning methodologies. This means that the country must redefine the profile of the civil engineer it wants, and adapt new educational thinking to suit local conditions. The following sections will explain the relationship between the current Colombian context and the transformations needed in civil engineering education.

# THE COLOMBIAN CONTEXT

Colombia is located in the northwestern corner of South America, with an area of 1,138,910 square kilometers and access to both the Pacific and the Atlantic oceans (coastlines of 1448 km and 1760 km, respectively). It is a country with approximately forty million inhabitants, the majority being Catholic and Spanish speaking. Approximately 80% of the population is settled in the central part of Colombia, the Andean Region, a mountainous, rugged terrain that is exposed to many natural hazards (e.g. earthquakes, floods, landslides). Currently, some 75% of the population lives in cities and about two million live as refugees under very difficult conditions. The socio-economic distribution of the population is approximately 55% percent low-income, 35% percent medium-income and 10% high-income, with a distribution of 52% female and 48% male.

Important figures related to the socio-economic situation of Colombia, are [2]: the unemployment level reached 18% in 2003; the current external debt is US$39 billion; 55% of the population lives under the poverty line; the Gross Domestic Product (GDP) per capita in 2002 was US$ 6300 (purchasing power parity, PPP); the current GDP composition is 19% agriculture, 26% industry and 55% services; and, the most important natural resources are: petroleum, natural gas, coal, iron ore, nickel, gold, copper, emeralds and hydropower energy.

Unfortunately, despite its strategic location and economic potential, Colombia has been ranked among the lowest (54th) on the scale of world competitiveness [1] and is one of the countries least attractive to foreign investment, due not only to the social conflict, but also to the poor and inefficient infrastructure.

In reference to its civil infrastructure, the critical figures from [3,4] are: (1) The kilometers of paved roads per hundred square kilometers are 2.5. This value is 90% smaller than in most European countries. (2) The kilometers of railroads per million inhabitants are 81.9. This value is about 7 to 10% of the corresponding figures for railroad networks in the United States and Canada. (3) The qualitative housing deficit is higher than 20%. Out of the existing 80%, 14.8% of the urban housing is overcrowded. (4) Only 46% of the housing available has water, sewer and electricity service; 20% does not receive any public services. (5) The most important cities in the country do not have efficient mass transit systems. Medellín, the second largest city in Colombia, is the only one with a notable rail system. In Bogotá, the capital city, a rapid transit bus system was installed in December 2000. Before that date, the transportation offered was poor in quality, insecure and inefficient in a city with more than 8 million inhabitants and where 70% of the population uses public transportation.

Studies show that there is a direct relationship between the quantity and quality of civil infrastructure and economical and social development [5-7]; consequently, civil engineers have a great responsibility.

Infrastructure development needs in Colombia face severe obstacles. Long term planning is almost non-existent and infrastructure is commonly developed on the basis of the immediate needs, or under severe political pressure. There is no consistent plan to integrate the country's policy for long-term competitiveness with industrial needs and investment in infrastructure. In addition, there is very weak technical support for the design and construction of such infrastructure. A significant part of the responsibility for these situations falls on the education system. Reports from industry and the government show that graduating civil engineers do not have the capability to face projects in a broad context that involves not only other disciplines (humanities, economics, social and politics sciences) but also abilities such as critical thinking, negotiation, leadership and effective communication. This situation supports the proposal of changing the current programs.

## CIVIL ENGINEERING EDUCATION IN THE COLOMBIAN CONTEXT

### Civil Engineering Programs

In 1996, the Association of Colombian Schools of Engineering (ACOFI) and the Colombian Institution for Higher Education (ICFES) developed a project for studying the current situation of civil engineering programs [8]. The results were discouraging. It showed that civil engineering programs are rigid, linear and over-designed. Furthermore, most programs do not instill the skills and attitudes required for the engineers of the 21st century [9,10].

Current civil engineering education in Colombia is basically the same as it was in the 1950s and 1960s. Traditional undergraduate programs were structured under a six-year scheme following the then existing European education system. During this period, the University of Los Andes opened a new program following the North-American perspective: more flexibility, fewer technical and more fundamental courses, the design of new integral education areas, and a five-year scheme including a credit system. Although all current academic programs are designed for five years, it takes almost six years for an average student to finish. In most universities, courses such as Structural Analysis I and II, Hydraulics I and II, Sanitary Engineering I and II, and Construction I and II, are almost the same as those taught in 1950s. A thorough review of civil engineering programs in Latin America shows that the situation is similar elsewhere [11].

Table 1 presents a comparison of five of the most important civil engineering programs in Colombia. In general, it can be concluded that a large number of technical courses are required with few electives in the humanistic and socioeconomic areas.

| | Colombian National University | Industrial University of Santander | Cauca University | Javeriana University | Los Andes University |
|---|---|---|---|---|---|
| Total courses* | 56 | 57 | 62 | 63 | 51 |
| Number of elective courses | 8 | 5 | 7 | 6 | 12 |
| Number of compulsory civil engineering courses | 26 | 30 | 31 | 25 | 20 |

* Without laboratories.

TABLE 1
SOME DIFFERENCES BETWEEN COLOMBIAN CIVIL ENGINEERING PROGRAMS

On the other hand, modifications of curricula at all universities have been superficial and have followed the same pattern: add, change or remove courses from the academic programs. Table 2 presents the evolution of the civil engineering program at the University of Los Andes during the last 30 years [12].

| | Period of time | | | |
|---|---|---|---|---|
| | 1970-1980 | 1980-1990 | 1990-2002 | 2004 |
| Total credits* | 170 | 170 | 164 | 158 |
| Total courses (without laboratories) | 57 | 57 | 55 | 50 |
| Total of credits in elective courses | 38 | 32 | 35 | 36 |
| Number of specific civil engineering compulsory courses | 22 | 25 | 22 | 20 |
| New compulsory courses or areas | ---- | Geometric design of roads | • Construction<br>• Pavements | ---- |

* A CREDIT IMPLIES 45 HOURS OF WORK, INCLUDING ATTENDANCE IN CLASSES AND INDIVIDUAL WORK. A NORMAL COURSE CONSTITUTES 3 CREDITS AND A LABORATORY COURSE CONSTITUTES 1 CREDIT).

TABLE 2

EVOLUTION OF THE CIVIL ENGINEERING UNDERGRADUATE PROGRAM, UNIVERSITY OF LOS ANDES

Comparison of the 1975 and the 2003 programs shows that most modifications since 1975 have been related to languages (Spanish and English), humanities, and elective courses. The basic cycle in sciences (physics, chemistry and mathematics) and the professional cycle (specific courses in geotechnics, structures, transportation systems and infrastructure, hydraulics, etc.) have not undergone significant changes. Modifications reported in specialized areas include construction management and a few isolated courses such as pavement and road design.

In conclusion, the essence of the program has not changed. Even though the courses now have a larger technology component, this component is still very limited. On the other hand, teaching methodologies, with few exceptions, do not include the active participation of the student as a principle element of the education process. Table 3 presents the main characteristics of the five years civil engineering undergraduate programs in Colombia.

Some of the weaknesses in the current civil engineering programs in the country identified by the author are:

- Programs are over-designed. The number of courses is excessive for an undergraduate program.
- There is no clear connection among the different areas. The students complete basic science courses, but these are not tied to the subjects that follow. By the time a student begins his advanced courses he realizes that he does not have the foundation for assimilating the new knowledge.
- Courses have a large lecture component. In some cases this means more than eight class-hours per week. This situation has two problems: (a) it does not encourage the students to do research; and (b) it limits the time the students have for individual or group activities.
- Most programs are rigid and linear. Students do not have an opportunity to be arbitrators of their own education development.
- Specific areas of knowledge have not been updated. While new technology is widely used, the programs are still tied to classical civil engineering concepts.

| | |
|---|---|
| Areas of knowledge | ▪ Basic sciences (mathematics, physics, chemistry).<br>▪ Social/cultural education (varies from one university to another)<br>▪ Basic engineering (courses common to all engineering programs)<br>▪ Basic civil engineering<br>▪ Advanced civil engineering |
| Areas in civil engineering | ▪ Structures<br>▪ Water resources (fluids, hydraulics, hydrology, etc)<br>▪ Geotechnics (soils, foundations)<br>▪ Transportation<br>▪ Road infrastructure<br>▪ Materials<br>▪ Construction management |

TABLE 3

GENERAL STRUCTURE OF CIVIL ENGINEERING PROGRAMS IN COLOMBIA

- Most programs do not contribute consistently to the development of skills and attitudes that are required of engineers by modern and future world conditions. (ABET 2003 Criteria, [13]).
- Most programs do not have the technological resources required by international standards. For instance, a typical civil engineering student does not have the opportunity to interact with specialized software and other technical resources.
- With the exception of certain universities, professors do not have a strong enough educational background.
- Research at all levels is very limited and is reduced to the particular interests of a few professors.

**Comparison between Colombian and North American Civil Engineering Programs**

A comprehensive comparison between Colombian and North American programs should include aspects such as: content and duration; organization of knowledge areas and their inter-relationship; skills and attitudes developed in students; philosophical concepts about knowledge, teaching and learning; and resources and tools that support the programs. Nevertheless, this kind of study is beyond the scope of this paper.

However, for the purpose of this work it is important to note that in American programs there is almost no separation among the six civil engineering areas [14]. These programs include not only a large number of elective courses, but also a new conception of civil engineering. For instance, traditional courses such as foundations and structural design have been replaced by courses like graphic communication in engineering; civil engineering systems; engineering modeling under uncertainty; engineering design; planning, design and management of civil engineering systems; technical and scientific communication; and development systems.

Likewise, the skills and attitudes developed in Colombian students are not sufficient to guarantee a successful professional performance. North American universities are very concerned about achieving the eleven outcomes (a-k) required by the ABET Criteria [13]. In this respect, it seems clear that civil engineers in Colombia have weaknesses in: (a) applying knowledge of mathematics, science and engineering, (b) functioning in multi-disciplinary teams, and (c) communicating effectively.

The development of these skills is closely related to teaching and assessment techniques, and to the resources that support the programs. New pedagogical concepts like active learning (constructivist philosophy), cooperative learning and peer learning have acquired wide acceptance in the United States over the last two decades. Most programs recognize the importance of placing the student at the center of the learning process, encouraging his active participation (e.g. Problem-Based Learning, PBL, and Hands-on) and moving away from traditional methodologies. The benefits of these concepts have been proven by innumerable experiments conducted worldwide. Some authors [15-18] state that teamwork allows the student to compare his worldview with his classmates, stimulating the consolidation of new knowledge. In this context the active participation of the students in their own personal growth seems to produce long-term benefits and could help to clarify the knowledge they have assimilated.

Most of the concepts discussed above are treated informally and are often overlooked in Colombian universities. Industry and the government when assessing the students once they become part of the labor market have observed the consequences of this situation.

## The Challenge: Achieving a New Concept of Civil Engineering Education

Nowadays, Colombia has favorable conditions for enhancing civil engineering education. Firstly, the government has set up a project to assess the quality standards of higher education. It has also promoted several research activities intended to improve engineering education. Secondly, the major universities have ongoing research programs investigating the possibility of re-designing their curriculum. Hopefully, these developments are favorable for the emergence of a new civil engineer who is more effective in the face of country's economic, social, technological and cultural realities.

However, to assure success, there are some tasks that should be considered:

- Develop a national consciousness of the importance of the civil engineering profession and define a profile of the civil engineer the country requires. The new civil engineer should respond to industry and government requirements; therefore, both government and industry must participate actively in the process. For that purpose, it is necessary to stimulate discussion about this subject (e.g. promote national and regional conferences, workshops and seminars).
- Re-think the academic concept of civil engineering programs. There should be a balance among the number of the courses and their contents. In order to achieve this goal, it is necessary to review the effectiveness of current curriculum structures. For example, it is time to ask if the basic natural sciences cycle of two years at the beginning of the program is giving students the tools they will need in future courses. Educational administrators should think about the possibility of extending the period of this cycle, allowing the professional cycle to begin earlier. Under this scheme, it is possible that students understand basic concepts when they need them, promoting a more effective and permanent learning processes.
- In order to design competitive programs, universities should explore new trends in civil engineering education worldwide (e.g. MIT with its "Conceive, Design,

Implement and Operate, CDIO," concept). It is also important to encourage faculty to attend international conferences in order to be in contact with new educational topics. This exploration should be considered a starting point in the long process of modifying educational schemes according to local and regional socio-economic conditions. In addition, it is time to study and include the impact that globalization processes will have on industry and education in developing countries. For example, Colombia and the United States are negotiating a free-trade agreement. This event will affect the industry and engineering profession, as it has been happening in Chile and Mexico.

- Comply with international standards of quality and excellence. This implies that Colombian students should be exposed to more technology and innovation processes.
- Involve civil engineering students in real projects and show them the responsibility that the profession demands in Colombia. Internships, international academic experiences, access to technology, participation in research projects, and academic projects among different programs should be real possibilities for engineering students. It is very important that universities develop a closer relationship with industry through partnership programs.
- Design and build infrastructure environments in accordance with new learning and teaching methodologies. Space and other related resources are important factors in the learning process and cannot be ignored. The infrastructure required should be not only physical, but also technological.
- Include, within the curriculum structure, engineering concepts such as reliability, uncertainty and optimization, which have shown to be essential in the training of future engineers. It is not necessary to create independent courses for these subjects; it seems more efficient to introduce them in technical courses through projects or case studies where students can recognize their importance and application.
- Adjust undergraduate programs to the already-existent graduate programs. New proposals should focus on shorter undergraduate programs with better connections with graduate programs (e.g. five years for bachelor plus master degrees).
- Design teacher-training programs in education. Teachers are fundamental actors in the education process. If universities want to change traditional teaching and learning methodologies, they should provide information, tools and support to their teachers. Universities should promote education research and should motivate their faculty to think about the effectiveness of their own pedagogic conceptions. Some authors [19-22] insist on the importance of this training as a tool for guaranteeing the accomplishment of learning objectives in courses and for the general success of a program.
- Finally, it is important to develop mechanisms to assess the effectiveness of new programs in the future. Well-designed programs that address the considerations presented above should encourage students to be constructors of and be responsible for their own education. Their professional performance should demonstrate that they truly have the required skills.

## CONCLUSIONS

Current civil engineering undergraduate programs in Colombia have important deficiencies. They are not only over-designed, but the current state of pedagogical methodologies does not encourage the acquisition of the long-term knowledge in the students. The country must think carefully about the profile of the civil engineer that it requires. Civil engineers should have attitudes such as: competence, integrity, commitment, tolerance, flexibility and reliability. They should also have specialized skills in basic sciences, engineering applications, computer science and technology. Furthermore, the civil engineer should develop and consolidate communication, social, presentation, teamwork, leadership and business management skills.

It is difficult to assure that a program can impart all of these characteristics to a person, but it is possible to design a program that helps a person achieve most of them by him/herself, during his/her professional development process. Specific issues like those described in this chapter are a first step in that direction. It is time to face the challenge of transforming the traditional idea of engineers as technicians to the idea of engineers as professionals.

The most important universities in Colombia have accepted this challenge. In a period of six years there will be new civil engineering programs and it will be possible to measure the effectiveness of such modifications.

Changing universities' educational model should be also considered by other developing countries. In particular, Latin American countries in the Andean Region where socio-economic conditions are very similar should start this process together. An educational model that includes common academic activities can contribute to the establishment of better relationships through partnerships that promote research and innovation in the region.

## References


1. World Economic Forum, *World Competitiveness Report*, Switzerland, 2001.
2. Instituto Geográfico Agustín Codazzi, Colombia, *Atlas de Colombia*, 2003.
3. CEPAL, *Un perfil del déficit de vivienda en Colombia*, Serie Financiamiento del desarrollo, Santiago de Chile, 1999.
4. Central Intelligence Agency, CIA, "Fact Book", found in the CIA website: http://www.cia.gov/cia/publications/factbook/index.html
5. Amsden, A., The Rise of "The Rest": Challenges to the West from Late-Industrializing Economies, Chapters 1-4, Oxford University Press, 2001.
6. Banister, D., and Joseph B., *Transport Investment and Economic Development*, Chapters 1-2, University College London Press, 2000.
7. Sánchez, M., and Caro, S., "Aceptabilidad del Riesgo como Criterio para la Competitividad," Proyecto financiado por el Banco de la República, Bogotá, Colombia, 2003.
8. Asociación Colombiana de Facultades de Ingeniería, ACOFI, "Actualización y modernización curricular en Ingeniería Civil," 1996.
9. Buckeridge, J., "A Y2K Imperative: the Globalization of Engineering Education," Global Journal of Engineering Education, Vol. 4, No. 1, 2000, pp. 19-24.
10. Nguyen, D.Q., "The Essential Skills and Attributes of an Engineer: A Comparative Study of Academics, Industry Personnel and Engineering Students," Global Journal of Engineering Education, Vol. 2, No. 1, 1998, pp. 65-75.

11. Universidad Católica de Chile, Chile, http://www.puc.cl/; Universidad de Buenos Aires, Argentina, http://www.uba.arg/homepage.html; Universidad Autónoma de México, México, http://www.unam.mx/

12. Universidad de Los Andes, Bogotá, Colombia, "Catálogo General", 1974-1975; 1983-1984; 1998; 2002.

13. Accreditation Board for Engineering and Technology, "Criteria for Accrediting Engineering Programs During the 2003-2004 Accreditation Cycle," http://www.abet.org.

14. Massachusetts Institute of Technology, http://web.mit.edu/civenv/; Berkeley University, http://www.ce.berkeley.edu/; Illinois University, http://www.engr.uiuc.edu/departments/; Georgia Institute of Technology, http://www.ce.gatech.edu/academics/.

15. Felder, R.M., and Brent, R., "Cooperative Learning in Technical Courses: Procedures, Pitfalls and Payoffs," ERIC Document Reproduction Service Report ED 377038, Supported by National Science Foundation, 1994.

16. Haller, C.R., Gallagher, V.J., and Felder, R.M., "Dynamics of Peer Education in Cooperative Learning Work Groups," Journal of Engineering Education, Vol. 89, No 3, 2000, pp. 285-293.

17. Perkins, D.N., "Technology Meets Constructivism: Do They Make a Marriage?" Educational Technology, Vol. 3, No. 5, 1991, pp. 18-22.

18. Felder, R., Woods, D., Stice, J., and Rugarcía, A., "The Future of Engineering Education II. Teaching Methods that Work," Journal of Chemical Engineering Education, Vol. 34, No.1, 2000, pp. 26-39.

19. Gibbs, G., "What Training of University Teachers Might Be Able to Achieve," Centre of Higher Education Practice, Open University, UK, 2002.

20. Randall, M., and Thortin, G., *Advising and Supporting Teachers*, Cambridge University Press, 2003.

21. Brent, R., and Felder, R., "Helping New Faculty Get Off to a Good Start," Proceeding of the 2000 Annual ASEE Meeting, St Louis, 2000.

22. Felder, R., "Teaching Teachers to Teach: The Case for Mentoring," Journal of Chemical Engineering Education, Vol. 27, No. 3, 1998, pp. 176-177.

**Silvia Caro** received her Civil Engineer degree from the University of Los Andes, Bogotá, Colombia in 2000. In 2001 she received her M.Sc. degree in Road and Civil Infrastructure Systems from the same University. Since 2001 she has been teaching and working as a research fellow in the Department of Civil and Environmental Engineering in the University of Los Andes. She belongs to the Geotechnical and Structure Research Group and to the Engineering Education Research Group (CHAICA) in the School of Engineering.

# Chapter 3

# The Interactive Knowledge Model: Knowledge Transmission And Cooperative Learning Environment

DOMINIQUE BOURGEOIS and PATRICK-YVES BADILLO
*Research Laboratory on Media, Information and Knowledge, University of Aix-Marseille, 21 rue Virgile Marron, 13392 Marseille Cedex 05, France. E-mail: dbourgeois@ejcm.univ-mrs.fr, badillo@romarin.univ-aix.fr*

*This paper proposes an analysis of the new pedagogical and knowledge paradigm, and especially of the cooperative learning environment based on Information and Communication Technology – ICT. First, we describe the new interactive knowledge model in which the learning process becomes more and more a common construction based on the cooperation between teachers, learners, and companies. We examine the role of ICT and the place of the teacher in this new learning process. We highlight an essential distinction between "telecommunications" transmission and "knowledge" transmission. We then analyze the question of knowledge transmission in a cooperative learning environment. We refer to the experiment of an important European project called Form-Ami. ICT can be considered as a catalyst of knowledge transmission, but the teaching act and the role of the pedagogue remain crucial.*

## INTRODUCTION

The harmonious development of the information society depends on the coherence of the "triptych": technology, education, and social diffusion. The technological push has made it possible to develop networks that store, manipulate and transmit increasingly large quantities of information. A new training paradigm appears with the rise of Information and Communication Technology – ICT. Traditional teaching principals must make a long and difficult adjustment as training and knowledge become accessible through new methods thanks to new technologies. Remote teaching becomes distributed teaching in which the learner plays an increasingly active role in interacting with the teacher.

Beyond technological developments, many significant breakthroughs have surfaced concerning knowledge and its diffusion [1]. According to T.H. Davenport and L. Prusak, knowledge constitutes "a fluid mix of framed experience, values, contextual information, and expert insight that provides a framework for evaluating and incorporating new experiences and information" ([2], p. xii). In short, knowledge is a key factor in the competitiveness of firms in the new information-based economy.

This paper analyses the new pedagogical and knowledge paradigm, and especially the cooperative learning environment based on ICT. First, we present the new pedagogical paradigm, which can be described as the "interactive knowledge model". We study the question of transmission and the role of technology. We then investigate the concept of the "cooperative learning environment" in the university framework. Finally, we argue, on the basis of an important European project called Form-Ami, that the pedagogue and the human relations between the teacher and the learners are at the heart of the teaching and knowledge transmission processes.

## The Interactive Knowledge Model, Knowledge Transmission and ICT

### The Pedagogical and Knowledge Paradigm Shift: The Interactive Knowledge Model

The old pedagogical paradigm can be defined as a knowledge push. We refer to the innovation theory to define the old pedagogical paradigm as a knowledge push model. In this model, which has been dominant since the beginning of the Gutenberg Galaxy, teachers are the holders of and have a monopoly on knowledge; they produce and diffuse knowledge from a specific place in society. But today this paradigm is being challenged by changes that affect universities. The university world today may be overturned tomorrow with the development of e-learning and the creation of new training tools.

A new interactive knowledge mediation paradigm is emerging. In the context of the new pedagogical paradigm, teachers share knowledge and produce knowledge in collaboration with students and companies or training organizations. In this new interactive knowledge mediation paradigm, the learning process increasingly becomes a common construction and collaboration. "Young people will be guiding their teachers to new kinds of learning models", predicts Mr. Juha Lipianen from Nokia (ED-MEDIA 2001, Hypermedia Laboratory, University of Tampere, June 2001). "Knowledge is not delivered; it grows in unpredictable ways because learners move through this space according to their own needs, intentionally connected to others. Participation in this space makes it grow for you, over time and through your use and interaction" [3]. In this new pedagogical paradigm, according to D. Jonassen [4], the learning process is efficient if it is organized as a "problem-solving" system; students are confronted with a problem and they have to solve it; the only professionally legitimate goal is "problem-solving" (decision-making, modeling).

This new teaching model can be interpreted as corresponding to the idea that the pupil is "at the center" of the education system but, according to L. Ferry ([5], pp. 46-47, our translation), "what is advisable to put at the center of the education system is not the pupil or knowledge only, but obviously the relationship between the pupil and knowledge". So it is of fundamental importance to take into account the role of the teacher and the work of the student in knowledge transmission. This analysis is very pertinent; in fact, the risk of focusing too much attention on the role of the learner is important because ICT has blurred our understanding of the educational process. With the new teaching paradigm was born the illusion that ICT would have a fundamental role in knowledge transmission. We will show that ICT is certainly useful from the point of view of "telecommunications" transmission, and can support an extremely interesting cooperative training but cannot, in any case, substitute for the fundamental teaching act performed by the professor, and for the work of the learner. In other words, ICT can be a positive catalyst in education, but cannot replace the links between the teacher and the learners.

## Information and Communication Technology and Knowledge Transmission

ICT has played a role in the paradigm shift. If one wants to understand this role it is necessary to reconsider the meaning of "information" and "communication" compared to knowledge transmission. C.E. Shannon, who was an engineer, originally developed the concept of information in "mechanistic" communication theory ([6]; see also Shannon and Weaver [7]). This approach was judged too "mechanistic" by many authors and in particular by the Palo Alto school which introduced, at the heart of its analysis, the communication "relation". (According to the Palo Alto School, and especially to Watzlawick [8], we are always communicating, since all human behaviour is a form of communication. The Palo Alto School relies on a systemic approach and insists on human relations.) Between the mechanics of information (Shannon theory) and the communication "relation" (Watzlawick), the ways become divergent. As Y. Winkin [9] underlines, "Shannon spoke initially about information, in the very specific meaning of 'quantitative measurement of the uncertainty of a message according to the degree of probability of each signal comprising this message'" (Petit Larousse). Weaver speaks rather about communication. Confusion between the two terms is widespread, which reflects the ambiguity of the terms "information theory" and "communication theory". Heinz von Foerster, one of the first researchers to use concepts of cybernetics in biology, has said of information that it is "the most vicious of the conceptual chameleons" (quoted by E. Morin [10], p. 290). In fact, it is necessary to speak of a menagerie of vicious chameleons ([9], pp. 31-32, our translation).

The theoretical disjoint alluded to by the image of the chameleon is at the center of the ambiguity which reigns in the information society. On the one hand, it is undeniable that the contribution of information technology is considerable, reducing space and offering new possibilities for collaboration, co-operation, construction and division of knowledge. On the other hand, knowledge transmission is not reducible to the mechanics of information.

### Knowledge Transmission or Telecommunications Transmission?

Following Shannon, telecommunications engineers have used the term transmission. In the French reference book on telecommunications written by F. Ducastel [11], a complete chapter is very naturally dedicated to the transmission function: "The transmission function corresponds to the remote transport of the transmission signals. The signals to be transmitted, coming from the local area networks, are initially treated, if an adaptation of the support appears necessary; they are then transmitted, in an electric or optical form, by using a guided or radiated transmission, from a transmitter towards one or more receivers" ([11], p. 213, our translation). It is clear that the term "transmission" is used here with the same meaning as telegraphy in the 19th century, from the Latin etymology "transmittere" which means "to send to the other side, to make go beyond", from which the word transmission ("way, crossing") is derived. The word "transmission" is used in this manner in various scientific and technical contexts.

The technical concept of transmission differs radically from another interpretation, again from the Latin word "transmittere," which means to pass from one being to another, to transmit life, to transmit a heritage, to bequeath, to give. It is obviously this definition that is privileged here, that of knowledge transmission, which goes beyond the diffusion of information. Thus, apart from the Shannon's mechanics of information and the communication relation of Palo Alto, there exists a field of analysis relating to knowledge transmission. Consequently, the question of the role of technology in this transmission arises.

### Technologies, Communication and Knowledge Transmission

According to R. Debray, transmission is a process that takes place over time and is based on institutions. By contrast, communication occurs in the short term, for example news in the media. "We will say that to communicate means to *transport information in space*, inside the same space-time sphere, and to transmit is to *transport information in time*, between different space-time spheres" ([12], p. 3). Communication is thus a necessary, but not sufficient, condition for transmission. The technical means of communication cannot ensure cultural transmission. It is precisely the "intoxication" ("*ivresse*") of communication which characterizes current society and which corresponds to the privilege granted to the "means of *space domestication* over the means of *time domestication*" ([12], p. 6).

Finally, there is on one side the "telecommunication" transmission and on the other the "heritage and knowledge diffusion" transmission. Yet, an illusion was born: many believed that "telecommunication" transmission automatically carried with it "knowledge" transmission, that it would be enough to develop technologies and "telecommunication" transmission that knowledge could be diffused in a systematic way. We will now examine the role of ICT in a cooperative learning environment. We will distinguish the catalyst and the reactive principle in knowledge transmission and we will refer to the experiment of an important European project.

## Knowledge Transmission and Cooperative Learning Environment

### The Form-Ami Project

Cooperative work and cooperative training tools have developed in the current educational world. According to the new pedagogical paradigm, the learners and the teachers jointly produce and share information and knowledge. We have tested this new pedagogical paradigm with a pilot European project, called Form-Ami ("**Form**ation en **A**rt et **M**écanique de l'**I**nformation").

Form-Ami was a project of the European Commission, Information Society Directorate-General, over the period 1999-2002, with a view to promoting new university training courses in the specialised skills required by information industries. A large European consortium of nineteen members including important academic institutions and information companies[1] implemented it. It included a new post-master diploma in "Art and Mechanics of Information", the realization of multimedia products, an international conference and ten videoconferences [13]. The post-Master training combined general theoretical training (information and telecommunications science and technology, design and ergonomics of multimedia information products, ICT economy and management), specialized training (three options: geographical information systems, security, pedagogy and multimedia) and practical realizations. It was an original training experiment, thanks to the intensive use of ICT with "présenciel" and remote teaching, the realization of multimedia products by the students on the basis of the courses, and a practical training period in industry.

Among 500 European candidates, 28 students finally obtained their diploma. The students were of 13 different nationalities; they largely represented the European Union, but non-European students were also represented. Five multimedia products linked to the courses were completed in the following subjects: theory of information, information market, security, geographical information systems, and pedagogy and multimedia. From a pedagogical point of view, a cooperative learning environment based on the Socratic teaching method was at the heart of the project. Of course, the pedagogy relied also on recent studies, particularly concerning the use of ICT in education.

## The Cooperative Learning Environment: Catalyst and Reactive Principle

The concepts of cooperative training and collaborative training are used in an undifferentiated way by many authors. However, these concepts clearly distinguish two differing teaching orientations. Within the framework of cooperative training, the teacher has a very important guiding role: he fixes a goal, for example the realization of a product, and implements a teaching strategy based on cooperation instead of competition. The group, under the direction of the teacher, cooperates in order to create the product.

Within the framework of the collaborative approach each participant is responsible for the realization of a part of the work and brings his contribution. The group then works together in a mutual fashion in which the authority of the teacher may be extremely modest. In the optimistic view of ICT in pedagogy, many advantages can be emphasized: suppression of the physical limits of the classroom, remote accessibility, increased exchanges among learners, high motivation of learners, more effective and individualized teaching, and great autonomy of the learners. In the Form-Ami project we adopted the cooperative approach with intensive use of ICT.

Within the framework of this interactive teaching model, technology plays a role in knowledge sharing and transmission. It is at this point of the approach that confusion often intervenes between the technical support, which accelerates the transfer of information, and the teaching reality, in which knowledge transmission is an ongoing

---

[1] The main members were: Université de la Méditerranée-EJCM (Ecole de Journalisme et de Communication de Marseille, France), ESIM (Ecole Supérieure d'Ingénieurs de Marseille, France), EMA (Ecole des Mines d'Alès, France), University of Tampere Hypermedia Laboratory (Finland), University of Koblenz-Landau (Germany), Fernando Pessoa University (Portugal), Cegetel, Grolier, Gemplus, Sun Microsystems.

process that relies on human relations. Various authors raised this aspect. "The principal problems will not be of technological nature, though their importance is significant, but rather of human nature, cultural and institutional" ([14], p. 171). Some authors even consider there to be no difference between traditional teaching and teaching through ICT. According to Russell [15], more than 355 publications confirm such an assertion.

Finally our working hypothesis is as follows. On the one hand, technologies cannot be used as a fundamental reactive agent of pedagogy. On the other hand, the role that they can play as a catalyst is undeniable. Of course, "The technology-as-magical-elixir approach is dangerous and unjustified by the evidence" ([16], pp. 230-231); clearly "technology is not a panacea for educational reform, but it can be a significant catalyst for change. To those looking for a simple innovative solution, technology is not the answer. To those looking for a powerful supporting collaborative environment, technology holds tremendous potential" ([15], p. 184). How does the teaching reaction function? The pedagogue is the fundamental reactive principle, and technology an interesting catalyst. We shall try to look further into this assertion through an analysis of the experiment in the Form-Ami European project.

## An Experiment in Cooperative Work: The "Socratic" Method and the Form-Ami Project

*"...research and knowledge are on the whole only reminiscence"*
*Plato, Menon, translation from Les Belles Lettres, tome III, 2, 1923*

Through the Form-Ami project, we will illustrate the preceding argumentation on the catalyst role of technologies. Indeed, twenty-eight students of thirteen different nationalities were selected and these students profited from a teaching environment supporting exchanges both between teachers and students and among the students themselves. There was an intensive use of ICT, with classroom teaching and remote teaching. In addition, the training in a numerical environment was founded on a "Socratic" teaching approach that ensured knowledge transmission.

Technological progress radically modifies the relationship between man and information and induces a considerable demand with regard to education systems. The virtues of a "Socratic school", based on four mainstays (to learn how to know, to learn how to make, to learn how to live together, to learn how to be; on this topic, see the publication of the European Commission [17]), must be accessible through the use of ICT. Indeed, ICT must help the teachers set up a teaching approach supporting not only training but also hypothetico-deductive reasoning modes which will permit the learners to acquire autonomy vis-à-vis life events.

The Socratic approach refers to the philosophy of education and to the teaching method presented by Plato. We refer in particular to the idea of the Socratic "maieutic", which is similar to the art of midwifery, and the conditions of knowledge learning. The role of the teacher is that of a guide and the learner is responsible for his learning, which cannot be an accumulation of knowledge, but rather a search for knowledge by reminiscence.

Within the framework of the Form-Ami project some ideas of C.S. Peirce are applied. We acquire our knowledge as participants and not as witnesses to whom knowledge would be "transmitted". The theory of "learning by doing" can also be quoted. According to Dewey [18], the act of knowledge always starts with the consciousness of a difficulty, and education rests on the resolution of problems. The Socratic approach was at the heart of the Form-Ami project; the students worked and learned together by doing research, in particular on the web, by questioning the teachers and by exchanging knowledge. Numerous discussions and meetings were organized between teachers and students so that the students could adopt and put into practice the principle of Socratic exercises. This pedagogy was developed in a numerical environment, with the students using the multimedia tools more and more and solving, by themselves, the problems they encountered. The Socratic method allowed the creation of multimedia products by the students themselves on the basis of the teachers' courses.

Form-Ami was a rich pedagogical experiment because the students, with help from the teachers, naturally, had to solve problems that deal with how to put the courses into practice, how to produce a multimedia product, and how to organize their practical training. Teachers also had to adapt their pedagogical approach and the educational organization to help the students solve the problems. This new pedagogical paradigm is also described, for example, by B. Collis [19]: "What is our first aim? Learning from experiences, from one's own and from those in an organization, and building upon these experiences".

The innovative image carried by ICT incontestably attracted the students at the time of their candidacies for the Form-Ami training. From this point of view, new technologies constitute, to some extent, a pretext that incites people to enroll in training. During the teaching process, technology was used to support research (on the web), exchanges (especially e-mail), and as practical achievements (development of multimedia products). However, the sharing of knowledge and the training rested on the Socratic method, and the extremely strong involvement of the teachers. On the basis of this pilot experiment, it is possible to affirm that technologies constitute a catalyst but not the reagent, which remains centered on the pedagogue. It is also noted that ICT was supportive within the framework of cooperative work and not of collaborative work; indeed, the autonomy of the students remains limited unless they are already sufficiently experienced in research, which is rather exceptional among students involved in traditional training.

## DISCUSSION AND SUMMARY

We have analyzed the question of knowledge transmission. In a first approach the application of the Shannon theory shows the existence of a transmitter (the teacher) and of a receiver (the learner). Pedagogy permits the transmission of knowledge, according to the following basic diagram:

T —— Pedagogy-Transmission ——▶ R

Teacher — Learner

In this basic diagram, technology is used to transmit knowledge. With the development of ICT the illusion has been that the technology of transmission is in itself the essence of pedagogy. There has been assimilation between the technology that is used to transmit (in the sense of telecommunications transmission) and the pedagogy that permits knowledge transmission. However, the virtues of new technologies are far from obvious. The specialists in cognition have indeed shown the risks of confusion and/or of cognitive overload. The analysis of the Form-Ami project has led us to conclude that the process of knowledge transmission rests, to a very large extent, on the teaching organization, the human relations, the motivation of the teachers and the learners rather than on technologies. Technologies seem to be a pretext or a catalyst but, in our opinion, do not appear fundamental to the training and knowledge transmission system. Of course, it is a refutable assertion; the generalization of this assertion is impossible to establish. However, we think it would be profitable to analyze the role of human relations in knowledge transmission compared to technology. The teaching act and the role of the pedagogue remain crucial. We even advance the idea that the more information and technologies are used, the more the role of the pedagogue will be decisive. ICT can favor knowledge transmission but at the same time increases the complexity of pedagogy. In this context, ICT is a catalyst of knowledge transmission, a catalyst because it aids the mediation between teachers and students but does not constitute the first factor, or even a reactive principle. ICT glosses over pedagogy but does not change the essence of knowledge transmission. What are absolutely decisive are the human environment and the quality of the mediation.

Obviously technologies have improved telecommunication transmission, and they can have a favorable role in knowledge transmission. Realizing this potential requires an essential action in terms of mediation. That action no longer concerns the field of technology, but rests in human relations. Such is the fundamental state of knowledge transmission.

## REFERENCES


1. Badillo, P.-Y., and Bourgeois, D., "Information Society and Network Economy: From Euphoria to Reality", Communication for "Economics for the Future," Major International Conference Organised by the Cambridge Journal of Economics in Celebration of the Centenary of Cambridge Economics, Cambridge, UK, 17-19 September, 2003.

2. Davenport, T. H., and Prusak, L., *Working Knowledge: How Organizations Manage What They Know*, Harvard University Press, Cambridge, MA, 1997.

3. Clifford, P., and Friesen, S., "Bringing Learning to Learners: The Galileo Educational Network," ED-MEDIA 2001, Hypermedia Laboratory, University of Tampere, June 2001.

4. Jonassen, D., "E-Learning to Solve Problems," ED-MEDIA 2001, Hypermedia Laboratory, University of Tampere, June 2001.

5. Ferry, L., Darcos, X., and Haigneret, C., *Lettre à tous ceux qui aiment l'Ecole*, Odile Jacob, Paris, 2003.

6. Shannon, C.E., "A Mathematical Theory of Communication," Bell System Technical Journal, Vol. 27, July and October 1948, pp. 379-423 and 623-656.

7. Shannon, C.E., and Weaver, W., *The Mathematical Theory of Communication,* University of Illinois Press, Urbana, 1949.

8. Watzlawick, P., Beavin, J., and Jackson, D., *Pragmatics of Human Communication,* Norton, New York, 1972.

9. Winkin, Y., *Anthropologie de la communication : de la théorie au terrain,* Editions Le Seuil, Paris, 2001.

10. Morin, E., *La Méthode : la nature de la nature* (tome1), Editions Le Seuil, Paris, 1977.

11. Ducastel, F. (Dir.), *Les Télécommunications*, V. A. Descours, Berger-Levrault International, Paris, 1993.

12. Debray, R., *Introduction à la médiologie*, PUF, Paris, 2000.

13. Badillo, P.-Y., "FORM-AMI, The Key for the Future: The New Pedagogical and Knowledge Mediation - A Strategy of Replication and Dissemination," Final Form-Ami Report for the European Commission, Marseille, July 2001.

14. Viens, J, "Environnements d'Apprentissages Collaboratifs à l'Université," in Karsenti, T., Larose, F., *Les TIC... Au Cœur Des Pédagogies Universitaires,* Presses de l'Université du Québec, Sainte Foy, 2001.

15. Russell, T. L., "The No Significant Difference Phenomenon," NCSU Office of Instructional Telecommunications, North Carolina, 1999.

16. Pepi, D., and Scheurman, G., "The Emperor's New Computer: A Critical Look at Our Appetite for Computer Technology," Journal of Teaching Education, Vol. 47, No. 3, 1996, pp. 229-236.

17. European Commission, *Teaching, Learning, Information: Towards an Open Socratic School,* Official Publications of the European Community, Luxembourg, 1998.

18. Dewey, J., *Experience and Education*, Touchstone, New York, 1938.

19. Collis, B., "Linking Organizational Knowledge and Learning," ED-MEDIA 2001, Hypermedia Laboratory, University of Tampere, June 2001.

**Patrick-Yves Badillo** is Professor at the Université de la Méditerranée – Aix-Marseille II and the Director of the Ecole de Journalisme et de Communication de Marseille (EJCM). He obtained an M.S. in Economics, Econometrics/Statistics from the Université de la Méditerranée Aix Marseille II, and an M.S. in Telecommunications from the Université de Paris IX Dauphine. He obtained his Ph.D. in 1987 in the field of the Theory of Innovation from Université de Paris I La Sorbonne. He first worked at France Télécom - Centre National d'Études des Télécommunications (CNET), as head of studies attached to the General Director of the CNET for Engineering and Economic Studies in the field of New Technologies (videocommunication networks, audiovisual industry and satellite systems). After his success at the Agrégation des Facultés de Droit et des Sciences Économiques in 1988, he had various responsibilities at the University. Since 2000 he has also been a Professor in Information and Communication Science. He is the Director of the M.S. degree program "Sciences et Technologies de l'Information et de la Communication et Médiation des Connaissances" and was the head of the FORM-AMI project (1999-2002). He is Director and founder of the Research Laboratory on Media, Information and Knowledge.

**Dominique Bourgeois** is Professor at the Université de la Méditerranée – Aix-Marseille II, France. She obtained her M.S. in Economics and languages from the University of Grenoble in 1975. She obtained a first Ph.D. in International Economics in 1978 (Université de la Méditerranée – Aix-Marseille II) and worked at the University of Abidjan, Ivory Coast, and the Université de la Méditerranée. She obtained a second Ph.D. in 1985 in Economics/Econometrics. After her success at the "Agrégation des Facultés de Droit et des Sciences Économiques" in 1985, she was Professor in Economics at the University of Angers and then at the Université de la Méditerranée. She produced different official reports for various institutions (Commissariat Général du Plan, etc.), especially in the field of econometric forecast. She is now Professor in Information and Communication Science and her research areas include information and communication technology, education and the development of electronic commerce and *e*-activities.

# Chapter 4

# Assessing Creativity Aspects in Engineering Students' Project Work

SHLOMO WAKS and MERDLER MOTI
*Technion - Israel Institute of Technology, Department of Education in Technology & Science, Haifa 32000, Israel. E-mail: waks@tx.technion.ac.il*

*Qualitative research methodologies have become more commonly applied in science and engineering education. There may be different reasons for implementing qualitative instruments of inquiry, ranging from ideological factors up to practical necessities whenever quantitative tools are inadequate. This article explores an array of both quantitative and qualitative methods for exploring creativity aspects by following students' activities on a comprehensive project and in producing a creditable artifact. We report on a study that has been carried out in the electronics and marketing disciplines. Findings indicate that "originality" is correlated to "system thinking" or "system approach" in the sense of using knowledge from other disciplines, while "innovation" depends on the extent to which the application of a new idea is adequate and effective. Nevertheless, both "originality" and "innovation" require sufficient disciplinary knowledge to work upon. Qualitative and quantitative methods of inquiry indicate similar and complementary explanatory factors related to creativity in engineering design.*

## INTRODUCTION

Creativity research, or rather the search for creativity, extends beyond content domains and disciplines and, as such, it has more commonly been studied in the last decade. Competition and the availability of advanced software and hardware have deployed a platform upon which creativity in engineering design could be attained on a daily basis. It is not surprising, perhaps, that creativity research emerged in engineering education, since patterns of creative behavior resemble those in developing artifacts. Nevertheless, valid and reliable methodological means of inquiry, regarding the phenomenon of creativity and its complexity, have not yet been established.

In the early nineteen-sixties, as the necessity for creativity emerged as a major issue during the era of space exploration, psychology researchers began to develop assumptions and theories about the nature of creativity. Normative means and tools for measuring creativity, especially creative personalities and creative thinking, were constructed and tested with extensive applications in education. In the late seventies, creativity research was divided into four categories: personality, process, product and place or environment. This categorization has been since then referred to as the 4-P's. Although creativity research in any category is essential for the understanding of creative behavior, it is still necessary to understand the relationships between the 4-P's.

While quantitative instruments and tests can estimate elements of the first two categories, quantitative measurement of the creative product cannot. Therefore, the logical approach of assigning the assessment of a creative product to experts in their field has been established [1]. Methodologically, this observation is important because the evaluation of a product or artifact has validity not only in engineering [2], but also ecologically [3]. Consequently, a product's embodied creativity and the expressed creativity of its producers are assessed by observations and interviews that often have phenomenological meaning. Still, it is essential to raise the question of what are the circumstances that are most likely to foster creativity, and whether they should be assessed. Since the essence of engineering is to design a useful product, creativity should be evaluated when working on a design project. Design has multi-farious aspects that could contribute to our understanding of how creativity can be fostered. Major factors whose influence could shed light on the ability to perform creatively include: motivation and personality; curricular and extra-curricular knowledge; dependence on the instructor; divergent or lateral thinking throughout the various stages of a project's development; and dealing with constraints of time and resources.

Engineering projects may have different scopes and purposes, any of which may be suitable for creativity research. For example, a certain group of high school students might engage in a selected topic during which a creativity fostering model, such as De-Bono's Cort model, could be examined for its usefulness [4]. Alternatively, students might work on a series of projects that gradually become less constrained [5], and the project might have the scope and freedom of a summative engineering project during which the effect of various creativity factors could be analyzed [6]. However, as we deal with many students engaged in wide-ranging and rather comprehensive projects, it is important to determine which cases should be thoroughly examined, presumably by qualitative in-depth interviews and project observations, so as to reveal more about the nature of creativity involved in engineering design.

Creative thinking that may pave the way to original ideas or solutions is insufficient in engineering design. Although an original solution to an engineering problem needs primarily sensitivity to gaps in information and knowledge, it must still be applicable and useful. This applicability is usually referred to as innovation; as such, it can be evaluated only when the designed product is completed and acknowledged by consensus to exhibit the novelty and adaptability which express creativity in a technological society [7]. Dieter [8] even suggests two alternative patterns of a creative product evolution. The rarest form is a sudden enlightenment, while more common is a targeted and gradual evolution, which, according to Dieter, demands well-based knowledge and precise thinking.

If we acknowledge working on a design project and producing an artifact to be an activity during which creativity is very likely to be expressed, then we should evaluate two aspects. The first is whether the idea or solution is original, i.e. that it expresses a new and unique approach to a problem when compared to what has commonly been produced during project work in that subject. The second is whether the solution is efficient and applicable.

## THE STUDY

We report the findings of a study examining aspects of creativity in the work of practical engineering students on their summative projects in two disciplines: 10 projects in electronics and 12 projects in marketing. The study was conducted without any intervention by the researchers. From that point of view, it has explored the nature of creativity as it is reflected in the field and circumstances. The study examines the assumption that originality can be obtained by a system approach to the topic or problem at hand, and that innovation requires compatibility and effectiveness. Findings indicate some bounds and conditions on which those assumptions are based.

| **Project work - study phase** | **Project work and product presentation** | **Introc ıction of video-recorded proje products and prese ation for evaluation by exper** | **Ic ntification of original aı l innovative projects and in ərviews of students and in tructors** |
|---|---|---|---|
| **Means of inquiry** | Nonparticipating video-recorded observations. | Experts evaluate creativity aspects by filling in a questionnaire while watching a video recording of the students introducing their products and being interviewed during the process. | In-depth interviews with students and their instructors after experts in the previous phase identify them as original and innovative. |
| **Inquiry objectives** | Description of presentation dynamics, focused on evaluators remarks which concern creativity aspects | Identify relationships between operational definitions of system approach, disciplinary knowledge, effectiveness and originality of idea/solution and innovation. | Reflective description of the project work by students and their instructors who produced a creative product. Students' self-evaluation of their creativity. |
| **Analysis** | Content analysis of the observation – categories classification of creativity aspects. | Reliability among evaluators by non- parametric test. Multi-variance analysis for original and innovative product as a dependent variable. Cluster analysis for identifying original and innovative projects. | Interview content analysis – categories classification of creativity aspects. Validation of categories by experts. |
| **Comments** | | | |

TABLE 1
THE STUDY ARRAY

Students that participated in the study learned practical engineering from 1998 to 2001 in a college located in northern Israel. Since the main goal of their study is practical, students were required to complete a major project. Once a student chose an instructor and began his project, the actual project activities would last approximately one year. The final product, which might be an artifact, instrument or a marketing plan, was then presented to evaluators. The presentation was usually conducted out in front of students, teachers and instructors. Although no questions or comments were permitted during the presentation, the situation was could still be quite stressful for the presenter. The process of working on the project was assessed for creativity aspects through qualitative and quantitative methods of inquiry as described in Table 1.

## RESULTS

In this paper, we describe the process of two experts in evaluating the originality and innovation reflected in project product presentations in the fields of electronics and marketing. We also examine presumed insights about the relationship between system approach and creativity.

Video-recorded presentations were displayed to two experts in every field. The experts filled in an 8-question questionnaire (see Appendix) for each presentation. The questionnaire included 3 taxonomic questions regarding the extent to which non-disciplinary knowledge was involved in the project (mentioning, using and successfully integrating non-disciplinary knowledge). Inter-correlations were examined between the experts as shown in Tables 2, 3.

| | exp 1-1 | exp 1-2 | exp 1-3 | exp 1-4 | exp 1-5 | exp 1-6 | exp 1-7 | exp 1-8 | exp 2-1 | exp 2-2 | exp 2-3 | exp 2-4 | exp 2-5 | exp 2-6 | exp 2-7 |
|---|---|---|---|---|---|---|---|---|---|---|---|---|---|---|---|
| exp1 - 2 | .683* | | | | | | | | | | | | | | |
| exp1 - 3 | .725* | .442 | | | | | | | | | | | | | |
| exp1 - 4 | .238 | .190 | .205 | | | | | | | | | | | | |
| exp1 - 5 | .551 | .854** | .477 | -.03 | | | | | | | | | | | |
| exp1 - 6 | .729* | .418 | .516 | -.11 | .227 | | | | | | | | | | |
| exp1 - 7 | .878** | .675* | .578 | .093 | .665* | .428 | | | | | | | | | |
| exp1 - 8 | .518 | .669* | .565 | -.14 | .679* | .514 | .547 | | | | | | | | |
| exp2-1 | .552 | .919** | .407 | .011 | .831** | .410 | .493 | .544 | | | | | | | |
| exp2-2 | .399 | .690* | .385 | -.09 | .727* | .212 | .566 | .840** | .599 | | | | | | |
| exp2-3 | .532 | .837** | .498 | -.12 | .730* | .342 | .587 | .787** | .768* | .779** | | | | | |
| exp2-4 | .275 | .591 | .423 | .065 | .587 | .082 | .441 | .758* | .514 | .866** | .655* | | | | |
| exp2-5 | .518 | .669* | .565 | -.14 | .679* | .514 | .547 | 1.0** | .544 | .840** | .787** | .76* | | | |
| exp2-6 | .703* | .590 | .471 | .084 | .348 | .859** | .485 | .724* | .478 | .557 | .548 | .39 | .724* | | |
| exp2-7 | .700* | .669* | .565 | .100 | .679* | .514 | .755* | .806** | .544 | .840** | .567 | .76* | .806** | .724* | |
| exp2-8 | .555 | .583 | .476 | .00 | .513 | .491 | .604 | .922** | .331 | .700* | .887* | .63* | .922** | .718* | .738* |

* Correlation is significant at the .05 level (2-tailed).

** Correlation is significant at the .01 level (2-tailed).

TABLE 2
SPEARMAN'S RHO (N=10) CORRELATIONS BETWEEN FACTORS AND BETWEEN TWO ELECTRONICS EXPERTS

**Key: exp 1** = the first expert, **exp 2** = the second expert, digit after dash indicates statement number.
**Example 1: exp 1-2** = second statement ("Student used knowledge from other fields.") of the first expert's evaluation questionnaire (see Appendix).
**Example 2: exp 2-7** = seventh statement ("Student applied the concept appropriately?) of the second experts' evaluation questionnaire for all projects.

| | exp 1-1 | exp 1-2 | exp1 -3 | exp 1-4 | exp 1-5 | exp 1-6 | exp 1-7 | exp 1-8 | exp 2-1 | exp 2-2 | exp 2-3 | exp 2-4 | exp 2-5 | exp 2-6 | exp 2-7 |
|---|---|---|---|---|---|---|---|---|---|---|---|---|---|---|---|
| exp1 - 2 | .756** | | | | | | | | | | | | | | |
| exp1 - 3 | .562 | .532 | | | | | | | | | | | | | |
| exp1 - 4 | .295 | .327 | .671* | | | | | | | | | | | | |
| exp1 - 5 | .594* | .307 | .479 | .461 | | | | | | | | | | | |
| exp1 - 6 | .541 | .292 | .532 | .653* | .679* | | | | | | | | | | |
| exp1 - 7 | .427 | .150 | .307 | .344 | .548 | .864** | | | | | | | | | |
| exp1 - 8 | .787** | .470 | .422 | .347 | .777** | .797** | .792** | | | | | | | | |
| exp2-1 | .710** | .582* | .663* | .206 | .392 | .533 | .395 | .576 | | | | | | | |
| exp2-2 | .602* | .602* | .312 | .430 | .266 | .461 | .229 | .478 | .647* | | | | | | |
| exp2-3 | .667* | .280 | .297 | .417 | .396 | .479 | .278 | .551 | .501 | .763** | | | | | |
| exp2-4 | .631* | .631* | .259 | .037 | .370 | .314 | .064 | .408 | .714** | .620* | .418 | | | | |
| exp2-5 | .762** | .515 | .555 | .500 | .691* | .745** | .582* | .738** | .772** | .788** | .749** | .579* | | | |
| exp2-6 | .760** | .531 | .415 | .128 | .562 | .498 | .356 | .706* | .865** | .733** | .677* | .764** | .823** | | |
| exp2-7 | .230 | .051 | .155 | .220 | .157 | .313 | .00 | .145 | .497 | .624* | .632* | .515 | .533 | .632* | |
| exp2-8 | .644* | .277 | .586* | .380 | .731** | .446 | .275 | .680* | .566 | .424 | .667* | .334 | .646* | .733** | .395 |

**. Correlation is significant at the .01 level (2-tailed).

*. Correlation is significant at the .05 level (2-tailed).

TABLE 3

SPEARMAN'S RHO (N=12) - CORRELATIONS BETWEEN FACTORS AND BETWEEN TWO MARKETING EXPERTS

A Wilcoxon Signed Ranks Test was performed in order to establish the concordance between the two experts. During the evaluation, the experts were interviewed to determine reliability and to find out the criteria by which they distinguish between low and high creativity aspects. After the evaluation, cluster analysis was conducted to decide which projects were most original and innovative. Based on this designation, students' presentations were analyzed and preliminary interviews were conducted with both students and their instructors. Linear regression and multi-variance analysis have been conducted to identify factors that influenced originality and innovation.

According to the Wilcoxon Signed Ranks Test, both experts concur to the degree that there are no statistically significant disagreements between them. Nevertheless, they disagreed on the need for disciplinary knowledge in originality. According to the multi-variance analysis, this may be explained by a 'threshold' effect, suggesting that disciplinary knowledge must be high enough to generate originality as observed in Figures 1 and 2. Therefore, every expert may perceive a different amount of disciplinary knowledge to be required for implementing a creative solution. According to the cluster analysis, the single most creative project was identified, one that dealt with computerized radio control of high voltage pylons. The students who carried out this project described in their interviews how specific experts, instructors and craftsmen had helped them on 'pinpointed' problems such that the main instructor (the one assigned to them) was almost irrelevant. These students also had previous experience with the environment and therefore had some applicable disciplinary knowledge.

The need for minimal disciplinary knowledge is essential but not sufficient for a creative project solution. As shown in the analysis, successful integration of non-disciplinary knowledge must accompany high disciplinary knowledge (var2*var3*var4).

Regarding innovative ideas (var6), the experts suggested that only an effective application of the project's idea (var7) leads to novelty.

Observations of the project presentations and student interviews provide supportive and complementary explanations.

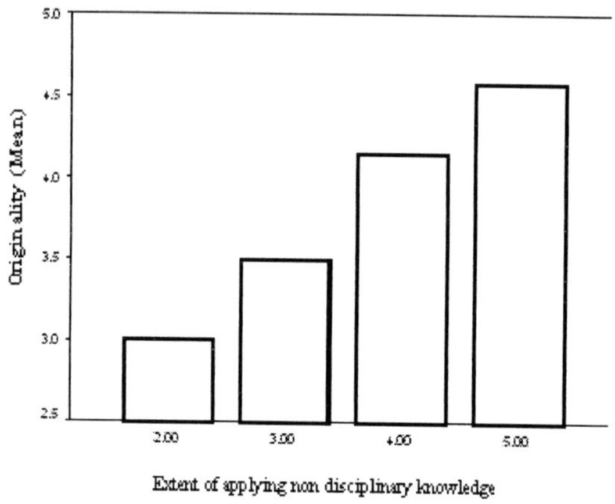


FIGURE 1
ORIGINALITY (MEAN) VS. NON-DISCIPLINARY KNOWLEDGE - ELECTRONICS

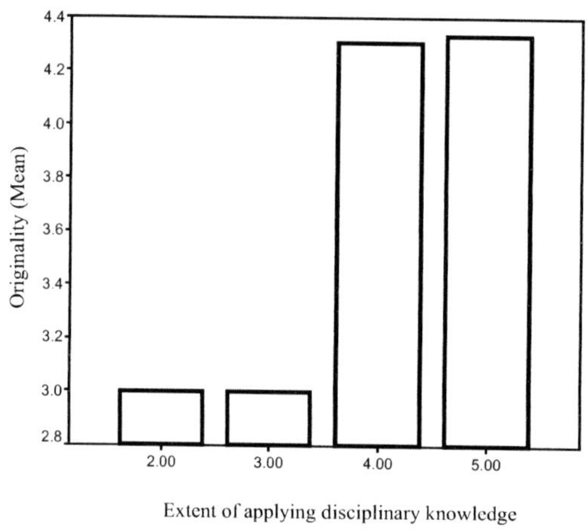


FIGURE 2
ORIGINALITY (MEAN) VS. DISCIPLINARY KNOWLEDGE - ELECTRONICS

**Supportive explanations**:

- The project examiners often referred to the complexity and challenging nature of the tasks that was undertaken by those identified as creative performers.
- Creative students expressed a need for extensive non-curricular knowledge, which was usually obtained by getting specific, "pinpoint" scholarly advice.
- Students described both past and present experience related to the topic at hand.

**Complementary explanations:**

- Creative students were anxious to talk about the ways in which they had overcome problems during the project while less creative students tend to blur their ways of resolving problems.
- Creative students stated their intention to work on new and unique ideas. This observation is important in the light of their disapproval of working on 'recycled' issues. They suggested a linkage between their desire to obtain uniqueness and the availability of 'pinpoint' instructors.

Surprisingly, the lack of agreement as to the role of disciplinary knowledge in encouraging originality has also emerged between the two marketing experts.

According to linear regression, it seems that while originality (var5) merely relates to mentioning the need for knowledge from other fields (var1), innovation in marketing is still a rather composite goal. In addition to a low, though significant, need for non-disciplinary knowledge, innovative marketing plans demand profound disciplinary knowledge (var4) and need to be usefully applied (var7) as described in Table 4.

| Model | R | R Square | Adjusted R Square | Std. Error of the Estimate | Change Statistics | | | | |
|---|---|---|---|---|---|---|---|---|---|
| | | | | | R Square Change | F Change | df1 | df2 | Sig. F Change |
| 1 | .672 | .451 | .426 | .68781 | .451 | 18.074 | 1 | 22 | .000 |
| 2 | .790 | .625 | .589 | .58222 | .174 | 9.704 | 1 | 21 | .005 |
| 3 | .843 | .711 | .668 | .52312 | .087 | 6.013 | 1 | 20 | .024 |

**ANOVA[d]**

| Model | | Sum of Squares | df | Mean Square | F | Sig. |
|---|---|---|---|---|---|---|
| 1 | Regression | 8.550 | 1 | 8.550 | 18.074 | .000[a] |
| | Residual | 10.408 | 22 | .473 | | |
| | Total | 18.958 | 23 | | | |
| 2 | Regression | 11.840 | 2 | 5.920 | 17.464 | .000[b] |
| | Residual | 7.119 | 21 | .339 | | |
| | Total | 18.958 | 23 | | | |
| 3 | Regression | 13.485 | 3 | 4.495 | 16.426 | .000[c] |
| | Residual | 5.473 | 20 | .274 | | |
| | Total | 18.958 | 23 | | | |

a. Predictors: (Constant), VAR1

b. Predictors: (Constant), VAR1, VAR7

c. Predictors: (Constant), VAR1, VAR7, VAR4

d. Dependent Variable: VAR6

TABLE 4

LINEAR REGRESSION (STEPWISE) OF FACTORS THAT AFFECT THE EMERGENCE OF INNOVATION

# CONCLUSIONS

One of the assumptions of this research was that a system approach to a problem or to a targeted product would facilitate opportunities to evoke original ideas and solutions. Although this assumption is corroborated by the analysis of the results, the results also highlight two major issues regarding the ways in which a system approach affects originality and innovation in engineering design. First, the systems approach is not a substitute for disciplinary knowledge, it fact it increases its importance. Second, the factors that trigger innovation appear to differ across disciplines. While the field of electronics innovation may be concerned only with the useful application of the artifact, marketing innovation may also require a comprehensive outlook involving sensitivity to other fields' impact on the marketing plan.

This preliminary but triangulated observation suggests two further research opportunities. The first is to clarify the reasons for the disagreement between experts as to the level of disciplinary knowledge required for the application of creative solutions.

One might reasonably question the validity of a specific expert's evaluation of interdisciplinary knowledge involvement in a particular project, but an evaluation of the disciplinary knowledge requirement in general should be less suspect. Therefore, this observation is somewhat surprising. The second research avenue is whether or not there exists similar 'threshold' and 'pinpoint' phenomena in other engineering fields.

## APPENDIX

| | To what extent... | Very high | ------- | | | Very low |
|---|---|---|---|---|---|---|
| Var1 | Student mentioned the need for knowledge from other fields? | 5 | 4 | 3 | 2 | 1 |
| Var2 | Student had used knowledge from other fields? | 5 | 4 | 3 | 2 | 1 |
| Var3 | Student integrated knowledge from other fields? | 5 | 4 | 3 | 2 | 1 |
| Var4 | Student had used knowledge from his/her field? | 5 | 4 | 3 | 2 | 1 |
| Var5 | Student expressed originality? | 5 | 4 | 3 | 2 | 1 |
| Var6 | Student introduced innovative ideas? | 5 | 4 | 3 | 2 | 1 |
| Var7 | Student had applied the idea appropriately? | 5 | 4 | 3 | 2 | 1 |
| Var8 | Student expressed creativity? | 5 | 4 | 3 | 2 | 1 |

EXPERTS' ASSESSMENT QUESTIONNAIRE

## REFERENCES


1. Amabile, T.M., *Creativity in Context: Update to Social Psychology of Creativity,* West View Press, 1996.

2. MacKinnon, D.W., In: *Research of Human Effectiveness: Identifying and Developing Creativity,* Creative Education Foundation, Buffalo, NY, 1978.

3. Isaksen, S.G., Puccio, G.J. and Treffinger, D.J., "An Ecological Approach to Creativity Research: Profiling for Creative Problem Solving," Journal of Creative Behavior, Vol. 27, No. 3, 1993.

4. Barak, M. and Doppelt, Y. "Integrating the cognitive research trust (Cort) program for creative thinking into a project – based technology curriculum.", Journal of Research in science & technology education," 17, 1999.

5. Matthews, J.M. and Shahriar, J.A., "Pedagogical Strategy for Enhancement of Creative Performance of the Students," European Journal of Engineering Education, Abingdon, Vol. 24, 1999.

6. Waks, S. and Merdler, M., "Creative Thinking of Practical Engineering Students During a Design Project," Research in Science & Technological Education, Vol. 21, No. 1, 2003.

7. Ripple, R.E., "Ordinary Creativity," Contemporary Educational Psychology, Vol. 14, 1989.

8. Dieter, G.E., *Engineering Design,* McGraw-Hill, New Jersey, 1991.


**Shlomo Waks** received his B.Sc. in Electrical Engineering in 1962, and his M.Sc. in Science Teaching in 1971 at Technion, the Israel Institute of Technology. He received his M.Sc. in Electrical Engineering and Ph.D. in Science & Math Teaching in 1973 at Michigan State University in the USA. He has been a Professor in the Technion Department of Education in Technology & Science, Haifa, Israel since 1973. He is a member of the Board of Editors of the International Journal of Engineering Education and the Journal Committee of the European Journal of Engineering Education. He is the author of "Curriculum Design - From an Art Towards a Science," 1995, TEMPUS Pub.

**Merdler Moti** received his B.Sc. in Aeronautical Engineering in 1987 and his M.Sc. in Technological and Science Education in 2002 at Technion, the Israel Institute of Technology. He has been a doctoral student since 2002. He carries out an extensive study of creativity in project-based learning at the Western Galilee College in Israel. He acts as an interdisciplinary instructor of students involved in project work.

# Chapter 5

# NeuroEngineering: The Integration of Neuroscience with Engineering

JACK W. JUDY and ALLAN J. TOBIN
*University of California, Los Angeles, California, 90095, USA.*
*E-mail: jjudy@ucla.edu*

*Neuroengineering is a new, exciting, challenging, and multi-disciplinary field of research that employs the latest engineering technologies to address problems having a neuroscientific basis. In order to accelerate progress in neuroengineering research, efforts beyond the teaming of experts and laboratories that specialize in engineering and neuroscience are necessary. Specifically, researchers with non-traditional backgrounds that combine neuroscience and selected fields of engineering could make major contributions. This article describes a neuroengineering research and training program developed at UCLA that targets development of such individuals. In addition to discussing the program, the multi-disciplinary research projects being pursued by the trainees are described and listed. The first three trainees have already graduated and are making an impact as Assistant Professors at universities in the United States and elsewhere in the world, as well as becoming leaders in industry at a neuroengineering start-up company. The future of the program is bright, and it is formally established as an integral part of UCLA.*

## INTRODUCTION

Revolutionary progress in treating neurodegenerative diseases (e.g., Alzheimer's, Parkinson's, Huntington's, etc.), in sensory and motor prostheses (e.g., cochlear, retina, muscle, etc.), and in dealing with spinal cord injuries (e.g., bladder, locomotion, etc.) require revolutionary approaches in multi-disciplinary research technologies, instrumentation, and graduate-level training. In order to address these needs, we have developed a specialized research and training program that reduces the gap that exists in the training that engineers and neuroscientists traditionally receive, the language they use (e.g., a "vector" means something very different to an engineer and a neuroscientist), the problem-solving methodologies they employ, and the technologies they leverage. This effort brings together departments from the life and medical sciences (e.g., neuroscience, neurology, neurosurgery, neurobiology, psychology, etc.) and the physical sciences and engineering (e.g., biomedical engineering, chemical engineering, computer science, electrical engineering, material science, mechanical engineering, applied mathematics, physics, etc.).

We seek to improve neuroscientific instrumentation to enable a better understanding of the nervous system as well as to develop advanced treatments and prostheses to better control aspects of the nervous system. The program described in this paper has been supported by the United States National Science Foundation with a funding opportunity specifically directed at supporting integrative graduate education and research training (IGERT) programs [1]. In many ways, what we are trying to do is create a new collaborative discipline, such as was done successfully with molecular and cellular biology and with materials science and engineering. Although the formulation of each of these collaborative disciplines took many years and encountered initial resistance, they became firmly established in the broader science and engineering curricula of many institutions. We believe that the collaborative discipline of neuroengineering will follow a similar path to establishment.

## GOALS AND OBJECTIVES

The goal of the UCLA NeuroEngineering Training (NET) Program is to prepare graduate students to be leaders in the revolutionary technological developments that will affect neuroscience in the 21st Century. By expanding the synergies between the UCLA Brain Research Institute (BRI) and the Henry Samueli School of Engineering and Applied Sciences (HSSEAS), the NET program promotes the application of new engineering technologies to neuroscience, including micromachining and microelectromechanical systems (MEMS). This training program provides the formal structure for collaborative neuroengineering research and training.

As one example of how engineering technologies can impact neuroscientific research, consider the revolutionary implications of MEMS technologies for neuroscience. The potential currently exists to develop arrays of microsystems that can be tailored to the physical and temporal dimensions of individual cells. Neuroscientists can now realistically envision sensing devices that allow real-time measurements at the cellular level of many different measurands (e.g., electrical, chemical, etc.). Information from such sensors could be monitored, analyzed, and used as a basis for experimental or medical intervention, again at a cellular level.

The objectives of the NET Program are (1) to enable students with a background in biological science to develop and execute projects that make use of state-of-the-art technology – including microelectromechanical systems (MEMS); in preparing students to use new technology, the NET Program also introduces them to basic concepts in engineering that are applicable to the study of systems neuroscience, including signal processing, communications, and information theory; (2) to enable students with a background in engineering to develop and execute projects that address problems that have a neuroscientific basis, including locomotion and pattern generation, central control of movement, and the processing of sensory information; and (3) to instill in all trainees the capacity for the multi-disciplinary teamwork that will be necessary for new scientific insights and dramatic technological progress.

Appreciating the difficulty of accomplishing this vision, faculty involved with the UCLA NeuroEngineering Training Program have established a curriculum and a structure that forces themselves and the trainees to confront both the theoretical and practical challenges of our shared goal. In addition to practical obstacles, substantial intellectual, organizational, and cultural impediments must also be overcome. For example, many micromachining and MEMS technologies have been derived from integrated-circuit (IC) manufacturing, and were developed by a group of engineers and physical scientists with little experience in the life sciences or biomedical engineering. Accomplishing the successful application of MEMS technologies to neuroscience will require an unusually strong emphasis on biological problems and their interactions with engineering solutions.

Unfortunately, almost all graduate programs in the life sciences prepare trainees to be academic scientists within traditional disciplines, using the traditional tools of that discipline. This educational conservatism limits the infusion of new technologies to the occasional chance encounter between researchers. With ever-increasing neuroscientific knowledge and ever more powerful technology, the NET Program faculty is convinced that we can prepare students for innovative work not only within universities but also within the more collaborative environments of industry, biotechnology, government laboratories, research institutes, and government agencies. Outside of biomedical engineering programs, few Ph.D. programs in biological science currently prepare students for these non-traditional careers and experiences.

## RESEARCH AND TRAINING STRATEGY

The basis of the UCLA NeuroEngineering Training Program has been our conviction that disciplinary, departmental, cultural, institutional, and social boundaries should not restrict the development of new concepts and technologies. The theme of the NET Program has been the collaborative search for: (1) applications of advanced engineering technology to the study of the nervous and neuromuscular systems; and (2) applications of neuroscientific knowledge to the design of new technologies with commercial and biomedical applications.

The UCLA NET Program Faculty in the HSSEAS have committed time and resources to establish the UCLA NET Program and to introduce neuroscience-oriented NET graduate students to advanced engineering concepts, including MEMS, signal processing, wireless communication, and information theory. Similarly, NET Program faculty within the BRI has committed time and resources to introduce engineering-oriented NET students to neuroscientific problems, particularly in the areas of sensory-information processing, movement control, and pattern generation. Perhaps most significantly, both the BRI and HSSEAS have a long-standing commitment to multi-disciplinary collaborations, though this is the first time that the two units have created a formal cross-school agreement. The physical juxtaposition of HSSEAS, BRI, the new Gonda (Goldschmied) Neuroscience and Genetics Research Center (which houses the BRI), and the School of Medicine further fosters this collaborative endeavor (all buildings are located within 100 m of each other).

The shared vision of the BRI and HSSEAS has been to form a center of educational excellence for multi-disciplinary neuroscience and its translation into engineering and technological applications. Formally, the program is a joint endeavor between the Interdepartmental Ph.D. Programs (IDP) for Neuroscience and for Biomedical

Engineering. An important part of our shared vision is a program that prepares our trainees to flourish in both traditional and nontraditional settings by taking the fullest advantage of the intellectual, cultural, and sociological diversity of our graduate programs and our faculty.

The starting point for the research strategy is the set of engineering technologies already developed by NET faculty. For example, ongoing HSSEAS MEMS research and the availability of these technologies has allowed us to identify new methods of addressing neuroscience problems, which provided an early focus for part of the NET Program. We are modifying, adapting, and customizing these technologies to address neuroscience problems first in an *in vitro* and ultimately in an *in vivo* manner. The results then provide a broad template for the development of new technologies in directions relevant to neuroscience. An illustration of this cyclical process is shown in Figure 1. While much of the early NET efforts have been in the adaptation phase, projects exist in all stages of the cycle. The NET Program strives to support a balanced mixture of projects in different phases.

Training a cohort of NET students together in classes co-taught by engineering and neuroscience faculty promotes and accelerates the appreciation and understanding of each field. For all the students, the unique feature of the Program has been the close interactions between engineers and neuroscientists – both students and faculty – in a wide variety of research and educational endeavors.

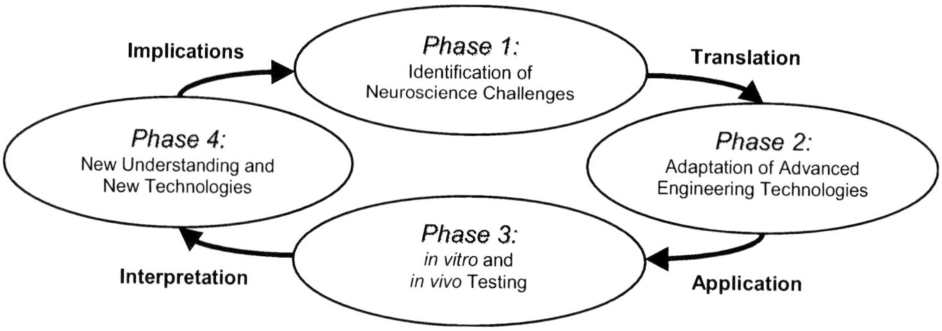


FIGURE 1
THE UCLA NET PROGRAM RESEARCH STRATEGY

## TRAINING PROGRAM

The NET Program draws students from disparate educational and social backgrounds. A major goal of the program has been to ensure that trainees from different backgrounds are conversant with the knowledge, concepts, vocabulary, and culture of both neuroscience and engineering.

### Core Courses

Both the Biomedical Engineering (BME) and Neuroscience (NS) Interdepartmental Programs (IDP) have developed a special track that provides an overview of both engineering and neuroscience. First-year NET students, regardless of whether they enter the BME or NS IDPs, take three core graduate courses in Neuroscience, normally

Neuroanatomy, Cellular Neurophysiology, and Systems Neuroscience, as well as a new HSSEAS course on NeuroEngineering.

Although trainees are carefully selected so that they enter the program well-prepared for the curriculum at the outset, students occasionally take an additional advanced undergraduate course in order to better prepare them for one or more of the core graduate-level courses. A particularly notable example is the fact that most trainees from engineering, who have never had an anatomy course, typically do not compete well when thrust directly into the graduate-level neuroanatomy course with students from neuroscience, biology, and medicine. Both the subject material (anatomy of the nervous system) and the pedagogical methods (massive memorization) are alien to these students. Fortunately, the graduate-level neuroanatomy course at our institution is offered in the winter quarter, allowing trainees to first take an undergraduate primer course on neuroanatomy. With this preparation, the engineering students compete very well in the graduate-level neuroanatomy course.

## Elective Courses

During their first and second years, NET Program students also choose electives from many departments (e.g., neuroscience, biomedical engineering, physiological sciences, electrical engineering, physiology, biophysics, etc.). A short list of some of the courses available to the students includes:

**Signals, Control Systems, and Information Processing:**

- Systems and Signals; Principles of Feedback Control
- Digital Signal Processing; Digital Image Processing
- Information Theory; Applied Bayesian Inference
- Cybernetics, Biomodeling, and Biomedical Computing
- Modeling in Neurobiology
- Neural Control of Physiological Systems
- Neural Systems for Motor Control
- Neural Basis of Memory; Computational Neuroscience

**Instrumentation Development and Application:**

- Methods in Cell Physiology; Bioelectricity
- Analog and Digital-Circuit Design
- Microelectromechanical Systems (MEMS) Fabrication
- Microsensors and Microinstruments; Microsciences

**Imaging Technology, Techniques, and Applications:**

- Biomedical Imaging
- Principles/Applications of Magnetic Resonance Imaging
- Neuroimaging and Brain Mapping
- Functional Neuroimaging: Techniques and Applications
- Physics and Chemistry of Nuclear Medicine
- Radiopharmaceutical Chemistry

**Processing and Engineering Cells, Tissues, and Materials:**

- Cellular and Molecular Neurochemistry
- Membrane Molecular Biology
- Biochemical Reaction Engineering
- Bioseparations and Bioprocess Engineering
- Introduction to Tissue Engineering
- Introduction to Biomaterials

## Seminars

All first-year NET students take a graduate seminar series that is co-taught by two faculty each quarter, with one instructor from HSSEAS and one from the BRI. Each seminar focuses on a single area of neuroengineering. Students are challenged both to develop critical skills in evaluating primary research papers and to design new approaches to current problems. Topics have included neural-electronic interfaces, engineering approaches to sensory signal processing, initiation and control of movement, microsensors, and neuroimaging.

## Responsible Conduct of Research

During the winter of the second year, students take a course in the responsible conduct of research. This course has already been developed for trainees in the UCLA Biotechnology Program. This course addresses problems in bringing scientific and technological discovery to practice. Topics will include issues of clinical trials, regulation (e.g., by the FDA), industrial partnerships, and entrepreneurship. Although most students enter this class with a preconception that breeds reluctance, all trainees have found this to be a very enlightening and fun course.

## Scientific Communication and Strategies for Translation Research

Since we expect that most graduates from this Ph.D. program will go on to careers in research, whether academic or commercial, it is vital that each student be able to effectively transmit their research accomplishments as well as understand the grant-writing process. This course addresses problems of scientific communication, ranging from the preparation of fellowship and grant applications to the presentation of scientific and technological news to the lay public.

## Laboratory Rotations

During the first year, each student is encouraged to complete two laboratory "rotations," one in a neuroscience laboratory and one in an engineering laboratory. While this is quite customary in the life science departments at our institution, it is not as common in the academic culture in the engineering school. Nevertheless, the wide range of research possibilities available to neuroengineering students behooves them to explore the laboratory realities of the topics they initially find the most interesting. Ideally each student undertakes a collaborative project that can be explored in both rotation laboratories, but even uncorrelated experience between the two lab rotations is extremely valuable. In addition to research work associated with different areas of investigation, the student-mentor compatibility is an important relationship to explore before committing to a particular laboratory.

## Fall Retreat

During the first quarter of each academic year, the NET Program sponsors a two-day retreat, which is attended by students, appropriate postdoctoral fellows, program faculty, and members of the industrial advisory board. It is at this event that the occasional culture clash between engineering and neuroscience has been the most evident. The participants in the first neuroengineering retreat, which was held in the fall of 1999, included trainees, mentors, other like-minded faculty, program administrators, and individuals invited from industry and federal agencies. Since those present felt that this meeting was extremely successful at kicking off our program, we sought to expand the audience and thus worked to combine it with the larger annual neuroscience retreat. Our first attempt at combining the retreats in Fall 2000 was not successful because many participants felt that too many compromises had been made. Indeed, few participants from neuroscience understood or valued the contribution of the neuroengineering components. We optimized the retreat format in subsequent years such that once again everyone viewed it as very successful. Important characteristics of the successful retreat format include:

1. Hold the neuroengineering and neuroscience retreats at the same time and at the same location.
2. Combine as many retreat activities as possible to facilitate social integration and exposure (e.g., meals, keynote speakers, social activities, ethics discussion, etc.).
3. Design a well-organized and clear joint program with parallel neuroscience and neuroengineering sessions so that participants can jump between the adjacent meeting rooms to listen to a talk with an intriguing title, increasing the chance for unexpected cross-fertilization.
4. Include a faculty and industry poster session. This event enables first-year and second-year students to see the research opportunities that exist in the labs of the presenting professors. It also enables students who are nearing graduation to examine the activities of the companies present to help them decide if an industrial career is right for them. The industry representatives have greatly appreciated the opportunity to meet a new crop of stellar prospective employees.

## Graduate Advising System

With a careful selection process, complete with comprehensive recruitment visits, self-directed and self-motivated students enter the program and rapidly find their niche within the UCLA neuroengineering community. It is absolutely critical that each student work hard to find both of his or her co-advisors (i.e., one in neuroscience or similar application area and one in engineering or similar technology field). Although in many cases an existing project and collaboration have been defined, and the trainees merely need to select the one they prefer most, some enterprising students with vision have more input in defining their project.

The most successful route for this approach has been for the students to first read through the research descriptions of the more than 200 involved faculty in the UCLA Brain Research Institute. After selecting the application areas they find most attractive, the trainees then ask each potential advisor how the research they are performing is technologically limited. This prompts the potential advisor, who in some instances is only vaguely familiar with the neuroengineering program, to think outside the traditional box of his or her discipline. Once the technological limitations are exposed, the trainees can interview the appropriate engineering faculty, sometimes exposing them for the first time to the potential neuroscientific application of their technology. The co-directors of the neuroengineering program facilitate meetings between trainees and the faculty involved to ensure all are aware of the exciting possibilities and practical constraints of the potential collaboration. The proportion of time that a student spends in each advisor's lab is determined by the project chosen. Efforts that require considerable technology development typically result in the student spending the majority of their dissertation research in the engineering or technology lab. Ideally, once the technology is ready, students will use it to demonstrate first-ever experiments in the neuroscience or application lab. The co-advisors then decide on the best selection of electives for each student given their research project. The NET Program monitors the progress and performance of each trainee and, in the event that the intended advising arrangement is not successful, prescribes remedies as needed.

## Administration And Evaluation

Day-to-day operations are the responsibility of two individuals (the Principal Investigators or "PIs") responsible for the program, who are also the authors of this article.

### Committees

The PIs consult regularly with a Steering Committee, which meets quarterly to discuss issues of policy and resource allocation. The whole training faculty regularly considers significant issues, either via e-mail or at semi-annual meetings. The Industrial and Government Advisory Committee develops relationships between the NET Program and government and industrial laboratories. They also supervise an in-depth and on-site external review of the NET Program. The Admissions and Recruitment Committee coordinates admissions and recruitment by formulating and evaluating recruitment strategies, especially for underrepresented groups. The Student Advising and Retention Committee advise individual students, identifying academic problems as they arise, and suggesting remedies. The committee evaluates the success of their efforts not only by numerical assessment of grades, student surveys, and retention rates, but also through discussions with individual students and groups of students at the annual retreat. The Curriculum Committee makes recommendations to the whole faculty concerning courses, course-content, and program requirements. They also provide feedback to instructors at the end of each quarter.

### Performance Assessment

Because the NET Program was an untried concept, both intellectually and administratively, it has been important for the Program to set performance standards and to evaluate the success of the Program in achieving them. In addition to evaluating the academic and professional success of our students, it has been important to have an external evaluation of the whole program.

The Industrial and Government Advisory Committee meets with the neuroengineering faculty annually to evaluate the program and provide suggestions for improvement. Additionally, a formal external review of the whole Program was performed once in the fourth year of the program (2003) and reported to the funding agency. In summary, the outside reviewers found our program to be very exciting and well conceived, despite the fact that it is academically challenging for both the students to complete and the faculty and university to effectively deliver. Their comments essentially called for improved balance in the overall program. Academically, this means potentially increasing the amount of required engineering course content (there is currently only one specifically required engineering course, neuroengineering; the rest are simply electives), increasing the enrollment of neuroscience students (most students currently enter through BME), and increasing the number of faculty from engineering and related fields (a majority of the faculty actively participating in the neuroengineering program come from the neuroscience or related application field). We are presently and actively implementing their recommendations to the best of our ability.

## Neuroengineering Research Projects

The neuroengineering research projects of current UCLA NET Program trainees cover a broad range of scientific questions and applications. Examples include research directed at making new technologies, tools, and devices that enable reliable and effective brain-computer or brain-instrument interfaces. The research seeks both a better understanding of the biology as well as the creation of devices directed at controlling neural function and restoring lost sensory and motor function. The following research projects have been selected as examples from the many active neuroengineering research projects underway at UCLA.

### Micromachined Probes for Deep-Brain Stimulation

A 22-mm-long multi-channel microprobe electrode array has been designed and fabricated for deep-brain stimulation (DBS) applications (Figure 2) [2]. The design attempts to minimize damage to surrounding tissue while providing sufficient stiffness to target small brain regions. The DBS microprobe is produced with a low-cost batch-fabrication process that includes a novel 3-D plating process that shapes the mechanical supporting layer of the probe into a tapered geometry similar to conventional microwires. Each microfabricated DBS probe has multiple electrodes that can be used to target and tailor the stimulus. The microprobes have the potential to significantly impact DBS research, such as the investigation of treatments for Parkinson's disease and its debilitating symptoms. In principle, this approach to multi-electrode microprobe fabrication could also be combined with integrated circuits in a batch-fabrication or hand-assembled process.

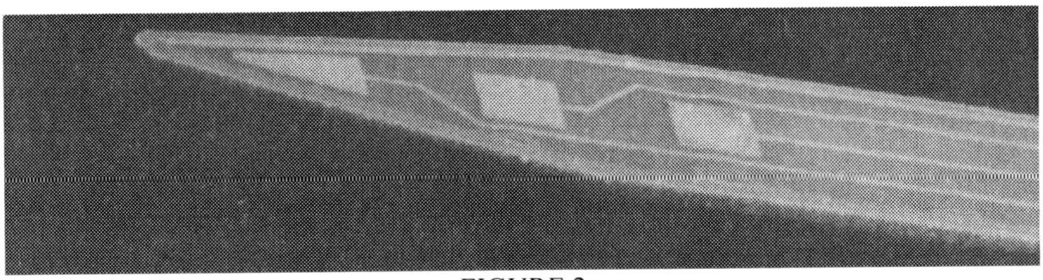

FIGURE 2
MICROMACHINED MULTI-ELECTRODE PROBES FOR DBS RESEARCH (VIEW OF UNDERSIDE) [2]

### Inductively Powered Wireless Neural Transceiver

The inability to perform electrophysiological experiments while a test subject is completely free to move and in a natural environment is currently hindering the study of some high-level neuroscience phenomena. A miniature, implantable, inductively powered and wireless single-channel neural recording device has been designed, fabricated, and tested to address this need (Figure 3) [3]. This device amplifies the recorded signal, which is then used to regulate a voltage-controlled oscillator about a 3.1-GHz carrier wave. This FM wave is sent through a power amplifier that drives a 50-Ω antenna load and transmits the signal into free space. The signal is picked up by a similar antenna and demodulated using off-the-shelf equipment. The demodulated signal has a high degree of correlation with the original input signal for inputs as small as 5 μV and as great as 1.5 mV. The device has a 50-cm transmit range that allows for continuous recordings from animals in their natural environments.

FIGURE 3
PACKAGED TRANSCEIVER TOP (LEFT) AND SIDE-VIEW (RIGHT) PHOTOGRAPHS [3]

### Real-Time *in vivo* Glutamate Measurement

Biochemical sensors have been developed that have the ability to detect in vivo glutamate concentrations in real-time using sol-gel glasses doped with enzymes or antibodies. The key feature of these sensors is the immobilization of highly selective biomolecules in a transparent, porous and inorganic glass matrix while the small analyte molecules are free to enter and diffuse through the porous network. The inorganic glass enables the sensors to be prepared in thin films that are compatible with micromachining technology, which allows them to be prepared in optical fiber or waveguide configurations. Additional work is being directed at making amperometric microsensors that use lipid layers on gold electrodes for the immobilization of membrane-bound redox enzymes.

### Fully Automated Planar Patch-Clamp Dose-Response Measurement System

We have designed, fabricated, integrated and tested microfabricated planar patch-clamp substrates and poly(dimethylsiloxane) (PDMS) microfluidic components suitable for use in an automated planar patch-clamp dose-response measurement system (Figure 4). Substrates with cell-patch-site apertures ranging from 300 nm to 12 μm were fabricated using standard MEMS techniques. The resistance of the cell-patch sites ranged from 0.2 to 47 MΩ for apertures ranging from 12 μm to 750 nm. The substrate capacitance was 17.2 pF per $mm^2$ of fluid contact. In addition, testing shows that the fluidic components are appropriate for driving human embryonic kidney cells (HEK 293) to patch apertures, for trapping cells on patch apertures, and for rapidly exchanging the extracellular fluid environment.

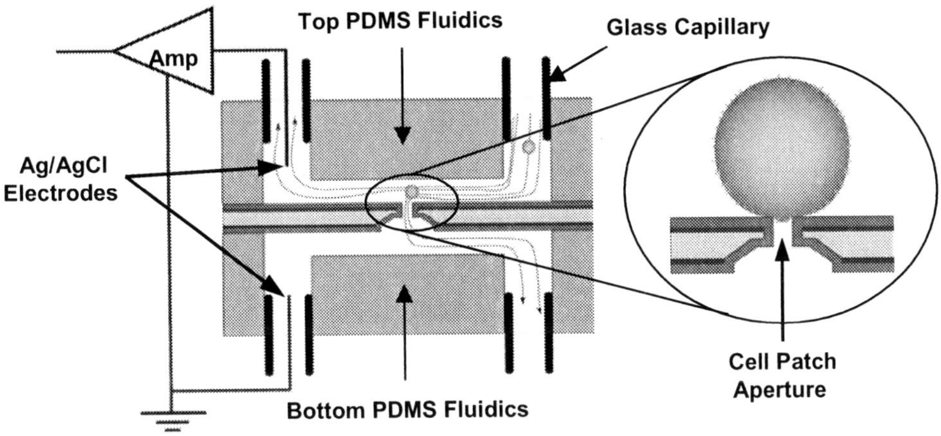


FIGURE 4
MICROFABRICATED PATCH CLAMP TECHNOLOGY [4]

## RECOVERY OF OCULAR MOTILITY FOLLOWING SIXTH-NERVE PALSY

Lesions to the sixth cranial nerve result in selective paralysis of the ipsilateral lateral rectus muscle, which causes convergent strabismus, a condition that nearly ten million people suffer from in the United States alone. Symptoms of sixth-nerve palsy include loss of horizontal eye control, convergent strabismus, and functional loss of sight in the affected eye. In moderate cases, the medial rectus is treated with Botulinum toxin to effectively balance horizontal eye muscle forces. In severe cases, the superior and inferior rectus muscles are surgically re-inserted laterally, followed by physical therapy and minimal recovery of horizontal eye motility. Since previous studies have indicated that a predictive relationship exists between horizontal rectus muscle force and eye position and velocity, our approach seeks to use an implantable stimulation device to exploit this relationship to reanimate the lateral rectus tissue and recover horizontal eye control.

## SUPPRESSION OF EPILEPTIC ACTIVITY BY DEEP-BRAIN STIMULATION

Electrical stimulation of the brain is an emerging alternative therapy for epilepsy, inspired by the failure of anticonvulsant drugs in 30% of those diagnosed with the disease. Our lab is investigating how deep-brain stimulation (DBS) can be employed to suppress epileptiform activity and/or the emergence of seizures. We are working in animal models of epilepsy to screen potential targets and develop effective stimulation strategies for application to clinical use. We are particularly interested in discerning how individual stimulation parameters (e.g., intensity, frequency, pulse-width, waveform) contribute to the suppression bought about by the electrical excitation or inhibition of brain cells using DBS (Figure 5) [5]. We are investigating two types of stimulation strategies. First, open-loop stimulation can be delivered at random or according to a specific schedule. Secondly, closed-loop strategies deliver stimulation in response to pre-defined brain activities (e.g., the beginning of a seizure or an epileptic spike on the EEG) detected by real-time electrophysiological recording. Our laboratory is conducting feasibility studies with fully implantable closed-loop devices in the rat model.

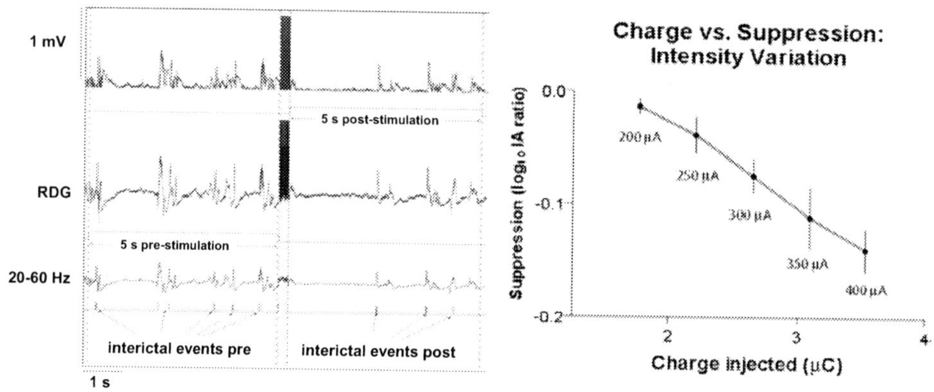


FIGURE 5

DBS-ELICITED SUPPRESSION OF EPILEPTIFORM ACTIVITY IN THE RIGHT DENTATE GYRUS (RDG) OF THE RAT HIPPOCAMPUS. EPILEPTIC SPIKING SUPPRESSED FOR 2 TO 3 SEC FOLLOWING STIMULATION (LEFT). RELATIONSHIP BETWEEN SIMULATION INTENSITY AND EPILEPTIC-SPIKING SUPPRESSION [5] (RIGHT).

### Closed-Loop Control for Energy Homeostasis

Since many diseases are associated with a particular change in activity of specific brain regions, our general objective is to design a closed-loop system to help control the symptoms of such diseases. As an initial study, we are designing a system to monitor and modify, in real-time, the energy homeostasis of the body. Such a system would provide a method of controlling related health problems, such as severe obesity and anorexia. We aim to transform the related known qualitative biological relationships into quantitative engineering time-and-frequency-dependent relationships. We will then use these relationships to build a closed-loop system to control the energy homeostasis of the body. Since our approach is generalizable, we hope it opens doors to new therapeutic approaches for many neurological problems and diseases.

### Micromachined Electrodes for High-Density Stimulation in Retinal Prostheses

High-resolution retinal prosthetic systems, such as that being developed by Second Sight Medical Products, Inc. (Figure 6, top), require high-density stimulation electrodes to restore the vision of blind patients suffering from retinitis pigmentosa and macular degeneration. To increase the effective area of the electrode, a micromachining process is used to add high-aspect-ratio structures to the electrode surface (Figure 6, bottom left and right) [6]. Since clearly the current density will not be uniform over the surface of the micromachined electrodes, modeling is needed to evaluate this complex metal-solution interface. Models have been developed using circuit simulation software (SPICE) that consist of networks of resistive and capacitive elements. By using the multi-dimensional circuit models, predictions of the dynamic current and voltage distribution on the surface of the micromachined electrode can be determined during each stimulation pulse, thereby enabling the quantification of overall electrode performance and longevity.

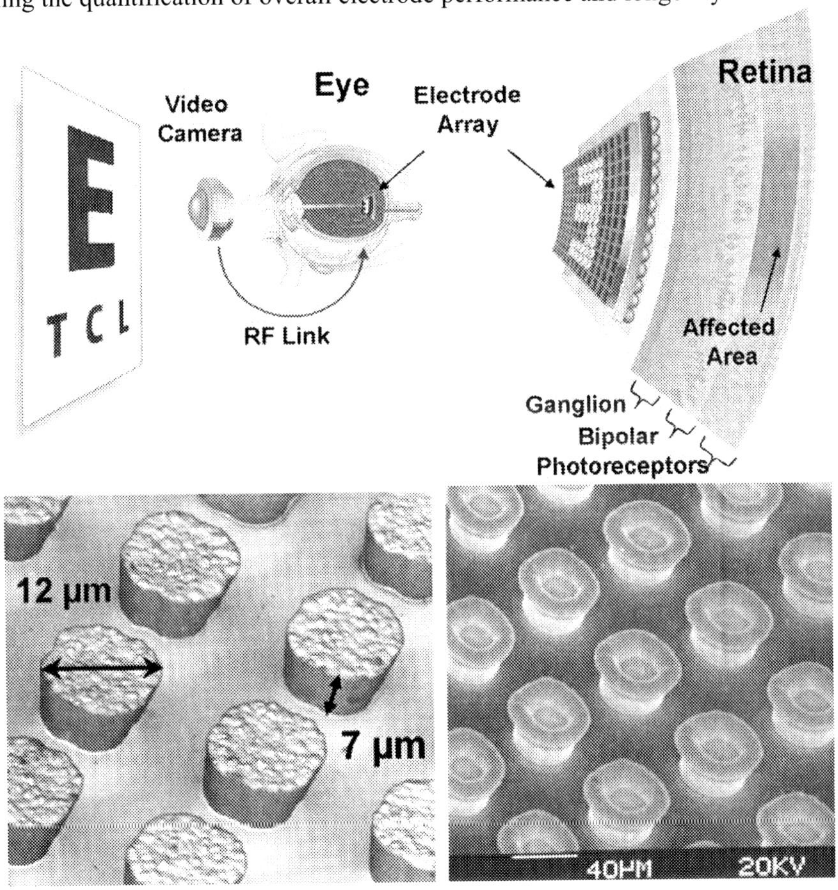


FIGURE 6

RETINAL PROSTHESIS SYSTEM FROM SECOND SIGHT MEDICAL PRODUCTS, INC. (TOP) AND MICROMACHINED ELECTRODES FOR HIGH-DENSITY NEURAL STIMULATION (BOTTOM, LEFT AND RIGHT) [7]

## MEMS-Enabled Ventricular Catheter for Hydrocephalus

One in every 500 newborns is afflicted by hydrocephalus, a condition characterized by an abnormal accumulation of cerebrospinal fluid (CSF) due to an imbalance between cerebrospinal fluid production and absorption [7]. Hydrocephalus may also be acquired later in life due to head trauma or tumor development. One of the most common treatments for patients with hydrocephalus is the surgical implantation of a cerebrospinal-fluid ventricular catheter. Unfortunately, the catheters have a relatively high failure rate and often need to be replaced within a few years, and sometimes within as little as several months. A malfunctioning (or obstructed) catheter can be a life-threatening condition [8].

The goal of this project is to design, fabricate, and test a smart ventricular catheter using MEMS technology. Microactuators will be integrated with the shunt to help resist occlusion due to cellular accumulation. Microsensors may also be added to monitor ventricle size, rate of occlusion, and cerebral pressure, which are parameters that can assist with patient monitoring. This project is currently in the exploratory stages and is focused on observing and quantifying the catheter obstruction process. Data obtained from this experiment will enable us to optimize the microactuator design.

## Neuronal Ensemble Recording and Stimulation of the Lumbar Spinal Circuits to Restore Locomotion

Interneuronal networks within the L1-L2 region of the spinal cord are intrinsically capable of coordinating the activation of multiple functional muscle groups as is required in locomotion (Figure 7). By tapping into this network using intraspinal microstimulation within the spinal cord, simple, natural and adaptable control of locomotion may be achievable. We are recording from the L1-L2 region of the spinal cord using high-density multi-electrode arrays during locomotion, and are correlating the spatiotemporal profile with electromyographic (EMG) data, videography, and ankle position/force measurements. Using principal component analysis and independent-component analysis, we will characterize the overall neural activity and identify neurons that are highly correlated with specific phases of the step cycle or rate of stepping. The results of this project will shed light on the organization of locomotor pattern generation and therapeutic targets. This work is motivated by the fact that spinal-cord injury affects upwards of 230,000 people and can cost between $500,000 and $2,000,000 per person.

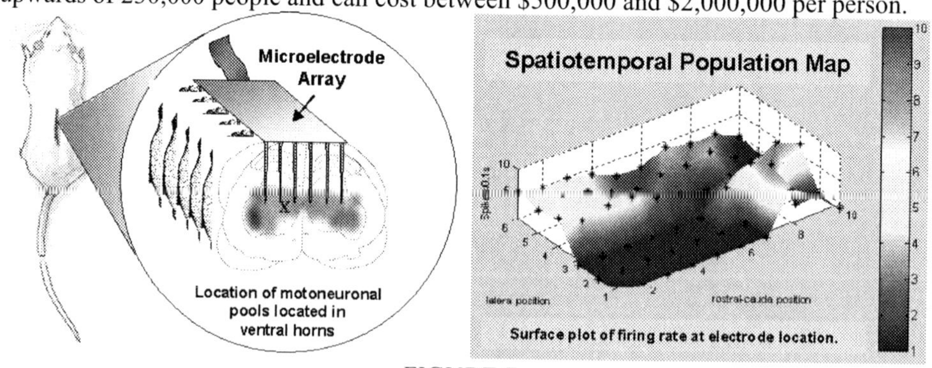


FIGURE 7

THE REGIONS OF THE SPINAL CORD ARE RECORDED WITH A MICROELECTRODE ARRAY TO DEVELOP ACTIVITY MAPS, CORRELATE THEM TO LOCOMOTION, AND PLAN THERAPEUTIC STIMULATION PATTERNS.

### Additional Neuroengineering Research Projects

The following is a list of other neuroengineering projects that are underway at UCLA:

- Characterization of Retinal-Ganglion-Cell Responses
- Computational Models of Working Memory Incorporating Neuronal Variability
- Coupling of Cerebral-Blood Volume and Flow in the Rat Somatosensory Cortex
- Amperometric Glutamate Biosensor
- Electromagnetic Diagnosis and Treatment of Neurological Disorders
- Glutamate Optical Biosensor
- Highly Specialized Microarray of Stimulus Electrodes
- Micro MRI/Depth Electrode
- Modeling Electrically Coupled Tissue
- Multi-modal and Multi-dimensional Atlas of the C57BL/6 Mouse Brain
- Noninvasive Therapeutic Gene Transfer for Neuropathic Pain
- Optical-Imaging-Aided Brain-Activity Sensor
- Polymer-Sphere Delivery of Proteins and Cytokines: in vitro Immunoactivation and in vivo Neuroscience Application
- Population Coding of Retinal-Ganglion Cells

## STATUS OF GRADUATES

Since the origin of the program in 1999, three students have graduated with Ph.D. degrees, two from biomedical engineering and one from neuroscience. Two were fully affiliated with the NET Program (i.e., completed all course requirements and participated in all program activities) and one student became strongly associated with the program (i.e., performed neuroengineering research, completed some but not all of the courses, and participated in some but not all of the program activities).

The first graduate, Jenna Rickus from the Neuroscience IDP, was co-advised by Allan J. Tobin (Neurology and Physiological Sciences) and Bruce Dunn (Materials Science and Engineering). Her dissertation research project was on glutamate optical biosensors. She is now an Assistant Professor at Purdue University in the Biological Engineering Department.

The second graduate from the program, Pedro Irazoqui-Pastor from the Biomedical Engineering IDP, was co-advised by Jack W. Judy (Electrical Engineering) and Istvan Mody (Neurology). His dissertation research project was on inductively powered wireless neural transceivers. He is now Vice President of IC Development at Triangle Biosystems, Inc. (www.tbsi.biz) and is leading their development of wireless neural recording and stimulation systems.

The third graduate, Naser Hamdi from the Biomedical Engineering IDP, was associated with the program but not a formal participant. He was co-advised by Harold Monbouquette (Chemical Engineering) and Marie-Francoise Chesselet (Neurology and Neurobiology). His dissertation research project was on amperometric glutamate biosensors. He is now as Assistant Professor at the Jordan University of Science and Technology in Jordan.

## Future of the Program

The NET Program has been quickly and broadly appreciated and supported on campus. As the NSF IGERT funds end, the continuation of the program is not in doubt – it will remain an integral part of the Biomedical Engineering and Neuroscience IDPs. However, the burden of financial support for the students and their participation in the retreat must be transferred from the IGERT to the co-advisors. Therefore, a modification of the program will be the prompt connection of a pair of faculty members with each student – something that has already happened to approximately 50% of the past and current trainees. In addition, we are applying to the NSF for a renewal.

## Summary

The UCLA NeuroEngineering Training (NET) Program has created a new training program for students with backgrounds in biological science and/or engineering to use and develop advanced engineering technologies to perform research that addresses problems in neuroscience. The program is fostering a new academic culture and multi-disciplinary teamwork needed for new scientific insights and dramatic technological progress in this new collaborative field of neuroengineering. Many exciting neuroengineering research projects are currently underway. The first three graduates now have positions of academic and corporate leadership in this new field.

## Acknowledgements

The success of the UCLA Neuroengineering Training Program is utterly dependent on the high-quality trainees that participate, the staff that administer the program (i.e., Barbara Cross, Cheri Smith, Suzie Vader, Jeanine Moreano, Larry Nadeau, and others), the faculty who must execute and manage the academic and technical efforts, the upper-level administrators of the university that appreciate the value and challenges associated with our unconventional program and effort, and lastly the National Science Foundation and IGERT Grant 9972802 that has funded this academic experiment. Additional resources from NIH have funded some of the different research thrust areas: Deep-Brain Stimulation for Parkinson's Disease Research (NIH NS38367), Development/Testing of Artificial Retinas for the Blind (NIH NEI HO021298), Brain-Computer Interfaces (Okawa Foundation), Neuronal Ensemble Recording and Stimulation of the Lumbar Spinal Circuits to Restore Locomotion (Roman Reed Foundation), and Fully Automated Planar Patch Clamp Dose-Response Measurement System (CMISE NASA Research Center NCC2-1364).

## References

1. National Science Foundation, IGERT: Integrative Graduate Education and Research Training Program: http://www.nsf.gov/home/crssprgm/igert/.

2. Motta, P., and Judy, J.W., "Micromachined Probes for Deep-Brain Stimulation", Proceedings of 2nd Annual International IEEE-EMBS Special Topic Conference on Microtechnologies in Medicine & Biology, Madison, Wisconsin, May 2-4, 2002, pp. 251-254.

3. Irazoqui-Pastor, P., Mody, I., and Judy, J.W., "Transcutaneous RF-Powered Neural Recording Device", Proceedings of 24th Annual International Conference of the IEEE Engineering in Medicine and Biology Society, Houston, TX, Oct 23-26, 2002.

4. Matthews, B., and Judy, J.W., "Towards a Fully Automated Planar Patch Clamp Dose-Response Measurement System", Proceedings of 7th International Conference on Miniaturized Chemical and BioChemical Analysis Systems, Squaw Valley, California, October 5-9, 2003.

5. Aldag, J.M., Bragin, A., Almajano, J., Engel Jr, J., Wilson, C.L., "Parameter Optimization of Electrical Stimulation to Suppress Interictal Spiking in a Chronic, Spontaneous-Seizure, Rodent Epilepsy Model", Epilepsia, Vol. 44S9, 2003, p. 320.

6. Hung, A., Zhou, D., Greenberg, R., and Judy, J.W., "Dynamic Electrochemical Simulation of Micromachined Electrodes for Neural-Stimulation Systems", Proceedings of 1st International IEEE EMBS Neural Engineering Conference, Capri, Italy, March 20-22, 2003.

7. National Institute of Neurological Disorders and Stroke, NINDS Hydrocephalus Information Page, http://www.ninds.nih.gov/health_and_medical/disorders/hydrocephalus.htm.

8. J.M. Drake, The Shunt Book, Blackwell Science, Cambridge, UK, 1995.

**Jack W. Judy** received his Ph.D. and M.S. degrees from the University of California, Berkeley, CA, in 1996 and 1994 respectively, as well as the B.S.E.E. degree, with summa cum laude honors, from the University of Minnesota, Minneapolis, MN, in 1989. He has also worked for Silicon Light Machines, Inc., Sunnyvale, CA, an optical-MEMS start-up company, from 1996 to 1997. He has been on the faculty of the Electrical Engineering Department at the University of California, Los Angeles, since 1997, where he is currently an Associate Professor. At UCLA he is also the Chair of the MEMS and Nanotechnology Major Field of the Electrical Engineering Department, and the co-director of the UCLA NeuroEngineering Training Program, an NSF-funded IGERT program sponsored jointly by the Biomedical Engineering Interdepartmental Program and the Brain Research Institute (BRI). His present research interests include additional novel ferromagnetic MEMS, electronic noses, and a variety of neuroengineering projects, such as microprobes for Parkinson's disease research, electrode arrays for retinal prosthetics, wireless neural transceivers, neural control systems for spinal-cord injury, ocular motility, and deep-brain stimulation.

**Allan J. Tobin** is Director of the UCLA Brain Research Institute and the Eleanor Leslie Chair in Neuroscience at UCLA. He holds appointments as Professor of Neurology and Professor of Physiological Science. He is also the Scientific Director of the Hereditary Disease Foundation (HDF). Dr. Tobin was the Chair of the Interdepartmental Ph.D. Program for Neuroscience from 1989-1995. He is the co-author of the textbooks Asking About Cells (with Richard E. Morel) and (with Jennie Dusheck) of Asking About Life, both published in 1997 by Saunders College Publishing. Dr. Tobin is the recipient of a Jacob Javits

Neuroscience Investigator Award from the National Institute of Neurological Disorders and Stroke. As Scientific Director of the HDF, he helped organize the consortium that identified the gene responsible for Huntington's disease. His laboratory studies the production and action of GABA, the brain's major inhibitory signal. These studies, he hopes, may eventually lead to new therapeutic approaches to epilepsy, Huntington's disease, and juvenile diabetes.

# Chapter 6

# The National Innovative Communication Engineering (NICE) Education Program of Taiwan

JIN-FU CHANG and SHYUE-WIN WEI
*National Chi-Nan University, Department of Electrical Engineering and Graduate Institute of Communication Engineering, Puli, Taiwan 545, ROC. E-mail: jfchang@ncnu.edu.tw*

*In 2002 Taiwan's Ministry of Education (MOE) launched a program initiating a new phase of reform in communication engineering education: National Innovative Communication Engineering (NICE) Education Program. This four-year NICE program seeks to produce high quality engineers for Taiwan's communication industries. NICE is partitioned into five technical areas: optical communication systems, network applications and services, communication devices, broadband Internet, and wireless networks. Five area centers are set up to coordinate and supervise the implementation of NICE. A program office is further established to ensure that NICE is successfully executed. Within the framework of NICE, state-of-the-art courses in telecommunication and networking are designed and experimented. These courses integrate theories with practices. Hands-on experiments are emphasized. Cross-university cooperation is given high priority and international collaboration is encouraged. NICE is a four-year program (2003-2006) with an annual budget of NT$80 million (approximately 2.28 million US) New Taiwan dollars. This program employs the following approaches: (i) partnership among universities, (ii) shared teaching laboratories, (iii) co-developed teaching materials, (iv) up-to-date regular and short courses, (v) cultivation of "seed lecturers," (vi) collaboration between universities and industries, and (vii) international students/scholars exchange. The detailed structure of NICE and its preliminary results are reported in the paper.*

## A Brief Historical Retrospect

Taiwan has for many years been an active and leading player in electronics and information technology (IT) in the world market. With this foundation, Taiwan wants to continue to be a front-runner in the booming global telecommunications market, particularly in the manufacturing sector. To maintain Taiwan's international competitiveness, one critical issue is the cultivation of not only sufficient but high quality human resources. Human resource cultivation relies on education and research training. In Taiwan, research training is implemented mainly through projects funded by the National Science Council (NSC), whereas education programs are in the hands of the Ministry of Education (MOE).

To achieve the goal of human resources development in IT, the MOE has implemented several programs in communication engineering education since 1993. The first four-year program was completed in 1996, while the second four-year program was carried out from 1997 to 2000 [1,2]. The budgets invested in these two phases were US$5 million and US$7.5 million, respectively. The first two programs focused on the building of infrastructure. Many fundamental teaching laboratories were established in these two stages. The third phase program was started in 2001. Course reorganization and key laboratories were emphasized in two particular areas: Communication Systems and Digital Processing, and Networking. This third phase was unfortunately funded for a brief one year only due to MOE's internal program reorganization, although it was originally designed to cover four years. Responding to continual appeal and to further enhance the overall quality of communication engineering education, Taiwan's MOE re-launched the current National Innovative Communication Engineering (NICE) education program in 2002 [3,4]. Based on the concept of partnership, NICE plans to merge and integrate resources from different branches of the education establishment in Taiwan. It should be emphasized that the funds allocated for these four national programs are directed at universities in Taiwan, as additions to their regular budget allocations from the MOE.

## Goals

In Taiwan, both the visibility and the quality of communication engineering education have been largely improved through the execution of prior phases of NICE. Although both hardware and software environments in each individual university or college have been improved significantly, these facilities need to be maintained and continuously upgraded to keep pace with the rapid changes of technologies. On the other hand, cross-university collaboration is a new dimension that needs further improvement and development. Thus, one emphasis of NICE is to integrate and share resources in different universities and colleges through partnerships. Based on the framework of partner schools, course materials, classrooms or laboratories developed at any one place can be shared through the Internet. Regional resource centers are set up to purchase and house expensive equipment to be shared by partner schools in the region. Training courses are designed and offered to cultivate seed instructors/lecturers. These courses include seminars or short courses taught by people invited from abroad. Every year NICE also plans to send university/college faculty members to attend short courses or visit industrial laboratories in foreign countries to learn new concepts and to broaden their visions.

To summarize, NICE is targeted at accomplishing the following: (1) cultivating seed instructors through regular/short courses and seminars; (2) sharing up-to-date course materials on the Internet; (3) sharing costly equipment through regional resource centers; (4) offering courses taught by industrial people; (5) co-sponsoring design competitions with industry; and (6) implementing international exchange or collaboration through visits and design competition.

## ARCHITECTURE AND FRAMEWORK

The organizational structure of NICE is shown in Figure 1. This program is partitioned into five key areas, including optical communication systems, network applications and services, communication devices, broadband Internet, and wireless networks. Table 1 shows the selected topics in each area. Five centers are set up to coordinate and supervise the implementation of NICE in their respective areas. These centers are placed at the following national universities: Taiwan, Sun-Yat-Sen, Central, Tsing-Hua, and Chiao-Tung. Each center is commissioned to a senior university professor. An executive secretary and a full-time assistant are funded to assist the center director in administrative matters. Executive secretaries are faculty members at the associate professor rank or above. The major tasks of an area center include course design and training, database establishment, set-up of teaching laboratories, international exchanges, etc. Each center has its own guidance committee to help ensure that the center proceeds on the right track. A guidance committee consists of experienced representatives from universities and industry. A program office is further established functionally on top of these area centers to ensure that NICE is successfully managed. The NICE program office is placed at the National Chi Nan University and run by Professor J.-F. Chang, President of the University. An advisory board is also organized by the MOE for guidance and consultation.

Partner schools are chosen through open calls and fair competition. Partner schools are responsible for helping their affiliated area center with the promotion of relevant activities. Among these five centers, the optical communication systems center and the communication devices center also arranged for their set up their regional laboratory to purchase and house expensive capital equipment. Partner schools share in the use of the equipment. These laboratories emphasize hands-on experiments. Through sharing of teaching materials and laboratory facilities, quality engineers can be fostered in large numbers. We believe that the concept of schools in partnership is an effective one to enhance the quality as well as the quantity of communication engineers in Taiwan.

The key tasks performed within the framework of NICE are discussed in the following sections.

## COURSE PLANNING AND TEACHING MATERIAL PRODUCTION

"Communications" cover not only a wide spectrum of topics, but also rapidly changing subjects. Therefore, the key is to start out with the foundations that universities in Taiwan already have to selectively choose and design new courses. This may be accomplished through the partner school architecture. Teaching materials, especially textbooks, are produced by partner schools in joint efforts. Textbooks are further tailored to suit the needs of students according to the nature of their colleges. In other words, course materials are produced on demand. To keep pace with technology, multimedia electronic and online teaching materials are preferred for easier dissemination.

## WEBSITE AND DATABASES

The Internet has become very powerful in interactive services and information dissemination. Taiwan is a nation with very high Internet penetration. For instance, universities in Taiwan are linked through TANET (Taiwan Academic Network) sponsored by the MOE. In NICE, each area center and the program office are required to build a website to collect teaching materials, including course information, handouts, transparencies, course or laboratory instructor manuals, etc. All materials and information will be carefully categorized and documented. Intellectual property rights are handled following government regulations.

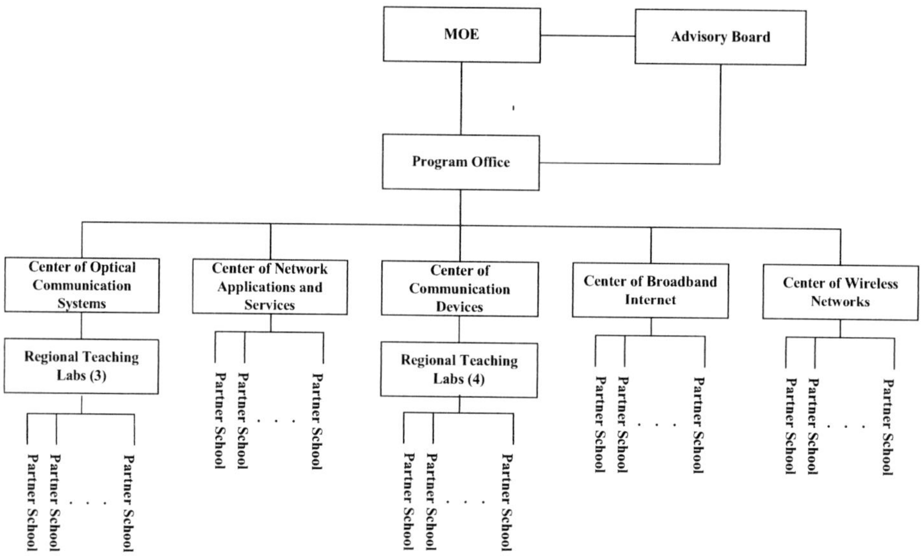


FIGURE 1
THE ARCHITECTURE OF NICE-PROGRAM IN TAIWAN

## REGIONAL SHARED TEACHING LABORATORIES

Equipment for optical communication systems and communication devices can be expensive; hence, two of the centers have decided to establish regional shared teaching laboratories to support the needs of schools in the region. The first, the Center of Optical Communication Systems, has opened three such laboratories in Taipei, Taichung, and Tainan. The second, the Center of Communication Devices, has set up four such laboratories in Chungli, Taichung, Tainan, and Kaoshiung.

## SHARED SOFTWARE CENTER

Simulation and design tools have become very important in recent years for the training of students. These software packages are generally expensive. A software center is set up to help negotiate reasonable education prices. This software center is designed also to offer training courses or programs to help students in their learning.

| Center | Technical Scope |
|---|---|
| Center of Optical Communication Systems | Optical signal transmission, Optical switching, Optical Routing, Photon devices and systems of optical communication |
| Center of Network Applications and Services | Multimedia communication technology and application systems, Network security and monitor, Worldwide information net technology, Mobile computing and applications |
| Center of Communication Devices | Radio frequency devices and IC design, photo devices and circuit design, Antenna design and electromagnetic wave measurement |
| Center of Broadband Internet | Advanced network switching technology, Internet security, Broadband network access technology, Embedded system core technology, New generation Internet (IPv6) integrated technology |
| Center of Wireless Networks | Wireless access networks, Mobile core networks, Wireless communication applications and services |

TABLE 1
THE TECHNICAL SCOPE OF EACH AREA CENTER

## SEED LECTURERS TRAINING

Seed instructor training is a very important task to be carried out in NICE. The communication industry in Taiwan has been growing fast and therefore urgently needs large numbers of high quality engineers. Seed instructors are trained at the area centers so that courses can be offered at these instructors' affiliated schools. NICE takes three approaches to fulfill this task. One is to support senior faculty members from reputable universities in offering courses at other universities. These senior faculty members are called "seed instructors" or "seed lecturers." The second approach is to select and send university faculty staff to attend overseas training courses to update their knowledge. Upon completion of these courses, they return to become seed lecturers. The third approach is to invite foreign experts to teach short courses designed for seed lecturers. NICE plans to train in four years approximately 300 seed lectures, many of them for trained private universities and technical universities/institutes.

## Exchange and Collaboration between Universities and Industries

There has been a steady increase in the collaboration between universities and industries in research and development work. The MOE has in recent years designed many incentives and mechanisms to nourish such a trend. It is always valuable to students to take courses taught by people from industry. On the other side of the coin, engineers working in the industrial sector need to keep their knowledge up to date. NICE will be offering courses to serve these purposes.

## Internationalization

Taiwan is aware of the need for internationalization if it does not want to become marginalized. The present administration in Taiwan has very much emphasized the importance of this issue. The NICE program makes no exception and is very eager to create a dialogue to interact with peers in other countries. This is why we proposed to the organizers of ICEE-2003 (Valencia, Spain, July, 2003) to grant a session to share our views and experiences with representatives from other countries. NICE has made possible greater international student and scholar exchanges, the organizing of websites in English, the co-production of teaching materials, and the developing of distance-learning environments. To collect up-to-date information, state-of-the-art knowledge, and evolving trends in other countries, NICE plans to send representatives to visit overseas universities, research establishments, and industrial enterprises.

## Communication Design Contest

The communication design contest was initiated by the MOE, Taiwan, in 1998 and it merged into the NICE program in 2003. The design contest serves as a very effective platform for students from different universities to interact. It also serves the purpose of demonstrating student skills in designing practical engineering systems. Participation from industry in terms of joint sponsorship and participation in review panels plays a key role in the success of such events. The design contest in NICE is offered to serve these purposes. As distinct from the past practices, NICE's design contest offers an additional opportunity for internationalization. Foreign teams are invited to participate in the competition.

## Performance Evaluation

Review or evaluation is a key step to help ensure the success of a program. In the NICE program, the area centers are required to file reports quarterly to the program office for evaluation. At the end of each year a review team will be dispatched to each center and regional laboratory to review progress. Poor performing projects face termination.

## The First Year Achievements

The four-year NICE program was planned in 2002 and officially started in 2003. Partner schools were organized through open calls. A total of 63 proposals were received out of which 35 were finally approved. Among the approved proposals approximately half were submitted from comprehensive universities while the remaining fifty percent came from technical universities/institutes. The Center of Optical Communication Systems comprises 13 partner schools, and planned to deliver five books and 6 (instruction) notes. National Taiwan University of Science and Technology, National Lien-Ho Institute of Technology, and National Cheng-Kung University were chosen to be the sites of regional shared laboratories. The Center of Communication Devices has 19 partner schools and planned to deliver two books and 7 notes. The four regional shared laboratories were placed at National Central University, Feng-Chia University, National Cheng-Kung University, and National Kaoshiung University. There are 11 partner schools for the Center of Network Applications and Services and 7 for each of the remaining two centers. The course materials to be delivered include the following: 2 books and 9 notes for Center of Network Applications and Services; 5 books and 6 notes for Center of Broadband Internet; and 2 books and 8 notes for Center of Wireless Networks.

There are already more than 60 individuals, including some from four industrial companies, participating in the first year of NICE and the number is expected to grow in the coming years. In the first year alone, 231 classes were offered with a total of 6601 students enrolled, exceeding the goal by a significant margin [3].

The first year budget allocated for NICE is 2.28 million US dollars and each area center takes a share of approximately 0.4 million. The first year budget was invested mostly in teaching material production and initial set-up of regional shared laboratories. The budget for international programs is 0.2 million of which seed instructor training consumes about 7 percent and will be weighted more in the coming years.

## SUMMARY

A new program in communications engineering education called NICE has been launched in Taiwan. This program is partitioned into five technical areas, including optical communication systems, network applications and services, communication devices, broadband Internet, and wireless networks. Five area centers are set up to coordinate and supervise the implementation of NICE. A program office is further established to oversee and coordinate, to ensure that the program is successfully executed. Based on the framework of partner schools, NICE nourishes a collaborative environment for universities in Taiwan. Key tasks include the following: instructional and laboratory materials design, production and distribution; seed instructor cultivation; design contest; university-industry collaboration; internationalization; and performance evaluation. NICE has a goal to help speed up the cultivation of good engineers in large quantities for Taiwan.

Although NICE is a continuation of past efforts, it has new designs. Encouraging progress has been seen during the first year. This program is expected to produce high quality human resources for Taiwan's communication industries.

## ACKNOWLEDGEMENTS

The NICE Program is supported by the Ministry of Education (MOE), Taiwan. The authors wish to thank Prof. J.L. Chen, National Taiwan University of Science and Technology, for her participation and support.

## References


1. Den, Chi-Fu, "Final Report of First Phase National Telecommunications Program (NTP)," NTP Office, National Science Council (NSC), Taiwan, September 2003.
2. Lan, M.-C., et al., "Education Enhancement of Communication Engineering in Taiwan," Proceedings of 2000 International Conference on Engineering Education (ICEE-2000), Taipei, Taiwan (2000).
3. Chang, J.-F., "Mid-Term Proposal of 4th Phase Innovative Communication Education Program," Ministry of Education (MOE), Taiwan, July 2002.
4. Chang, J.-F., "Annual Report of 4th Phase Innovative Communication Education Program," Ministry of Education (MOE), Taiwan, December 2002.



**Jin-Fu Chang** is presently President of the National Chi Nan University, Puli, Taiwan. He received his B.S.E.E. from the National Taiwan University in 1970 and Ph.D. in electrical engineering and computer sciences from the University of California, Berkeley in 1977. In the past years he has held teaching positions at the National Taiwan University, National Central University, and National Chi Nan University. He also worked several years at the Ministry of Education and the National Science Council, Taiwan.

**Shyue-Win Wei** received the Ph.D. degree in electronics from National Chiao-Tung University, Taiwan, R.O.C., in 1990. Then he joined Telecommunication Laboratories, R.O.C., where he worked on the development of a high-bit-rate digital subscriber line transmission system. From 1992-1997 he was an Associate Professor at the Department of Electrical Engineering, Chung-Hua University. He was appointed Professor 1998, and moved to the National Chi-Nan University in 2000. His research interests include digital transmission system, digital subscriber lines, coding theory, and related VLSI circuits' design.

# Chapter 7

# An Innovative Teaching Program for Network Applications and Services

REN-HUNG HWANG[1] and CHUNG-MING HUANG[2]
*[1]National Chung-Cheng University, Chia-Yi, Taiwan 621, ROC. E-mail: rhhwang@cs.ccu.edu.tw. [2]National Cheng-Kung University, Tainan, Taiwan 701, ROC. E-mail: huangcm@locust.csie.ncku.edu.tw*

*In this paper, we present the plan for the education program in network applications and services within the National Innovative Communication Education Program (NICE) of Taiwan. The major goal of our education program is to organize teaching partnerships among universities to develop outstanding teaching materials for advanced courses in the network applications and services area. We will present the infrastructure of our teaching partnerships among universities and the courses for which we have chosen to develop courseware. Four categories in the network applications and services area are identified to be the most important for Taiwan, namely, multimedia communication techniques and applications, network security and monitoring, Internet and web techniques, and mobile computing applications. We have launched the development of courseware, including textbooks and teaching materials (such as PowerPoint slides), at the beginning of year 2003 and expect to run the program for four years (2003 to 2006). An e-Learning platform is also developed for selected courses to experience the new learning style. Besides developing courseware, several activities will be arranged to extend the program to have more universities participate and use the developed courseware. These activities include arranging regular, short-term, and promotion courses to incubate seed lecturers from universities nationwide, collaborating with industries to develop courseware according to their needs, arranging short-term courses for professionals from industries, international student/professor exchange programs, international contests on network applications, and tours to well-known international companies to help students and scholars in Taiwan learn state-of-the-art technologies in the network applications and services area. Finally, this program will be evaluated annually to assure the effectiveness of the proposed activities.*

## INTRODUCTION

Recently, many countries have started to reform their higher education systems [1,2]. As described by Chang and Wei [3] in the previous chapter in this volume, in 2002, the Ministry of Education (MOE) of Taiwan launched a new program of reform in communications engineering education: National Innovative Communication Education Program (NICE-Program). The communication technologies are classified into five categories, including optical communication systems, network applications and services, communication devices, broadband Internet, and wireless networks. In this paper, we report the program for network applications and services of the NICE-program.

In the 21st century, Internet and mobile communication will play a major role in our daily life. Several network-related services, such as e-Learning, e-government, e-campus, and e-life, have become more and more mature. As a consequence, we expect that the market for network applications and services will grow steadily in the next decade. According to Datamonitor's report [4], investment in mobile communication applications was \$0.5 billion in 2001, but will reach \$2.1 billion in 2006. E-commerce on mobile devices, such as cellular phones and PDAs, will be the major market drivers. Due to the volume of E-commerce activity, Internet security becomes more and more important. Thus, the market for network security will also grow rapidly in the next few years. According to Dataquest's prediction [5], the market value for network and software security will reach \$6.7 billion in 2004.

Advances in optical communication and switching technology also result in rapid bandwidth growth. A direct consequence of high Internet bandwidth is the spread of multimedia applications over the Internet, such as Internet phone, video on demand, video conferencing, and interactive TV. Multimedia streaming technology is identified as the key technology for these applications. Furthermore, Internet technologies and applications, such as search engine, portal, distance learning, distance medication, XML applications, data mining, and Internet computing are also expected to have a growing market. Therefore, in this program, we divide the network applications and services area into four categories: multimedia communication techniques and applications, network security and monitoring, Internet and web techniques, and mobile computing applications.

## GOALS OF THE NETWORK APPLICATIONS AND SERVICES PROGRAM

The ultimate goal of this program is to produce a large number of high quality engineers with experience in state-of-the-art technologies in the network applications and services area in order to promote Taiwan's international competition in this industry. To achieve this goal, several activities are planned for the next four years. The activities include the following:

- Establish a course-promotion center and partner schools.
- Reform existing courses.
- Develop high quality courseware.
- Arrange regular, short-term, and promotion courses to incubate seed lecturers from universities nationwide.
- Collaborate with industry to develop more industry-oriented courseware.
- Develop short-term courses for industry professionals.

- Establish international student/professor exchange programs.
- Conduct international contests on network applications.
- Provide tours to well-known international companies.

Through these activities, the explicit goals of this program are summarized as follows:

1. Educate adequate high quality engineers to meet industry's needs.
2. Establish a teaching center for network applications and services with twelve partner schools, including public and private universities, colleges, and institutes of technology.
3. Develop high quality courseware for sixteen courses and three textbooks that cover the four most important areas in network applications and services.
4. Organize two training workshops each year to incubate at least eighty seed lecturers from thirty-seven schools. The best twenty seed lecturers will be given an additional budget to run a related course.
5. Hold at least ten distinguished lectures on advanced topics in the network applications and services area.
6. Organize at least one short-term training course for professionals from industry each year.
7. Create a website as the communication center as well as the repository for courseware developed by this program, such that courseware can be accessed via the Internet.
8. Encourage collaboration between universities and industry, such as promotion courses, seminars on new technologies and new services, communication contests, software donations, intern jobs, etc.
9. Internationalize the program to import state-of-the-art technologies and courseware from overseas. For each year, the program arranges at least two tutorials with well-known foreign speakers, one trip for a professor to go overseas to take short courses on advanced topics, one international contest on network applications, and one overseas visit to well-known international companies.
10. Evaluate the performance of partner schools quarterly, conduct one on-site visit at each partner school yearly, and hold an exhibition of courseware developed at the end of each year.

## ARCHITECTURE AND FRAMEWORK OF THE NICE-PROGRAM

The architecture of the network applications and services program is shown in Figure 1. The teaching center is located at National Cheng Kung University and chaired by Dr. Chung-Ming Huang, a professor in the department of Computer Science and Information Engineering. Twelve partner schools are chosen specifically to include public and private universities, colleges, and institutes of technology such that all types of schools have an equal chance to join the program. As shown in Figure 1, partner schools in the multimedia communication techniques and applications area include National Sun Yat-Sen University (NSYSU), National Chung Cheng University (CCU), National Dong Hwa University (NDHU), and Southern Taiwan University of Science and Technology (STUT). Partner schools in mobile computing and applications include Feng Chia University (FCU), National HuWei Institute of Technology (NHIT), and National Pingtung University of Science and Technology (NPUST). Partner schools in Internet

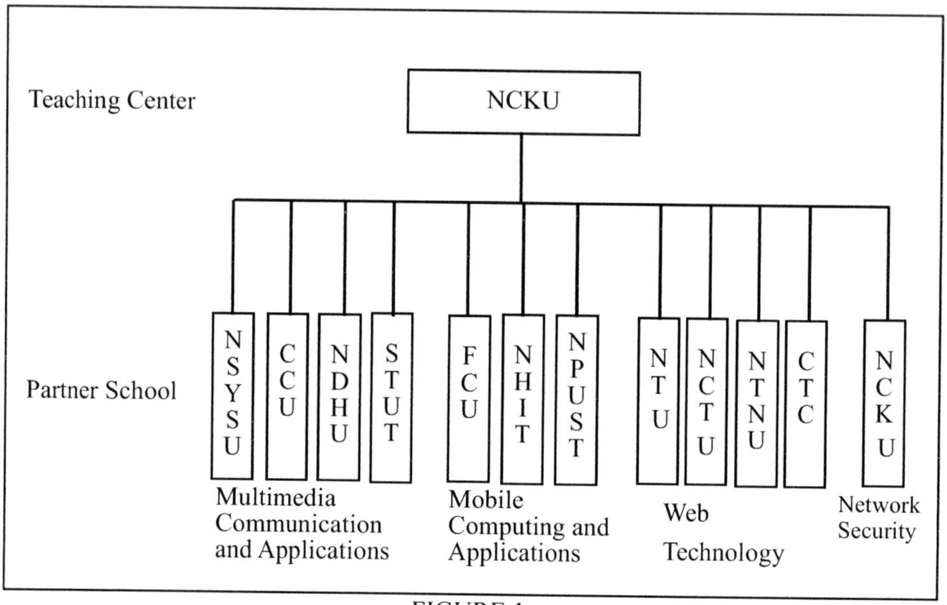


FIGURE 1
ARCHITECTURE OF THE NETWORK APPLICATIONS AND SERVICES PROGRAM

and web techniques include National Taiwan University (NTU), National Chao Tung University (NCTU), National Taiwan Normal University (NTNU), and Chungtai Institute of Health Sciences and Technology (CTC). Finally, the network security and monitoring area has only one partner school, namely, National Cheng Kung University (NCKU).

An open call for proposal was conducted in the summer of 2002. Eleven courses spread over three areas were selected as the seed courses for the first year from a pool of approximately fifty courses. The selections were based on following criteria:

1. Courses that are designed with state-of-the-art technologies and that are in high demand by industry.
2. Courses that appeal to a wide population.
3. Fundamental courses are excluded as their course materials were already developed in previous programs.
4. Courses that are overly specific are not considered unless the technology covered is a key future technology.

Table 1 shows the eleven courses that were selected. (A new course, "Advanced Topics in Internet Security", will be launched a year later to cover technologies in the network security area.) In general, each partner school is responsible for developing one set of courseware or a textbook. Each partner school is asked collaborate with other schools in developing the courseware so that it will be suitable for general universities as well as institutes of technology. Collaborative schools are also listed in Table 1. Course materials and textbooks developed by partner schools will be delivered to all universities, colleges and institutes of science and technology in Taiwan free of charge so that a large number of students and engineers can be trained. It is expected that the framework of partner and collaborative schools will be an efficient way to increase the quality as well as quantity of professionals in the networking industry in Taiwan.

| Area | Partner School | Course | Collaborative Schools |
|---|---|---|---|
| Multimedia communication techniques and applications | NDHU | Multimedia network applications for next generation Internet | Chung Yuan Christian University (CYCU) |
| | NSYSU | Multimedia streaming technology | National Kaohsiung University of Applied Sciences (KUAS)<br>Tajen Institute of Technology (TIOT)<br>Nan Jeon Institute of Technology (NJTC) |
| | CCU | Design technology for network multimedia e-book | Chaoyang University of Technology (CYUT)<br>National Yunlin University of Science & Technology (NYUST) |
| | STUT | Network digital media | NDHU |
| Mobile computing and applications | FCU | Technology for mobile agent | NanKai College (NKC)<br>Taichung Healthcare & Management University (THMU) |
| | NHIT | Mobile e-commerce | National Taichung Institute OF Technology (NTIT)<br>CCU<br>Chung Hua University (CHU)<br>Mingchi Institute of Technology (MIT) |
| | NPUST | Multimedia application system for handheld devices | National Central University (NCU) |
| Internet and web techniques | NTU | Internet computing | NCKU<br>National Chengchi University (NCCU) |
| | NCTU | Internet data mining (Textbook) | NTU<br>MingHsin University of Science & Technology (MUST)<br>Chung Hua University (CHU) |
| | NTNU | Network computing and XML (Textbook) | National Taiwan University of Science & Technology (NTUST)<br>National Taipei Teachers College (NTPTC) |
| | CTC | Technology for transmitting medical images and documents | Chung Shan Medical University (CSMU)<br>National ChungHsing University (NCHU)<br>Taipei Medical University (TMU) |

TABLE 2
COURSES DEVELOPED IN THE FIRST YEAR

## STRATEGIES TO EFFECTIVELY DEPLOY THE NETWORK APPLICATIONS AND SERVICES PROGRAM

The following strategies are adopted to speed up the promotion of the network applications and services program:

### Course Planning and Teaching Materials Production

As mentioned above, eleven courses are selected as pioneer courses for this program. Among the eleven courses, two of them will emphasize writing textbooks while the remainder will focus on developing teaching materials such as presentation slides, documents for hands-on experiments, etc. Textbooks and teaching materials will be co-designed by instructors from collaborative schools. To be adaptive to the disparate needs of different kinds of schools (e.g. general universities vs. institutes of technology) different versions of textbooks and teaching materials may be developed. Furthermore, teaching materials in multimedia and electronic formats are preferred because they are easier to deliver and promote. One or two courses will also be selected for presentation via the e-Learning environment (the first course chosen to experiment in Spring, 2004 is "Multimedia Network Applications for Next Generation Internet", developed by NDHU. This course will be taught cooperatively with professors from Taiwan as well as from the USA). A SCORM-compliant LMS has been developed, and the courseware developed for these courses will also be SCORM-compliant [6]. The LMS also provides collaborative functions to allow students from different universities to collaborate on a project. Various functions, such as e-Learning tracking, student portfolio, student management, assessment tracking and management, and collaboration environment will be provided for effective and adaptive learning.

### Establishment of Website and Database

A website, located at the URL http://www.pcnase.csie.ncku.edu.tw/, is created to collect all teaching materials, including teaching handouts, transparencies, course information, teaching experience sharing supplies, teaching/learning assistance materials, and so on. All materials and information are categorized and filed. All documents on the website will be translated to Adobe Acrobat PDF format for easy access. We will also encourage partner schools to prepare courseware in the SCORM-compliant format for easy management and exchange.

### Training Seed Lecturers

Due to the urgent requirement of manpower and engineers in the networking industry, we have to recruit and train more lecturers to educate more students in a short time. Several workshops will be arranged each year to promote the courseware developed by this program. Professors who are interested in using the courseware will be invited to attend the training program and become the seed lecturers. In 2003, three workshops were held at FCU, NCKU, and NTU on July 11, 16, and August 1, respectively. These three universities are located in the central, southern, and northern parts of Taiwan, respectively. At the end of 2003, twelve attendees were selected to become seed lecturers. They will be given additional budget in 2004 to attempt to use the courseware developed in this program.

## Collaboration between Universities and Industry

One of the major goals for the NICE-program is to enhance students' background and practical experiences so that they do not experience a huge practical shock on entering the workforce [7]. Two approaches are adopted to achieve this goal. First, when developing courseware, we will collaborate with industry to develop courseware according to their needs, and arrange short-term courses for professionals from industry to train them in new technologies. Second, we will arrange for students to do internships with industry or to engage in hands-on experiments in the laboratories of some networking companies during the summer. We will also encourage industry to donate software and hardware to our partner schools. In March 2003, Microsoft Taiwan donated ten copies of MSDN AA (Microsoft Developer Network Academic Alliance) to this program.

## Internationalization

Internationalization is an important concern in introducing state-of-the-art technologies into the program. Three approaches are adopted. First, we will invite at least three overseas experts to teach short courses for the seed lecturers each year. In 2003, we invited experts from IBM, Hong Kong University of Science and Technology, and University of California, Davis, California to give short courses on data mining, peer to peer (P2P) applications, pervasive computing, and Internet security, respectively, during the summer. Second, each year, we will select at least one seed lecturer to attend overseas training courses to learn state-of-the-art technologies. Two professors were selected in 2003 to attend training courses held in the USA in areas of mobile communication and wireless networks, and distributed simulation technology. Finally, we will arrange for seed lecturers to visit well-known international companies in the networking industry. In October 2003, we visited the Information Sharing Laboratories of NTT and Waseda University, Japan, to discuss topics such as Mobile IP/Internet, IPv6, and audio/video/multimedia communication technologies.

## Communication Contest

In past years, the MOE of Taiwan selected a university to organize a national communication contest. Starting in 2003, the contest is divided into five areas with each teaching center of the NICE-program in charge of one area (the contest was cancelled in 2003 due to SARS). In 2004, the contest will begin in April and end in June. Winners of the contest will have the chance to go overseas for an international contest or to visit well-known companies. Starting next year, we plan to open the contest to international participants in order to broaden students' view and to have international competition.

## Distinguished Lectures

One way to expose the teaching experience of senior professors in well-developed universities to young professors is to organize seminars or workshops and invite those senior professors as distinguished speakers. We call these events distinguished lectures. Making use of well-designed courses, well-written teaching materials, and senior professors' teaching experiences can quickly convey the state-of-the-art technologies and good teaching styles to junior professors. In 2003, we conducted seventeen distinguished lectures that cover the four major areas in network applications and services. Approximately twenty distinguished lectures are planned for 2004.

### Performance Evaluation

The success of this program relies on the cooperation of partner schools and collaborative schools. Therefore, a bi-monthly meeting is called by the teaching center to discuss the progress of various activities, and to exchange ideas and comments for improvement. Each partner school is required to submit a quarterly report so the teaching center can keep track of their progress. The teaching center will arrange at least one on-site visit to each partner school each year. A workshop will be arranged by the teaching center to show the accomplishment of the teaching center and its partner schools. Besides the connection between the teaching center and the partner schools, the chair and executive secretary of each teaching center also attend regular meetings arranged by the NICE-Program office to report their progress.

The outcomes of 2003 were very successful. Courseware developed in this program has spread to universities, colleges, and industry nationwide through our promotion strategies. Tutorials and distinguished lectures are very popular and help domestic and junior teachers to learn state-of-the-art technologies. One courseware bundle developed in 2003 has both Chinese and English versions. We feel this is a good trend and will encourage more partner schools to develop courseware in both Chinese and English.

## Summary

We have described the education program in network applications and services within the National Innovative Communication Education Program (NICE) of Taiwan. Our work deals with network applications and services, one of five areas under NICE that also include optical communication systems, communication devices, broadband Internet, and wireless networks. We have presented the current working status for the network applications and services area. In particular, we have established twelve partner schools. During the program's four years, we plan to develop high quality courseware containing sixteen courses and three textbooks. Various activities, including distinguished lectures, tutorials, and short overseas courses are conducted to incubate seed lecturers such that the courseware developed by this program can be widely delivered to universities, colleges, and institutes of technology.

## Acknowledgements

The Network Applications and Services Program of the NICE-Program is supported by the Ministry of Education (MOE), Taiwan, R.O.C.

## References

1. Prakash, S., "Reformation of Science Education in Indian Universities," Current Science, Vol. 78, No. 12, June 2000, p. 25.

2. Tsuruta, Y., "On-Going Changes to Higher Education in Japan and Some Key Issues," Inside the Higher Education Institution in Japan: Responding to Change and Reforms Seminar, the Daiwa Anglo-Japanese Foundation, 26 November 2003.

3. Chang, J.F., and Wei, S.W., "The National Innovative Communication Engineering Education (NICE) Program of Taiwan," Chapter 6, Aung, W. et al. (eds.), *Innovations 2004: World Innovations in Engineering Education and Research*, Int. Network for Eng. Ed. and Res. (iNEER), Arlington, VA, 2004, pp. 55-62.

4. http://www.datamonitor.com/

5. http:// www.dataquest.com/

6. ADL SCORM. Specifications, http://www.adlnet.org/

7. Eyerer, P., Hefer, B., and Krause, D., "The Reformation of Technical Education through Project-Orientated Education (TheoPrax)," Global J. of Engng. Educ., Vol. 4, No. 3, 2000, pp. 281-286.

**Ren-Hung Hwang** received his B.S. degree in Computer Science and Information Engineering from National Taiwan University, Taipei, Taiwan, in 1985, and M.S. and Ph.D. degrees in Computer Science from the University of Massachusetts, Amherst, Massachusetts, USA, in 1989 and 1993, respectively. He joined the Department of Computer Science and Information Engineering, National Chung Cheng University, Chia-Yi, Taiwan, in 1993, where he is now a Professor and Chair of the Department of Communications Engineering. He is also the Executive Secretary of the Network Applications and Services Program of the NICE-Program. His research interests include peer-to-peer applications, ad hoc networks, e-Learning, and 3G.

**Chung-Ming Huang** received his B.S. degree in Electrical Engineering from National Taiwan University, Taipei, Taiwan, in 1984, and M.S. and Ph.D. degrees in Computer and Information Science from Ohio State University, USA, in 1987 and 1991, respectively. He joined the Department of Computer Science and Information Engineering, National Cheng Kung University, Tainan, Taiwan, in 1991, where he is now a Professor. He is also the Director of the Network Applications and Services Program of the NICE-program. His research interests include broadband Internet, mobile IPv4/IPv6, next generation mobile communication protocols and software, interactive multimedia systems, and audio/video/multimedia streaming.

# Chapter 8

# Chipskills Program - A Unique Partnership Between University, Industry and Government

JUGDUTT SINGH and AKHTAR KALAM
*Victoria University, Melbourne, Australia. E-mail: jack.singh@vu.edu.au*

*This paper describes the "Chipskills" program that forms a key component in revitalising the microelectronics industry in Australia. The program is a partnership between universities, industry and the State Government of Victoria to provide the skill needs of the microelectronics industry and thus support the Government's economic strategy to attract the semiconductor industry to Victoria. Seed funding of A$7.1 million has been provided by the Government for the development of curriculum, learning modules, and staff development to meet the spectrum of education and skills training requirements of the microchip industry's workforce. The program is unique in Australia in that it has been developed and is presented by staff from partner universities and industry. By cooperating in this way, the participating universities can achieve economies of scale that would otherwise be impossible in the Australian microelectronics arena and should allow Australian engineers to develop the skills necessary to contribute to the global microelectronics design market.*

## INTRODUCTION

Microelectronics is the key driver of applications across a diverse range of industries and over the last thirty years has been one of the world's fastest growing industries. The global sale for the microelectronics industry is forecast to be US$140 billion. However, Australia currently has very little perceived presence in this industry. A common misconception is that it requires billions of dollars of investment in integrated circuit manufacturing facilities to get into the microelectronics business. This is not true at all. The future powerhouses of the microelectronics industry will be those without fabrication facilities. They will have the best people that design the integrated circuits – a corporation's greatest asset remains its employees. The employment market is so competitive in the USA and Europe that the brain drain in the microelectronics industry will only continue to get worse unless Australia starts competing. There are opportunities in the global microelectronics industry. Key high-growth areas are:

- Wireless Technologies
- Internet Technologies
- Embedded Systems
- Automotive and Transportation
- Bio-Technology and Bio-Medical
- Defense

Microelectronics is the enabling technology underpinning all of these areas and all nations are keen to attract and develop the microelectronics industry for the expected economic benefits it brings. Microelectronics technology was highlighted as a key component in the Information and Communication Technologies (ICT) industry in a recent report to the Prime Minister's Science, Engineering and Innovation Council (PMSEIC) [1].

Overall, Australia is well positioned to take part in the microelectronics revolution. Australia has good tertiary education infrastructure, economic and political stability and work ethic similar to USA.

## WHY IS THE MICROELECTRONICS INDUSTRY IMPORTANT?

Microelectronics, along with software, is a key enabling technology of the information society. While software allows us to perform computing tasks microelectronics provides the hardware platforms on which software runs.

The enabling character of microelectronics underpins the development of many sectors of the economy. This is nowhere more evident than in the impact of microelectronics on the Information Economy and its role as the driver of the Internet. The design and production of microchips is highly research intensive and requires skilled engineers, designers, software programmers and other technical skills. The microelectronics industry is therefore a key knowledge industry requiring highly skilled people.

Nations are keen to attract and develop the microelectronics industry for the expected economic benefits it brings. These benefits can be summarised as follows:

- it is highly knowledge intensive, employing highly skilled, highly paid people
- its progress relies on continuous R&D activity, involving universities
- it attracts investment and jobs to communities
- its products have global implications and can boost export values

There are several key characteristics that influence the development of microelectronics industry [2]. The degree to which these are present in a community will influence ability of the local microelectronics community to prosper and grow over time. These characteristics are:

- the availability of skilled people,
- the presence of universities with research programs in microelectronics
- a cluster of microelectronics companies that can take advantages of the above characteristics.

Microelectronics companies will be attracted to a location if the first and second of these characteristics exist. Access to a pool of skills is fairly self-evident. The level of skill and training required, whether for microchip design or fabrication, is specialised and not readily found. Microchip design is an activity that is becoming increasingly mobile, and companies whose business it is to design microchips are prepared to establish operations where a pool of microchip design skills may be available. The Chipskills program seeks to benefit from this emerging trend [3].

Microelectronics is a fast changing field of knowledge. The pressure to have products ready for the market is growing. In order to keep up, companies need access to well-trained people who are current in the latest technologies. Universities also need access to advanced technology to train highly qualified staff. Microelectronics companies therefore look to universities to carry out research and development tasks that will drive the future development of the industry.

## Current State of the Microelectronics Industry in Victoria

A few industry players, notably Ericsson, NEC, Agilent and Fujitsu, have operated microelectronics design centres in Melbourne for a number of years. More recently, Semiconductor Technologies Australia and Bandspeed Pty Ltd have also established local design centres. Further, Melbourne has recently become the location for the only large-scale microchip fabrication facility in Victoria, and only second in Australia, with A$33 million investment by Robert Bosch in a wafer making facility. In October 2000 NEC announced that it would inject an additional A$70M over three years into 3G mobile research and development. The company intends to recruit 150 engineers to undertake the additional R&D at the Mulgrave Technology Centre in Melbourne. Most of the recruitment is in the areas of microelectronics engineering (ASIC, RF, hardware, DSP and embedded systems). Even more recently Analog Devices has established their presence in Melbourne through the acquisition of the local technology company Signal Processing Associates. Companies such as Infineon and Motorola have also shown interest in establishing their microelectronics R&D centre in Victoria. Development activity of this type will prevent Victoria from becoming, in the words of the Allen report "only a spectator in the microelectronics industry [4].

There are a few issues that need to be addressed in order to further develop the microelectronics cluster and develop Victoria as a Centre of Excellence in microelectronics. They are shortage of talent, the prohibitive cost of Electronic Design Automation (EDA) tools, research and development, and the right venture capital environment.

The Victorian Government has had the foresight to recognise the need to support the growth of the microelectronics cluster in Victoria. It already provided funding to implement postgraduate programs in Microelectronic Engineering - the "Chipskills" program, encouraged industry-university R&D collaboration, established network with the microelectronics industry in Victoria and overseas [5]. This goes a long way in addressing some of the issues highlighted above.

The Victorian Government through Multimedia Victoria and the Chipskills program is continuing its support in further developing Victoria as the Centre of Excellence in microelectronic design and testing thus encouraging investment and start-ups in microelectronics. The purpose of Chipskills program is to support the growth of graduates who will be available to the industry in Victoria and to allow for qualified engineers to choose a new career path as well as upgrade their qualifications.

## CHIPSKILLS PROGRAM

The Information and Communications Technologies (ICT) industry is a key priority industry for the Government. One sector of the ICT industry, microelectronics, has been targeted as having the potential to grow in the State. This growth is the result of changes in technology in which microelectronics is becoming increasingly important in shaping the development of many other industries, such as medical, agriculture, automotive and manufacturing.

The existing microelectronics industry in Australia is small, but with the appropriate effort, it can grow. The State Government, through the Chipskills program, committed funds to assist in the development of industry-relevant curricula, which will support the growth of a skilled workforce, a critical requirement if the industry is to grow in the State [6]. The courses developed as part of the Chipskills program include, Master of Engineering – Microelectronics Engineering, Graduate Diploma in Microelectronics Engineering and Graduate Certificate in Microelectronics Engineering.

## MASTER OF ENGINEERING - MICROELECTRONICS ENGINEERING

The Master of Engineering – Microelectronics Engineering degree course [7], or M.Eng. for short, has a number of features that are unique amongst Australian coursework postgraduate programs. Firstly, it has been developed cooperatively by staff of four universities in the inner Melbourne area - Victoria University, RMIT University, Latrobe University, and Swinburne University of Technology and is the first time that this level of cooperation has been attempted and achieved in an Australian context. The course is separately accredited by the Chipskills partner universities with common structures, rules and unit syllabus definitions.

Although students are asked to select a "home" campus (from which they will ultimately graduate), they travel between the various university campuses to attend classes. Even though the partnering universities are in within easy commutable distance to one another, the course delivery is programmed so that students do not have to travel to different university on the same day. This contrasts with programs such as **MIDAS** [8], **Alba** [9] and **Socware** [10] that tend to focus on one central point of delivery. It has the advantage of permitting the various universities to leverage their existing resources (such as laboratory space and computing infrastructure) and support skills.

The students have access to high-end EDA tools, such as Cadence [11] and Synopsys [12] used by industry. They are required to use these tools in all the appropriate units offered in the course and also on industry-backed research and development projects. EDA tools of this calibre are normally vastly too expensive for Australian universities to purchase. At present, the State Government of Victoria underwrites the cost of the tool suite. Their availability represents a major attraction to potential students.

The program is comprehensively supported by industry [13]. Companies support the program in a range of ways – by including it as part of their staff development activities, by funding student scholarships and placements, by assisting in the curriculum development and modification, by assisting in the curriculum delivery and assessment, and by offering and supervising thesis and project work. This level of direct industry involvement is not typical of coursework postgraduate programs in Australia.

## PHILOSOPHY AND AIMS OF THE COURSE

The microelectronics engineer today is faced with many challenges brought about by the rapid advances in computer, multimedia and telecommunication technology. The M.Eng. course in Microelectronic Engineering addresses all aspects of this Technology, from high-level specification of microelectronic systems, through implementation alternatives, and the effective use of design tools, to realisation of integrated circuits. The course aims to produce engineers with the necessary skills and practical experience to satisfy the requirements of the microelectronics industry. An important feature of the course is the opportunity it provides for the students to design their own integrated circuits. The general aims of the course are to provide graduates with:

- high levels of both logical and lateral thinking development so that the graduates can lead constructive change through innovation;
- the ability to use a multi-disciplinary engineering philosophy towards the synthesis, design and integration of solutions; and
- a level of professional development in confidence, judgment and experience such that the implementation of proposed solutions proceeds successfully.

The specific aims of the course are to:

- develop integrated circuit design expertise in embedded systems, digital, mixed signal and system-on-chip systems design and verification.
- develop a basic understanding of the device physics, the fabrication process and the testing to the level needed by IC designers.
- develop the advanced technical and algorithmic skills necessary to master state-of-the-art microelectronic technology;
- develop research skills necessary to obtain specialist knowledge of issues pertinent to integrated circuit design;
- cultivate logical and lateral thinking that leads to creation and innovation in the pursuit of solutions to engineering problems.

## STRUCTURE OF THE COURSE

The structure of the Master of Engineering degree course is presented in Figure 1. The course is structured to allow students to exit at different academic levels with either a Graduate Certificate, Graduate Diploma or Master of Engineering qualifications. The completion of a Graduate Certificate in Microelectronic Engineering requires successful completion of four units; the Graduate Diploma in Microelectronic Engineering requires successful completion of either eight units or six units and minor project, while the Master of Engineering in Microelectronic Engineering requires successful completion of either eight units and major project or ten units and minor project.

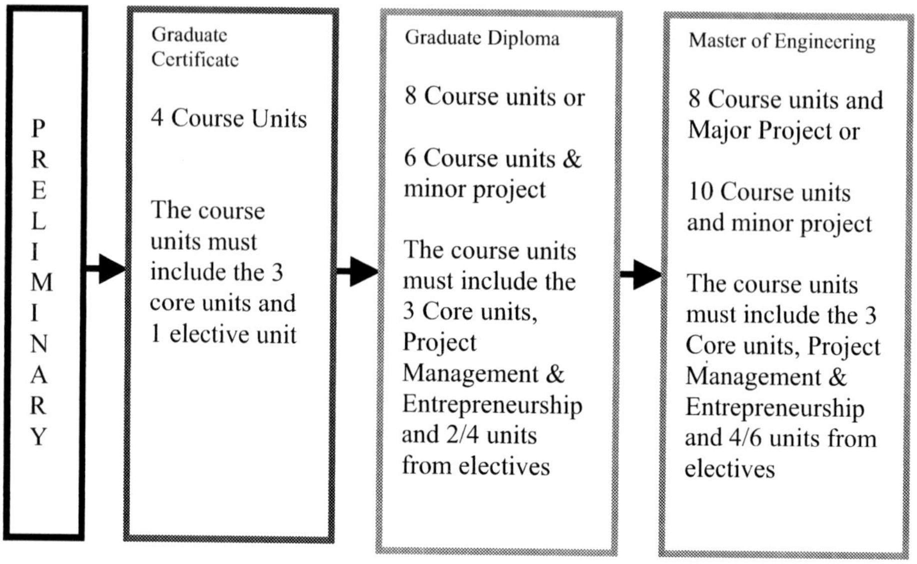


FIGURE 1
COURSE STRUCTURE – GRADUATE CERTIFICATE, GRADUATE DIPLOMA AND MASTER OF ENGINEERING

For the coursework component, a full-time student will take 4 units per semester. This comfortably caters for part-time students, especially when the units are taught in the evening.

- Normally it will take 1 semester to complete Graduate Certificate, 2 semesters to complete Graduate Diploma and 3 semesters to complete Master of Engineering on full-time basis.
- Where feasible, the program will be structured as 2 semesters of coursework, followed by a 1 semester of project. This means that students will be available to the industry after only one-year (perhaps on a 'cadet' or trial basis for 1 semester), if they are permitted to undertake their project as part of their employment.
- All units are structured to be of equivalent credit, allowing for the modularity necessary in such a collaborative venture.

## ENTRY REQUIREMENTS

Admission to the course normally requires a four-year Bachelor of Engineering degree in Electronic Engineering, Computer Engineering, Communication/Telecommunication Engineering, or a four-year Bachelor of Science (Honours) degree in an appropriate field, or an equivalent qualification.

Applicants with a three year Bachelor of Science degree (in appropriate field) or a Bachelor of Engineering degree in another field may also be considered for admission on the condition that they may be required to take additional (preliminary) subjects that will strengthen their knowledge and skills in digital systems, analog electronics and microprocessor systems.

Full-fee paying international students are required to have qualifications equivalent to those above, and in addition, they must provide evidence of proficiency in English Language, as assessed by International English Language Testing, with an overall band score of 6+, subject to individual profile.

A panel comprising of academics from each of the partner universities will carry out student selection into this course.

## UNIT DETAILS

### Core Units

| Unit Name | Credit Points |
|---|---|
| HDL and High Level Synthesis | 15 |
| Integrated Circuit Design | 15 |
| EDA tools and Design Methodology | 15 |

### Electives Units

| Unit Name | Credit Points |
|---|---|
| Digital System Design | 15 |
| Embedded Systems Design | 15 |
| Emerging Topics in IC Design | 15 |
| Advanced VLSI Design | 15 |
| VLSI Digital Signal Processing Sys. | 15 |
| Reliability and Testability in IC Dgn. | 15 |
| Introduction to Microsystem Tech. | 15 |
| Introduction to Semiconductor Device Fabrication | 15 |
| Semiconductor Device Physics | 15 |
| Project Management & Entrepreneurship | 15 |
| RF and Mixed Signal Design | 15 |
| Minor Project | 30 |
| Major Project | 60 |

## MAJOR PROJECT

The Major Project is equivalent to one semester of full time study. This time includes all aspects of preliminary research, consultation, design,, implementation, testing and reporting. Many of the projects are based with industry partners. The advantages of this include:

- immediate relevance
- improved probability of a commercially relevant outcome
- safe and positive recruitment process
- strengthened industry – academia networking

The projects undertaken are industry initiated, candidate initiated or university initiated. A three-way process between industry, universities and the project candidates is implemented to bring together the best available team for a particular project. All projects have a senior supervisor from one of the cooperating universities. The students are enrolled in the project unit at the university of the senior supervisor. The project supervising university is the one from which the student graduates. Supervisors are responsible for: providing advice and monitoring the project solution alternatives; monitoring progress; advising about reporting and assessment and providing a list for possible project examiners.

Each of the cooperating universities has in place regulations that apply to the assessment of the postgraduate course work qualifications with project reports. A typical project assessment includes: an oral presentation, preferably public but subject to commercial non-disclosure conditions, and a written report. Two examiners, at least one of them being external to the supervising university, examine the written report.

## Project Intellectual Property

In the development of a project solution there may be a number of parties that can legitimately claim partial rights to intellectual property generated by the project. There may also be background intellectual property that is owned by the involved parties. Typical parties to the project intellectual property are: the project sponsoring company; the supervising university; the student; and the project supervisor. All the intellectual property issues are resolved and an agreement is seen and signed by all parties before the commencement of the project.

The projects are often “Commercial-in-Confidence” and subject to “Non-disclosure Agreements” by parties that need to know in order that the project progresses. The “Non-disclosure Agreements” are signed at an early discussion phase of the project to legally protect the commercial interest of the project source.

## Program Implementation

The Master of Engineering in Microelectronics was first introduced in 2002 with an intake of approximately 15 students. The demand for the course has grown dramatically since its introduction and at present approximately 130 new students joins the course each semester.

After completing two semesters of course work program, each student is required to undertake either a major project or a minor project with an additional two units. The majority of the projects are sponsored by the partner companies such as: Agere, Semiconductor Technologies Australia, Swip, Analog Devices, Motorola, Redcentre, Dynamic Hearing, etc. As part of the collaborative project, the industry partners assist in the delivery and assessment of the curriculum.

## Conclusion

Although Australia is a small player in the microelectronics arena, it has a number of positive things going for it. In particular, it is economically and politically stable and has a work ethic similar to the U.S. and Europe. Relative to other countries, R&D costs in Australia are quite low and the quality of its information infrastructure is world-class. Furthermore, the research culture in Australia is very similar to that of the U.S. and Europe and is somewhat unique within the Asian region [14]. These features make it likely that Australia can be a significant player in microelectronics design.

The Chipskills Program is a unique partnership of universities, industry and government to provide the skill needs of the microelectronics industry and thus support the Government's economic strategy to attract the semiconductor industry to Victoria, Australia. The program is funded by the State Government of Victoria and supported by the ICT industry. Seed funding of A$7.1 million has been provided by the Government for the development of curriculum, learning modules, and staff development to meet the spectrum of education and skills training requirements of the microchip industry's workforce.

While the Chipskills postgraduate programs in microelectronics will stimulate a supply of engineers with a high-level of microelectronic design skills, including knowledge of the leading EDA tools and design flows used by industry, the programs represent a long-term solution to industry skill needs. The Government also recognises that industry has a range of short-term skill needs. Industry can't always wait eighteen months to get a graduate with new skills and, smaller companies particularly, can't always afford to travel to relevant training courses and workshops overseas. The Chipskills project aims to address these more immediate skill needs through a series of specially targeted short courses that will bring international expertise to within the reach of all Victorian microelectronics companies and confirm Victoria's position as the centre for microelectronics design in Australia.

## REFERENCES


1. Australia's Information and Communications Technology (ICT) Research Base. Driving the "New Economy". www.isr.gov.au/science/pmseic/6thmeeting.html

2. M. A. Rosen, "Challenges and Opportunities in Engineering Education: A Canadian Perspective," Chapter 4, Aung, W. et al. (eds.), *Innovations 2003: World Innovations in Engineering Education and Research,* Int. Network for Eng. Ed. and Res. (iNEER), Arlington, VA, 2003, pp. 25-35.

3. J. Singh, A. Kalam, A. Zayegh, P. Beckett and L. Ismahil, "Chipskills Project – A Unique Partnership Between University, Industry and Government," Proceedings of 2003 International Conference on Engineering Education (ICEE-2003), Paper No. 2543, Valencia, Spain, 2003.

4. Spectator or Serious Player? Competitiveness of Australia's Information Industries. Allen Consortium, 24 April 1997. http://www.dist.gov.au/itt/tskforce/allen/index.html

5. Lan, Man-Chi; Chang, Yao-Wen; Chen, Liang-Gee; and Jen, Chein-Wei, "The VLSI Circuits and Systems Educational Program in Taiwan," Proceedings of 2003 International Conference on Engineering Education (ICEE-2003), Paper No. 3441, Valencia, Spain, 2003.

6. Jeng, Jeng-Ywan; Wanbg, A.B., Huang, L. S., Lee, C.K., Chang, M., and Chen, I-P., "Manpower Development of the Op-Mechatronics Education Program for Optoelectronic Industry in Taiwan," Proceedings of 2003 International Conference on Engineering Education (ICEE-2003), Paper No. 2932, Valencia, Spain, 2003.

7. Chipskills Website – Growing Microelectronics Design Industry in Victoria, http://www.vlsi.vu.edu.au

8. The Microelectronics Industry Design Association of Ireland (MIDAS), http://www.midasireland.ie

9. The Alba Centre, http://www.albacentre.com/

10. Socware, http:// www.socware.com

11. Cadence Design Systems, http://www.cadence.com

12. Synopsys – Electronics Design Automation Solutions and Services, http://www.synopsys.com

13. Victorian Microelectronics Designer's Network, http://www.mmv.vic.gov.au/microelectronics.

14. C. Nicol, "The State of Research in Australia: Brain Drain, University Research Funding and the Microelectronics Industry," Prime Minister's Science, Engineering and Innovation Council, (PMSEIC), Canberra, Australia, June 2000.

**Jugdutt Singh** received his B.Sc. in Electronics Engineering from University of Brighton, UK and M.Sc. in Electronics Engineering from University of Alberta, Canada in 1978 and 1986, respectively. He completed his Ph.D. at Victoria University, Australia in 1997. He has held teaching appointment at Fiji Institute of Technology. Since 1989 he has been at Victoria University, Melbourne, Australia. Currently he is a Professor in the Centre for Telecommunications and Microelectronics at Victoria University. His major areas of research interests are in RF and mixed signal design, reconfigurable architectures, low-power VLSI circuits and systems design. He has published a number of articles in education and research microelectronics area.

**Akhtar Kalam** received his B.Sc. and B.Sc. Engineering from Calcutta University and Aligarh Muslim University, India in 1969 and 1973 respectively. He completed his MS and Ph.D. at the University of Oklahoma, USA and the University of Bath, UK in 1975 and 1981, respectively. He has worked with Ingersoll Rand and other electrical manufacturers. He has held teaching appointments at the University of Technology, Baghdad, Iraq and Capricornia Institute of Advanced Education, Rockhampton, Queensland, Australia. Since 1985 he has been at Victoria University of Technology, Melbourne, Australia. Currently he is the Deputy Dean of the Faculty of Engineering and Science and is the Head of Werribee Group of Campuses. His major areas of interests are power system analysis, control, protection and cogeneration systems. He has conducted research, provided industrial consultancy and published over 220 articles on energy systems and written over 15 books in the area. Professor Kalam is a Fellow of IEAust, IEE and a member of IEEE.

# Chapter 9

# Use of Web-Based Review of Prior Course Material to Increase the Intellectual Challenge of an Engineering Course

KAREN E. SCHMAHL
*Miami University, Oxford, OH 45056, USA. E-mail: schmahke@muohio.edu*

*Mathematics and science courses provide the foundation upon which engineering science is built. An engineering course, therefore, typically will require completion of specified mathematics and science courses before students are admitted. Often a professor in a higher-level course will experience frustration at students' lack of retention of the prior course material. Professors may accommodate the students by spending class time to review material, or reducing the rigor of the course topic material. An engineering course in Statistical Quality Control requires a fundamental knowledge of statistics to provide the basis of a sound understanding of the theory and application of quality control principles. This paper establishes a need for review of prerequisite course material and proposes use of web-based review material outside of normal classroom time. Review materials include online lectures, simulations, and reference to numerous statistics sites.*

## INTRODUCTION

A major source of frustration for engineering professors teaching upper level courses is lack of student retention of prior course material. The lack of retention of important mathematical and scientific knowledge needed for upper level courses can come about for many reasons. A student may not have learned the material well when the prerequisite course was taken. The prerequisite course may have been taken several semesters prior to the course where the knowledge is needed and the student has forgotten important aspects. Students may not have realized the relevance of material to the future courses and so did not keep good course notes or even keep the textbook.

Whatever the reason for the lack of retention, the professor is faced with the dilemma of how to handle the deficit. Sometimes valuable class time is spent reviewing prior material. Other times, students are expected to review material on their own. A professor may even reduce the rigor of a course to compensate for the low level of student ability.

The professor must consider reasonable expectations for student retention of prior course material and reasonable effort in the current course. It is the responsibility of the professor to know what was taught in the prior course and what was expected of the students at that time. It is certainly reasonable to expect students to have kept the textbook and notes from the prior course and to perform a review on their own of the material. Many engineering textbooks also provide a chapter for brief review of pertinent information from the typical prerequisite courses, and students can use this as a resource. Where the success of a current course is dependent on the students' prior knowledge, however, it is often in best interest of both professor and student for the professor to provide some review assistance.

Use of web-based material provides a new mechanism for providing review material for prior course work. Review materials may include online lectures, simulations, and reference to web sites that contain additional review material. By providing such material in a logical manner, students can be quickly guided in a review process outside of regular class time. Students can go through the material at their own pace. Those students who have had the course recently, or learned the material very well, should be able to quickly go through the material. Students who need more time can take the time. Either way, all students would benefit and the expectation for the level of knowledge required will have been established. A course can then proceed with students prepared to face a rigorous course content.

Use of web-based material for review of statistics has enhanced the teaching of a course in statistical quality control in the Manufacturing and Mechanical Engineering program at Miami University in Oxford, Ohio, USA. The Quality Planning and Control (QPC) course requires a fundamental knowledge of statistics to provide the basis of a sound understanding of the theory and application of quality control principles. Successful completion of a course in calculus-based statistics is required for entry into the QPC course. This paper establishes a need for review of prerequisite course material and proposes considerable use of web-based material outside of normal classroom time. Computer-based materials include online lectures, simulations, statistical analysis software, and reference to numerous statistics sites. The efforts in increasing the statistical rigor of the course by providing the review has improved student understanding of quality control principles as can be shown in the successful addition of complex case studies in a pilot class.

## NEED FOR REVIEW

A course in statistical quality control can actually be taught with several different levels of statistical rigor. For example, a person can learn to apply statistical process control charts without understanding the underlying statistics. One could make the argument that proper application of quality control techniques is more important than theoretical understanding. However, what differentiates an engineer from a technologist is that deeper theoretical understanding and ability to synthesize a situation. Engineers should answer why as well as how. An understanding of the basic assumptions behind an analysis will allow the critical assessment of a manufacturing process problem that a quality engineer should be able to perform. Therefore, a rigorous statistics course is generally required before any study of quality engineering.

Weakness in engineering mathematical skills was noted in a study by Townend [1]. The study tested students' mathematical knowledge at a foundation level and results suggested that knowledge previously taken for granted, such as elementary trigonometry and factorialization, could no longer be relied upon by professors. The Townend study then focused on improving initial learning of mathematics skills.

Students at Miami University following the Manufacturing Engineering and Engineering Management curriculums are scheduled to take a calculus-based statistics course in the spring semester of the sophomore (second) year of study. The Quality Planning and Control course is scheduled for the spring semester of the junior (third) year. Most students follow the standard curriculum, but often students take courses earlier or later in their program. Pre-requisites are strictly enforced and no student is allowed to stay in the QPC class without having had statistics in a previous semester.

The professor's subjective assessment was that the students were not as strong in their statistics as they should be when they reached the Quality Planning and Control Course. A major contributor to the problem was the time lapse between the statistics course and the QPC course. Given other constraints of the program, it was not possible to modify the standard curriculum to move the courses to consecutive semesters.

Over many years of teaching the course, several approaches had been taken to compensate for this. An "inherited" course syllabus, one that had been used in prior semesters, was used the first time the author taught the course. This syllabus devoted three weeks to review of statistics before starting new material and prompted several student complaints on the final course evaluation. The next semester, the professor gave a brief survey to students asking what level of coverage they had had previously, when they had taken the statistics course, and what instructor they had had. From that survey data, it would appear that students with the same instructor the same semester received very different levels of instruction. The statistics review was reduced to one lecture and one lab period with some hands-on activities to reinforce key concepts. The expectation was also that students would review more on their own as needed. The students were assigned homework problems from the statistics review chapter in the textbook. The lectures then assumed that the students had a grasp of the applicable concepts. However, assessment of the student work, primarily through testing, still focused on proper application of the quality control tools, rather than understanding of the concepts behind the tools. Students could do reasonably well in the class without having a thorough understanding of the application of statistics in quality engineering, or the theoretical underpinnings of many of the quality tools.

That this situation is not unusual is evident in the 1997 revision of the Engineering Council UK Standards and Routes to Registration (SATOR), which noted weakness of engineering students in that they may appear to demonstrate reasonable level of competency as measured by their mathematics assessment, but they are often unable to apply knowledge to mainstream engineering [2]. The Townsend project previously mentioned was in response to the 1997 SATOR revision [1].

Further development of the QPC course hinged on the students' ability in statistics. In an effort to gauge this ability, students were asked to complete a survey regarding how well prepared they thought they were with statistics. The survey was administered the first week of class. Students were asked a series of five questions to gauge their level of confidence in working statistics problems. Problems were posed on normal distribution, Central Limit Theorem, Poisson distribution and regression analysis. The students were asked to check off a response to being asked the question, not to work out the problem itself. (This approach was taken to minimize the amount of valuable class time taken to gather data.) A "retention" score between zero and four was then calculated based on their responses.

A sample survey "retention" question is as follows: A sample of 40 parts is taken from a process. The sample mean is 23. The sample standard deviation is 1.1. What is the 90% confidence interval for the true mean of the population if normality is assumed? Possible responses were:

- Give me a normal table and I will work the problem.
- I could work this problem with a brief refresher of how to read the normal table.
- I could work this problem if I looked back at my stats notes/book for a few minutes.
- It would be best if someone sat down with me and helped me work through this.
- It would take considerable refreshing for me to work this.

Students were also asked to indicate level of agreement from strongly agree (5) to strongly disagree (1) with the following statement: I feel well prepared to apply statistics that I previously learned to the Quality Planning and Control course. This is considered the students "preparedness" score.

The retention scores versus the preparedness scores are plotted in Figure 1 for the eighteen students in the spring offering of the Quality Planning and Control course. As would be expected, the students that answered that they could work the statistics problems described also felt better prepared to use the statistics in the current course. As also expected, students that took the statistics course in the previous semester generally scored themselves higher in retention and preparedness than students who had taken the class a year before (two semesters previous).

Only 33% of the class agreed with the statement that they were well prepared to use their statistics. Approximately 22% did not agree with the statement and the remaining 45% were neutral in their response. Overall the survey showed that the students did not have a high level of confidence in their statistical ability.

To increase the sophistication of the application of statistics in the QPC course, most of the students would need an aggressive review of their statistics and the course would need to be revised to incorporate additional statistics emphasis.

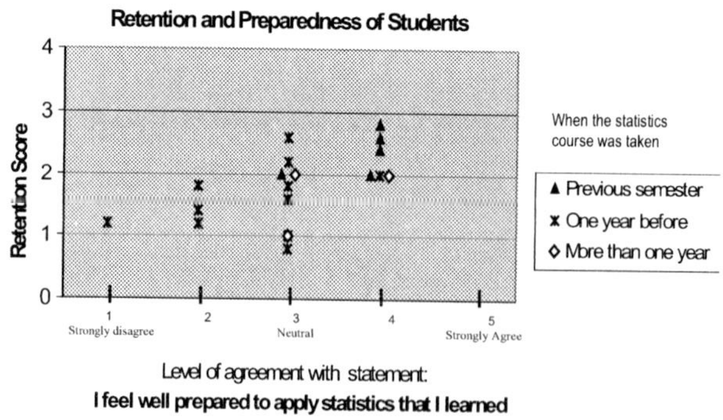


FIGURE 1

RETENTION AND PREPAREDNESS OF STUDENTS

## THE WEB-BASED STATISTICS REVIEW

Several challenges became quickly apparent in preparing to develop a statistics review. One challenge in reviewing statistics was to guide the students sufficiently but not to use valuable class time. Another was to make it adaptable to the different levels of student ability.

The review should not be a repeat of their statistics course and should relate directly to quality control issues.

Also, the review ideally should not be boring and should be designed to enhance knowledge retention – so a multi-faceted review with interactive components would be best. Research has shown that limited knowledge is retained in passive learning, such as with reading, but increases with visual stimulation and is greatest with participative elements [3].

To meet these challenges, web-based and other computer tools provided a solution. Search of the literature showed numerous instances of effective teaching and learning through Internet-based tools applied to math, science and engineering courses. Conference papers abound and are too numerous to mention. Journal papers reviewed included [4,5,6,7,8,9,10,11]. Among the literature, some papers discussed using tools to enhance existing courses while others discussed fully web-based courses. The application of web-based materials as a means to review prerequisite knowledge was not found in the literature.

The opportunity to develop the materials came about when a web-based version of the SPC course was to be offered. For the web-based course, a series of four statistics review modules were developed. Each began with a web-based lecture covering theory and working examples. In addition, several web sites where the students could get additional information or example problems were linked to the course web site. Textbook problems were assigned and additional follow-up exercises were developed incorporating statistical simulation software tools and Minitab Statistical Software [12].

The four web-based review lectures were initially developed as PowerPoint lectures with voice clips. These were then transformed to HTML pages and access was provided to them through the course web site. The first three lectures on continuous distributions provide the basis for theoretical understanding needed early in the course. The fourth lecture over discrete distribution topics correspond to later course material. The idea is that students can spend as much or as little time listening to the lectures as they need.

The course web site provides links to other web-based statistics material. Search of the web for appropriate links found numerous sites. Some excellent online statistics books are offered by StatSoft Inc. [13], the National Institute of Standards and Technology [14], Rice University [15] and by the University of Alabama Huntsville[16].

While students are not required to listen to the lectures or access the other sites, they are required to complete several problem sets and follow-on exercises. Although the solutions are provided for them on the course web site, students are required to include their solutions in a notebook submitted near the end of the course.

Computer-based simulations provide an interactive opportunity for the follow-on exercises with each lecture. The simulated quincunx and sampling bowls of commercial software, The Bean Machine [17], are used in two review exercises and also for later course instruction. Free simulation applets are offered in numerous web sites and were also used for class exercises. For example, in the Rice Virtual Lab in Statistics [15], the applet titled "Normal Approximation of Binomial" is used in an exercise to reinforce when it is appropriate to use the approximation. The exercises in which the simulations are used are typically short, typically completed in fifteen to thirty minutes. The intent is to provide some variety in instruction and quickly reinforce a concept or idea. The students are not required to turn the exercises in for grading, but are to include them in their course notebooks.

The simulation exercises plus the online course discussion board also provide a mechanism to ensure the students are on track with the review. They are required to post messages regarding several exercises by specified deadlines. For example: A normal distribution was generated with the simulated quincunx. (A quincunx is a device that allows beads to drop through an array of pins placed equally spaced in a board. When the bead hits a pin it is equally likely to fall to the left or the right creating the normal distribution.) The students calculate the natural process limits, considered to be plus or minus three standard deviations. The funnel which drops the bead is then moved and a small number of beads dropped. Based on the sample the students predict what percent of a large number of beads would fall outside the original process limits. The students are required to post their original and shifted process data and assess how well they were able to predict the number outside the limits. Because each student's data is simulated, every posting is different. Allowing the students to look at each other's data lets them know whether they are on track. The instructor monitors the postings as well.

The required simulation exercises were comparable to the statistics review labs previously performed with actual quincunxes and sampling bowls in groups of four students. While, in general, hands-on exercises would be preferred in a laboratory, for these exercises considerable time appeared to be inefficiently utilized by the teams and the exercises took considerably longer than the instructor felt they should have. Time seemed to be wasted on getting into groups, deciding who was going to do what, and just tinkering with the devices. Also, too often, a student in a group would sit back and let the others do the work, merely copying the data and calculations from a teammate's worksheet. The online simulations, while not quite as "fun," more efficiently utilize the students' time and require that each student complete the work with his or her own set of data. Students can still, however, compare data with each other and discuss the lab.

Minitab Statistical Software [12] is a tool used throughout the course. For each module of the statistics review, the students are provided with worksheets of problems to solve manually, then given the Minitab instructions to work the problems and thus check their own work. Only two of these Minitab review exercises were required to allow the students to become familiar with the software. Three others were developed to give students an opportunity for additional practice in a given area if they felt they needed it or to get more experience with the Minitab software. Review of regression analysis was covered in a Minitab exercise and handout (no lecture) later in the course.

Relative to learning retention, there are many benefits of the web-based lectures and associated activities over relying on students to read the text review chapters. With the lectures, the students are listening, looking at pictures and diagrams along with reading the slides. The simulation exercises provide an interactive learning environment that, as noted in [18], facilitates learner inquiry, discussion, discovery, and general contribution to their own learning. Such simulations have been found to improve student learning [11] by increasing understanding of abstract components [19]. The discussion board postings help to develop reciprocity and cooperation among the students, an important practice in undergraduate education [20,21].

## TESTING OF REVIEW MODULES & MODIFICATION OF COURSE MATERIAL

The review material was piloted with a group of 6 students taking the course in a new, completely web-based offering. These students were expected to complete required assignments in the first three review modules, along with two non-review modules, by the end of the second week of the course. The fourth review module was assigned later, as was the regression review.

The initial reaction of several students in the pilot group to the first three review modules seemed to be, "Why do we have to do all this statistics work that the regular class does not have to do?" The required elements (textbook problems and labs) of the review were essentially the same as the "regular" class. However, it appears that by giving the students the lectures and additional problems, the students felt the increased level of expectation of statistics knowledge. All six students did use the review lectures, but only three did any of the optional exercises.

Now that the expectation was set, the challenge for the professor was to improve the course content and assessment by incorporating the statistics more aggressively. This was looked at from two viewpoints: application of previously learned statistical techniques to quality control problems, and increased theoretical understanding of the new quality control techniques learned.

While the review material focused on quality control issues, opportunity was taken to incorporate increased application of previously learned statistical techniques throughout the semester in graded assignments. For example, a case was modified to require a hypothesis test to make a judgment as to whether the variance of a process was actually improved. Another case now requires interpretation of the correlation between a process variable and part characteristic.

Lectures had already contained the basic theoretical background to the statistical quality tools. With the intensive review, however, a shift in the approach to presenting the content was made. The lectures built upon the review and occasionally took less time to present concepts. In other areas, a deeper level of presentation was possible. For example, the recent review of confidence intervals allowed a discussion of confidence intervals for process capability ratios, a topic not previously broached.

Increasing the intellectual challenge of a course can be looked at with respect to Bloom's Taxonomy of Educational Objectives [22] and relative to the assessment methods used in the course. The original Taxonomy identified six levels of intellectual skill ability in the cognitive domain: knowledge, comprehension, application, analysis, synthesis, and evaluation. The primary method of assessment of student learning for the course, prior to changes, was through in-class testing. Classroom exams can most easily test factual knowledge and application of techniques. While questions can be formulated, it is not as easy to test true comprehension of concepts. Time constraints will generally limit the use of questions that are complex enough to evaluate analysis, synthesis, and evaluation abilities. Therefore, four cases were developed incorporating greater levels of statistics emphasis, providing for application and analysis using new tools, and requiring synthesis of relatively complex situations to make evaluations/recommendations for future action. In terms of the Revised Taxonomy [23], the cases developed approach the highest levels of intellectual thinking: creative thinking using metacognitive knowledge.

In the Townend project, it was also noted that students' examinations were predominately procedural in nature rather than conceptual. Their results in going to a case-based approach also suggested that that case use increases mathematical confidence and competence, while also increasing ability to use math in an engineering context [1].

For the pilot web-based course, 50% of a student's grade in the course was based on the cases. A comprehensive final exam was still given, accounting for 30% of the grade. Homework problems, labs and exercises, all completed and organized in a course notebook, are the basis of the remaining portion of the grade. Previously, the greatest portion of the student's grade for the course was through testing.

With the success of using web-based review materials in the pilot course, the classroom version of the course will now be modified. The review modules will be made available to the students. Course content and assessment will be modified as appropriate.

## Study Limitations

A limitation of this study is in the small number of students who participated in the pilot project and lack of statistically viable measurements to discern levels of learning compared to the standard class. Final exam average scores were slightly higher for the pilot students than the class-based students. Pedagogically, the web-based review materials were superior to student review with the textbook. But classroom time is taken in the traditional class for review and the professor is available for additional help during office hours. Note should be made that numerous studies have shown factual learning to be comparable for online and traditional in-class instruction [24,25,26,27,28].

## Concluding Remarks

The web-based review of statistics provided an effective means to overcome students' lack of retention of the statistics knowledge required for a challenging Quality Planning and Control course. Providing the review lets the students know the expectations for the course. The computer-based format allowed students to proceed with the review at their own pace and intensity that they required. The review allowed the course to include more in-depth coverage of the topic material that will be of benefit to students in their engineering careers.

Now that the material has been developed it can be used every semester with little additional effort. However, the effort to initially develop the material was substantial. The web-based review lectures for this course averaged around forty hours each to prepare. This time included research to find appropriate examples using the statistics in quality control, preparing the PowerPoint slides, recording the "script" to go with each slide, putting the material in HTML format, and loading it to the course web-site.

As also noted in [9], after the initial time consuming effort to develop web-based materials, the instructor was able to provide high-quality and well-structured pedagogic material, integrate online references that enrich learning experiences, facilitate self-learning tasks and student responsibility, support student collaborative efforts with the discussion boards, and provide learning-by-doing tools such as simulations and virtual laboratories. In addition, having more time available for new material and having students prepared to handle more challenging material enriched the course.

From the user point of view, the web-based review material was available as needed, was self-paced, and provided feedback mechanisms. Every student understood what the expected knowledge from the prerequisite course was, and the structured review gave the student the opportunity to review pertinent material in an effective manner.

Using a web-based approach to reviewing prior course material could be applied to any course where prior content is pertinent. This pilot offering of web-based review showed that such an approach can be successful in increasing the intellectual challenge of an engineering course.

## Acknowledgements

Funding for development of the web-based Quality Planning and Control course was provided in part by the Ohio Learning Network/Engineering Management Consortium.

## References

1. Townend, M.S., "Integrating Case Studies in Engineering Mathematics: A Response to SARTOR 3," Teaching in Higher Education, Vol. 6, No. 2, 2001, pp. 203-215.

2. Engineering Council UK, Standards and Routes to Registration, SATOR 3rd Ed, London, 1997.

3. Dale, Edgar, *Audio-Visual Methods in Teaching, 3rd Edition*, Holt Rinehart and Winston, New York, 1969.

4. Cameron, B.H., "Effectiveness of Simulation in a Hybrid and Online Networking Course," The Quarterly Review of Distance Education, Vol. 4, No. 1, 2003, pp. 51-55.

5. Baillie, C., and Percoco, G., "A Study of Present Use and Usefulness of Computer-Based Learning at a Technical University," European Journal of Engineering Education, Vol. 25, No. 1, 2000, pp. 33-44.

6. Fuks, H., Gerosa, M.A., and deLucena, C.J.P., "The Development and Application of Distance Learning Courses on the Internet," Open Learning, Vol. 17, No. 1, 2002, pp. 23-38.

7. Khalifa, M., and Lam, R., "Web-Based Learning: Effects on Learning Process and Outcome," IEEE Transactions on Education, Vol. 45, No. 4, 2002, pp. 350-356.

8. Korwin, A.R., and Jones, R.E., "Do Hands-On, Technology Based Activities Enhance Learning by Reinforcing Cognitive Knowledge and Retention?" Journal of Technology Education, Vol. 1, No. 2, 1990.

9. Michau, F., Gentil, S., and Barrault, M., "Expected Benefits of Web-Based Learning for Engineering Education: Examples in Control Engineering," European Journal of Engineering Education, Vol. 26, No. 2, 2001, pp. 151-168.

10. Seng, L, and Mohamad, F.S., "Online Learning, Is It Meant for Science Courses?" The Internet and Higher Education, Vol. 5, No. 2, 2002, pp. 109-118.

11. White, B.E., "A Microworld-Based Approach to Science Education," Scanlon, E. et al. (eds.), *New Directions in Educational Technology*, Springer, New York, 1994.

12. MINITAB Statistical Software, Release 13. Minitab, Inc.

13. StatSoft, Inc., Electronic Statistics Textbook, StatSoft, Tulsa, OK, 2002, URL: http://www.statsoft.com/textbook/stathome.html

14. NIST/SEMATECH e-Handbook of Statistical Methods, 2002, URL: http://www.itl.nist.gov/div898/handbook/

15. Rice Virtual Lab in Statistics, Rice University, Houston TX, URL: http://www.ruf.rice.edu/~lane/rvls.html

16. Virtual Laboratories in Probability and Statistics, University of Alabama Huntsville, 2002, URL: http://www.math.uah.edu/stat/index.xml

17. Imboden, J., The Bean Machine for Windows (software), 2002.

18. James, J., "Practical Issues in Interactive Multimedia Design," Proceedings of ED-MEDIA/ED-TELECOM98 World Conference on Educational Multimedia and Hypermedia and World Conference on Educational Telecommunications, Freiburg, Germany, 1988.

19. Reiber, L.P., "Animation as Feedback in a Computer Based Simulation: Differential Effects on Self-Efficacy and Achievement," Journal of Computer-Based Simulations, Vol. 16, No. 1, 1996, pp. 1-10.

20. American Association of Higher Education, "Seven Principles of Good Practice in Undergraduate Education," Seven Principles Resource Center, Winona State University, Winona, MN, 1987.

21. Newlin, M.H., and Wang, A.Y., "Integrating Technology and Pedagogy: Web Instruction and the Seven Principles of Undergraduate Education," Teaching of Psychology, Vol. 29, No. 4, 2002, pp. 325-330.

22. Bloom, B.S. (ed.), Taxonomy of Educational Objectives: The Classification of Educational Goals. Handbook 1: Cognitive Domain, David McKay, New York, 1956.

23. Anderson, L.W. et al. (eds.), A Taxonomy for Learning, Teaching and Assessing: A Revision of Bloom's Taxonomy of Educational Objectives, Longman, New York, 2001.

24. Johnson, S.D., Aragon, S.R., Shaik, N., and Palma-Rivas, N., "Comparative Analysis of Learner Satisfaction and Learning Outcomes in Online and Face-to-Face Learning Environments," Journal of Interactive Learning Research, Vol. 11, No. 1, 2000, pp. 29-49.

25. Kekkonen-Moneta, S., and Moneta, G.B., "E-Learning in Hong Kong: Comparing Learning Outcomes in Online Multimedia and Lecture Versions of an Introductory Computing Course," British Journal of Educational Technology, Vol. 33, No. 4, 2002, pp. 423-433.

26. Nichols, J., Shaffer, B., and Shockey, K., "Changing the Face of Instruction: Is Online or In-Class More Effective?" College & Research Libraries, Vol. 64, No. 5, 2003, pp. 378-388.

27. Parker, D., and Gemino, A, "Inside Online Learning: Comparing Conceptual and Technique Learning Performance in Place-Based and ALN Formats," Journal of Asynchronous Learning Networks, Vol. 5, No. 2, 2001, pp. 64-74.

28. Schulman, A.H., and Sims, R.L., "Learning in an Online Format Versus an In-Class Format: An Experimental Study," T H E Journal, Vol. 26, No. 11, 1999, pp. 54-57.

**Karen Schmahl** received the B.S. and M.E. degrees in Industrial Engineering from Texas A&M University. She spent 13 years in industry at Rockwell International, E-Systems and General Electric Aircraft Engines prior to returning to school to complete a Ph.D. at the University of Cincinnati. She began at Miami University in Oxford, Ohio, USA in 1994 and is currently an Associate Professor of Manufacturing and Mechanical Engineering. Her research interests are in the areas of quality management systems and lean manufacturing.

# Chapter 10

# A Web-Accessible Heat Exchanger Experiment

CLARK K. COLTON, MARC KNIGHT, RUBAIYAT A. KHAN, SARAH IBRAHIM and RICHARD WEST
*Massachusetts Institute of Technology, Department of Chemical Engineering, Cambridge, MA 0213, USA. E-mail: ckcolton@mit.edu*

*We have developed a remotely controlled heat exchanger experiment as part of the I-Lab project at MIT. It has been used on the Internet by students in a variety of chemical engineering courses, and the response of students to the online experiment has been assessed. The hardware consists of a general service unit, custom designed and fabricated for us (HT-30XC, Armfield, Ltd., England), on which a heat exchanger is mounted (flat plate, shell and tube, or double pipe). Cold water flows through the exchanger in a single pass, whereas hot water is recycled to a heated reservoir. Both fluid flow rates, their direction (cocurrent or countercurrent), and the inlet hot temperature are monitored and are under the students' control. All other inlet and outlet temperatures are monitored. Monitoring and control is performed with a computer/web server using LabVIEW 6.1 software (National Instruments, USA). Data is published to web-accessible LabVIEW graphical user interfaces (GUIs) or via a Data Socket Server (National Instruments) to a Java2 GUI. A database (Microsoft SQL) is used for registering, authentication, and scheduling (ASP.NET) and for collaboration management software (Java2), which provides chat capabilities and the ability to pass local control between team members who are collaborating on carrying out the experiment from computers in different locations. The experiment has been used successfully in various ways in courses. Student response has been generally favorable, and students appreciate the ability to take data from real equipment, especially in engineering science courses that otherwise contain no laboratory component. Student response improves with increasing sophistication of the GUI and is very high when everything works as intended. Conversely, student response can be very negative when problems in the software or in the control or operation of the hardware occur. These findings suggest that web-accessible, remotely controlled laboratory experiments can be valuable in a variety of settings, and that the value of the experience, as perceived by the students, depends strongly upon the quality, richness, and reliability of the website and GUI.*

## INTRODUCTION

Web-accessible laboratory experiments emerged in the mid 1990s and are gaining popularity because they provide students with remote access to measurements with real equipment while making efficient use of resources. The first web-accessible experiment [1], an industrial robot that could be operated with Internet browsers, was developed using custom software written in C, C++, and Java, and notable efforts continue using Java [2,3]. LabVIEW™ (National Instruments, Austin, TX), a flexible and easy-to-use graphical programming language, was initially used for web-accessible experiments [4] at about the same time. The rapid expansion in web-accessible experiments using LabVIEW has been reviewed through 2000 [5] in a journal issue [6] devoted entirely to the use of LabVIEW in engineering education, including web laboratories. According to one tabulation [7], as of early 2004 there are about 60 or more remotely controllable devices accessible on the Internet.

We have developed a remotely controlled heat exchanger experiment as part of the I-Lab project at MIT [8,9]. Students in a variety of chemical engineering courses have used it on the Internet, and the response of students to the online experiment has been assessed. Initially we used Java to develop a graphical user interface (GUI) because the then-current version of LabVIEW (6i) could not display it on the web unless the client had exclusive control of the server. During the course of our work, LabVIEW version 6.1, which did not suffer from this limitation, was released, thereby giving us the opportunity to develop a LabVIEW 6.1 GUI for comparison with the Java version. In this paper we first describe the heat exchanger equipment and the way in which dependent and independent variables are monitored and controlled. There follow descriptions of the overall software architecture, the Java and LabVIEW client GUIs, and the user management, authorization, and scheduling. We close with a brief initial assessment of student responses regarding their experiences in performing assignments on the web (an area largely neglected by the literature in this field), and their preferences for using either the LabVIEW or Java GUI.

## HEAT EXCHANGER SYSTEM OVERVIEW

We worked with Armfield Ltd. (Hampshire, England) to develop a remotely controlled experiment. Armfield markets a range of small-scale heat exchangers that represent the common types found in industry for indirect transfer of heat from one fluid stream to another. We acquired three heat exchangers, double-pipe or tubular (HT31), flat plate (HT32) and shell-and-tube (HT33), and one service unit (HT30XC), which was modified by Armfield for remote control by a computer.

The service unit (HT30XC) is a bench-top apparatus onto which one of the heat exchangers may be individually mounted. The Service Unit provides necessary services and measurement facilities for investigating and comparing the different heat exchanger working principles and operating characteristics. These services include providing streams of hot water (heating fluid) and cold water (process fluid) at variable flowrates to the heat exchanger under evaluation. The service unit can also control flow direction by operating in either cocurrent or countercurrent mode. The ability to change the type of exchanger quickly, without the use of tools, and the rapid response of the system under optimal control to changes in water flow rate or temperature allow the steady-state experiments to be carried out in a relatively short period of time.

The HT30XC is connected to the computer through a universal serial bus (USB) port and comes equipped with a software driver for the Windows operating system. It is designed to operate from a Windows computer, and all the parameters are computer-controlled with no manual intervention required other than setting up the equipment and switching it on. The software bundled with the HT30XC was not designed to be web-accessible, so we designed our own software architecture to put the experiment on the web.

Figure 1 is a schematic diagram of the heat exchanger flow circuit on the service unit and its interaction with computer software for monitoring and control. A pressure regulator controls the single pass cold water stream, and the recirculated hot water stream is controlled by a bi-directional gear pump that allows the exchanger to be configured for cocurrent or countercurrent operation. Four k-type thermocouples monitor the inlet and outlet temperatures of the hot and cold water streams. There are also two flow meters that measure the flow rates of the hot and cold water streams. The thermocouples, flow meters, pressure regulator, and pump all interface with the computer via the software driver. Additionally, feedback mechanisms use input from the thermocouples and flow meters to control the flowrates and the inlet hot water temperature by heating the hot water reservoir.

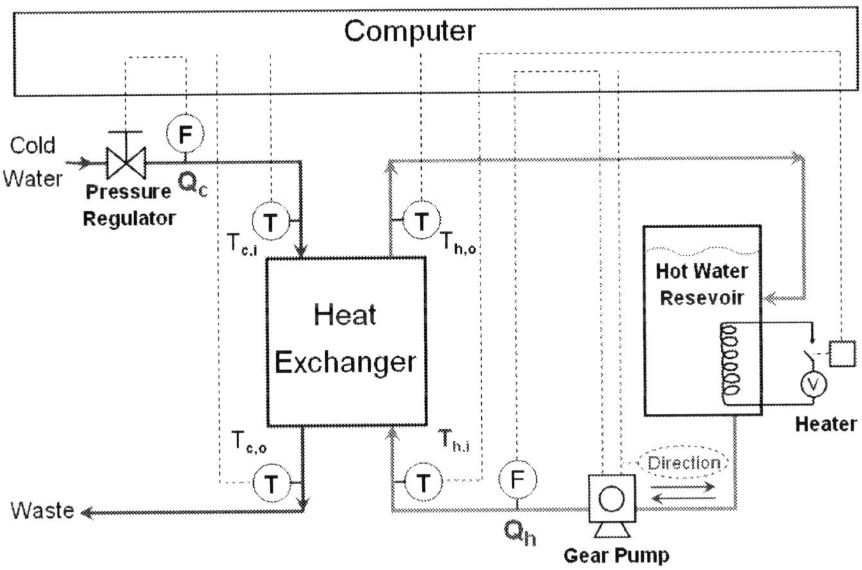


FIGURE 1
I-LAB HEAT EXCHANGER SCHEMATIC DIAGRAM

An important aspect of the I-Lab Heat Exchanger is its real-time nature. A change in one of the flowrate values or the hot water setpoint leads to a transient response on the data output, and the desired steady state value should be reached as soon as possible. To achieve this goal, control algorithms were implemented for the hot and cold water flow rates and the hot water temperature. The hot water temperature was controlled by placing algorithms that determined the on/off operation of the heater, and the cold and hot water flow rates were controlled through algorithms that determined pump and valve settings.

All the control algorithms were based upon proportional-integral-derivative (PID) control. Once the algorithms were in place, the Ziegler-Nichols closed-loop tuning method was used to find the optimal PID parameters. The parameters so obtained for the cold and hot water flow rates were applicable over the entire useful range (cold 0.5 to 5.0 L/min, hot 0.5 to 5.0 L/min); however, optimal parameters for the hot water temperature control varied as a function of the hot water flow rate. The optimal parameters are very sensitive to the physical characteristics of the heat exchanger including tube lengths and diameter. Any change to the arrangement of the exchanger and tubing will render the PID parameters sub-optimal. The optimal PID parameters were calculated for each of the heat exchangers (shell and tube, flat plate, and double pipe); in addition, a different set of parameters was obtained for each mode of operation (countercurrent and cocurrent).

## SOFTWARE ARCHITECTURE

Figure 2 depicts the software architecture used for the I-Lab Heat Exchanger Project. At the core of the architecture is the LabVIEW 6.1 software that communicates with the heat exchanger and makes data accessible to users of the system. The architecture also performs user registration, authentication, and scheduling and is flexible enough to allow students to work in teams on an experiment.

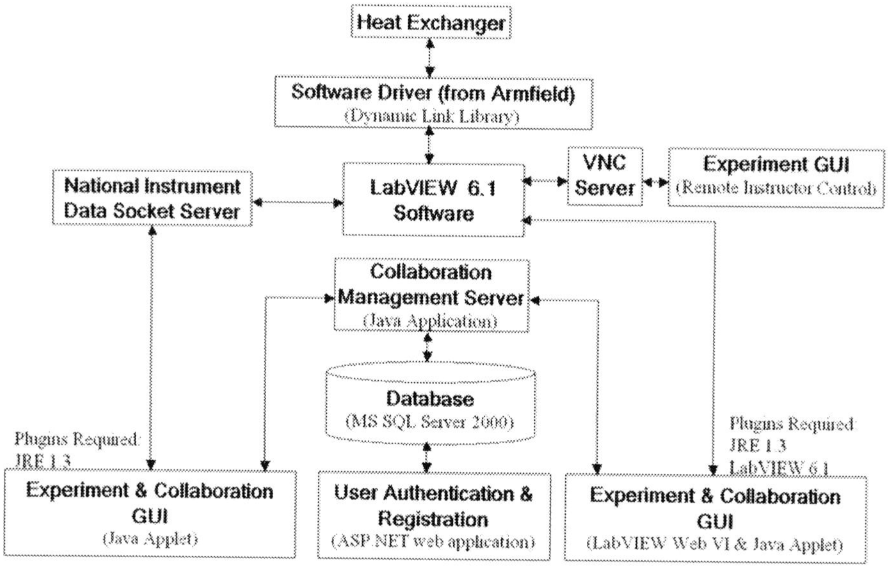


FIGURE 2
I-LAB HEAT EXCHANGER PROJECT SOFTWARE OVERVIEW

The HT30XC ships with software that allow the USB interface to be controlled from a computer. Included with this software is a Dynamic Link Library file (DLL), which provides a means to access the input and output parameters of the HT30XC Heat Exchanger Service Unit from any Windows program. Performing the appropriate function call to the DLL returns the I/O data.

To perform the function calls to the DLL file, we originally used LabVIEW 6i and later switched to LabVIEW 6.1 when it became available. In LabVIEW, block diagrams or VIs (Virtual Instruments) are used to develop a Graphical User Interface (GUI) to monitor and control parameters. This programming language contains a very large library of Graphical Instrument control tools, such as knobs, dials, and charts, which make creating user GUIs very easy. It also has built in functions that allow for easy use of the DLL file for the HT30XC service unit. In LabVIEW, this is accomplished via the Call Library Function, which can provide access to a particular function in a specified DLL. The Call Library Function block diagram is shown in Figure 3. Through the use of this feature and the DLL file, the LabVIEW software is able to read from and write to both the analog and digital channels of the HT30XC service unit. Analog channels are used for transmitting temperature and flowrate data whereas digital channels are used to turn on and enable the service unit.

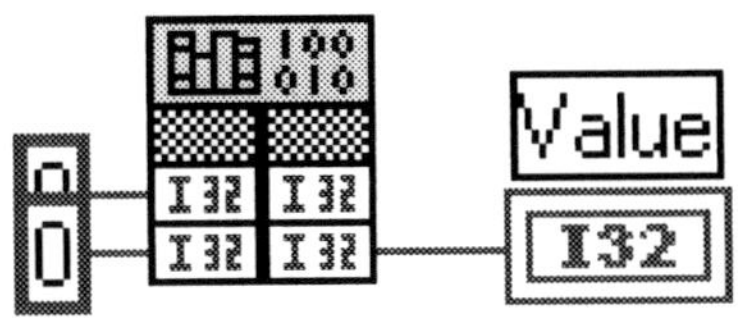


FIGURE 3
CALL LIBRARY FUNCTION BLOCK DIAGRAM

The I-Lab Heat Exchanger Project was designed to have two types of web-accessible interfaces, one type for instructors and another for students. To facilitate remote instructor control, a VNC server was used. VNC is a product from AT&T Laboratories, Cambridge, England, that allows the instructor to remotely view the desktop of the server computer, essentially allowing the instructor to have the same control as if he/she were seated at the server console. For remote student control, two types of interfaces were developed. With LabVIEW 6i, one interface was developed using the Java language. The second interface was developed once LabVIEW 6.1 became available, which permitted remote control of the GUI on the web. The Java interface included collaboration features, whereas the LabVIEW 6.1 interface was viewed from a webpage that contained a Java Chat window in addition to the interface.

## JAVA CLIENT INTERFACE

In implementing the Java interface, a DataSocket server was used to serve as an intermediary between the Java interface and the underlying LabVIEW software. The DataSocket server is a software product developed by National Instruments that enables real-time sharing of data among a variety of software clients developed in different programming languages.

The DataSocket server uses a publish-subscribe model to share the data. The server publishes real-time data to the Internet at a specific address via a DSTP protocol (Data Sockets Transfer Protocol). Each data point binds to a specific URI (with a prefix dstp://). In our case, the LabVIEW GUI sends data to the URIs, and the Java client interfaces subscribe to the server using these specific data addresses to listen for updates. Conversely, the Java interface can also publish parameter control values to the

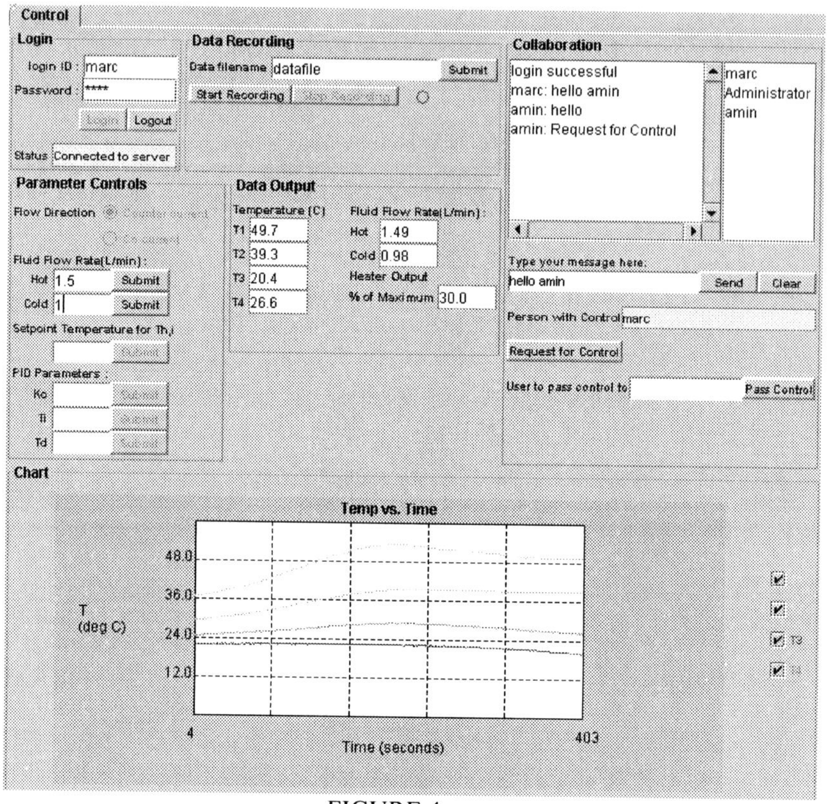


FIGURE 4

JAVA CLIENT INTERFACE

DataSocket server while the LabVIEW software listens for updates. Both of these operations were necessary to provide proper control via the Java Interface.

The Java interface is divided into many sections as depicted in Figure 4. The login panel handles user authentication, and relevant data can be viewed on the chart or in the data output panel. All the relevant experiment parameters are contained in the parameter controls panel, and the collaboration window panel contains all elements of the system's collaboration capabilities.

In the login panel, a user must enter his/her Login ID and Password and press the "Login" button. Every user is assigned a Login ID and Password when they first register to use the system. Requiring this information provides an additional level of authentication to ensure that the user is authorized to perform the experiment. When the "Login" button is pressed, the application creates a socket connection with the collaboration server and passes to it a Login message. Once a user is logged in, he/she will be able to view temperatures and flow rates, and their Login ID will appear in the collaboration panel.

Students can change the experiment parameters by submitting values via the parameter controls panel. In the parameter controls panel, the student may have control over the:

- Flow direction (countercurrent or cocurrent)
- Hot fluid flow rate
- Cold fluid flow rate
- Temperature set point
- PID (Proportional-Integral-Derivative) parameters

Students must click the Submit button to update the parameter values. Depending on the assignment, students may not have control over all the parameters. If a student does not have control of the parameter, the corresponding button will appear grey.

Shown in the data output panel are the actual temperature and flow values. The temperature values are received from thermocouples mounted at different locations in the heat exchanger. For the flat plate HT32 and the shell-and-tube HT33 heat exchanger, the thermocouple locations are:

- Hot fluid inlet (T1)
- Hot fluid outlet (T2)
- Cold fluid inlet (T3)
- Cold fluid outlet (T4)

For the double pipe HT31 heat exchanger, two additional thermocouples are located at the midpoint between two sections (Hot fluid midpoint, T5; Cold fluid midpoint, T6).

The chart panel contains a scrolling graph that can display these four temperatures. Users can select which combination of temperatures they wish to view by selecting the appropriate checkboxes on the side.

The data recording panel allows students to save their data in Excel format in a folder on the server. There is a text box to input the filename and a button that submits the filename to the server. The student can press the start recording button only after the submit button has been pressed. A green light will appear when recording starts after which the student can press the Stop Recording button to stop the data recording.

The collaboration panel is divided into two distinct components. Starting at the top is the chat component. On the left is a list of all users currently logged in and on the right is the chat window, where users can view messages. Below these two boxes is a field where users can enter messages and send them to other users.

Below this field is the "Pass the Token" component that controls which user can control the system parameters shown on the left hand side of the interface. If a user does not have control over a parameter, the Submit button next to it will appear grey. The name of the user that currently has control over the parameters appears next to the label "Person with control." The first user to log into the system is automatically assigned control of the parameters, and his/her Login ID will appear in this box. If a user wants to gain control of the parameters, he/she must press the button labeled "Request for Control", which broadcasts a message to all users who have logged in via the chat window. The user who has control of the parameters can transfer control to the user who made the request by typing the user's username into the "Pass Control" field and pressing the appropriate button.

On the server side, there is a collaboration server running which enables these features. Its basic job is to receive a message from one client interface, interpret it, and broadcast the appropriate message to the other clients connected to the server. For example, if a student enters a comment in their chat window, a message is sent to the server that then broadcasts the comment to all the other clients so that it appears in the other client interfaces. Additionally, the server can broadcast messages about who should have control of the equipment, which would enable/disable control over the parameters.

## LabVIEW Client Interface

LabVIEW 6.1 was the first version of LabVIEW to offer web-based control of a LabVIEW VI. As a result, we decided to develop a client GUI using LabVIEW 6.1 for two main reasons:

1. To explore the possibility of reducing system development time, as compared to Java programming, if additional pieces of equipment were to be made web-accessible.
2. To provide students with a more user-friendly interface which more closely resembles a control panel that they might find on a laboratory instrument.

It was our hypothesis that students would prefer to use the LabVIEW interface since the controls included knobs, buttons, and real-time, scaleable scrolling graphs. Figure 5 depicts the LabVIEW Client Interface that was developed. This GUI is embedded into a webpage that also contains a Java chat window.

The GUI is divided into two pages. On the page shown in Figure 5, there are controls and indicators divided into three distinct sections: Parameter Controls, Temperature Data, and Flow Rate Data. On the second page, there is a real time data table that contains temperature and flow rate data from the heat exchanger.

In the upper right corner of the interface is the section labeled Parameter Controls that contains all the input parameters for the heat exchanger. There are two knobs that control the hot and cold flow rates. Alternatively, a user can enter a number for the flow

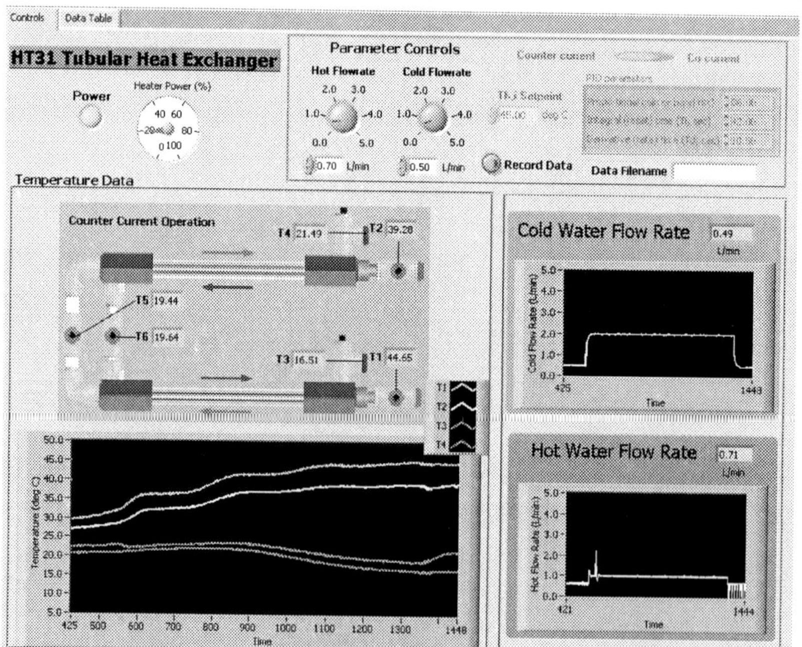


FIGURE 5

LABVIEW CLIENT INTERFACE

rate in the boxes below the knobs. The flow rates can vary between 0.5 and 5 liters per minute. A user can switch between countercurrent and cocurrent operation by dragging the selector switch to the appropriate side. There also is an input panel for the PID parameters and an input box for the hot water setpoint temperature. If a user wishes to record data, he/she can type a filename in the appropriate box and press the adjacent button to start data recording. The button will turn green once data recording is initiated and the data is stored on the server.

One of the weaknesses of the Java Client Interface was the way it represented the temperature data from the heat exchanger. In the Temperature Data section of the LabVIEW interface, both numerical and graphical data are presented so that students can see how the temperatures change over time. There is a diagram of the HT31 double-pipe heat exchanger, and arrows represent the direction of fluid flow when configured for countercurrent operation. Overlaid on this image are the six temperature readings from thermocouples placed at six points on the exchanger. Looking at the image, students can see the direction of fluid flow, the location where temperature readings are taken, and can read the numerical temperatures at those points. The graph shows the inlet and outlet temperatures for the hot and cold fluid streams. The temperature graph scrolls constantly while the system is running and shows the last five minutes of temperature data. Students can scale the temperature axis (ordinate) by clicking on the maximum or minimum values and entering a new value. The temperature graph is helpful in determining when the heat exchanger has reached steady state so that students can begin taking measurements.

The Flow Rate Data section contains two scrolling graphs, one for the hot flowrate and one for the cold flowrate. Each graph panel also contains a numerical indicator of the flowrate. The flowrate axis (ordinate) of both graphs is scaleable by clicking on the maximum or minimum values and entering a new value. The flow rate graphs show data for the last three minutes. These graphs are helpful in detecting when the flowrate has reached steady state. In addition, these graphs also detect irregularities in the control algorithm, which can occur at very low values of the hot water flow rate. This is associated with the operating range limit of the pump, which causes the flow rate to jump between values such as 0.2 and 0.6 L/min when the flow rate is set at 0.4 L/min.

The second page of the GUI contains is a data table that contains temperature and flowrate values recorded since the system was turned on. This feature was designed to give students a better estimate of the data that they would be recording to an Excel file. It also enables them to better understand the precision of the measurements they will be making.

## User Management, Authentication, and Scheduling

On the I-Lab Heat Exchanger Project website (http://heatex.mit.edu), we make use of the Microsoft .NET architecture for web services. The features of .NET make it very simple to collect, manipulate, and display data. The individual pages are ASP.NET web forms, and C# was used as the programming language. In order to provide database functionality for the website, Microsoft SQL Server 2000 was used as the database server. Because of constraints on the equipment available, this database server was forced to reside on the same machine as the web server. Access to the SQL server database was performed using ADO.NET.

Users must register for the system before they are allowed to perform an experiment. At the time of registration, the user is prompted to choose a login ID and password, and to enter their e-mail address and phone number. Users can also select the team number they belong to at this point. This information is stored in the SQL database. Before a user can perform the experiment, he/she must enter his/her login ID and password, which will be checked against existing values in the database.

Students can also sign up for times to perform an experiment on the web. There is a schedule page that shows all the timeslots and who has the slot reserved, if anyone. When a student selects an available timeslot, their name and team will appear on the schedule page.

## STUDENT ASSESSMENT

The heat exchanger has been successfully used in a variety of classes and for other purposes at MIT. It has been used in (1) Transport Processes classes to demonstrate principles of heat transfer, (2) Process Dynamics and Control classes to demonstrate how to tune PID parameters with equipment in real-time, quantify a response to disturbances, and illustrate the general notion of how a real piece of equipment may be abstracted in diagrams and equations, and (3) Chemical Engineering Project Laboratory classes to provide experience in data collection, analysis and presentation, and technical report writing. In all cases, students carried out the web experiment without supervision.

The students filled out assessments after each class to evaluate usability, fulfillment of educational objectives, suggested improvements, how beneficial the learning experience was, and how much fun the experiment was to perform. The surveys show favorable responses in general and demonstrate that the I-Lab experience has improved considerably over time as the interface has become more sophisticated, as more tools and options have been added, and as software and hardware problems have been fixed. An example of this improvement can be seen in Figure 6. The assessments show that students have very low tolerance for imperfections in the computer interface or for software and hardware problems that may arise during the course of an experiment. Without having the heat exchanger equipment directly in front of them, the students find it harder to fathom that equipment is prone to failure and malfunction.

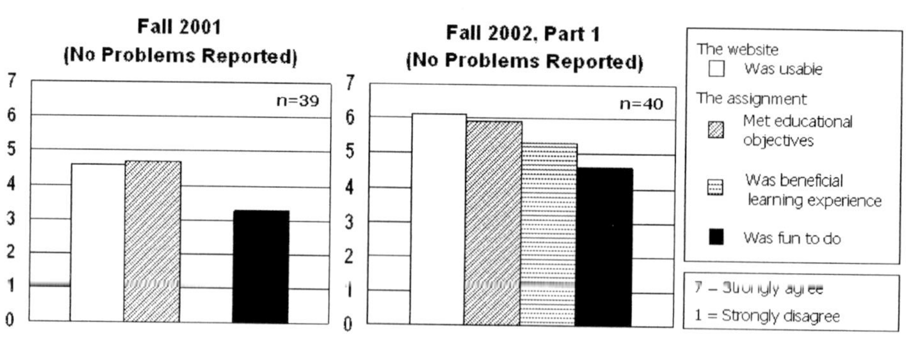


FIGURE 6
STUDENT ASSESSMENT OF I-LAB HEAT EXCHANGER: COMPARISON OF 10.302 (TRANSFER PROCESSES) IN FALL 2001 AND FALL 2002 USING THE JAVA 2 INTERFACE

There is compelling evidence that students' experience with the experiment is a strong function of the context in which the experiment is performed. Students who performed the experiment in the Chemical Project Laboratory class rated their experiences as less beneficial and less fun than those who performed it in other classes. This can be clearly seen in Figure 7. One explanation is that the purpose of the experiment in the Chemical Project Laboratory class is mainly to allow students to practice taking data, analyzing it, and then writing a technical report on their findings rather than explore or test a chemical engineering concept. Therefore, the purpose and context in which the experiment is performed is vitally important to the success of the experiment. Some students who did not read the instructions had difficulties; others could not obtain stable values of $T_{h,i}$ and $Q_h$ because of unexpected hardware problems and/or inadequacies in the control scheme employed. Students who had difficulties gave lower scores even though the problems they experienced were solved.

| | **Reported Problems** | |
|---|---|---|
| | Yes | No |
| Usability on the web: | 3.9 | 5.4 |
| Comparison between GUIs: | | |
| I liked Java 2 (Part 1): | 6.0 | 5.4 |
| I liked LabVIEW (Part 2): | 5.1 | 5.3 |
| If I had to choose one, I prefer: | | |
| Java 2: | 5 | 10 |
| LabVIEW: | 2 | 17 |
| The chat facility for communication was useful: | 5.5 | 5.5 |
| | n = 7 | n = 27 |

TABLE I

STUDENT ASSESSMENT OF I-LAB HEAT EXCHANGER: COMPARISON OF ASSESSMENTS OF DIFFERENT GUIS

Lastly, the students carried out a two-part assignment, one with the Java GUI and the other with the LabVIEW GUI. Students liked both GUIs (see Table I), but, when asked to pick one, they preferred the LabVIEW interface to the Java interface (when comparison is made between students who had no problems with either interface). Students who had training in Java but had no previous experience with LabVIEW, created the GUIs [8,9]. After learning it while creating the LabVIEW GUI, both agreed that they would choose LabVIEW if they were to create another GUI because it reduces development time.

These findings suggest that web-accessible, remotely controlled laboratory experiments can be valuable in a variety of settings and that the value of the experience, as perceived by the students, depends strongly upon the quality, richness, and reliability of the website and GUI. In the case of complex experiments, students seem to have a lower tolerance for equipment problems than they might have with a hands-on experiment.

## ACKNOWLEDGEMENTS

This work was supported in part through the I-Campus Initiative at MIT by a grant from the Microsoft Corporation. Conversations with James Trevelyan (Univ. of Western Australia) and Jim Henry (Univ. of Tennessee) were helpful in development of the website.

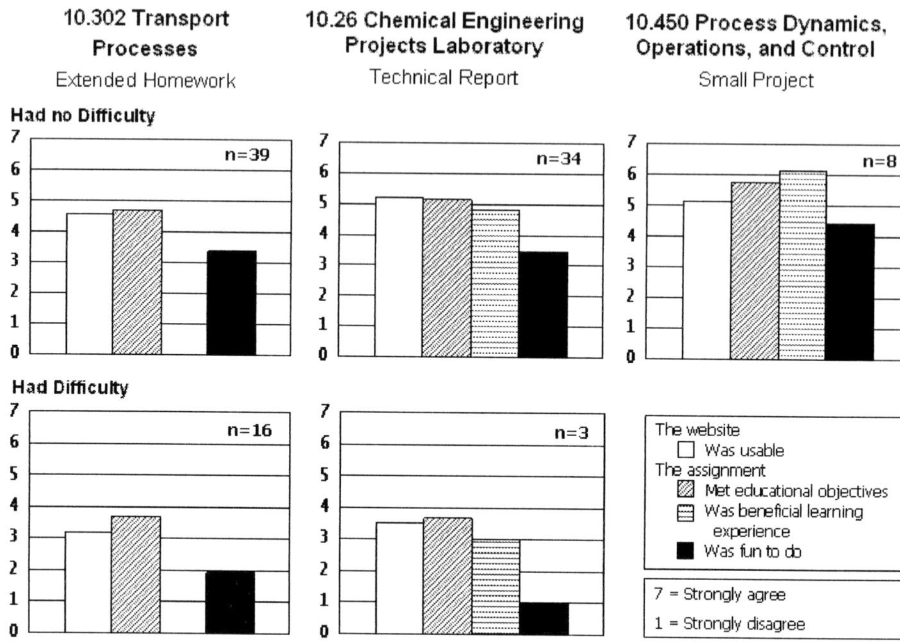


FIGURE 7

STUDENT ASSESSMENT OF I-LAB HEAT EXCHANGER: COMPARISON OF EVALUATIONS IN DIFFERENT CLASSES

## REFERENCES


1. Taylor, K., and J.P. Trevelyan "Australia's Telerobot on the Web," 26th Symposium on Industrial Robotics, Singapore, Oct. 1995, pp. 39-44.

2. del Alamo, J.A., Brooks, L., Mclean, C., Hardison, J., Mishuris, G., Chang, V., and Hui, L., "The MIT Microelectronics WebLab: a Web-Enabled Remote Laboratory for Microelectronic Device Characterization," Proceedings of Networked Learning 2002, Berlin, Germany, May 1-4, 2002, Paper # 10029-03-JD-047.

3. del Alamo, J. A., Hardison, J., Mishuris, G., Brooks, L., McLean, C., Chang, V., and Hui, L., "Educational Experiments with an Online Microelectronics Characterization Laboratory," Proceedings of 2002 International Conference on Engineering Education, ICEE-2002, Manchester, UK, Aug. 19-21, 2002, Paper 0102.

4. Henry, J., "Running Laboratory Experiments via the World Wide Web," ASEE Annual Conference, Session 3513, 1996. Available from National Instruments Academic Resources CD_ROM 2001 Edition.

5. Ertugrul, N., "Towards Virtual Laboratories: a Survey of LabVIEW-based Teaching/Learning Tools and Future Trends," Int. J. Eng. Ed., 2000, Vol. 16, No. 3, pp. 171-180.

6. Ertugrul, N. and Henry, J. (guest editors), "LabVIEW Applications in Engineering," Special Issue, Int. J. Eng. Educ., Vol. 16, No. 3, 2000.

7. http://www.mech.uwa.edu.au/jpt/tele/remote-labs.html

8. Khan, R.A., "Software Architecture for Web-Accessible Heat Exchanger Experiment," SM Thesis, MIT, Cambridge, MA, 2002.

9. Knight, M., "Connecting and Teaching Students via Web Services for an Online Laboratory." M.Eng. Thesis, MIT, Cambridge, MA, 2003.

**Clark K. Colton** received the B.Ch.E. degree from Cornell University in 1964 and the Ph.D. degree from Massachusetts Institute of Technology in 1969. He joined the faculty of MIT thereafter, becoming full professor in 1976. He was Deputy Head of Chemical Engineering in 1977-1978 and Bayer Professor of Chemical Engineering in 1980-1986. His research interests are in bioengineering.

**Marc Knight** received his S.B. degrees in Electrical Engineering and Computer Sciences and in Management Sciences in 2002, and the M.Eng. degree in 2003, all from Massachusetts Institute of Technology. He is currently working at Morgan Stanley in New York City as an Analyst in the Fixed Income Division.

**Rubaiyat A. Khan** received his B.S. degree in Civil and Environmental Engineering from Bangladesh University of Engineering and Technology, Dhaka, Bangladesh in 2000 and his S.M. degree in Civil and Environmental Engineering from Massachusetts Institute of Technology in 2002. He is currently a consultant at Barra, Inc. in Berkeley, CA in the area of financial risk management solutions for investment managers.

**Sarah Ibrahim** received her S.B. degree from Massachusetts Institute of Technology and is currently enrolled for an S.M. degree from the Biological Engineering Division of MIT. She participated in this project under the Undergraduate Research Opportunities Program.

**Richard West** is enrolled in his fourth year in Chemical Engineering at Cambridge University, Cambridge, UK. He participated in this project under the Undergraduate Research Opportunities Program as an exchange student at MIT in 2002-2003.

# Chapter 11

# A Virtual Company Multimedia-Based Learning Environment for Teaching Students in Manufacturing

J. D. TEDFORD, M. G. MCCARTHY and R. H. A. SEIDEL
*Department of Mechanical Engineering, The University of Auckland, Auckland, New Zealand. E-mail: d.tedford@auckland.ac.nz*

*The development of a novel approach to teaching Manufacturing Systems to third year engineering students at the University of Auckland is presented. This is achieved through the use of a multimedia-based learning system that delivers immersive projects via a virtual company environment called "INFOstation Limited." The INFOstation concept is described and the multimedia-based modules representing typical departments within the medium-sized engineering company are explained. The virtual company concept attempts to provide students with an appreciation for the importance of time in an engineer's decision-making process, and to promote a comprehensive understanding of the process of design, planning and manufacturing as integrated entities. A detailed description is given of a particular real-life ergonomics problem in which students are required, by the company's "Engineering Manager," to analyse a materials handling operation for ergonomic impact, to complete a professional quality report on the situation and to recommend any required workspace changes. The results of student feedback on the INFOstation exercise are presented and planned future developments are described. Part of the ongoing work is the development of an assessment instrument to gauge the reliability and effectiveness of the overall "virtual company" concept and of individual modules within it. The intention is to discover if some topics and techniques are more suitable for this learning approach than others and to see if any significant variations exist for students who learned English as a second language.*

## INTRODUCTION

The University of Auckland's Mechanical Engineering Department in 1998 introduced a Manufacturing Systems course into its four-year undergraduate degree programme. This compulsory course for all penultimate year students has proved to be demanding on the available resources in the Department due to increasing enrolments. As a result, contact hours with students have become restricted, and academic staff involved in the teaching of this course find it difficult to adequately expose students to the necessary core manufacturing systems knowledge required by engineers while at the same time ensuring that the students obtain a clear and comprehensive view of the topic.

Of concern to the educators involved with these students, who are studying manufacturing systems with little or no workplace experience, was the students' understanding of the importance of the dual issues of time and integration [1]. In practice, manufacturing engineers deal daily with production targets, maintenance schedules, resource limitations and financial constraints, usually under severe time pressures. Furthermore, the inevitable interaction between the departments in a manufacturing company need to be considered in any decision-making process. It is therefore essential that the importance of the effective use of "Time" is impressed on students, as well as introducing them to the concept of "Systems Integration."

Although professionals in most engineering disciplines are subject to time constraints, none, we believe, feel the pressure more strongly than manufacturing engineers. In practice, manufacturing engineers are constantly under pressure to meet production targets and delivery schedules, and to minimise disruptions to normal production activities arising from maintenance and commissioning activities.

Today's manufacturing environment requires that a wide range of physical resources from stand-alone machine tools to continuous flow machines, be properly integrated. Less tangible systems such as planning, quality control and human resources also must be seamlessly integrated into the organisation's operations. In successful manufacturing industries, the processes of product design, systems/project control, and the management of manufacturing operations are interactive, dynamic and interrelated. Thus, systems integration when effectively carried out is a key factor in the operation of a successful and profitable manufacturing organisation.

Most topics that mechanical engineering students study in a typical undergraduate programme, such as mechanics and thermodynamics, concentrate on scientific theories that are relatively easy to comprehend by the use of numerical examples. The issues of time and systems integration in a manufacturing systems course, however, are difficult to demonstrate or explore in a conventional lecture or laboratory session.

Although undergraduate students are given a solid background in the use of resources such as materials and space, they have much less guidance in understanding the crucial importance of their use of time and the consequences their decisions. There is a need, therefore, to give students the capability to grasp these concepts through the use of a teaching methodology that gives them a global view of how each individual decision combines with others to form an integrated solution to a problem.

The "INFOstation Company" concept [2] described here attempts to provide students with an appreciation of the importance of time in a manufacturing engineer's decision-making process, and with the opportunity to gain a comprehensive understanding of how the design, planning and processing phases of a manufacturing environment form a complete and integrated system.

## TEACHING METHODOLOGY

Several years ago, the Manufacturing Systems Group within the Mechanical Engineering Department adopted a project-based learning approach for its final year elective course on Technology Management. This was done in order to allow students to gain first-hand experience and a feel for actual manufacturing systems and processes whilst at university. It was also an attempt to improve the students' understanding of underlying theory and to demonstrate to them that collaboration and teamwork were important attributes for their full development as engineers [3,4,5].

Students, in groups of three or four, were given actual industrial problems to investigate and solve. This was done with the co-operation of local manufacturing companies that participated in the programme. The results of this initiative were encouraging, with positive student feedback as well as significant gains in their understanding of the processes of problem-solving in dynamic, real-life situations.

Whilst this project-based approach was very successful, it did not entirely solve the problems associated with providing students with the best possible learning experience. Although students had the opportunity to discuss their assigned project with managers and other personnel within the host company and, in the process, learned about the organisation's structure, communication networks etc., they did not have time to explore the full range of activities in the organisation and to fully appreciate the complex interaction of job functions and processes within the company.

It was also generally not possible to impress on students the importance of time as a manufacturing constraint during these projects. This was because the ideal student project is one that deals with a problem that the company needs to address, but one that is not critical for the company's short-term success. The projects generally tended to be those that had been postponed because of a lack of time, staffing or finance within the organisation. Because of this, the projects rarely had a critical time constraints associated with them other than those imposed by the usual university course schedule. Thus, the need to make appropriate use of time by students was not as obvious as is normally found when dealing with real-life production emergencies.

Of more concern, however, was the fact that this industry-based project approach was obviously unworkable if applied to an introductory course in Manufacturing Systems, which typically had an enrolment of around one hundred students each year.

The primary requirement for the introductory course is to deliver essential knowledge about manufacturing systems topics that are needed to provide an adequate foundation for later study of the subject. It is necessary, therefore, to deliver the material in a manner that emphasises to students the complexities and interactions between departments within a typical manufacturing organisation.

In recent years, researchers in the field of engineering education have investigated the application of multimedia technologies to the education of manufacturing engineers. Many of these developments have included the use of "manufacturing games" [6]. These games can range in complexity from following a simple set of decision rules incorporated into a spreadsheet to real-time management games where the students must maintain ongoing contact with the simulation in order to deal with the evolving progress of the scenario. These simulations often change the ground rules or parameters throughout the exercise to constantly challenge the student's decision-making ability. Another common approach is to simulate individual shop floor machines by computer "stand-ins" [7,8] often using proprietary simulation software such as Quest® by Deneb or Arena® by Rockwell Software Ltd.

In order to meet the requirements described above, and with the constraints of large student numbers and limited available contact time, the INFOstation concept was developed [9]. The concept was designed not to mimic the powerful layout/process simulation and 3D graphics abilities of proprietary software such as Arena® or Quest® or to have the student work through a single-issue simulation. Instead, INFOStation was designed to provide a basic virtual company scenario that would concentrate on emphasising the interconnectivity of many tasks and processes within a manufacturing organisation. For example, the INFOstation simulation includes Administration and Financial Departments and their associated systems as well as the more usual Design and Manufacturing Departments.

When fully developed, it is intended that INFOstation will consist of a number of multimedia-based "modules," each of which will cover a specific aspect of manufacturing. Each module will consist of a multimedia-based virtual project that will provide students with the necessary parameters to analyse and solve open-ended problems, and will provide comprehensive background resource information. To emphasise the criticality of time to the students, they are required to respond periodically throughout the project with interim reports indicating their progress to date. This forces the students to plan and structure their analysis and solution to the problem rather than leaving everything to the last minute, as tends to happen with traditional assignments. The nature of the projects also generally requires students to work collectively to arrive at practical solutions. Thus, the need for teamwork, leadership, communication skills and an understanding of ethical constraints form a part of their decision-making process. [10,11] As a typical example of the type of immersive project presented to students within INFOstation, an "Ergonomics Project" is described later.

## THE VIRTUAL COMPANY

In the third year of their engineering degree, students are first introduced to the concept of the virtual company in an assignment designed to reinforce and apply their knowledge of ergonomic principles to workplace design. The name of the virtual company is INFOstation Limited.

INFOstation Ltd. is the virtual manufacturing organisation that will be the basis of ongoing development. It is a medium-sized manufacturing organisation, with a virtual workforce of two hundred people, half of whom are engaged in production, the tool room or maintenance. Eighty staff members are administrative, including those dealing with accounts, sales and marketing. There are twenty staff in the engineering function which includes designers, manufacturing engineers and QA specialists. In future development, it is intended that a live connection be made, via the Internet, with a real local manufacturer, as suggested by Dessouky and Verma [12].

INFOstation Ltd. has a web site on the School of Engineering's Intranet. Its home page contains background information about the company as well as links that connect the student "employee" to the web pages of other company departments which they may need to "visit" and/or obtain information from. Some of these are shown in Figure 1.

The company currently encompasses four departments - the Design Office, the Planning Office, the Quality Assurance Department and the Administration Office.

Further developments will see the addition of several more functional areas to complete the enterprise.

## ERGONOMICS PROJECT

One exercise that students are required to complete is an ergonomics assignment. In it, the students are told, through the company's online Staff Manual, that the Planning Office is responsible for the efficient planning, maintenance and review of the handling, assembly and machining tasks within the organisation. They are also told that this responsibility includes ensuring that staff are not required to carry out tasks which may be dangerous, stressful, or detrimental to their health. In order to meet this responsibility, the Planning Office engineers must be familiar with the theory and practice of ergonomics as well as the relevant health and safety requirements and regulations.

After joining the Planning Department as a virtual employee, the student receives a memorandum from their employer, the Manufacturing Manager, asking them to complete an ergonomic investigation into a handling operation that is causing some concern to both the employees and their trade union representative.

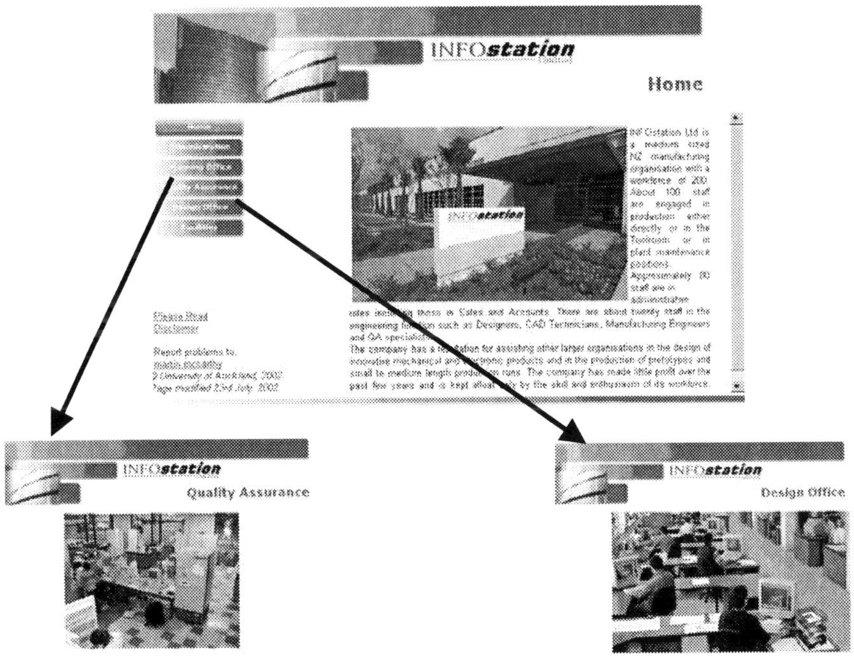


FIGURE 1
THE VIRTUAL COMPANY'S HOME AND DEPARTMENT PAGES ON THE INTRANET

The project scenario is based on a real process from industry in which an operator performs a sequence of repetitive tasks. The job consists of the operator removing packs of empty drink cans from a pallet, placing them on a conveyor and unbundling them in preparation for filling. The total job cycle included removing the packs from a six-layer-high stack on a pallet, so reaching from eye height to almost floor level was involved. Since the packs were three deep across the pallet, the task also involved different severities of horizontal reach. The analysis of the task included unstacking a complete pallet load to allow the total severity of the job to be established. For the assignment, however, the students were asked to analyse only the most severe motions for the operator during the work cycle.

The objectives of the exercise were to:

- Make students aware of ergonomic issues in the workplace.
- Reinforce and extend the material covered in lectures.
- Give students hands-on experience in industrial problem-solving and productivity improvement.
- Give students practice in learning and working with professional computer-based analysis tools.
- Practice the professional skills of communication and report writing.

Students accessed a short video clip of the operation online (Figure 2) and some data collected by earlier investigators from the web sites of the appropriate departments within the organisation. This data included the length of a work shift, the body weight of the operator, the weight of the load handled and industry work practice standards.

FIGURE 2
A VIDEO FRAME FROM THE ERGONOMICS PROJECT

Students used a commercial software package called ErgoEASE©, which they accessed from the Planning Office website, to perform the ergonomic analysis. The program has a graphical interface that allows students to input the results of a detailed methods analysis together with general task parameters to produce a range of reports on the ergonomic impact of the operation. The program includes detailed graphical help screens to assist students who are unfamiliar with ergonomics terminology and acronyms.

## PROJECT FORMAT

In the assignment, students were sent a memorandum indicating that a detailed ergonomic analysis of the job was required. They were also told that they should periodically update the manufacturing manager on their progress and submit a final report on completion of their study.

It was indicated that their report should include, among other things:

a. An analysis of the employees' energy expenditure in performing the task.
b. Comments on the ergonomic impact of the job of unloading the pallet.

c. A complete Rapid Upper Limb Assessment (RULA) analysis on what they regarded as the most ergonomically sensitive parts of the work cycle.
d. Recommendations on a suitable illumination level for the workspace based on relevant industry standards.

It was suggested to students that, as a competent INFOstation employee, their solution should consider all of the usual relevant commercial issues and constraints, such as: the cost of implementing any recommended changes, anticipated effectiveness of their solution, pay-back period, downtime, likelihood of staff/union acceptance and compliance with health and safety regulations.

## STUDENT FEEDBACK

The results from an anonymous feedback survey completed by the participating students was collected and analyzed at the end of the project. The questionnaire covered responses to both the specific ergonomics project and the overall virtual company approach to learning. The questionnaire made ten specific statements, covering both aspects, and the students responded in one of six ways. The percentage of each response to the statements is shown in Table 1.

Key: SA - Strongly Agree A - Agree U - Undecided

D - Disagree SD - Strongly Disagree

| SA | A | U | D | SD |
|---|---|---|---|---|
| The INFOstation web pages were clear and uncluttered | | | | |
| 14% | 74% | 10% | 2% | 0% |
| The number of hyperlinks per page was about right | | | | |
| 8% | 60% | 24% | 8% | 0% |
| The hyperlinks on the web pages are clearly identifiable | | | | |
| 10% | 54% | 23% | 13% | 0% |
| Important data on the ErgoEASE© project was easy to find | | | | |
| 4% | 56% | 28% | 12% | 0% |
| Navigating my way around the INFOstation site was easy | | | | |
| 24% | 54% | 16% | 4% | 2% |
| The instructions in the manager's memo were easy to interpret | | | | |
| 13% | 55% | 21% | 9% | 2% |
| I could obtain all the information needed to complete the project | | | | |
| 10% | 52% | 24% | 6% | 8% |
| The use of a real industry scenario added interest to the project | | | | |
| 20% | 48% | 24% | 2% | 6% |
| The use of an industry scenario added relevancy to the project | | | | |
| 13% | 52% | 23% | 10% | 2% |
| I would recommend that this concept be extended to other topics | | | | |
| 14% | 64% | 20% | 2% | 0% |

TABLE 1

RESPONSES TO STUDENT FEEDBACK SURVEY

In addition to responding to the statements by circling one of the six options available to them, students were also encouraged to complete an open-ended section which solicited their comments on the assignment and its method of presentation.

Overall, the responses were very encouraging and indicated a general acceptance of both the virtual company concept and the realistic nature of the project assigned to them. The feedback also indicated that there were some significant pedagogical issues raised within the exercise itself. Some students had difficulty in understanding what was required of them. This appeared to be because they were not used to receiving instructions, or data, in a narrative form – in this case in the format of a typical company inter-departmental memorandum. Another problem concerned the use of the ErgoEASE software itself. Although not difficult to use, students had little time to familiarise themselves with it. As a consequence, some students attempted to use all of the program's analysis modules when this was not required for the purpose of the project. This led to some students producing analyses which were not required by the memorandum of instructions.

In general, however, students felt that the INFOstation format made the ergonomics assignment more interesting and life-like. The particular students who were personally interviewed following the exercise indicated that they were keen to have other topics delivered in a similar manner.

## Proposed Future Developments

In the near future, the virtual company concept is being extended to include the financial operations of the company. Students will participate in a project that will involve the analysis of INFOstation's cash flow and financial status, together with a requirement for them to make some financially and technically sound decisions about much-needed capital investment for the organisation. Once logged-on to the INFOstation environment, students will view a short video clip of the "Manufacturing Manager" outlining a specific manufacturing problem. Students will then be able to run a video of three presentations by equipment suppliers' representatives outlining the technical and financial benefits of their competing products. Students will be required to analyse the presentations, take notes of important parameters and use the data to produce a report containing a recommendation for purchase and an analysis of the expected return on investment. This exercise will be a part of a general Engineering Management course in which Manufacturing Systems students participate.

Final year students involved in a Design course will also use INFOstation to participate in a "design for assembly" exercise in which they will access a knowledge-based expert system, formulated on the Boothroyd/Dewhurst methodology [13], from INFOstation's Design Office. This will assist them in carrying out a re-engineering analysis on a specific product that has been identified by the production manager as requiring improvements aimed at increasing ease of assembly. In addition, the project concept will be extended further to include topics such as design for "x," quality assurance, resource utilisation and environmental issues.

To support the development of the INFOstation initiative, research will also be conducted on producing an assessment instrument to gauge the reliability and effectiveness of the overall concept and of the individual modules within it. There is a need to discover which topics and assessment techniques are most suitable for this "virtual company" approach. The development of the assessment instrument will be based on the concepts proposed by the CDIO Initiative [14].

## CONCLUSIONS

The staff involved in the development of the INFOstation concept and its application in the teaching of specific topics to engineering students believes that the ergonomics exercise was successful. By delivering the topic in a more immersive fashion, students felt that it was more realistic and more interesting than run-of-the-mill assignments. The completed reports were generally of a high quality with some students really entering into the spirit of the exercise and presenting their results and commentary in the format, style and language they would be expected to use in a workplace technical report.

This project, like all new course design initiatives, requires some development time by the staff involved. But the time investment is not unreasonable and is expected to be repaid when the exercise is repeated in following years.

From the feedback gained, it would appear that students learned a significant amount about extracting data and instructions from a typical industry memorandum. This area will also be investigated in more detail to determine if any correlation exists between students who have difficulty with this aspect of the assignment and those for whom English is a second language. There is a desire among academic staff to investigate this area of narrative scene setting to discover if there is a saturation point where information overload becomes a problem and students loose sight of the task at hand.

## REFERENCES


1. McCarthy, M.G., "A Manufacturing Simulation Project - The Virtual Factory, Exemplary Models in Education," APEC-HURDIT Network Project, Toronto, Canada, 1996.
2. Seidel, R., "Project-Based and Hypermedia-Supported learning approaches in Manufacturing Engineering," Proc. of 17th Australasian Association for Engineering Education Annual Conference, Brisbane, Australia, 2001, pp. 415-456.
3. CDIO Initiative web site, http://www.cdio.org/
4. Tedford, J.D., "Developing a New Learning Environment for Engineers," Proc. 23rd. International Conference on Improving University Learning and Teaching, Dublin, Ireland, Vol. 2,, July 1998, pp. 757-767.
5. Terenzini, P.T., et-al, "Collaborative Learning vs. Lecture/Discussion: Students' Reported Learning Gains," Journal of Engineering Education, Jan 2001,90,1, pp.123-130.
6. Jackson, P.l., Muckstadt., et-al, Course Material for Manufacturing System Design, http://www.orie.cornell.edu~jackson.asee.html, Cornell University, Ithaca, USA, 1994.
7. Tufekci, S. Quoted from: http://gurgun.ise.ufl.edu. University of Florida, Gainsville, USA, 1999.
8. Musbah, M.M. and Tharmaseelan, T., "Using Simulation as a Tool in Implementing Cellular Manufacturing in a Project-Based Organisation," Proceedings of the 14th National Conference on Manufacturing Research, Derby, UK, 1998, pp. 405-411.

9. Seidel, R. and Sitha, K., "A Novel Hypermedia Supported Teaching Approach," Proceedings of the 2nd UICEE Annual Conference on Engineering Education, Auckland, New Zealand, 1999.

10. Dessouky, M. and Verma, S., "A Methodology for Developing a Web-Based Factory Simulator for Manufacturing Education," HE Transactions Vol. 33, 2001, pp. 167–180.

11. Ilic, V., "Sustainability and Ethics in Engineering," Proceedings of 2003 International Conference on Engineering Education, ICEE-2003, Paper No. 2046, Valencia, Spain, 2003.

12. Moskalski, N.M., "Teaching Teamwork," Chap. 3, Aung, W. et al. (eds.), *Innovations 2003: World Innovations in Engineering Education and Research*, Int. Network for Eng. Ed. and Res. (iNEER), Arlington, VA, 2003, pp. 17-23.

13. Boothroyd, G. and Dewhurst, P., *Product Design for Assembly*, Wakefield, RI, Boothroyd Dewhurst Inc., 1989.

14. Berggren, K-F., Brodeur, D., et-al, "CDIO: An International Initiative for Reforming Engineering Education," World Transactions on Engineering and Technology Education, Vol. 2, No. 1, 2003.

**Des Tedford** received a B.Sc. and Ph.D. in Mechanical Engineering from the Queen's University of Belfast. He has had academic appointments at the Queen's University of Belfast, the University of Rhode Island, the University of Wollongong, and the University of Auckland. Throughout this time he has been active in the continuing professional development of engineers, which has led to a significant number of publications and in-house training programmes on engineering management related topics. He is Course Director for the Master of Engineering Management programme at the University of Auckland.

**Martin McCarthy** has a Masters Degree in Engineering Management from the University of Auckland and is currently completing a Ph.D. He is a is a Senior Tutor in the Department of Mechanical Engineering at the University of Auckland and is a Chartered Engineer by profession, with many years experience in mechanical and electronics product design, manufacturing systems and fire prevention. Martin's current interests include research into the effective teaching of engineering design and manufacturing with emphasis on the use of multimedia tools and simulation.

**Rainer Seidel** is a Senior Lecturer in the Department of Mechanical Engineering at the University of Auckland. He holds the equivalent of a Masters Degree in Engineering and Commerce from Karlsruhe University (Germany) where he specialised in Industrial Engineering, Business Management and Accounting, Information Technology, and Operations Research; and a Ph.D. in Engineering from the University of Auckland. His main research, teaching and consulting activities are in the areas of technology management, organisational learning and engineering education.

# Chapter 12

# Online-Based Teaching Can Facilitate Skills Development for Today's Global Manufacturing Environment

ERIK BOHEMIA
*University of Western Sydney, School of Engineering and Industrial Design, Locked Bag 1797, Penrith South Distribution Centre, NSW 1797, Australia. E-mail: e.bohemia@uws.edu.au*

*In 2001, the School of Engineering and Industrial Design undertook an initiative to conduct a pilot study of the online delivery of the 'Design Management: Organisational Skills for Designers' subject. This subject is one of the core subjects offered by the Industrial Design program to industrial design and engineering students. The subject has been developed to address industry needs identified in previous research. One of the key learning outcomes of the subject was for students to develop distance communication and virtual teamwork skills, which are becoming increasingly important in new product development as a result of the changing global manufacturing environment. The online component of the subject was structured to stimulate students' independent learning and to develop their critical thinking by arousing their curiosity in the topics by incorporating the seven processes of critical thinking and writing. Student evaluation data has been used to evaluate the effectiveness of the approaches introduced in this subject.*

## INTRODUCTION

The increased globalisation of manufacturing is forcing organisations to increase flexibility and at the same time reduce product development time. Therefore, organisations are incorporating strategies that enable them to tackle both of these aims. However, these strategies, such as cross-functional teams and overlapping product development stages, require organisational members, including industrial designers and engineers, to have skill sets that enable them to operate successfully in this new setting. The aim of this paper is to explore the usefulness of online teaching methods for developing skills that are relevant in contemporary workplaces, particularly those using global manufacturing practices.

## CHANGING MANUFACTURING ENVIRONMENT

Increased global competition is changing manufacturing environments and forcing organisations to introduce new products on domestic and international markets at an increasingly faster pace [1-3], resulting in *'Time Base Competition'* [4], which has become a critical measure of business performance [2,5,6].

It is proposed that this new emphasis on speed and flexibility requires a new and different approach for managing new product development [e.g. 4,7,8-13]. It is suggested that a holistic *concurrent* product development approach might be better aligned with today's goals of maximum speed and flexibility than traditional sequential product development.

*Concurrent engineering,* is a relatively new approach [14], and as opposed to a sequential process, uses overlapping design stages with fewer 'gates', and *cross-functional* teams to shorten the development time of a project [15-28]. The major advantage is that a number of critical issues concerned with product development can be looked at and dealt with simultaneously in the early stages of the design development [29], so that any immediate and future requirements can be taken into account [24, 30]. It is argued that *concurrent engineering* fosters collaboration, communication and mutual appreciation and understanding between various team members from different disciplines; it breaks down the boundaries between various disciplines by utilising *cross-functional* teams, consisting of key personnel from various functions [22,24,25,31-33].

Tomiyama [23] and Shemas [24] state that good communication is essential for the successful implementation of *concurrent engineering.* Song, Neeley and Yuzhen [34] identified that effective communication fosters integration between departments, and this in turn leads to successful new product development. In a classical manufacturing model the progression of communication and interaction flow is mostly one-way (serial). Feedback is also one way and occurs with very little consultation. This type of communication and interaction generally occurs in hierarchical (functional) structures [35] and it can compromise design integrity [36]. As Furbershaw explains: "In the past, industrial designers would do nice renderings of what the product should look like, then they would pass these over to engineering. The engineers might examine it and say it's stupid and couldn't be made that way. They would change it to make it manufacturable, but they wouldn't care about the aesthetics or human factors of the product. Then the engineers would pass it on to manufacturing, which would have its own concerns. There was no product continuity" [31, p. 39].

On the other hand, the flow of communication and interaction between various functional departments in the *concurrent engineering* environment allows relevant information to be passed on to any functional department within the organisation. According to Maccoby, the above communication takes place in a *'heterarchy'* team structure. Maccoby has proposed that the team members need to "learn the principles of effective dialogue" in order to successfully operate in this structure [35, p. 60].

It is acknowledged that well-integrated teams are critically important for a successful product development process [31,37-41]. Maccoby states that team members should have *cross-functional* capabilities in order to function appropriately in teams [42]. These include: understanding of the project/company vision; solid grounding in their discipline so that they can communicate any developments from their knowledge area; respect for others and their views; a functional empowerment; team skills that include "brainstorming, listening, asking clarifying questions, seeking consensus"; and the proper attitude to help others to succeed rather than focusing "on their own interest" [42, pp 57-58]. This would suggest that team members have to be 'team-players' with well-developed team-skills. According to Oxman, it is important "to establish a knowledge base related to team behaviour in order to prepare designers for interaction in design teams" [33, p. 282].

*Concurrent engineering* is not an "out of the box product", a fixed methodology, or just teamwork; it is a culture, philosophy and way of thinking [43,44]. This means that behavioural and mental constructs have to be altered [18]. Thus, organisations have to re-examine "the way products are designed, manufactured, [and] serviced" [44].

Curriculum development in the higher education of industrial designers and engineers must take into account changes in contemporary manufacturing organisations and provide students with skill sets for operating effectively in these settings.

## SUBJECT IMPLEMENTATION

The University of Western Sydney services a vast area of Greater Western Sydney through its 6 major campuses. These campuses are spread over a geographical area of more than 2,000 square kilometres or almost three-quarters of Sydney [45,46]. During 2000, the University underwent a major restructuring. As a result of this, it was decided that beginning in 2002, the Industrial Design degree would be offered on two campuses, Penrith and Campbelltown. These campuses are about a one-hour drive apart. To minimise excessive travel by lecturers between campuses, various possibilities were explored, one of which was the online delivery of some subjects [47]. Therefore, in 2001 the School of Engineering and Industrial Design undertook an initiative to conduct a pilot of the online delivery of the Design Management: Organisational Skills for Designers subject. An evaluation of the pilot by students revealed positive learning outcomes, and it was decided to implement the online delivery component of this subject on an ongoing basis [47].

Now in its third year, this subject is one of the core subjects offered by the Industrial Design program to senior students enrolled in: Design & Technology, Industrial Design, Industrial Design Engineering and Master of Engineering (Industrial Design). The subject has been developed to address industry needs identified in previous research [48-53]. Key learning outcomes include:

i. For students to understand manufacturing paradigms and their impact on the product development process and the design process.
ii. For students to understand the impact of organisational structures, strategies and processes on the design process.
iii. To provide students with an understanding of key skills that will enable them to work successfully with various organisational members in the product development process. These skills include teamwork, decision-making and communication, analysis and problem-solving.
iv. The online component of this subject provides students with the opportunity to develop distance communication and virtual teamwork skills, skills that are becoming increasingly important in new product development.

The above subject's aims address a number of Engineering Graduate Attributes [54] such as: the ability to function effectively as an individual and in multi-disciplinary and multi-cultural teams, with the capacity to be a leader or manager as well as an effective team member; the social and cultural responsibilities of professional engineer; effective communication (both written and verbal); and the awareness of the need to undertake lifelong learning, and capacity to do so.

This paper focuses on the learning outcomes outlined in section (iv) above. Virtual workgroups have been incorporated into the course structure in order to provide students with practical exposure to group work-related skills, such as virtual teamwork, decision-making, and long distance communication. The group work aims to expose students, in a practical setting, to the benefits and limitations of online group communication and collaboration [47].

## DESCRIPTION OF TUTORIAL STRUCTURE

In this subject, the tutorial activities are delivered in an online mode while the lectures are delivered in a face-to-face mode at both campuses. The tutorials have been organised in a way that promotes online communication and virtual teamwork skills.

This has been achieved as follows. Students from the two campuses are grouped into seminar groups of 3 to 4 members, with each group containing members from both campuses (i.e. Penrith and Campbelltown). Eight seminar groups are then combined to form a larger tutorial group of approximately 25 students. Each tutorial group has their own discussion board using the WebCT software. In addition, each seminar group has their own "coffee room" (a non-moderated discussion forum) to facilitate their online group participation in order to accomplish their online group assignments.

FIGURE 1
SCHEMATIC PRESENTATION OF THE ACTIVITIES

Each of the seminar groups selects one of the weekly lecture topics on which to prepare a seminar paper. The seminar group's task is to prepare a discussion paper on the lecture topic and post it to the online discussion board. The rest of the "non-presenting" seminar groups in each of the respective tutorial groups provide a response to the weekly seminar paper. A final summary is provided by the presenting seminar group, which addresses points/issues raised by the "non-presenting" groups. Finally, the tutor provides online feedback to each of the seminar groups each week.

A structured timetable for the presentation of seminar papers is provided to the students at the outset of the course. For example, seminar papers need to be posted for tutorial group discussion at a particular time each week[2] (see Figure 1).

## DEVELOPMENT OF INDEPENDENT LEARNING SKILLS

While one of the key aims for the subject was acquisition of virtual teamwork skills, other learning outcomes also needed to be incorporated into the subject design. For example, critical thinking skills and writing skills are also important in a global manufacturing organisation.

Students' independent learning and critical thinking skills are stimulated by arousing their curiosity in the topics through the seven processes of critical thinking and writing, as outlined by James, Scoufis, Farrell and Carmichael [55]. These are: (i) questioning and analysing, (ii) immersing, (iii) questioning, (iv) linking, (v) viewing, (vi) understanding frameworks, and (vii) positioning. In the Design Management: Organisational Skills for Designers subject, this is accomplished by the following activities and the subject structure:

i. The first step of the learning process is for the students to read required texts and accompanying case study notes prior the lecture. In this step students should be 'questioning and analysing' what they have read.
ii. During the second step of the learning process students attend a lecture where they are exposed to various views and a debate on the topic, which should lead to an 'immersion' in the topic.
iii. At this point, the students are required to prepare a group seminar presentation paper in which they link theory and practice through analysis of a case study. At this stage students should be 'questioning' the issues presented in the lecture and readings.
iv. This stage comprises the production and submission of the 'final' seminar paper. At this stage students should be able to demonstrate 'linking' between what was covered in the lecture(s) and readings.
v. During this activity, students who have reviewed the seminar paper prepare and submit their feedback. This results in a number of varied 'views' on that weekly topic (seminar paper). Thus students learn to be assessed on an ongoing basis [56].
vi. The presetting students review their colleagues' and tutor's feedback. At this stage, the feedback encourages the students to develop good 'understanding frameworks'.
vii. At this stage, the presenting student groups prepare and submit their seminar summary papers in which they address any issues raised in their peers' and their lecturer's feedback. During this activity students should reach the last stage of the critical thinking process, and that is *'positioning'*.

The above activities not only encompass the seven processes of the critical thinking as outlined by James, Scoufis, Farrell, and Carmichael [55], but also prepare students for their other assessments (e.g. essay writing and end-of-the-semester exam).

---

[2]Five days after the lecture on the same topic

## Methods Used to Evaluate Student Experiences

Feedback on the students' experiences and learning outcomes was collected by two methods. The university's standard education evaluation survey, SEEQ (Student Evaluation of Educational Quality) was administered at the end of semester, prior to the final examination. Students complete this survey anonymously. Completion of the survey is voluntary. In addition, a specifically designed questionnaire was developed and administered, also at the end of the semester. This questionnaire was prepared in collaboration with the Educational Development Centre at UWS. The questionnaire focused on the student's experience with the online discussion tasks, and asked them to compare their online and face-to-face tutorial experiences. Students were also given the opportunity to comment and to provide additional clarification on any of their responses. Key categories were developed in QSR NVivo from the open-ended questions and additional comments, and the quantitative results were analysed using SPSS.

## Results from the Subject Evaluation

Results from the student survey administered at the end of the subject showed that:

- 93% of students indicated that this subject had helped them to develop virtual team skills
- 78% of students indicated that their writing skills have improved as a result of activities undertaken in this subject
- 75% of the students felt that the online discussion has encouraged collaboration between their group members
- 53% of students perceived that the online activities in this subject has helped them in developing a deeper understanding of the subject

These results suggest that most of the students perceived the subject was facilitating their learning and especially in the intended areas of development of virtual teamwork and written communication skills.

## Importance of Providing Feedback in Online Delivery

Appropriate rapid feedback plays an important role in students' learning and development [56,57]. Cronin and Sparrow identified the following as the main characteristics of effective written feedback: specific and clear comments, comprehensive constructive suggestions, encouragement, prompt return of assignments, justification for grades, respectful presentation and tone [58]. Therefore, students were provided with feedback, which encouraged their learning and their participation, as is evident from this message that was posted into one of the groups 'coffee room':

- *well i'd like to congratulate to all of us. having been mentioned by [the lecturer] on a good response. we should try to keep it up.*

It is important to note that nearly two-thirds (64%) of students indicated that they have received **more feedback** than they would have normally received in other face-to-face only subjects.

As stated previously, the use of online technologies was intended to enhance the student's learning experience. More importantly, however, the aim was to present students with a "whole" learning experience. Thus, students' learning experiences were enhanced through several avenues. These are listed below and are followed by example comments submitted by students on their end-of-the-semester subject survey (in italics):

1. Encourage and motivate students from the very start.
    - *[Lecturer] was very approachable. Even before the subject had begun, email communication was very prompt and helpful. The way the subject was structured was extremely effective. The on-line component combined with face-to-face lecture was helpful and very effective in teaching the subject. Overall, excellent.*
2. Define well-structured activities and clear learning outcomes.
    - *Well structured, approachable, motivational, but firm lecturer. You knew what was expected of you from the beginning and it didn't change.*
    - *The lecturer was very organized so there was no ambiguity once we grasped the system he had set in place.*
3. Demonstrate passion and knowledge for the profession (leading by example).
    - *Knowledgeable: he has a great knowledge about design and enjoys sharing his knowledge.*
    - *Enthusiasm – he likes the subject and helps us appreciate it too. Sincerity – he is sincere in wanting us to learn and improve and gives us his time to help us improve.*
    - *Organisation – always prepared, organised. Assignments marked quickly, he is always on time.*
4. Exhibit genuine interest in students' learning and well-being.
    - *[Lecturer is] willing to help students through email, phone, [on-line discussion board]. Very quick to returning calls, email etc. Good – very good teaching skills and methods. I enjoyed learning from [this lecturer].*
    - *[Lecturer] seemed involved in lectures and showed high interest in what we took away from lectures.*
5. Provide an environment in which students are comfortable asking questions and having an open discussion.
    - *Comfortable classroom – always willing to answer questions. e.g. [lecturer] was very good as a lecturer, encouraging open communication and providing a large amount of class material.*
    - *[Lecturer has] also involved us all in discussions each week which gave us a clear understanding of the course.*
6. Use humour to engage students.
    - *Very approachable, humorous and he knows his stuff up to date.*
7. Vary teaching styles and source of instruction.
    - *Analysis of each weeks articles in depth. Different media besides the reader etc. Video was good.*
    - *Varies teaching style. Always different and new – kept everyone interested.*
8. Link theory to practice.
    - *Ability to divergently discuss topics from many points of view with practical examples relative to topic and issues.*
9. Structure the lectures, tutorials, and other activities to encourage students' participation.
    - *Encouraged participation through weekly activities from reader.*

   - *Very open and willing to communicate or discuss issues numerous times.*
10. Encourage open discussions between students.
   - *The use of on-line discussion for group participation. Making the lectures student-involved with open discussion being used.*
11. Provide prompt, critical feedback that motivates students.
   - *Feedback and explanation from lecturer on topics of discussion and assignments.*
   - *Does give useful advice and criticise not useless command.*
12. Develop students' teamwork skills and involve students in a team environment.
   - *The teamwork tutorials were a good initiation into working in a team.*

## CONCLUSION

The students' feedback indicates that the Design Management: Organisational Skills for Designers subject was well received and the aims of the subject have been achieved. Creating assignments that encourage the use of online communication forums amongst the group and tutorial members, which subsequently promoted development of virtual teamwork skills, have achieved this positive outcome. Incorporating the seven processes of critical thinking, and the aspects of good teaching/learning practices has also facilitated this learning process. It can be concluded that the use of distance learning technologies in the subject provides new opportunities for students to develop skills that are useful in global manufacturing environments.

Last year, the School of Engineering and Industrial Design, in the interest of rationalisation and cost savings, made the decision that beginning in 2005 it would operate only on one campus, Penrith. This decision, to deliver all the subjects on only one campus, will remove the physical distance that currently exists amongst the group members. The elimination of this physical distance will remove the necessity for students and teams to use the online communication forums to complete their group work.

However, as has been illustrated in this paper, the online component of the subject provides valuable skills for designers and engineers in contemporary manufacturing workplaces and also contributes to the development of learning outcomes that might not be developed in a face-to-face setting. Therefore, the author seeks to partner with external educational institutions that are interested in research and development of long distance virtual teamwork skills, and in providing an opportunity for their students to be exposed to this type of environment. Interested institutions are encouraged to contact the author.

## ACKNOWLEDGEMENT

I would like to thank Kerry Harman whose input while redeveloping this subject was invaluable, and germinated a vast number of ideas. It is also important to thank Associate Professor Heinz Luettringhaus and Professor Steven Riley for giving me the opportunity to redevelop the above subject. Finally, I would like to thank the staff from the Educational Development Centre who have been important in developing the above subject.

# REFERENCES


1. Schilling, M.A., and Hill, C.W.L., "Managing the New Product Development Process: Strategic Imperative," Academy of Management Executive, Vol. 12, No. 3, 1998, pp. 67-81.

2. Dröge, C., and Calantone, R., "New Product Strategy, and Performance in Two Environments," Industrial Marketing Management, Vol. 25, No. 6, 1996, pp. 555-566.

3. Biemans, W.G., "Interval and External Networks in Product Development: A Case for Integration," in *Product Development: Meeting the Challenge of the Design-Marketing Interface*, Bruce, M. and Biemans, W. G., Eds. Hoboken, NJ: John Wiley & Sons, 1995, pp. 137-159.

4. Stalk, G., Jr. and Hount, T.M., *Competing against Time: How Time-Based Competition Is Reshaping Global Markets,* The Free Press, New York, NY, USA, 1990.

5. Flynn, B. B., "The Relationship between Quality Management Practice, Infrastructure and Fast Product Innovation," Benchmarking for Quality Management & Technology, Vol. 1, No. 1, 1994, pp. 48-64.

6. Thamhain, H.J., "Concurrent Engineering: Criteria for Effective Implementation," Industrial Management, Vol. 36, No. 6, 1994, pp. 29-32.

7. Takeuchi, H. and Nonaka, I., "The New Product Development Game: Stop Running the Relay Race and Take up Rugby," Harvard Business Review, Vol. 64, No. 1, 1986, pp. 137-146.

8. Gupta, A.K. and Souder, W.E., "Key Drivers of Reduced Cycle Time," Research Technology Management, Vol. 41, No. 4, 1998, pp. 38-43.

9. Vesey, J.T., "The New Competitors Think in Terms of Speed-to-Market," Sam Advanced Management Journal, Vol. 56, No. 4, 1991, pp. 26-33.

10. Womack, J.P., Jones, D.T., and Roos, D., *The Machine That Changed the World,* Rawson Associates, New York, NY, USA, 1990.

11. Hansen, J.G., "Design for the Decision Point," Design Management Journal, Vol. 5, No. 1, 1994, pp. 47-51.

12. Gupta, A.K., and Wilemon, D., "Changing Patterns in Industrial R&D Management," The Journal of Product Innovation Management, Vol. 13, 1996, pp. 497-511.

13. Womack, J.P., and Jones, D.T., *Lean Thinking: Banish Waste and Create Wealth in Your Corporation.* Simon and Schuster, New York, NY, USA, 1996.

14. Swink, M.L., Sandving, J.C., and Mabert, V.A., "Customizing Concurrent Engineering Processes: Five Case Studies," Journal of Product Innovation Management, Vol. 13, 1996, pp. 229-224.

15. Cook, B.M., "Design Gets a Face Lift," Industry Week, Vol. 240, No. 22, 1991, pp. 46-50.

16. Reinertsen, D.G., *Managing the Design Factory: The Product Developer's Toolkit,* Free Press, New York, NY, USA, 1997.

17. Cooper, R.G. and Kleinschmidt, E.J., "Determinants of Timeliness in Product Development," The Journal of Product Innovation Management, Vol. 11, 1994, pp. 381-396.

18. Franze, H.A., "Design for Environmental Capability of Automobiles: New Life-Cycle Management Tools in the BMW Product Development Process," Proceedings of International

Conference on Engineering Design 97: World Class Design by World Class Methods, Tampere, Finland, 1997, pp. 23-35.

19. Hsiao, S.-W., "A Concurrent Engineering Based Method for Developing a Baby Carriage," International Journal of Advanced Manufacturing Technology, Vol. 12, 1996, pp. 455-462.

20. Liker, J.K., Ettlie, J. E., and Campbell, J.C., "Engineered in Japan: Japanese Technology-Management Practices," in *Japan Business and Economics Series*. New York, NY, USA: Oxford University Press, 1995.

21. Whitney, D.E., "Nippondenso Co. Ltd.: A Case Study of Strategic Product Design," in *Engineered in Japan: Japanese Technology-Management Practices, Japan Business and Economics Series*, Liker, J.K., Ettlie, J.E., and Campbell, J.C., Eds. New York, NY, USA: Oxford University Press, 1995, pp. 115-151.

22. Elshennawy, A.K., Krishnaswamy, G.M., and Mollaghasemi, M., "Concurrent Engineering Deployment: A Virtual Reality Approach," Integrated Manufacturing Systems, Vol. 4, No. 4, 1993, pp. 24-28.

23. Tomiyama, T., "A Note on Research Directions of Design Studies," Proceeding of International Conference on Engineering Design 97: World Class Design by World Class Methods, Tampere, Finland, 1997, pp. 29-34.

24. Shemas, D.G., and Derakhshan, S., "Organizational Approach to the Implementation of Simultaneous Engineering," International Journal of Operations and Production Management, Vol. 14, No. 10, 1994, pp. 30-43.

25. Flynn, B.B., Sakakibara, S., and Schroeder, R.G., "Relationship between JIT and TQM: Practices and Performance," Academy of Management Journal, Vol. 38, No. 5, 1995, pp. 1325-1360.

26. Business Victoria Department of State Development, Yates, L., Cooper, B., Scoular, R., Ieronimo, N., et al., "Design for Agility: Creating a New Competitive Strength for Australia's Automotive Industry," Business Victoria Department of State Development, Melbourne, Vic, Australia, September 1996.

27. Mortimer, J. and Hartley, J., "Simultaneous Engineering," Department of Trade and Industry, London, UK, bookletJune 1991.

28. Teng, S.-H. G. and Ho, S.-Y., "Failure Mode and Effects Analysis: An Integrated Approach for Product Design and Process Control," International Journal of Quality and Reliability Management, Vol. 13, No. 5, 1996, pp. 8-26.

29. Anderson, D.M., *Agile Product Development for Mass Customization: How to Develop and Deliver Products for Mass Customization, Niche Markets, JIT, Build-to-Order and Flexible Manufacturing,* Irwin, Chicago, USA, 1997.

30. Sarkis, J., "Manufacturing Strategy and Environmental Consciousness," Technovation, Vol. 15, No. 2, 1995, pp. 79-97.

31. Braham, J., "The Wizards of Industrial Design," Machine Design, Vol. 64, No. 23, 1992, pp. 37-41.

32. Harkins, J., "A Means to Our Ends: 3D CAD Tools Help to Accomplish Design Goals in the 90s," Design Management Journal, Vol. 5, No. 2, 1994, pp. 79-82.

33. Oxman, R., "Observing the Observers: Research Issues in Analysing Design Activity," Design Studies, Vol. 16, No. 2, 1995, pp. 275-283.

34. Song, M.X., Neeley, S.M., and Yuzhen, Z., "Managing R&D-Marketing Integration in the New Product Development Process," Industrial Marketing Management, Vol. 25,1996, pp. 545-553.

35. Maccoby, M., "Creating Network Competence," Research • Technology Management, Vol. 43, No. 3, 2000, pp. 59-60.

36. Clark, K.B. and Fujimoto, T., "The Power of Product Integrity," Harvard Business Review, Vol. 68, No. 6, 1990, pp. 107-118.

37. Karbhari, V.M., Burns, J.S., and Wilkins, D.J., "Total Quality Design: An Approach for Customer Satisfaction in Critical Advanced Technologies," Benchmarking for Quality Management & Technology, Vol. 1, No. 1, 1994, pp. 65-88.

38. Song, M.X., Montoya-Weiss, M.M., and Schmidt, J., "Antecedents and Consequences of Cross-Functional Cooperation: A Comparison of R&D Manufacturing and Marketing Perspective," Journal of Product Innovation Management, Vol. 14, 1997, pp. 35-47.

39. Song, M.X., Souder, W.E., and Dyer, B., "A Casual Model of the Impact of Skills, Synergy, and Design Sensitivity on New Product Performance," Journal of Product Innovation Management, Vol. 14, 1997, pp. 88-101.

40. Pahl, G., Badke-Schaub, P., and Frankenberger, E., "Résumé of 12 Years Interdisciplinary Empirical Studies of Engineering Design in Germany," Design Studies, Vol. 20, No. 5, 1999, pp. 481-494.

41. Beckman, S.L. and Mowery, D., "Getting the Right Products to Market: A Study of Product Definition in the Electronic Industry," Design Management Journal, Vol. 4, No. 2, 1993, pp. 54-61.

42. Maccoby, M., "Building Cross-Functional Capability: What It Really Takes," Research • Technology Management, Vol. 42, No. 3, 1999, pp. 56-58.

43. Brooks, B. M., "Concurrent Engineering - What Is It?" Proceedings of IEE Colloquium on Concurrent Engineering, London, UK, 1993, pp. 1-3.

44. Fisher, R. A., "Can Technology Improve Product Development?" Machine Design, Vol. 65, No. 10, 1993, pp. 202.

45. Contractor, A., "Uni Amid Swinging Voters May Get Extra $10m," in *The Sydney Morning Herald.* Sydney, NSW, 2003, pp. 5.

46. Contractor, A., "PM Helps Uni Straddling Crucial Seats," in *The Sydney Morning Herald.* Sydney, NSW, 2003, pp. 3.

47. Bohemia, E., and Harman, K., "Comparative Study of On-Line and Face-to-Face Subject Delivery in Industrial Design," Proceedings of Exploring Emerging Design Paradigm - ICSID Educational Seminar 2001, Seongnam City, Korea, 2001, pp. 218-227.

48. Bohemia, E., "Designer as Integrator: Reality or Rhetoric?" The Design Journal, Vol. 5, No. 2, 2002, pp. 23-34.

49. Bohemia, E., "Performance of Industrial Designers: Australian Perspective," Proceeding of International Conference on Design Education: Re-inventing Design Education in the University, Curtin University of Technology, 2000, pp. 103-114.

50. Bohemia, E., "Lean Manufacturing in Australia and Its Impact on Industrial Design," Proceeding of International Conference on Engineering Design 97: World Class Design by World Class Methods, Tampere, Finland, 1997, pp. 101-104.

51. Bohemia, E., "Suitability of Industrial Designers to Manage a Product Development Group: Australian Perspective," Design Management Journal: Academic Review, Vol. 1, No. 1, 2000, pp. 40-54.

52. Lawrence, P., Barlow, W., Conrad, T., Denham, P., et al., "Winning by Design: Technical Papers; the Corporate Role of Design in Australian Industry," The Warren Centre for Advanced Engineering, University of Sydney, Sydney, NSW, Australia, Report Volume 2, 0-86758-231-1, 1987.

53. Freeman, D., Goodman, J., Axarlis, S., et al., "Competing by Design: The National Design Review Report," National Design Review Steering Committee, The Australian Academy of Design, St Leonards, NSW, Australia, March 1995.

54. IEAust, "Manual for the Accreditation of Professional Engineering Programs," The Institution of Engineers, Australia, 1999.

55. James, B., Scoufis, M., Farrell, H., and Carmichael, E., "Unravelling the Mysteries of Critical Thinking," University of Western Sydney, Nepean & Video Education Australasia Pty Ltd, Sydney, NSW, Australia, accompanying video tape notes 1-86341-840-7, 1999.

56. Lazerson, M., Wagner, U., and Shumanis, N., "What Makes a Revolution?" Change, Vol. 32, No. 3, 2000, pp. 12-19.

57. Petre, M., Carswell, L., Blaine, P., and Thomas, P., "Innovations in Large-Scale Supported Distance Teaching: Transformation for the Internet, Not Just Translation," Journal of Engineering Education, Vol. 87, No. 4, 1998, pp. 423-432.

58. Cronin, M., and Sparrow, H., "Focusing on Feedback: An Examination of Effective Written Feedback to Promote Learning for External Students," Proceedings of 8th Annual Teaching Learning Forum - Teaching in the Disciplines/Learning in Context, The University of Western Australia, Perth, WA, 1999, pp. 87-92.


**Erik Bohemia** is a Senior Lecturer and Head of Industrial Design Engineering and Industrial Design programs at the School of Engineering and Industrial Design at the University of Western Sydney. Erik's research focus is on changes taking place in the areas of manufacturing methods, new product development processes, quality issues, business management and operational management, and how these impact the industrial design profession. The results from his research have been used to guide the development of a curriculum in design management subjects at the University of Western Sydney so that future graduates may more effectively fulfill industry requirements. Erik's research has been published in international journals and conferences on Design Management and Design and Engineering Education.

# Chapter 13

## Mobile Phones in the Lecture Theatre - Using Wireless Technology as a Pedagogical Aid

FRODE EIKA SANDNES and OLAV TALBERG
*Faculty of Engineering, Oslo University College, P.O. Box 4, St. Olavs plass, N-0130 Oslo, Norway. E-mail: frodes@iu.hio.no*

*In most engineering colleges, giving lectures for 300 or 400 students is a practical necessity. Most lecture theatres are large and, when filled up, are intimidating to the average student and do not invite students to be active participants during lectures. These theatres also restrict the instructor's opportunities to interact with the students. In this paper we describe a "proof-of-concept" system developed at Oslo University College that exploits the use of mobile phones in the lecture theatre with the intention of enhancing learning through increased teacher-student interaction.*

### MOBILE TECHNOLOGY AND LEARNING

The idea of incorporating mobile technology into the learning process is already an established thought [1,2, 4-17]. Universities and colleges worldwide have invested in wireless campus networks allowing students to work freely, anywhere, anytime.

Until recently, the main efforts in mobile learning have focused on different ways of using laptop computers with wireless network connections [2,3]. Students possess personal laptop computers that they carry around campus, using them in the dormitory, the library, etc., for individual work, collaboration in groups or even to take notes in the lecture theatre. Laptop computers, however, are relatively large and hence cumbersome to carry around, and it takes some time to reactivate them once they are in sleep mode. The utility of laptop computers, therefore, is limited.

Another recent technical innovation is the PDA [18], the personal digital assistant. PDAs are small, easy to transport and are to some extent a fashion accessory. Many students use PDAs for note taking during lectures. As they take a very short time to start up, PDAs are suitable for quick, urgent and even impulsive actions. Still, owing to their relative high costs and other reasons, not all students possess PDAs. One level further down the device hierarchy is the mobile phone. The obvious advantages of mobile phones are their low price and wide availability. Nearly all students own a mobile phone, and they use them actively.

Mobile phones do have a potential as a pedagogical aid in learning. SMS (Short Messaging Service) has become a de-facto lightweight means of asynchronous communication among youths. Students can exchange meaningful and subject-related messages during and outside of lectures. For example, engineering degree programmes frequently rely on group work involving meetings. At Oslo University College students frequently negotiate meeting place and time for these meetings via SMS [19].

Mobile phones, however, are generally viewed as a nuisance in the context of lectures and classroom presentations. Incoming calls with loud annoying ring-tones and students rushing out of the room to answer the call frequently disturb and draw attention away from lectures. Even though a humorous ring-tone may lighten up a monotonous lecture, it generally takes the focus away from the lecture. Even worse, students may be tempted to play games with their handsets during lectures. The overall common attitude among teaching staff is that mobile phones should be banned from classrooms and lectures. So why should mobile phones be taken seriously as a positive pedagogical aid during lectures?

## RATIONALE

Modern pedagogy clearly separates lecturing from learning; lecturing is a one-way communication situation – that of delivery of content. In spite of the fact that most teaching staff acknowledge that lectures do not necessarily assist learning, many engineering colleges are dependent on this form of teaching due to large student numbers and scarce resources. Thus, though not pedagogically optimal, lecturing is a practical necessity. Most lecture theatres are large and may easily accommodate 300 - 400 students. When filled up, theatres are intimidating to the average student and do not invite him or her to actively interact with the teacher during the lecture. Research by Banham and Elender [20], however, suggests that learning can be stimulated in large classes by increasing the range of work in which students are actively involved. Felder [21] recommends getting students actively involved and helping them develop a sense of community.

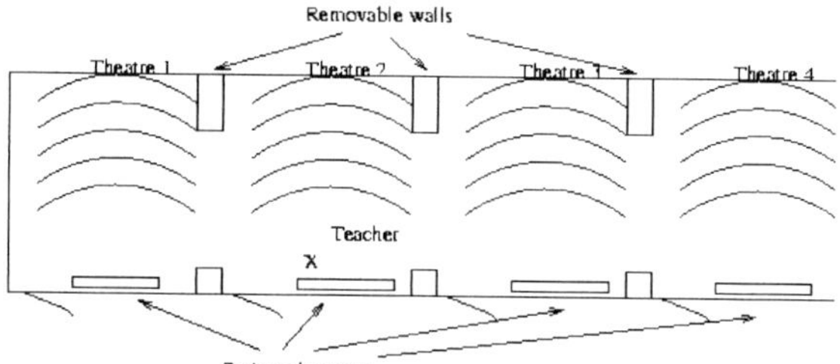


FIGURE 1
THE LAYOUT OF THE LECTURE THEATRE(S) AT OUC

At the Faculty of Engineering at Oslo University College, frequently one teacher serves 400 students. Our lecture facilities are organised into four adjoining lecture theatres, each with approximately 100 seats (see Figures 1 and 2). Each theatre is equipped with a projector connected to the teacher's computer. The four theatres are separated by removable walls, i.e. the rooms can be configured into four theatres with 100 students each, or 2 theatres with 200 students each or with 100 students in one and 300 in the other, or one theatre with the capacity of 400 students.

When the facilities are configured into four separate lecture theatres the lecturing environment is acceptable. A good amount of teacher-student interaction can take place, and the teacher is free to use the blackboard instead of the projector providing a more relaxed means of presentation. But as soon as two or more theatres are joined together, the lecturing situation changes drastically. First, the teacher no longer can rely on his or her own voice but is required to use either the stationary microphone positioned at the teacher's desk or a mobile wireless microphone. Secondly, the teacher is unable to use the blackboard efficiently since only the students in the same room are able to read the writing. Thirdly, the teacher loses the vital visual contact with some of the students, owing to distance and obstructing pillars, as many students may be seated behind supporting pillars located in the theatre. Naturally, this reduces the overall effectiveness of the lectures, since the teacher can no longer rely on students' body language and other cues in order to adjust the delivery of the presentation [22]. Consequently, teachers feel uncomfortable and the lectures become sub-optimal.

FIGURE 2
LECTURE THEATRES AT OUC

Obviously it is not only the teachers who find this environment problematic. Students in the adjacent lecture theatres who have limited visual contact with the teacher must rely on the projection screen and the amplified audio from the speakers. These students are unable to see all the teacher's gestures - gestures that intentionally or unintentionally might be essential for comprehension of the material presented. Even more important, when sitting in a lecture theatre that is separated from the teacher's, it is physically difficult for a student to ask questions and make comments simply because he/she may be neither seen nor heard by fellow students and/or the teacher. On the other hand, students in the teacher's field of view often feel too intimidated to ask questions or to comment, simply because of the sheer number of students. Research has shown that class size affects academic performance (see the classic study by Glass and Smith [23]). Further, research by Goldstein and Benassi [24] shows that students' perception of a good lecturer is directly related to what they perceive as a good discussion leader.

The problem we face by our faculty is simply this: "How can the student-teacher interaction be improved in large lecture theatres?" The explicit research questions addressed in this study are: (1) How can commonly available wireless technology be used to engage students and improve student-teacher interaction in large lecture theatres? and (2) How should such a system be realised in practice?

## INCREASING INTERACTIVITY IN THE CLASSROOM

The current project attempts to address this problem, and our approach is to let the students use their mobile handsets to interact with the lecturer. The students log into the course website via their phones and access the appropriate software. The software allows the students to respond to polls, either planned or spontaneous, and make comments and ask questions anonymously in real-time. Comments and questions as well as poll results are displayed on the projection screen in the lecture theatres for everyone to see, and the teacher can therefore respond to the information appearing. For instance, the teacher may respond to a question asked by a student via the system. One apparent benefit is that even very humble and shy students can participate in the interaction. Another advantage is that both the students and the teacher have access to a log of the entire session, and the log can be studied after the lecture. The system can also be used to perform short in-class-examinations or tests with some degree of security.

One important advantage of using mobile phones is that existing infrastructure is exploited. New educational technology often fails because institutions are unable to invest in the required infrastructure [25]; however, most students possess a handset – at least in most northern European countries. Although other technologies such as PDAs (personal digital assistants) might be more suitable from a technical point of view, they are impractical as PDAs are not commonly used or owned by the average student. Most students are also quite skilled using mobile technology – although they usually have cumbersome 12-key numeric keypads, new text entry technology such as Tegic's T9 system allows students to type text comfortably with impressive speed. Many students have acquired such training from frequent use of SMS messages [26].

In this paper we discuss a prototype system developed in the Faculty of Engineering, Oslo University College, that allows Internet-enabled Java phones to be used in lecture theatres as part of teaching and learning. The system is intended for lecture theatres with more than 60 students, a class size large enough to be intimidating for students, resulting in less student-teacher interaction than has been observed for smaller classes.

The current prototype has three main functions: (a) performing real-time polls, in which the teacher acquires general opinions regarding various issues, or holds a quiz; (b) uploading queries, in which students post requests or questions anonymously in real-time from mobile handsets; and (c) providing feedback, in which students send anonymous hints in real-time to the teacher, such as when they do not understand a particular part of the lecture, or when the teacher is advancing too quickly or speaking too softly. The information is assimilated by a server and accessed by a web-client and presented on a large projection screen in the lecture theatre that can be seen by students and teacher.

## Wireless Technology

As mobile phones and other portable devices become more powerful with wireless Internet access and the ability to run custom wireless applications using Micro Java, mobile phone technology also becomes more interesting from a pedagogical point of view. Within the next few years most mobile phones will be equipped with these applications and a majority of students will be accordingly empowered.

Currently, most mobile phones on sale in Europe are Internet enabled (GPRS) [27], and allow simple web browsing of WML (Wireless Markup Language) and more recently XHTML-basic (Extended HyperText Markup Language) documents [28]. Further, many handhelds come equipped with a Java virtual machine designed to run J2ME Java programs [29]. The J2ME platform is a reduced version of standard Java. For example, only a subset of the standard libraries are available and there is no floating point number support. In addition, J2ME comes with platform-specific libraries designed for the limited user interfaces and operating environments of the mobile handsets. In addition, there are vendor-specific libraries that allow Java programs to control model-specific features such as vibration. Java programs for mobile phones - usually called MIDlets (Mobile Information Device) - can exploit basic Internet connectivity, simple persistent storage known as records and simple mono- and multi-colour graphics. The advantage of MIDlets is that they can be designed to run on all Java-enabled mobile phones. All students with a Java-enabled phone should be able to run the programs, irrespective of handset model and manufacturer. Currently, the J2ME platform is predominantly used for games and entertainment.

## THE GAMESHOW TEACHING SYSTEM - AN OVERVIEW

We have chosen to name our prototype teaching system the 'GameShow' with reference to the audience participation aspect observed in some television game shows. The GameShow teaching system comprises three main parts – the students' phone clients, the teacher's client and the server. Students possess Internet- and Java-enabled mobile phones. Their handsets run instances of the GameShow mobile client MIDlet software, communicating directly with the closest base station operated by the mobile network operator. The mobile network provides a gateway to the Internet and thus the departmental GameShow application server. This server centrally coordinates all the interactions. At the front of the lecture theatre is the teacher's computer, either a stationary desktop or a laptop computer connected to the projectors. This audio-visual setup is usually used for displaying PowerPoint presentations or performing other computer-based demonstrations. The lecturer accesses the teacher's client using a standard web browser as the teaching portal has a web-based interface.

## THE SERVER

The GameShow server also controls all the student-teacher interactions. Our prototype is implemented on the Java platform using a mix of servlets and JSP pages [30,31]. A relational database (MySQL) is used for storing persistent data, for example prepared poll questions, poll results and entire lecture sessions. The server is administered through a web interface using a standard HTML web browser, and the special-purpose mobile client interacts with the server via the HTTP protocol.

## THE TEACHER CLIENT

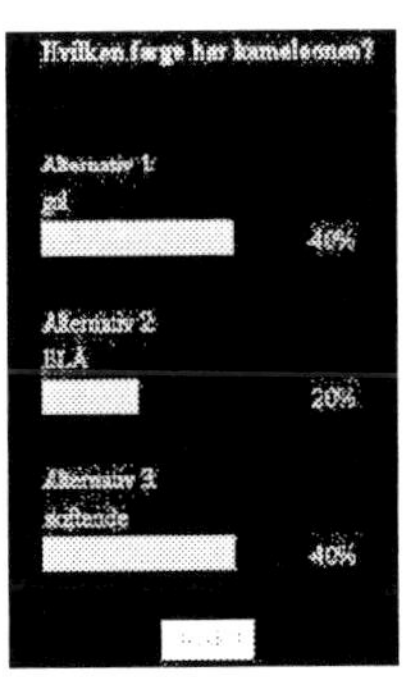


FIGURE 3
QUIZ STATISTICS

A standard web browser is used for accessing the GameShow software. The students in a "live" mode can use GameShow during lectures or in an "offline" mode during pre-lecture preparation or post-lecture analysis.

The teacher has two options. First, she can prepare general questions, requests and messages that students may want to ask during the lecture. With this option, to pass a message or question to the teacher the student simply activates the phone, selects the question or request, and presses the appropriate button. Second, the teacher can prepare polls, quizzes or review questions that the students will be asked to answer during the lecture. After the students have answered the questions, the statistics can be shown on the projected screen instantly (see Figure 3).

FIGURE 4
THE TEACHERS' CLIENT LOGIN

To make modifications the teacher must be authenticated (see Figure 4). Teachers can register themselves. Once logged into the system, the teacher is given a number of options (see Figure 5).

If the teacher starts "live" mode, the web browser will allow incoming textual and iconic messages to be displayed. The display is updated every 3 seconds, and the last 4 messages are always shown (see Figure 6).

FIGURE 5
MAIN MENU

Using the questionnaire editor (see Figure 7), the teacher may also use a spontaneous generic test template to prepare poll questions and define answer alternatives prior to the lecture. The questions are associated with a particular lecture and the teacher must specify for which lecture the questions are to be used. There are three templates - one each for questions with two, three and four answer alternatives. In these instances, the teacher's client only displays the text "Select one of … ", and the teacher speaks the question orally. The options "A", "B", "C" and "D" are displayed on the screen while the teacher orally specifies what answers the letters correspond to.

The "perform poll" or "examination" function is activated from the main menu. This page displays a random security code (see Figure 8) that is used by the students present in the lecture to post their answers. A continuously updated counter shows the number of replies received. The security code is only displayed when the teacher client is accessed from the designated computer in the lecture theatre or other authorised computers. This prevents external users from tampering with the voting results. External users, such as students at home accessing the teacher client, will not see the security code and hence will be unable to vote.

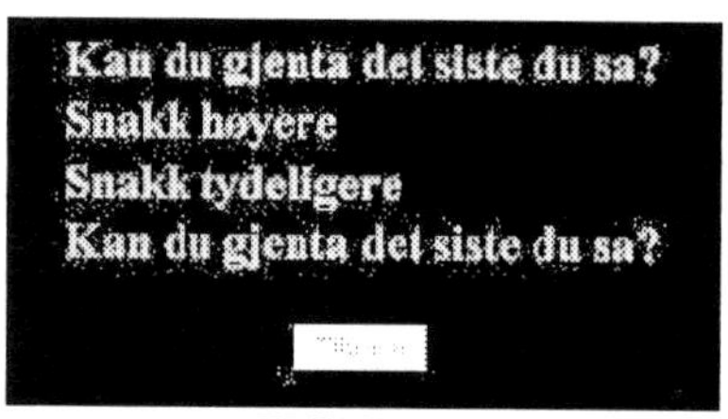


FIGURE 6
LIVE MODE – STUDENTS' COMMENTS

The poll runs for as long as the teacher desires, and is stopped by the teacher. The results are shown on a summary page providing statistics, i.e. how many percent of the voters chose each option.

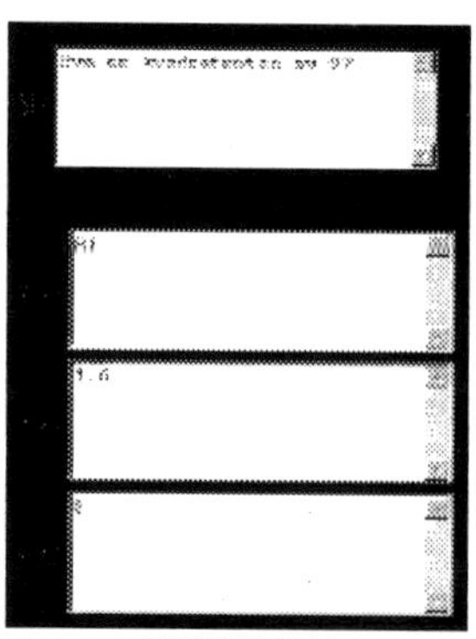

FIGURE 7
AUTHORING QUESTIONNAIRES

The students can also access the teacher's client in read-only mode, with no authentication required. Students unable to

FIGURE 8
SECURITY ACCESS CODE DISPLAYED IN THE LECTURE THEATRE

attend the lecture can then observe remotely what is happening in the lecture. Furthermore, after the lecture, students can revise questions that arose during the lecture.

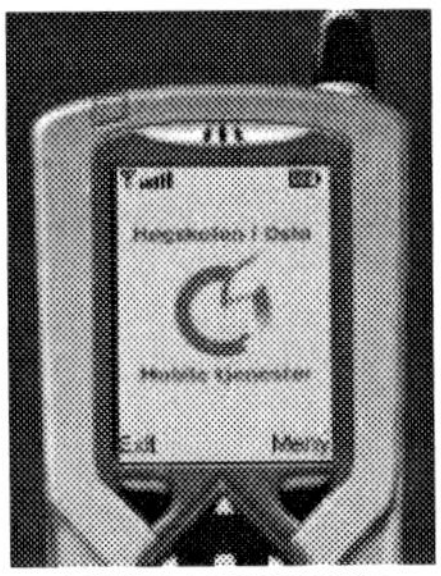

(A) SPLASH SCREEN

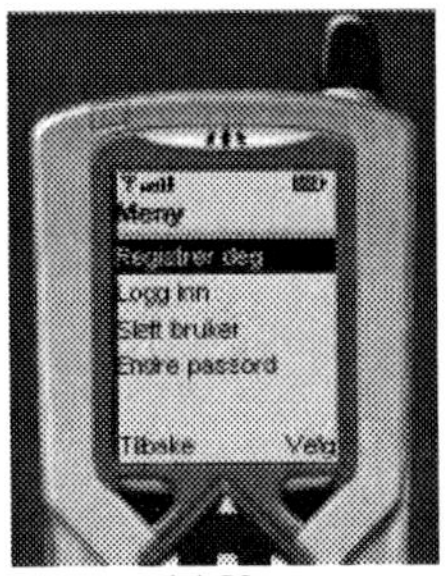


(B) USER ADMINISTRATION

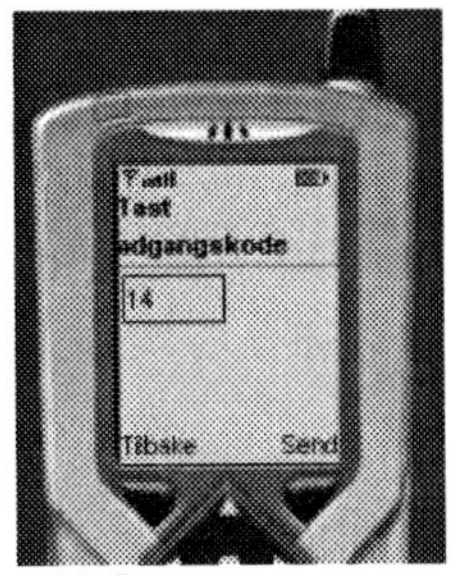


(C) STUDENT LOGGING INTO THE SYSTEM

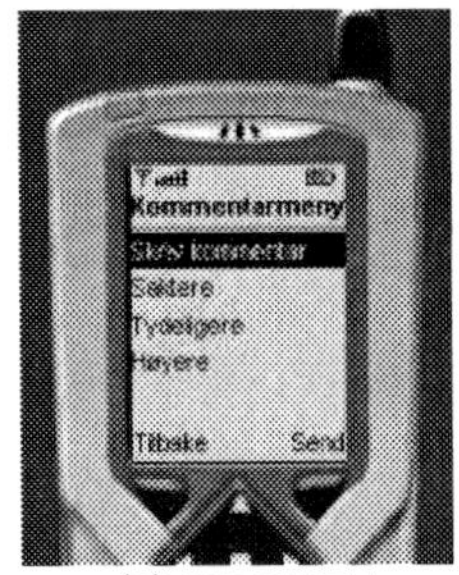


(D) CHOOSING A PREPARED COMMENT

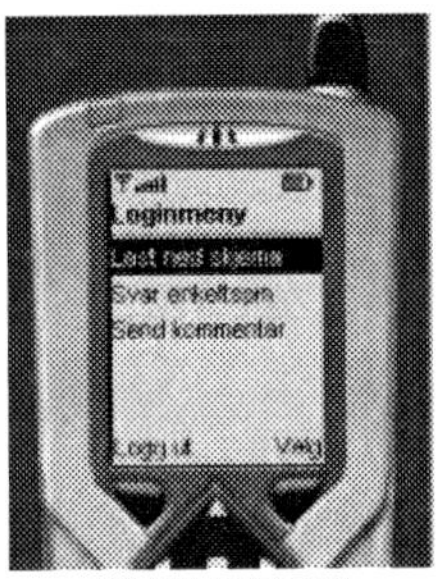


(E) MAIN MENU

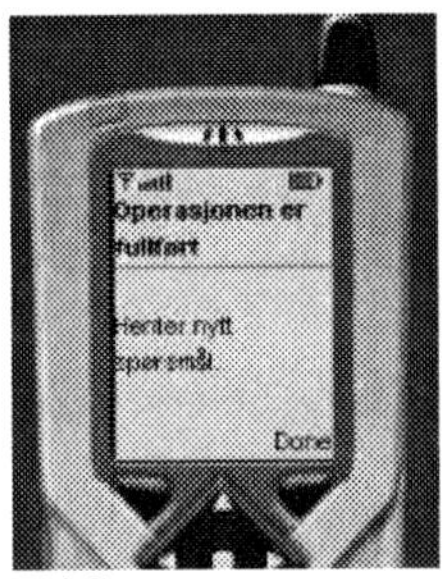


(F) STUDENT RETRIEVING QUESTIONNAIRE

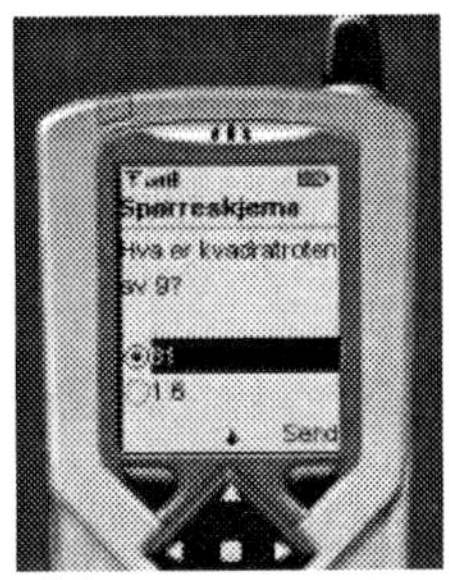


(G) RESPONDING TO A QUESTION

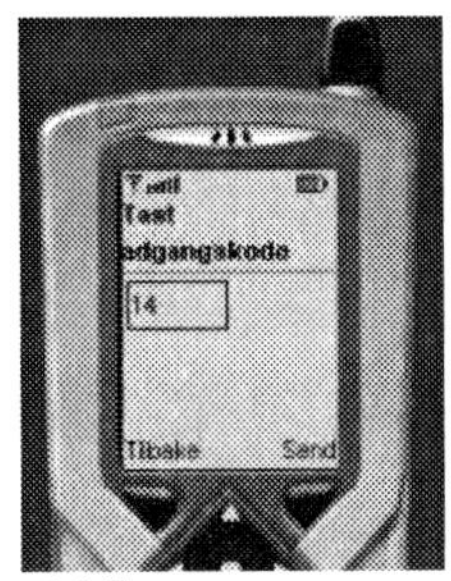


(H) STUDENT LOGGING INTO THE SYSTEM

FIGURE 9
STUDENTS' CLIENT

## THE STUDENT CLIENT

The students' MIDlet application must be downloaded and set up prior to the lecture. This is only done once. The teacher instructs the students to do so. The MIDlet is available from the departmental web site, and is set up by registering the user the first time the MIDlet is run (see Figures 9a and 9b). The registration procedure consists of entering personal information (name etc.) and selecting a password. Once registered, the student can log into the system using the chosen password for authentication (see Figure 9c).

The student is given four choices (see Figure 9e): (1) Download the questions or quizzes for the lecture (see Figure 9f); (2) Respond to a single question posted by the teacher (see Figure 9g); (3) Send a question or a request to the teacher (see Figure 9h); or (4) Log out. To download the questions the student must provide the security code. The single question function is chosen to access and respond to a single question posted by the teacher during the lecture. The "send question" function is used by the student to post a request, question or comment to the server. The request is subsequently displayed on the teacher's client if it is in "live" mode. Currently, the students have to type their question or request, but future versions could also provide a list of preset comments, questions and iconic messages for commonly needed phrases to make the system more efficient to use.

## FICTITIOUS GAMESHOW SCENARIOS

Emma is a second year computer science student doing a compulsory course in statistics. Consequently, the lecture hall is always full. Emma is a bright student but statistics is slightly outside her main field of study. One of her teachers, John, is a mathematician who loves the beauty of mathematics. One day John is covering the topic "analysis of variance", and is covering the theory behind this analysis technique. Towards the end of the lecture, John quickly summarises his main topic and moves on to the subject "two-factor anova." Emma notices that there is a change in topic but missed the conclusions, and felt that John did not cover the applications of the technique properly. She retrieves her mobile phone, logs into the system via the MIDlet, and selects the "send question" option. She enters the question "Can u rep conclusion, plx" and presses submit. Two seconds later the text appears on the projected screen and on the console in front of the teacher. John does not notice it immediately, but 15 seconds later he sees the message. Meanwhile, additional messages of support have arrived from other students – "Agree, don't understand a thing." John gets the point and goes back to the conclusion to give it a more in-depth revision.

Birger is a computer science lecturer teaching computer security for a class of 170 students. Towards the end of the course he is teaching biometrics. He is curious about what the students think about the future of biometrics. He therefore asks the students the question "Will biometrics become commercially viable? Select A for yes and B for no!" Then he selects the two-question template question from the console and starts the poll. He tells the students to start. It takes about 10 seconds for the students to retrieve their mobile phones from their pockets, log into the system and locate the question. Then the replies arrive. After about 40 seconds he sees that 140 students have cast their vote and that not all the students have brought their mobile phones. He therefore concludes the poll and displays the results: 63% of the students think that biometrics will take off while 37% are sceptical.

The results of the poll have potential for further follow-up by John, which he will of course consider at leisure in his office; but it's obvious that the use of the mobile phone has increased student participation and also given John ideas of how to follow-up the activities during the lecture so as to increase the students' learning.

## PRACTICAL DIFFICULTIES

A number of factors have prevented us from deploying the system with large groups of students. Firstly, the mobile coverage in the lecture theatre is poor due to the shielding of the thick walls and the distance to the closest mobile base station. Secondly, a single base station may not reliably handle the traffic from 400 mobile users simultaneously. Thirdly, although the technology is available, only a minority of students has already invested in the new mobile-technology. Finally, it is currently not trivial to configure Internet-enabled phones so they can use the Internet.

Computer students should be able to perform such a configuration quite easily, but less computer-literate students from other departments may have problems doing so; however, it is only a matter of time before such technology is commonly available and in common use. Moreover, base stations are gradually becoming more powerful and more numerous in inner-city areas and should shortly be able to handle a large volume of low-intensity traffic, such as the packets transmitted by the students' GRPS handsets. Another possibility is the use of the school's own WLAN or Bluetooth. A number of competing, complementary, exciting and promising technologies are ready to come on stream.

## Current and Future Work

We are currently improving the system by exploiting the Bluetooth technology available in many new mobile handsets, which allows mobile phones to communicate with other mobile devices across free ad-hoc networks without utilising the costly mobile operators' networks, as cost is an important factor for students. Another avenue of cost-reduction research would be to exploit WLAN technology (IEEE 802.11b/g) as educational institutions are increasingly offering such wireless infrastructure. Another ongoing project is the development is a real-time question classification unit that will allow a large number of simultaneous questions to be automatically sorted into related categories in real-time. This will simplify the job of responding to a large number of students' in-class queries for the teacher as many questions are related or similar. There are also unresolved issues regarding the user interaction design (ergonomic aspects) – for example, efficient text entry [26].

## Conclusions

An experimental "proof-of-concept" system - the GameShow – has been introduced. The system is designed with the objective of strengthening the student-teacher interaction during large class lectures by means of commodity mobile phones. The implementation of this or similar systems may change our view on mobile phones and their potential as pedagogical aids. Further development of the technology and its infrastructure along with further testing with regard to its pedagogical aspects may in the future give academia a new learning and teaching aid. The present system is available for free download from the Oslo University College website [19].

## Acknowledgements

The authors are grateful to Kjetil Haraldstad and Daniel Rustad for implementing the system, Tore Øfsdahl for providing the lecture theatre photograph, and the seven anonymous reviewers for their helpful comments and suggestions for this manuscript.

## References

1. Srivastava, M., Muntz, R. and Potkonjak, M., "Smart Kindergarten: Sensor-Based Wireless Networks for Smart Developmental Problem-solving Enviroments," Proceedings of the Seventh Annual International Conference on Mobile Computing and Networking, ACM Press, 2001, pp. 132-138.

2. Campbell, A.B. and Pargas, R.P., "Departmental Concerns: Laptops in the Classroom," Proceedings of the 34th Technical Symposium on Computer Science Education, ACM Press, 2003, pp. 98-102.

3. Cisco Systems, "Carolina Using Wireless with New Freshman Laptop Requirement to Transform Learning," January 24, 2001. http://newsroom.cisco.com/dlls/fsp-nisapi5a77.html.

4. Olsen, F., "The Wireless Revolution," The Chronicle of Higher Education, October 9, 2002, http://chronicle.com/free/2000/10/2000100901t.htm.

5. Olsen, F., "Carnegie Mellon Works to Make Computers Invisible and Pervasive," The Chronicle of Higher Education, October 12, 2000, http://chronicle.com/free/2000/10/2000101202t.htm.

6. Sharples, M., Corlett, D. and Westmancott, O., "The Design and Implementation of a Mobile Learning Resource," Personal and Ubiquitous Computing, Springer Verlag, Vol. 6, No. 3, 2002, pp. 220-234.

7. Najjar, L.J., Thompson, C. and Ockerman, J.J., "Using a Wearable Computer for Continuous Learning and Support," Mobile Networks and Applications, ACM Press, Vol. 4, No. 1, 1999, pp. 69-74.

8. Grant, W.C, "Wireless Coyote: A Computer-Supported Field Trip," Communications of the ACM, Vol. 36, No. 5, 1993, pp. 57-59.

9. Liu, T-Z, Wang, H-Y, Liang, J-K, Chan, T-W and Jiang, J-C, "Applying Wireless Technology to Build a Highly Interactive Learning Environment," Proceedings of the IEEE International Workshop on Wireless and Mobile Technologies in Education (WMTE'02), 2002.

10. Jerez, H., Abdullah, C., Christodoulou, C., Reenen, J. V. and Jordan, R, "Wireless Enabling Education and Research Activities," Proceedings of the 2003 International Conference on Engineering Education (ICEE-2003), Ref. No. 3301, Valencia, Spain, 2003.

11. Smørdal, O., Gregory, J. and Langseth, K. J., "PDAs in Medical Education and Practice," Proceedings of the IEEE International Workshop on Wireless and Mobile Technologies in Education (WMTE'02), 2002.

12. Chang, C-Y and Sheu, J-P, "Design and Implementation of Ad-Hoc Classroom and eSchoolbag System for Ubiquitous Learning," Proceedings of the IEEE International Workshop on Wireless and Mobile Technologies in Education (WMTE'02), 2002.

13. Holmes, A. and Smith, K.J., "Do Mobile and Wireless Technologies Add Value to Higher Education," Proceedings of the ASEE/IEEE Frontiers of Education Conference, Boston, Massachuets, USA, November 2002, T3F-1...T3F-4.

14. Shotsberger, P.G., Vetter R., "Teaching and Learning in the Wireless Classroom," IEEE Computer, March 2001, pp. 110-111.

15. Shields, J. and Poftak, A. "A Report Card on Handheld Computing," Technology and Learning, Vol. 22, No. 7, 2002, pp. 24-36.

16. Nybo, V., "Will Handheld Computers Work in the Classroom?" Syllabus, November, 2001, pp. 36-37.

17. Griffioen, J., Seales, W. and Lumpp, J., "Teaching in Real-time Wireless Classrooms," Journal of Engineering Education, October 1999, pp. 397-402.

18. Thunhelle, T., Nydahl, A. and Sollie, R.S., "Trådløs Kommunikasjon i Undervisning," Gjøvik University College, http://hovedprosjekter.hig.no/v2003/data/gruppe5/index.html, 2003.

19. Haraldstad, K., and Rustad, D., "Mobile Læringstjenester," Oslo University College, http://student.iu.hio.no/hovedprosjekter/2002/data/33/, 2002.

20. Bankham, J. and Elender, F., "Applying Person-Centered Principles to Teaching Large Classes," British Journal of Guidance and Counselling, Vol. 23, No. 2, 1995.

21. Felder, R.M., "Beating the Numbers Game: Effective Teaching in Large Classes," Proceedings of ASEE Annual Conference, Milwaukee, June 1997.

22. Cashin, W.E., "Improving Lectures," IDEA Paper 14, Kansas State University, 1985.

23. Glass, G. V. and Smith, M. K., "Meta-Analysis of Research on Class Size and Achievement," Educational Evaluation and Policy Analysis, Vol. 1, 1979, pp. 2-16.

24. Goldstein, G., & Benassi, V., "Students' Perceptions of Excellent Lecturers and Discussion Leaders," Journal on Excellence in College Teaching, Vol. 7, No. 2, 1996, pp. 81-97.

25. Dede, C., "Six Challenges for Educational Technology," *Learning with Technology,* Dede, C. (ed.), ASCD Yearbook, 1998.

26. Sandnes, F.E., Thorkildssen, H.W., Arvei, A. and Buverud, J.O., "Techniques for Fast and Easy Mobile Text-Entry with Three-Keys," Proceedings of HCISS'37 Hawaiian International Conference on System Science – Minitrack on Distributed and Mobile Information Systems (CD-ROM), IEEE Computer Society Press, January 2004.

27. Beaulieu, M., *Wireless Internet: Applications and Architecture: Building Professional Wireless Applications Worldwide,* Addison Wesley, 2002.

28. Burkhardt, J., Henn, H., Hepper, S., Rintdoorff, K. and Schak, T., *Pervasive Computing - Technology and Architecture of Mobile Internet Applications,* Addison Wesley, 2002.

29. Muchow, J.W., *Core J2ME Technology & MIDP,* Prentice Hall, 2002.

30. Sandnes, F.E, "EazyTagz - An Environment for Building Powerful Interactive Teaching Portals for Non-programmers," Proceedings of the 2002 International Conference on Engineering Education (ICEE-2002), Paper 0211, Manchester, United Kingdom, August, 2002.

31. Sandnes, F.E., *Modern Application Development: Thin Clients and Fat Servers,* Tapir Academic Press, Tronheim, Norway, 2002.

**Frode Eika Sandnes** received a B.Sc. in computer science from the University of Newcastle upon Tyne, U.K. in 1993 and a Ph.D. in computer science from the University of Reading, U.K. in 1997. Dr. Sandnes is currently an associate professor in the department of Computer Science at Oslo University College. Dr Sandnes' research interests include parallel processing, error-correction, mobile user interfaces and university-level pedagogy.

**Olav Talberg** has an MA degree in English (1971) from the University of Bergen, Norway. Talberg has written a series of textbooks on Technical English for upper secondary schools and colleges. Talberg is currently an assistant professor in the department of Computer Science at Oslo University College and is currently doing a research project in collaborative authoring at the engineering faculty. His main interests are within the pedagogical implementation of ICTs in various learning environments.

# Chapter 14

## Teaching a Large Class Using an Adaptive Technique

YONG LIAN
*National University of Singapore, Department of Electrical and Computer Engineering, 10 Kent Ridge Crescent, Singapore 119260.*
*E-mail: eleliany@nus.edu.sg*

*This chapter presents an innovative teaching method that effectively integrates information technology (IT) into teaching and learning. The new method uses an adaptive technique to trace learners' performance during and after a lecture so that the educator is able to quickly adjust course contents to tailor them to the needs of the learners. The success of the method is based on two feedback loops established between learners and educator. The in-class feedback is created by a SMS (Short Message Service) message-based quiz system and the after-class feedback is realized by an Educator-to-Learner Portal, e-Learning Hub [1], which provides information on learners' progress after a lecture. The intelligence of the e-Learning Hub greatly reduces the workload of educators when dealing with a large number of learners. The new method brings a new learning experience to students in a learner-centered environment.*

### INTRODUCTION

Universities around the world always encourage developing innovative teaching methods for higher education. An area that is of concern to many institutions is the development of innovative teaching methods for large classes, driven by increasing enrollments. Teaching an overcrowded class is a very challenging task for educators due to many constraints imposed on such classes. In Singapore, a class of more than 100 students is considered a large class. According to a survey done in Australia [2], some large classes in Australia have more than 1000 students. The question that an educator faces is how to create a learner-centered environment and promote active learning among all students when there are so many students to deal with. The simplest answer is to break a large class into small classes. This is quite difficult due to economic constraints. The real answer lies in supporting a serious attempt to examine alternative solutions in the face of economic realities.

Over the years, much effort has been made to understand the difficulties faced by educators in teaching large classes. The challenges include encouraging class attendance; delivering well-balanced course contents to maintain students' interest on the subject; understanding the students well enough to create interactive classes; identifying and making time to help students who are at risk; and getting/giving feedback and managing student inquiries. Many methods [3-6] have been suggested to address these issues. With the rapid development in communication and information technology, especially the Internet, it is possible to address these difficulties with the help of various existing technologies.

The proper use of communication and information technology not only enhances student-learning experience, but also maximizes student productivity. There are many existing web-based teaching methods [7-18]. Among them, the adaptive teaching method [16-18] is very interesting as it provides a mechanism to personalize teaching for each student. The advantage of adaptive teaching is its ability to change the course contents according to a student's needs, and is very useful for teaching engineering students.

Engineering students learn better if the course contents match the knowledge level of the individual student. A large number of mathematical equations are normally used to model real world problems, and students need to put in some effort to gain an appreciation of the underlying concepts. Although visualizing concepts using multimedia technology can help students to understand certain concepts, the ability of an educator to deliver the right amount of information to students is the key to successful teaching and learning. In a large class, an educator will be able to teach more effectively if he knows the difficulties confronting the students. To this end, feedback loops must be in place to keep track of students' progress during and after each lecture.

In this chapter, we present an adaptive teaching method that utilizes an intelligent learning portal to facilitate learning and to maximize the effectiveness of teaching a large class. This method uses a teaching and learning strategy comprised of two elements: in-class feedback and after-class feedback. The in-class feedback loop creates an active learning environment within a classroom and the after-class feedback loop promotes other learning activities, especially knowledge transfer. These two elements allow the educator to rapidly adjust course contents and to make classroom activities more efficient and more closely tuned to students' needs.

## The Adaptive Teaching Method

Adaptive teaching is a fairly new technique that utilizes the most recent information technology to enhance the learning experience. It may be the optimal instruction method that effectively meets the individual needs of the students, targeting students' different aptitude and traits, including intellectual abilities, personalities, and cognitive styles of learning [16]. It addresses these differences by implementing diagnostic and prescriptive instructional systems designed to make learning more effective, efficient and meaningful. The adaptive technique is widely used in control and digital signal processing systems to enhance performance of the systems. An example of such an application is adaptive filter, which improves system performance by adjusting the governing coefficients constantly to minimize the difference between the desired output and actual results.

Such an approach is very effective for performance enhancement in a dynamic system. The general principle of the adaptive technique is to monitor the system closely and to adjust the system whenever there is a need to keep the system in the best condition. This principle is equally applicable to the teaching and learning process in order to improve the efficiency of the process and the quality of the results.

The adaptive teaching method exploits interactions between learners and educator. Two feedback loops, i.e. an in-class and an after-class loop, are established between educator and learner to facilitate such interactions. The role of feedback in any learning environment is clearly important. It provides two main functions; first, it helps the educator to know what difficulties students may face in their learning process. With such information, the educator will be in a good position to help students overcome their difficulties. Second, it allows students to discover their weaknesses, and to become engaged in the learning process.

To implement the method, an educator needs to prepare two sets of carefully constructed assignments, one for an in-class test and another for after-class activities. The aim of the in-class quizzes is to test students' understanding during a lecture while the after-class activities engage the student in other learning activities. The original lecture notes have to be adjusted for use in the adaptive teaching method. For example, additional materials for difficult concepts can be included as hidden slides in a PowerPoint file. Such additional slides will be covered only if the in-class quiz indicates that more explanations are needed for a difficult concept. Other materials such as brief review of prerequisite and mathematical formulas are very helpful for classes with diverse backgrounds.

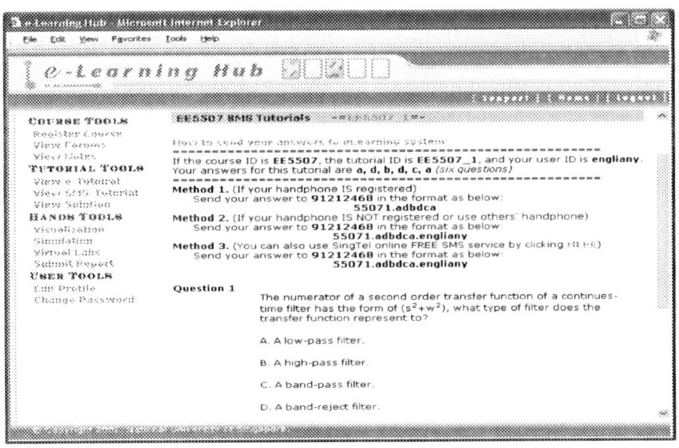


FIGURE 1

AN EXAMPLE OF THE SMS QUIZ

During a lecture, an educator gives students a quiz whenever there is a need to check their understanding on a given topic. Students answer the quiz by Short Message Service (SMS) available to mobile phone users. Students send the SMS messages to an instant feedback system that automatically grades students' answers and generates a set of statistical bar charts for viewing by the educator. From the bar charts, the educator is able to identify the students' difficulties immediately and adjust the course contents accordingly to suit students' needs. The details of the instant feedback system will be presented in the next section. Figures 1, 2, and 3 show examples of a SMS quiz and the reports generated by the feedback system, respectively. The quiz in Figure 1 was used at the beginning of one of the postgraduate courses to test students' background on analog filters. The purpose of the quiz was to find out whether filter fundamentals needed to be reviewed before new filter design techniques were introduced, since the students came from different universities. Although they met the prerequisites of the course, they could have very different backgrounds on filter design. Figure 2 indicates how many students entered the correct answers. It was very surprising to find out that only 11% of the students were properly prepared for the very basic question on filter transfer function. Figure 3 revealed more details on students' inputs to the question. From the figures, the educator may conclude that the majority of the students did not have a clear conception on the relationship between a filter and its transfer function. He recalled several hidden slides from the PowerPoint file and covered this topic. Reviewing additional background information allowed the students to more easily understand the new materials.

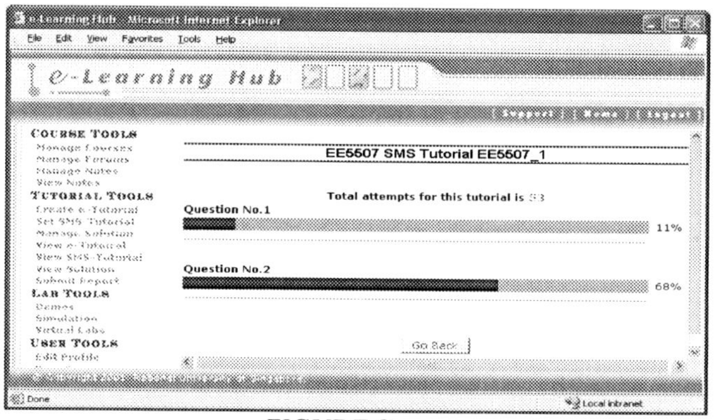


FIGURE 2
BAR CHART SHOWING CORRECT ANSWERS FROM STUDENTS

The in-class feedback loop serves the purpose of information collection that helps the educator to address students' immediate concerns. It has three advantages. First, it provides a clear picture on where students stand. Oftentimes we teach a course based on our experience and use previous batches of students as a reference. This gives us a rough idea on students' learning patterns, but a new batch of students may learn differently. We found that students from computer engineering learn quite differently from electrical engineering students, just as junior and senior students are different. With increased diversity of the student population, not only in race and culture, but also in age, our past experience may not work well. The in-class feedback loop helps the educator to calibrate the students more accurately, so (s)he can address the problems instantaneously. Second, it enables students to know their problems immediately. Often, students seem to understand a new concept, but they are at a loss when asked to apply it. Third, the system acts as a tracking mechanism to monitor students' performance. Each time a student responds to a quiz, the system creates a record for him. At any time, the system is able to generate reports for each student, which includes the performance of the student during and after each class. The educator can set a threshold to trigger the system to alert him on weak students.

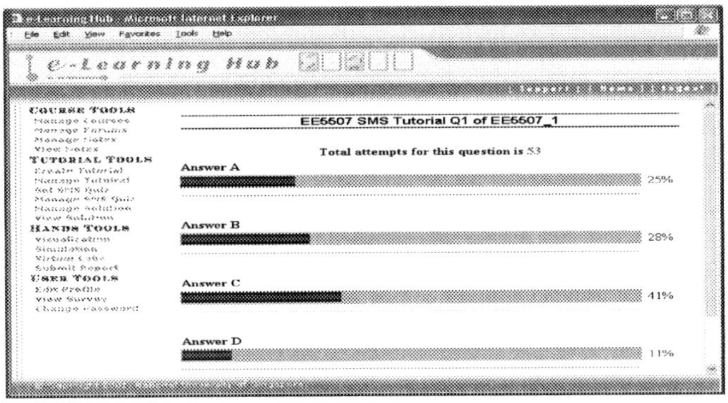


FIGURE 3

THE STATISTICAL BAR CHART SHOWING DETAILS OF THE STUDENTS' INPUTS

Very often in a large class we do not give students something to do during a two-hour lecture session, except note taking. To facilitate active learning, we must get students involved in the class. In-class feedback is one of the learning activities that helps student to reflect on what they have learned. Our experience shows that despite some initial hesitation among students, participation level increases very rapidly.

After the lecture, students will log on to an adaptive teaching web site to complete other learning activities, such as going through online tutorials, online virtual experiments, and viewing multimedia demonstrations, and submit their results two to three hours before the next lecture. Based on the statistical reports, the educator adjusts the contents and activities for the next lecture. The problems that surface during the after-class activities are addressed before new contents are presented. This will maximize the gain for students. Figures 4(a) and 4(b) show an example of one of the online tutorial questions with symbolic answers. Figure 4(a) is the workspace for the student and Figure 4(b) shows the automatic diagnosis provided by the *e-Learning Hub* (see below). It is obvious that adaptive teaching will bring the benefits to both learner and educator. Students come to class prepared and already engaged with the material as the after-class feedback loop helps them to identify their weakness. Furthermore, the educator knows exactly where the students are and what problems they have.

The after-class feedback loop plays another important role in the learning process. If we classify a learning process into six levels [19], i.e. knowledge, comprehension, application, analysis, synthesis, and evaluation, then the instant feedback loop addresses the issues in the first two learning levels. The after-class feedback loop extends learning to the next two levels, i.e., application and analysis. This is done through an enhanced intelligent Educator-to-Learner portal [1,20], *e-Learning Hub*. The *e-Learning Hub* solves two major problems faced by educators: tracking of students' performance and knowledge transfer. Since adaptive teaching requires an educator to know students' progress between lectures, it puts a lot of pressure on the educator as he has to analyze all feedback in a very short time frame, normally a couple of hours before the lecture. The task becomes mission-impossible if a class involves several hundreds of students. One of the ways to solve this problem is to limit tutorial/assignment to multiple-choice questions (MCQs) or numerical-answer questions so that a computer program can be used to mark the solutions and generate the statistical data on time. Unfortunately, this method is not quite suitable for many engineering courses, especially for advanced undergraduate courses and graduate courses. This is because the tutorial/assignment in advanced engineering courses tends to use symbolic or graphics-based answers. With the *e-Learning Hub*, the symbolic and simple graphic answers can be verified automatically by the system. The performance of the students are recorded by the *e-Learning Hub* and presented to the educator at any time. The workload of educators is thus greatly reduced.

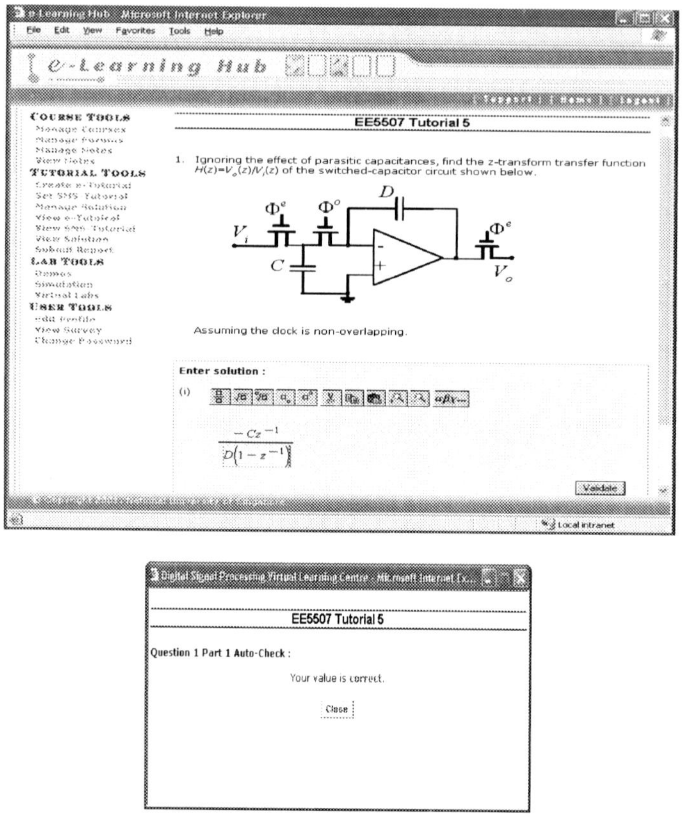


FIGURE 4

INTELLIGENT TUTORIAL SYSTEM, (TOP) ANSWER ENTERING INTERFACE, (BOTTOM) AUTO-DIAGNOSIS WINDOW

Knowledge transfer is another important element in engineering education. It not only trains students on application and analysis skills but also stimulates students' creativity. While a single sentence and mathematical answer is always possible for a specific tutorial question, it is highly desirable for students to carry out a reasoned simulation before answering the question. MATLAB is a tool that is capable of such a simulation for many electrical engineering topics, but the learning curve involved in understanding the tool can distract a student from adopting it. To give students more hands-on experience and to cultivate students' interest in a subject, a web-based virtual simulation tool that allows students to access it from anywhere at any time is included in *e-Learning Hub*. An example is shown in Figure 5. This example allows students to explore the detailed frequency responses of frequency-response masking filters. In a digital signal processing course, the ability to develop digital signal processing algorithms is important for students as they go deeper into the world of digital signal processing. A web-based virtual laboratory was developed for this purpose as shown in Figure 6. This lab trains students in developing digital signal processing algorithms.

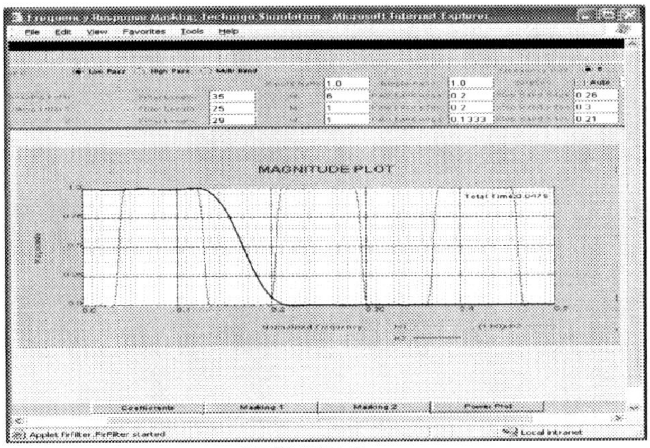


FIGURE 5

A VIRTUAL SIMULATION FOR FREQUENCY-RESPONSE MASKING FILTER DESIGNS

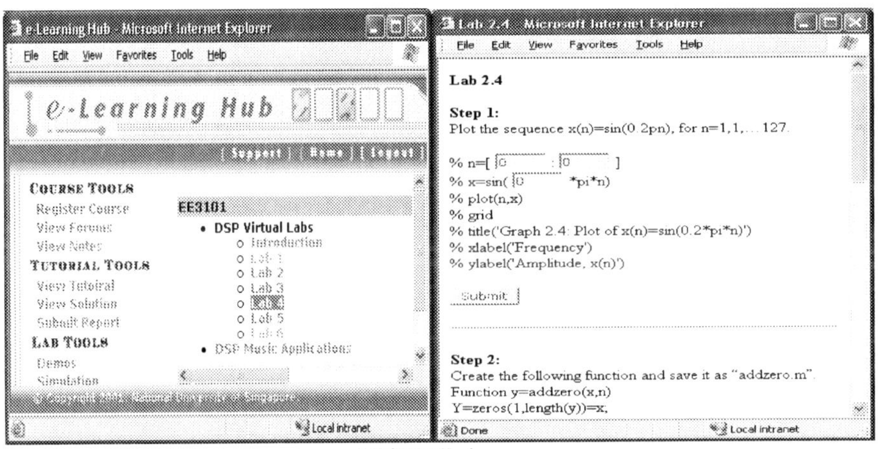


FIGURE 6

VIRTUAL LAB USER INTERFACE AND AN EXAMPLE OF GUIDED LAB

Until now, we have presented the essential part of the adaptive teaching method. To make it successful in a large class, a supporting tool is absolutely necessary. An Educator-to-Learner portal is developed for this purpose.

## THE SUPPORTING TOOLS

As stated previously, two types of feedback loops are needed in the adaptive teaching method: an in-class and an after-class feedback loop. In-class feedback is achieved by using SMS provided by mobile phone. The selection of SMS as an instant feedback tool has two advantages. Firstly, it is the most cost effective way to gather instant feedback and is easy to integrate with a web-based system. The service is highly reliable. Other rapid feedback systems require heavy investment on hardware, such as voting systems.

These systems lack flexibility as they are difficult to move around once installed. Although a wireless-based voting system provides a flexible solution, its cost is too high. With the increasing popularity of the mobile phone, we found that more than half of our students bring mobile phone to the class. This makes SMS-based instant feedback feasible. Secondly, students are very familiar with SMS service, and are very excited when such a service is applied to learning. Our observation shows that students are much more active in using SMS service than traditional in-class probing. The drawback of the SMS system is the requirement of a mobile phone for each student. To address this problem, small discussion groups of 2 – 4 students are formed at the beginning of the course with each group having at least one mobile phone. A think-pair-share approach [21] is adopted when an in-class quiz is conducted. Students work on the quiz individually and then pair up to compare and improve their answers before they submit a final one to the SMS system. Such an approach allows every student to participate in the SMS quiz. The instant feedback system is able to publish two types of quizzes: multiple choice and numeric answer quizzes. User interfaces for quizzes are designed for user-friendliness and the What-You-See-Is-What-You-Get effect. To reduce the learning curve, the quiz creation interface adopted is similar to the text editor used in word processors, as shown in Figure 7.

The after-class feedback loop is established through the use of an intelligent Educator-to-Learner portal, *e-Learning Hub*. It contains the following tools:

- multimedia demonstration tools for difficult concepts;
- visualization tools for exploring difficult concepts;
- an intelligent tutorial system that is capable of grading students' answers automatically. Currently it supports four types of questions: multiple-choice questions, numeric questions, symbolic questions, and questions with simple graphic answer. For symbolic answers, the system is capable of telling whether a student's answer is correct or wrong if it is in a format different from the standard answer, e.g. if the standard answer to a question is $\frac{s^2-2s+1}{s^2+2s+1}$, the system will indicate that the following answers are all correct: $\frac{1-2s+s^2}{1+2s+s^2}, \frac{1-s(2-s)}{1+2s+s^2}, \frac{(1-s)^2}{(1+s)^2}$;
- virtual simulation tools that allow students to carry out simulations at any time from anywhere;
- virtual lab tools that allow students to conduct experiments using a web browser.

Additional information about the *e-Learning Hub* can be found in [1] and [20].

## EVALUATION OF THE NEW METHOD

The adaptive teaching method has been used in teaching undergraduate and post-graduate students in the past two years. The courses covered are Digital Signal Processing Fundamentals for Level 3 full-time and part-time undergraduate students, and Integrated Circuits Design and Analysis for full-time and part-time post-graduate students. The class sizes range from 50 – 400 students. The full-time undergraduate students have very similar academic backgrounds and most of them are in the same age group as they enter the university directly from the junior colleges. Most part-time undergraduate students are local students with diverse academic backgrounds and ages.

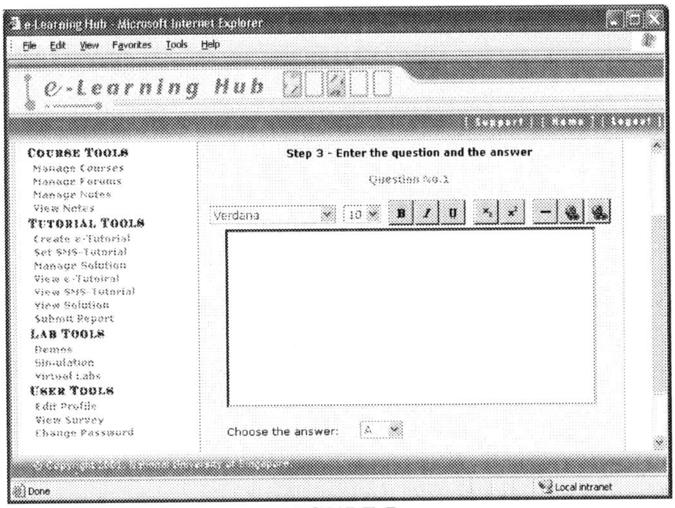


FIGURE 7
SMS QUIZ CREATION INTERFACE

All postgraduate students have great diversities in academic background, race, age, and culture. They range from Electrical Engineering graduates to Physics graduates, from company managers to fresh graduates, and from Master Degree holders to Diploma holders. Many of them are foreign students.

In evaluating the effectiveness of the new method, the following measurements were used: (1) the perceived cognitive outcomes; (2) the academic performance of the student; and (3) feedback from students. On cognition, the observation showed that students need less time to develop a full comprehension on the difficult concepts, and their thinking skill is enhanced. This could be the combined result of in-class and after-class activities. The in-class one removes learning obstacles for students and stimulates their interests in learning. The after-class one provides them with guided practice to develop their problem-solving skills. Sixty-seven percent of full-time students and 71% of part-time students surveyed agreed that the in-class and after-class feedback helped them understand and learn the module effectively. Sixty-two percent of full-time and 91% of part-time students agreed that the new method helped them think more deeply into the problems beyond what was taught in class. Eighty-seven percent of full-time and 93% of part-time students rate the new method between good and excellent. It is interesting to note that the responses from part-time students are more positive than those from full-time students. This is expected as part-time students are from very different backgrounds and they benefit most from the new method. It is not surprising that this group of students shows major improvement on their academic performance.

On academic performance, the improvement on grades is notable. A student's grade depends on results of the final examination and continuous assessments. Figure 8 shows the final examination results of one of the Master level courses for two different batches of students, where the x-axis is the students' grades and the y-axis is the number of students. The level of the examination questions are kept as close as possible. Comparing the right figure to the left, the improvement on students' marks after the introduction of the method is very encouraging.

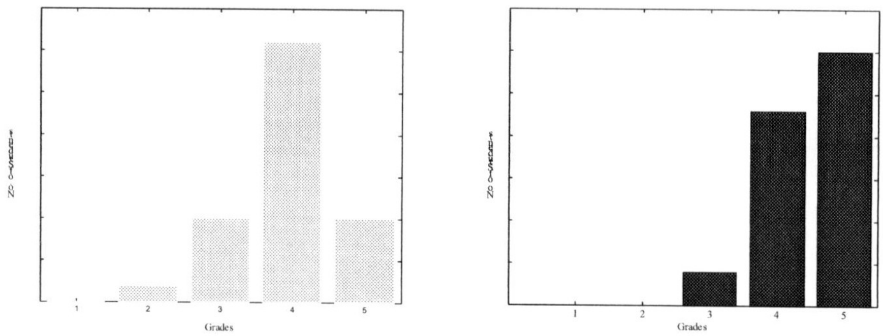


FIGURE 8

STUDENTS' MARKS BEFORE (LEFT) AND AFTER (RIGHT) THE INTRODUCTION OF THE METHOD

On student feedback, the following comments were taken from the university teaching evaluation system on modules taught using the adaptive teaching method.

1. "Helps me to understand concepts effectively and makes me more interested in the subject."
2. "Inculcates interest in students and effective teaching ability."
3. "Online tutorial allows students to do tutorials and spot mistakes in solutions immediately. Ability to bring ideas and concepts across clearly."
4. "Wonderful teaching style. Immediate feedback."
5. "Stirs the interest in me to study this difficult module."
6. "Wonderful teaching style, increases interest in this course."
7. "Interested, proactive approach to learning with the innovative methods."
8. "Explains very well, and concerned with students' understanding level."
9. "Very concerned about the student's progress."
10. "Makes good use of technology."

## CONCLUSIONS

We have presented an adaptive teaching method that improves the effectiveness of learning in large classes. The proposed method is based on two feedback loops that continuously monitor learners' progress during and after a lecture. In-class feedback is achieved using SMS provided by mobile phones and the after-class feedback is done through an enhanced Education-to-Learner portal, *e-Learning Hub*. With two feedback loops in place, the educator is able to deliver well-balanced course contents that match the students' knowledge level. Doing so makes learning more effective, efficient and meaningful. Although the initial target of the method is for large classes, it is equally applicable to any class size. For a small class, the instant feedback loop can be established using other forms such as student probing, think-pair-share, etc.

## Acknowledgements

The author gratefully acknowledge contributions made by his students, Y.H. Tan, K.S. Chua, C.L. Boh, and X.Q. Zhang, for the development of *e-Learning Hub*. This work was supported by the National University of Singapore Academic Research Fund R263-000-227-112.

## References

1. Lian, Y., and Tan, Y.H., "An Educator-to-Learner Portal for Just-in-Time Teaching," Second Symposium on Teaching and Learning in Higher Education, 2002, pp. 420-425.
2. Herbert, D., and Hannam, R., "Report on Survey of Large Class Teaching in Australia," http://www.tedi.uq.edu.au/largeclasses/pdfs/Resources_2ReportonSurvey.pdf, Teaching and Educational Development Institute, The University of Queensland, Australia.
3. Teaching Large Classes, Part V, Suggestion for Further Reading, http://www.uwo.ca/edo/tlc/lc_part5.html, Educational Development Office, The University of Western Ontario.
4. Johnson, R.N, Enerson, D.M. and Plank, K.M., "Selected Bibliography for Teaching Large Classes," http://www.psu.edu/celt/largeclass/lcbib.html, Center for Excellence in Learning and Teaching, The Pennsylvania State University.
5. Gibbs, G., and Jenkins, A., *Teaching Large Classes in Higher Education,* Kegan Paul, London, 1992.
6. Carbone, E., *Teaching Large Classes: Tools and Strategies,* SAGE Publications, 1998.
7. Lian, Y., "The Joy of Learning DSP in a Large Class," Proceeding of 2003 IEEE International Conference on Acoustics, Speech, and Signal Processing, Vol. 3, 2003, pp.753-756.
8. Gillet, D., "Towards Flexible Learning in Engineering," Chap. 11, Aung, W., et al. (ed.), *Innovations 2003: World Innovations in Engineering Education and Research,* Int. Network for Eng. Ed. and Res. (iNEER), Arlington, VA, 2003, pp. 95-102.
9. Woolf, B., and Poli, C., "Best Practice in Building Multimedia Tutors: Multimedia Instruction for Engineering Education," Chap. 17, Aung, W., et al. (ed.), *Innovations 2003: World Innovations in Engineering Education and Research,* Int. Network for Eng. Ed. and Res. (iNEER), Arlington, VA, 2003, pp. 159-174.
10. Tuschák, R., Vajk, I., Bars, R., Hetthéssy, J., Barta, A., and Charaf, H., "Teaching Basic Control Theory: A Novel Approach Using WEB-MatLab Based Teaching Materials," Chap. 19, Aung, W., et al. (ed.), *Innovations 2003: World Innovations in Engineering Education and Research,* Int. Network for Eng. Ed. and Res. (iNEER), Arlington, VA, 2003, pp. 185-194.
11. Daku, B., and Diefes-Dux, H., "An Effective System for Implementing Self-Learning, Online Instruction for Engineering Students," Chap. 31, Aung, W., et al. (ed.), *Innovations 2003: World Innovations in Engineering Education and Research,* Int. Network for Eng. Ed. and Res. (iNEER), Arlington, VA, 2003, pp. 307-318.
12. Lian, Y., "Learning Digital Signal Processing at Six Degrees of Freedom," Proceedings of Second Symposium on Teaching and Learning in Higher Education, 2002, pp. 444-449.

13. Novak, G.M., Patterson, E.T., Gavrin, A.D., and Christian, W., *Just-in-Time Teaching: Blending Active Learning with Web Technology,* Prentice Hall, 1999.

14. CoLos Main Entry Page [Internet]. http://www.colos.ec-lyon.fr

15. Shaffer, J., Hamaker, J., and Picone, J., "Visualization of Signal Processing Concepts," The Proceedings of 1998 International Conference on Acoustics, Speech, and Signal Processing, Vol. 3, 1998, pp. 1853-1856.

16. Srisethanil, C., "ITS-Engineering: Providing Adaptive Teaching in the Engineering Tutor," Frontiers in Education: Engineering Education for the 21st Century, November 1995.

17. Specht, M., and Oppermann, R., "ATS-Adaptive Teaching System a WWW-Based ITS," The Proceedings of Workshop Adaptivitt und Benutzermodellierung in Interaktiven Softwaresystemen, 1998.

18. Barbara, H.R., "Adaptive Learning Guides," The Proceedings of the IASTED Conference on Computers and Advanced Technology in Education, Banff, Canada, June, (2001).

19. Wills, S., McNaught, C., "Evaluation of Computer-Based Learning in Higher Education," Journal of Computing in Higher Education, Vol.7, No.2, 1996, pp. 106-128.

20. Lian, Y., and Hu, B., "An Intelligent Learning Portal for Distance Learning," The Proceedings of 2003 International Conference on Engineering Education (ICEE-2003), Valencia, Spain, 21-25 July 2003.

21. Lochhead, J., and Whimbey, A., "Teaching Analytical Reasoning through Thinking Aloud Pair Problem Solving," J.E. Stice (Ed.), *Developing Critical Thinking and Problem-Solving Abilities, New Directions for Teaching and Learning*, Jossey-Bass, 1987.

**Yong Lian** is an Associate Professor in the Department of Electrical and Computer Engineering at the National University of Singapore. His research interests include digital filter design, VLSI implementation of high-speed digital systems, Biomedical instrumentations, RF IC design, and Internet-based learning. Dr. Lian received the 1996 IEEE Circuits and Systems Society's Guillemin-Cauer Award. He currently serves as an Associate Editor for the IEEE Transactions on Circuits and Systems Part I and has been an Associate Editor for Circuits, Systems and Signal Processing since 2000. He was an Associate Editor of IEEE Transactions on Circuits and Systems Part II from 2002 to 2003, and was the Guest Editor of the "Special Issue on Frequency-Response Masking Technique and Its Applications in Circuits, Systems and Signal Processing" in 2003. Dr. Lian serves as Secretary of IEEE Circuits and Systems Society Biomedical Circuits and Systems Technical Committee and is a member of the Digital Signal Processing Technical Committee of the IEEE Circuits and Systems Society.

# Chapter 15

## An Online Microelectronics Device Characterization Laboratory with a Circuit-Like User Interface

JESÚS A. DEL ALAMO, VICTOR CHANG, JAMES HARDISON, DAVID ZYCH and LISA HUI
*Massachusetts Institute of Technology, Cambridge, MA 02139, USA.*
*E-mail: alamo@mit.edu*

*We have developed and deployed a new graphical user interface for an online microelectronics characterization laboratory that we call the MIT Microelectronics WebLab (or "WebLab" for short). WebLab allows the characterization of transistors and other microelectronic devices in real time through the Internet. The new graphical interface consists of a Java applet that uses the circuit language of electrical engineering to specify the experiment to be performed. The latest release of WebLab features a new remote management system designed to allow the efficient deployment of the system in multiple subjects with large numbers of students. Since its release in mid 2002, this version of WebLab has been used by nearly 700 students to complete assignments for credit in six different subjects.*

### INTRODUCTION

In the teaching of microelectronic device physics, hands-on characterization of transistors and other devices substantially enhances the educational experience. However, for a variety of logistical and economic reasons, conventional microelectronics courses are rarely able to include a laboratory component. A remote laboratory that allows the characterization of microelectronic devices without the need for the user to be in front of the experimental setup solves many of these logistical concerns while largely preserving the educational experience. The use of web technology allows the creation of such a laboratory while imposing minimal requirements on the remote user. Many institutions in different fields have explored this concept of an online laboratory [1-7].

Over the last few years, we have been experimenting at MIT with a system called the MIT Microelectronics WebLab, or WebLab for short. This system allows microelectronics device characterization through the web [8,9]. There are similar efforts throughout the world, many using different software solutions to achieve the same goal [10-15]. While a complete architectural discussion of the WebLab system is beyond the scope of this paper, general software tools have been used to create a system that is flexible, highly functional and educationally sound. Through WebLab, students can take current-voltage measurements on transistors and other microelectronics devices in real time from anywhere and at any time. WebLab was first deployed in 1998, and since then it has been used for graded course assignments in several graduate and undergraduate subjects at several different universities. The basic architecture of the system and its use in a variety of educational settings were reported in [7]. A prototype collaboration system for WebLab was described in [16].

In this work, we describe WebLab 5.0, released in mid-2002 and used in several educational settings during the academic year 2002-2003. WebLab 5.0 features two key improvements over previous versions of the system. First, it employs a graphical user interface (GUI) that uses the circuit language of electrical engineering to specify the experiment to be performed. This is much more intuitive than the textual, form-based user interface that was utilized in previous versions of WebLab. The text-based approach, though very similar in nature to the instrument's own interface, was found to be intimidating and to represent a significant barrier to first-time users. The new graphical interface improves usability by reducing clutter and hiding unnecessary details from the user.

The second major improvement in this latest release of WebLab is a new remote management system that has been designed to allow the efficient deployment of WebLab in multiple subjects with large numbers of students. Through this system, the WebLab administrator can manage user accounts, user groups, device specifications, and device access permissions, as well as examine a variety of system usage records. These are all important tasks that can become a significant administrative burden as the number of courses and the number of students that use the lab increase.

WebLab 5.0 was deployed mid-2002 and extensively utilized during the academic year 2002-2003. This paper describes the key improvements in WebLab 5.0 and summarizes the educational experiments that have been performed. We begin by providing some technical background on WebLab. Next, we describe the new graphical user interface, followed by presentation of the new remote management system. Finally, we summarize the most significant educational experiments performed with WebLab 5.0.

## BACKGROUND

At the heart of WebLab is an Agilent 4155B Semiconductor Parameter Analyzer, a state-of-the-art instrument for measuring the current-voltage characteristics of microelectronics devices and small circuits with up to eight terminals. The analyzer is connected to an Agilent E5250A Switching Matrix, a multiplexer which, in its current configuration, allows up to eight different devices to be connected to the system at any given time. Both the analyzer and the switching matrix are controlled by a GPIB (IEEE 488) interface card on the WebLab server computer, which is currently a Dell PowerEdge 4600 running the Windows 2000 Server operating system. The devices under test are mounted in a packaged form onto standard test fixtures that are connected to the switching matrix.

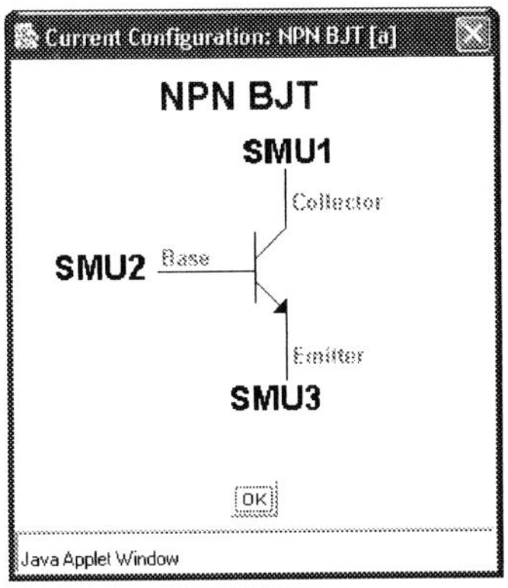


FIGURE 1
WEBLAB 4.2 CLASSICAL INTERFACE: DEVICE CONFIGURATION FRAME

The user interface for WebLab is a Java applet which duplicates the essential functionality of the analyzer's console, allowing the user to set up a measurement for one of the devices that is currently connected to the system (the necessary information about these devices is provided by the server when the applet loads). When the user is ready to execute a measurement, the applet sends the measurement specifications to the server. The server then sets the switching matrix to the appropriate device, instructs the analyzer to run the measurement, obtains the results from the analyzer, and finally transmits the data back to the applet, where they can be graphed and otherwise analyzed by the user. More details of the system's operation are given in [8,9].

Since its original release in 1988 WebLab, through its versions 1.0 through 4.x, has provided a textual, form-based user interface (now referred to as the "Classical" Interface) that closely mimics the front panel of the analyzer itself. This approach was based on our judgment that the designers of the instrument would have already put a great deal of thought into creating an optimal user interface for performing device characterization measurements.

The Classical Interface consists of three frames. The *device configuration frame*, shown in Figure 1, illustrates the connections between the ports of the analyzer and the terminals of the microelectronics device under test. The *measurement specification frame*, shown in Figure 2, allows the user to specify the voltages or currents to be applied to each terminal, and optionally to specify simple algebraic functions on the data which will be computed by the analyzer as the experiment is being performed. The *measurement results frame*, shown in Figure 3, displays a customizable graph of the data returned by the analyzer, and also allows the user to download a copy of this data for offline processing using an external spreadsheet or

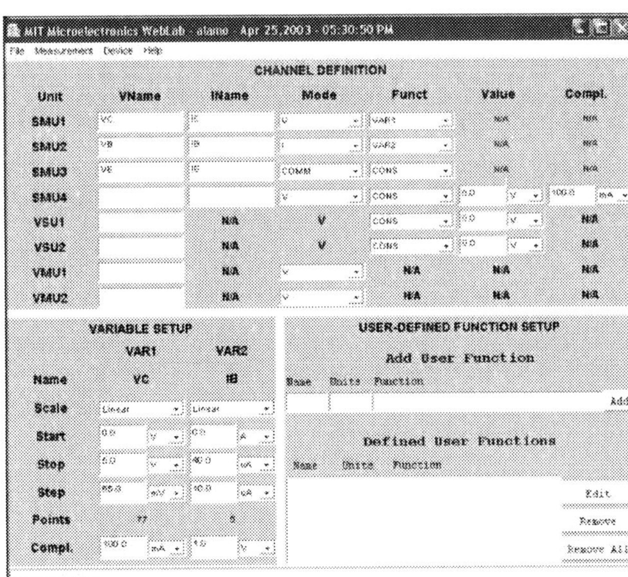


FIGURE 2
WEBLAB 4.2 CLASSICAL INTERFACE: MEASUREMENT SPECIFICATION FRAME

mathematics software application.

The advantage of the Classical Interface is that it is extremely simple, and consequently runs very reliably on any standard Java-enabled web browser. Also, the Classical version of the applet downloads quickly because the code archive is very small, about 58 KB. We have therefore continued to maintain the Classical applet in WebLab 5.0 to accommodate those users for whom compatibility issues and/or network bandwidth constraints make it difficult or impossible to use the new Graphical applet.

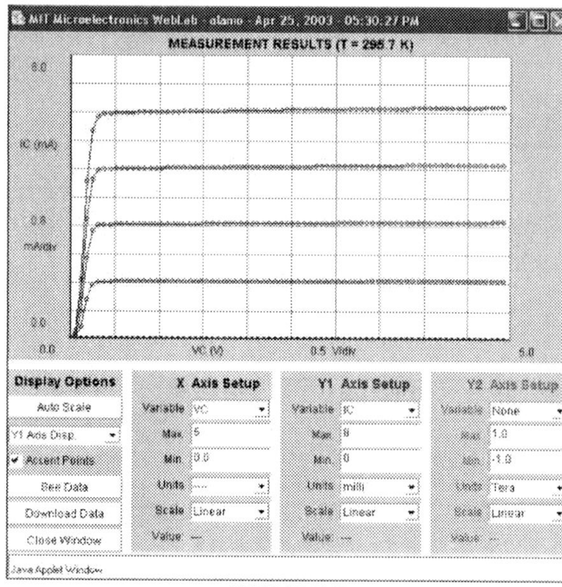


FIGURE 3
WEBLAB 4.2 CLASSICAL INTERFACE: MEASUREMENT RESULTS FRAME

However, the Classical Interface has two major problems. First, the text-based measurement specification frame is highly intimidating and not particularly intuitive. This has always represented a significant psychological barrier for first-time users of the system. Note that this effect is not limited to first-time users who are students; it also applies to teachers of microelectronics who are evaluating WebLab for possible adoption into their courses.

The second problem with the Classical Interface is the clutter created by the use of multiple frames. This becomes especially troublesome when considering the collaborative use of WebLab. A prototype collaboration system for WebLab was presented in [16]. The key concept behind this system is the “dual-domain interface”, in which each user is given essentially two copies of a standard WebLab interface: an individual domain which belongs to that user alone, and a group domain that is controlled by only one member of the group at a time but seen by all members of the group. In this way, individual and group work can proceed simultaneously but separately. However, a dual-domain interface based on the Classical applet requires a large number of frames: two measurement specification frames (one for each domain), two measurement results frames, and at least one additional frame for listing the composition of the group and transferring control of the group domain between members. Chatting among group members (either as part of the WebLab client, or using a different third-party messaging application) requires yet another frame. At best, having all of these frames open at once results in a great deal of clutter and confusion; at worst, they may not fit onto the user’s display at all.

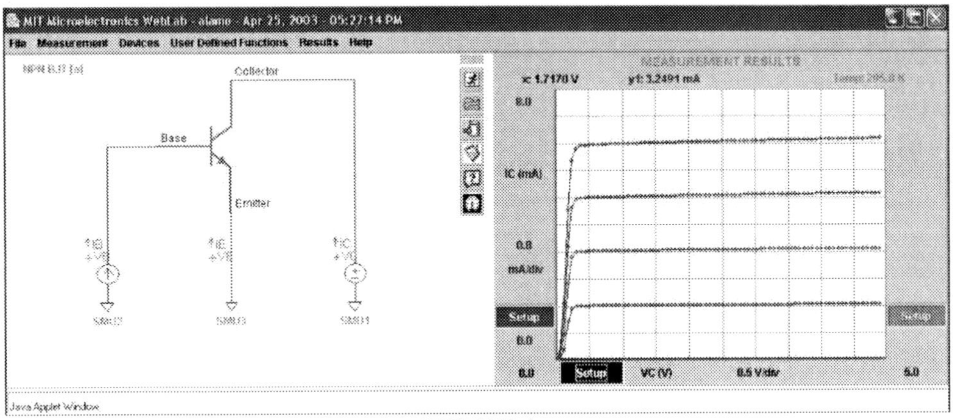


FIGURE 4

WEBLAB 5.0 GRAPHICAL INTERFACE: MAIN WINDOW

## A GRAPHICAL USER INTERFACE FOR WEBLAB 5.0

The new Graphical Interface featured in WebLab 5.0, shown in Figure 4, was designed to address the deficiencies of the Classical Interface. It uses the standard circuit language of electrical engineering to represent the experiment setup, making the user experience much more intuitive. It also reduces desktop clutter by combining the measurement specification panel and the measurement results panel into a single window. This unified interface greatly simplifies the look and feel of the dual-domain collaboration system [17].

The measurement specification panel of WebLab 5.0, on the left in Figure 4, displays a circuit schematic that illustrates the connections between the device under test and the various ports of the analyzer. As each analyzer port is configured, its icon in the schematic changes to reflect the port's new function and behavior. For example, Figure 4 shows an experiment setup for a bipolar transistor. The port connected to the emitter is programmed to be a direct ground connection, the port connected to the base is a variable current source, and the port connected to the collector is a variable voltage source. The WebLab server provides a bitmap image for each device attached to the system and specifies the pixel coordinates of the terminals; the remaining pieces of the schematic diagram (the terminal labels, port icons, and connecting wires) are assembled dynamically by the interface.

To make the user experience less intimidating, many features of the interface have been moved out of the main window and into dialog boxes and menu items. The analyzer ports are configured through dialogs like the one shown in Figure 5, which the user opens by clicking on a port icon in the schematic diagram. Similar dialog boxes are used to configure each axis of the measurement results graph. Several other features, such as user-defined functions, viewing data, and downloading data, are now accessed through the applet's menu bar. The net effect of all of these changes is that the main window of the interface is much less cluttered than before; instead of being confronted simultaneously with every available piece of detailed information (as was the case in the Classical Interface), the user is initially presented with an intuitive visual representation of the measurement. All of the details are still present in the interface, but the user can examine them a few at a time by opening and closing the various dialog boxes.

The Graphical Interface does have a few disadvantages. The applet code for the Graphical Interface makes substantial use of the Swing toolkit, a sophisticated library of GUI components for Java that is much more versatile and powerful than the older Abstract Window Toolkit (AWT). However, because the Swing toolkit is more recent, it is not directly supported by the Java virtual machines in standard Java-enabled web browsers. WebLab users must therefore download and install the Java Plug-in from Sun in order to use the Graphical applet. Unfortunately, the Java Plug-in itself is not yet free of bugs, and certain operations in the Graphical applet that run flawlessly on one client machine may cause the applet to behave erratically or even crash when run on a second client machine that uses a different version of the Java Plug-in (or even the same version of the Java Plug-in but for a different operating system). Debugging the Graphical applet has therefore been a complicated, ongoing process. While the majority of our students are able to use the system without ever encountering technical problems, each new WebLab assignment usually results in at least one bug report that we are eventually able to trace back to an obscure inconsistency in the behavior of the Java Plug-in. Additionally, there are some platforms (Windows 95 and 98, and Mac OS 8.x and 9.x) on which we have never been able to successfully run the Graphical applet at all due to inadequate Java support. Finally, because the Graphical Interface is substantially more complicated than the Classical Interface, the code archive is about 261 KB and the applet takes longer to download and longer to initialize. However, for users with compatible system configurations and enough network bandwidth to support the longer load time, the intuitive feel of the new Graphical Interface has dramatically improved the overall quality of the user experience.

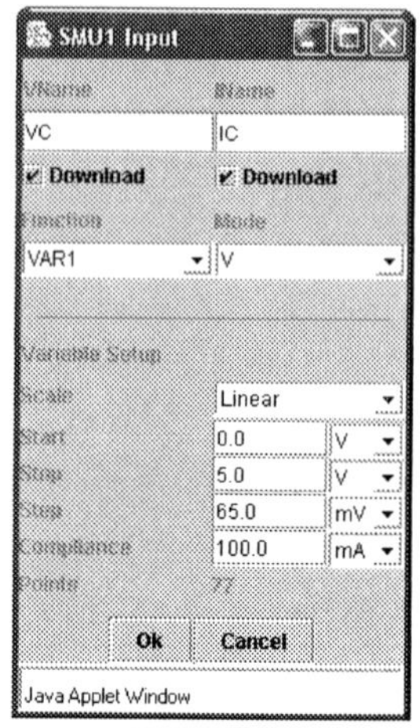


FIGURE 5
WEBLAB 5.0 GRAPHICAL INTERFACE: PORT CONFIGURATION DIALOG

## A REMOTE MANAGEMENT SYSTEM FOR WEBLAB 5.0

The second significant improvement in WebLab 5.0 is the introduction of a remote management system. Prior to the v. 5.0 release, most of the administrative tasks had to be performed while at the lab. This meant that a common task, such as adding a user account, could only be performed by a lab manager who was physically present in the lab. As the number of WebLab users and the overall demand on the system increased, the need for a more efficient method of system administration also grew. By taking advantage of Microsoft Active Server Pages (ASP) technology and a dedicated SQL database, lab managers using WebLab 5.0 are able to perform the bulk of their tasks over the Internet.

Technologically speaking, ASP is an established way of incorporating functional logic into online forms or documents. In particular, ASP allows for this functional logic to utilize server components such as a database. In this way, information can be programmatically stored and retrieved by the functions written into a set of online forms. The technology involved is rather simple, but the value added is significant. In the case of

WebLab, it allows administrative tasks that were previously time-consuming and inconvenient to become efficient, highly available and, in some cases, automatic.

For instance, the new management system streamlines common user management tasks. In the current implementation, users are responsible for registering their own accounts. These registrations are then either approved or rejected by a lab manager at the WebLab site. Realizing that the lab manager and user may not be logged onto the site at the same time, e-mail alerts are sent to the appropriate groups at each step.

From this same web interface, a lab manager may add a new user to a custom user group and grant the user access to resources on the system, such as physical devices or software functions. Again, all of this functionality is available to a lab manager at any Internet-enabled workstation regardless of time or location. The end result of this is that students are able to log onto the system and perform experiments within hours or even minutes of requesting their accounts where, prior to version 5.0, it could take days.

From the WebLab site, lab managers may also edit the resources available on the system as well as the permissions associated with those resources. Of particular import is that the representations of the physical devices connected to the lab can be edited through this interface. In fact, the only portion of the lab device setup process that cannot be performed via the Internet is physically connecting the device to the system. This means that there is minimal lab setup work necessary to accommodate a new class or even multiple classes simultaneously.

Finally, a searchable records system allows lab managers to track user logons and job executions. This, coupled with new bug reporting and system notice interfaces, allows problem tracking and resolution to occur much faster than before. By adding an easy means of communication between lab users and managers, students who are having problems may report those problems to a manager, who then can view system logs and correspond with the user in order to isolate and solve the problem. Also, the resolution can then be quickly posted to the site, making the information available to all users of the system.

The end result of the new functionality is that the system experiences less downtime because the resistance to addressing potential issues is lowered. This means that the technology involved in the lab becomes more transparent to the student and they are able to focus their experience more on the educational aspect of the system.

## EDUCATIONAL EXPERIMENTS USING WEBLAB 5.0

Since its deployment in mid 2002, WebLab 5.0 has been used in several subjects for credit-bearing assignments. In the Fall of 2002, WebLab was used in three courses: *6.012 "Microelectronic Devices and Circuits"*, a junior-level course at MIT with about 120 students; *6.720 "Integrated Microelectronics Devices*", a graduate-level subject at MIT with about 50 students; and *SMA5104 "Fundamentals of Semiconductor Device Physics"*, a graduate-level subject in the Singapore-MIT Alliance with about 20 students.

In the Spring of 2003, WebLab was used again in another edition of 6.012 at MIT (with about 120 undergraduate students) and also in FFY171/ETI145 "Semiconductor Devices," a sophomore level subject at Chalmers University in Sweden, with 350 students. This has been the largest deployment of WebLab to date. All together, nearly 700 students have used WebLab 5.0 for lab projects for credit since its deployment.

Since the architecture of WebLab 5.0 is the same as that of earlier versions, the scaling behavior of WebLab should be unaffected. The Chalmers University experiment allowed us to explore the capabilities of the system beyond our previous peak [9]. On Feb 25, 2003, between noon and 1 PM EST, 134 characterization experiments were executed in a single hour (on average, that means a job every 27 seconds). This represents a 35% increase over our previous record.

In each of the above courses, students were also asked to complete a survey detailing their experiences using WebLab 5.0. This survey typically requests feedback regarding the usability and effectiveness of the system as an educational tool. The user feedback we have received suggests that the new Graphical Interface indeed provides for a pleasant and intuitive working environment. Here is the opinion of a 6.012 student in the Spring of 2003: "The WebLab interface worked flawlessly for me. The applet was very intuitive to use and I was able to complete the data collection portion of the assignment easily." Here is another: "I found WebLab to be very user friendly and a great tool. It beats walking all the way to the lab to perform these measurements."

## CONCLUSIONS

In conclusion, we have developed a new Graphical User Interface for an online microelectronics device characterization laboratory. The new interface uses the circuit language of electrical engineering to specify an experiment. Students report that the new interface is intuitive and feels familiar. On the negative side, the graphical nature of the interface requires a Java Plug-in which makes debugging more complicated. We have also introduced a new management system that allows remote user management and system management through the web. This has allowed us to support courses with as many as 350 students. WebLab 5.0 has now been used by nearly 700 students in different subjects from academic institutions on three continents.

## ACKNOWLEDGEMENT

This project is funded by Microsoft through iCampus, the MIT-Microsoft Alliance. The instruments used in WebLab were donated by Agilent Technologies.

## REFERENCES

1. Cesnik, C.E.S., and Radcliffe, R., "Mechanical Structures Interactive Lab," Proceedings of 2002 International Conference on Engineering Education, ICEE-2002, Manchester, UK, 2002.

2. Amaratunga, K. and Sudarshan, R., "A Virtual Laboratory for Real-Time Monitoring of Civil Engineering Infrastructure," Proceedings of 2002 International Conference on Engineering Education, ICEE-2002, Manchester, UK, 2002.

3. Biesenbach, R. and Pautzke, F., "Remote Control of Industrial Robots via Internet - Procedures and Techniques of Tele-Learning in the Virtual Laboratory Amberg," Proceedings of 2002 International Conference on Engineering Education, ICEE-2002, Manchester, UK, 2002.

4. Trevelyan, J., "Towards Cost Effective On-Line Laboratories," World Congress on Networked Learning in a Global Environment, Berlin, Germany, 2002.

5. Penarrocha, V. Rodrigo, "Virtual Laboratories: A Reference Model and Practical Experiences," World Congress on Networked Learning in a Global Environment, Berlin, Germany, 2002.

6. Ko, C.C., Chen, B.M., Chen, J., Zhuang Y., and Tan, K.C., "Development of a Web-Based Laboratory for Control Experiments on a Coupled Tank Apparatus," IEEE Transactions on Education, Vol. 44, No. 1, February 2001, pp. 76-86.

7. Henry, J., "Running Laboratory Experiments via the World Wide Web," ASEE Annual Conference, Session 3513, 1996. Available from National Instruments Academic Resources CD-ROM 2001 Edition.

8. del Alamo, J.A., Brooks, L., McLean, C., Hardison, J., Mishuris, G., Chang V., and Hui, L., "The MIT Microelectronics WebLab: a Web-Enabled Remote Laboratory for Microelectronic Device Characterization," World Congress on Networked Learning in a Global Environment, Berlin, Germany, 2002.

9. del Alamo, J.A., Hardison, J., Mishuris, G., Brooks, L., McLean, C., Chang V., and Hui, L., "Educational Experiments with an Online Microelectronics Laboratory," Proceedings of 2002 International Conference on Engineering Education, ICEE-2002, Manchester, UK, 2002.

10. Fjeldly, T.A., Strandman J.O., and Berntzen, R., "LAB-on-WEB – A Comprehensive Electronic Device Laboratory on a Chip Accessible Via Internet," Proceedings of 2002 International Conference on Engineering Education, ICEE-2002, Manchester, UK, 2002.

11. Shen, H., Xu, Z., Dalager, B., Kristiansen, V., Strom, O., Shur, M., Fjeldly, T., Lü, J.Q., and Ytterdal, T., "Conducting Laboratory Experiments over the Internet," IEEE Trans. on Education, August 1999.

12. Fjeldly, T.A. and Shur, M.S., Editors, *LAB-on-the WEB. Running Real Electronics Experiments via the Internet*, John Wiley and Sons, ISBN: 0-471-41375-5, 2003.

13. Gómez, F., Cervera, M. and Martínez, J., "A World Wide Web Based Architecture for the Implementation of a Virtual Laboratory," Proceedings of the 2000 Euromicro Conference, Vol. 2, Maasstricht, Netherlands, September 2000, pp 56-62.

14. Söderlund, A., Ingvarson, F., Lundgren, P. and Jeppson, K.O., "The Remote Laboratory – A New Complement in Engineering Education," Proceedings of 2002 International Conference on Engineering Education, ICEE-2002, Manchester, United Kingdom, 2002.

15. Cervera, M., Gómez, F. and Martínez, J., "Remote Measurement System using Internet," Proceedings of the 1999 WebNet Conference, Hawaii, October 1999.

16. Chang, V. and del Alamo, J.A., "Collaborative WebLab: Enabling Collaboration in an Online Laboratory." World Congress on Networked Learning in a Global Environment, Berlin (Germany), 2002.

17. Lin, Y., "A Collaboration System and a Graphical Interface for the MIT Microelectronics WebLab, " MIT Master of Engineering Thesis, May 2002.

**Jesús A. del Alamo** received the degree of Telecommunications Engineer from the Polytechnic University of Madrid in 1980, and the M.S. and Ph.D. degrees in Electrical Engineering from Stanford University in 1983 and 1985, respectively. From 1985 to 1988 he was a research engineer with NTT LSI Laboratories in Atsugi (Japan). Since 1988 he has been with the Department of Electrical Engineering and Computer Science of Massachusetts Institute of Technology where he is currently a Professor. His current research interests are on Gigahertz power transistors: Si LDMOS on SOI, RF power CMOS, GaAs PHEMTs and InP HEMTs. He is also active in online laboratories for science and engineering education.

**Victor Chang** graduated with a S.B. in Electrical Engineering and Computer Science from the Massachusetts Institute of Technology in 2000 and joined the WebLab group. He started as a graduate research assistant, investigating methods of online collaboration for WebLab. After receiving his M.Eng. in EE/CS from MIT, he became WebLab's project manager and led the development of WebLab 5.0.

**James Hardison** received the S.B. degree in Electrical Engineering and Computer Science from the Massachusetts Institute of Technology in 2002. Since 2000, he has contributed to the WebLab group as System Manager and Internet Application Developer. Currently, he works with the WebLab group as a Research Engineer with the Microsystems Technology Labs at MIT.

**David Zych** received the degrees of S.B. in Computer Science and M.Eng. in Electrical Engineering and Computer Science from MIT in 2000 and 2001, respectively, and has been working with the Microelectronics Weblab project since April 2002. He also teaches introductory physics for the Concourse program, an independent learning community within MIT (neither elite nor remedial) that allows 60 freshmen each year to enjoy many of the advantages of a small school environment while still retaining the opportunities offered by the Institute as a whole.

**Lisa Hui** graduated in 2002 with a S.B. degree in Computer Science from Massachusetts Institute of Technology.

# Chapter 16

# Traditional Laboratory Exercises by Remote Experimentations in Electrical Engineering Education

INGVAR GUSTAVSSON
*Blekinge Institute of Technology, SE-372 25, Ronneby, Sweden.*
*E-mail: ingvar.gustavsson@bth.se*

*Laboratory (lab) work is recognized as an efficient method for students to assimilate knowledge and develop skills for solving real world problems. A traditional undergraduate electronics laboratory at Blekinge Institute of Technology (BTH), Sweden, provides eight identical lab stations. Each station is equipped with a breadboard, some desktop instruments and a power supply. The Internet provides new opportunities for remote experimentation. Laboratory exercises in electrical engineering courses such as circuit theory and basic electronics can be performed remotely using real equipment. Experiments on electrical circuits have been conducted over the Internet using experimental hardware located in a small closed room at BTH. This tiny laboratory provides one lab station equipped with computer-based instruments and a remotely controlled switch matrix to replace the traditional breadboard. The matrix makes it possible to make all the necessary connections to form a circuit and to connect test probes. The aim is to emulate a traditional laboratory. This paper describes the remote laboratory implemented and compares it with a traditional one at BTH.*

## INTRODUCTION

During the last decades there has been a trend towards increased use of simulation in engineering education. One reason for this is that physical experiments are expensive both to implement and maintain. Another obvious reason is the belief that simulators can replace physical experiments [1]. A few years ago the lab sessions in the circuit analysis course at Blekinge Institute of Technology (BTH), Sweden, were replaced by simulated experiments due to budget cuts. One traditional session and three remote ones are now provided.

The Greek philosopher Aristotle (ca. 340 B.C.) asserted that a body when falling increases speed for a brief period; it then falls at a constant speed which is proportional to its weight. Thus, a body that has a weight twice that of another takes half as long to fall from a given height. For Aristotle, no experiment was necessary to substantiate the theory. Almost two thousand years later, in 1586, the Dutch mathematician Simon Stevin performed an experiment to test the ancient "natural law." He dropped two lead balls, one ten times the weight of the other and noted that their impact on a board on the ground produced "a single sensation of sound" [2]. Experiments are indispensable for students if they are to trust physical laws and be aware of their limitations. Practical projects and experiments also provide the framework for students to learn to cope with real world problems. Simulations cannot replace experiments. Simulator programs are based on calculation methods emanating from the physical laws or models. On the other hand, many laws and models are complicated, and the related calculation methods are difficult to use; where this is the case, simulations are a useful and convenient tool for solving problems if you know that the simulator used covers all aspects of the problems to be solved.

The traditional way of conducting experiments is to go to a university laboratory where students work in teams and receive tutorial help from teachers. There is no doubt that nothing can replace synchronous learning through face-to-face interaction, but it is not always feasible for students to attend conventional classes. Models for using information technology to enhance the learning experience for students who are asynchronous in time or/and space and which are also suitable for on-campus students have been presented earlier [3]. A number of so-called remote laboratories have been set up by a number of universities around the world. These offer remote access to laboratory equipment and experimental setups [4]. Conventional laboratory exercises have been used for many decades with good results to teach science and engineering. To design a remote laboratory where students can perform traditional laboratory exercises from home or elsewhere is a good starting point from which new teaching methods may emerge.

A lab station with Internet access is set up in a small closed room at BTH [5]. The desktop instruments are replaced by computer-based ones with virtual front panels, and the breadboard is replaced by a switch matrix which can be controlled remotely. A number of people located in different places around the globe can perform experiments simultaneously using client PCs connected to the BTH laboratory server via the Internet. They download client software from the laboratory web site. This is one way to use equipment and premises efficiently. The cost of the equipment and its maintenance can be reduced if the number of lab stations is reduced and/or the laboratory is used outside ordinary working hours. It is possible to emulate a whole traditional laboratory equipped with 8 lab stations using time sharing and the single lab station in the closed room.

The remote laboratory project at BTH started in 1999 and the first version, a lab station with Internet access providing some fixed experiments in circuit analysis, was set up in a small closed room and released in the year 2000. It was tested in regular campus courses [6]. The second version released in the year 2002 represented a major step forward towards an electronics laboratory capable of emulating a traditional one [7]. In this current version the experiments are not fixed but the teacher can bring his own circuit analysis experiments into the laboratory. In the next section, sessions in a traditional laboratory at BTH are scrutinized. The remote laboratory and its performance are then discussed. Finally, the two types of laboratories are compared.

## EXERCISES IN A TRADITIONAL LABORATORY

Students perform traditional laboratory exercises in local laboratories supervised by instructors. The laboratories are open during organized laboratory sessions only. The main reasons are as follows:

- A laboratory with electrical experiment is a hazardous place due to the high voltage. At BTH, students must pass a test before they are allowed to enter the laboratories.
- Some instruments are very expensive and can be damaged by improper use.
- Equipment can be stolen.

A traditional laboratory session in undergraduate education requires a laboratory where there are several identical lab stations allowing a number of students to perform experiments simultaneously. Two to three students work in a team and share the same lab station.

Most laboratory sessions at BTH are four hours long and each session is supervised by one instructor only. In the traditional undergraduate electronics laboratory there are eight identical lab stations. At each station there is a lab box with a white plastic breadboard and some voltage sources and instruments as shown in Figure 1. There is no standard laboratory for undergraduate education in circuit analysis and basic electronics but most laboratories at universities in Sweden have similar equipment. The breadboard is large enough to accommodate circuits with many components and nodes. It is also possible to assemble compact circuits providing relatively high bandwidth. The instructor provides the components necessary for each laboratory session. The engineer who was previously responsible for maintaining the laboratory equipment has become an instructor in order to cut costs. Nowadays some students may use a significant part of the time available to troubleshoot connecting failures or other faults arising from a lack of maintenance. All experiments are described in an instruction manual. The normal procedure for performing a single experiment is as follows:

- Every student team forms the circuit specified in the laboratory instruction manual using a voltage source, the breadboard, and some of the components provided. A circuit is a collection of electrical elements interconnected in some way. Examples of elements used are resistors, inductors, capacitors and sources [8]. At least one of the instruments is connected to test points in the circuit.
- The instructor checks each circuit formed to avoid possible damage. If the circuit is safe the student team is allowed to continue by activating the source. When the source is activated transients will appear in the circuit. The duration of these transients is usually short in comparison with the time required for manual operation of the instruments, and the students will not notice them.
- The students read the instruments and evaluate the results. If they are acceptable they record them in a laboratory report. Where this is not the case, some troubleshooting must be carried out if necessary, with the support of the instructor.

Even if it is impossible to see the electrical current with the naked eye, it is easy for a student to see that the electrical experiment is genuine and not a simulation if desktop instruments are used. It is more difficult when computer-based instruments are used because a computer is involved. Is it possible, for example, that a result displayed on a virtual front panel on the computer screen originates from a simulator program running in

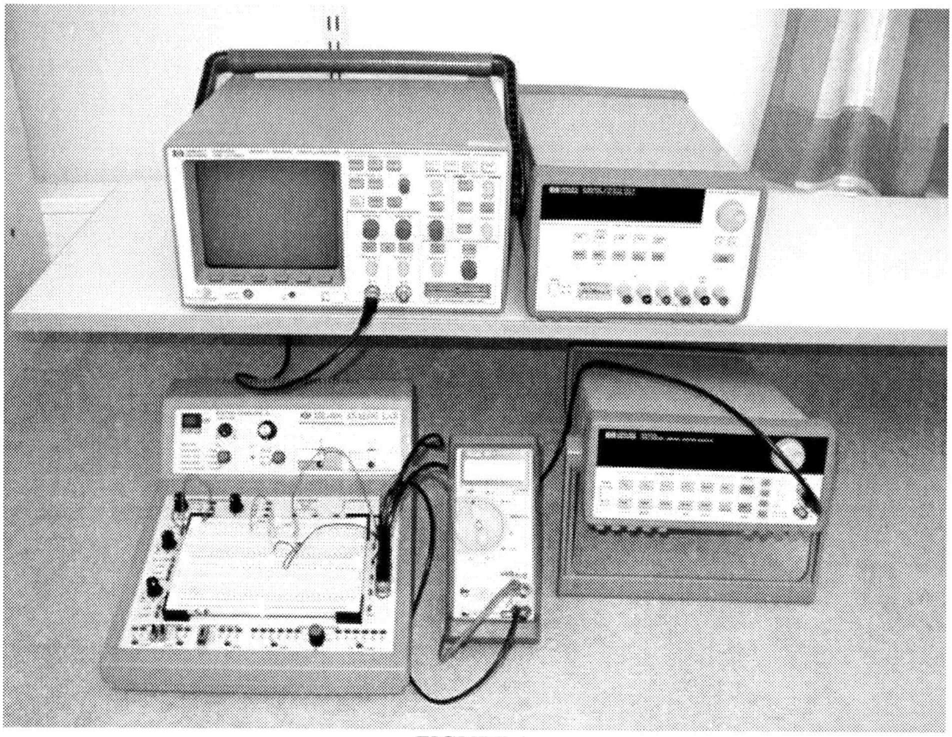

FIGURE 1
A TRADITIONAL LAB STATION IN A LABORATORY FOR UNDERGRADUATE EDUCATION IN ELECTRICAL ENGINEERING AT BTH

the computer? No, such a program needs a complete description of the circuit, and only a limited amount of data concerning the circuit and the electrical phenomena occurring in it can be fed into the computer via the instruments. It is always easy to modify the circuit, for example, to remove a critical wire from the circuit and observe that the output from the instruments will change. A circuit simulation does not depend on something in the physical world.

## A Remote Laboratory for Traditional Electrical and Electronics Experiments

The experiment server in the closed room at BTH is shown in Figure 2. To the left there is a PXI (PCI Extensions for Instrumentation) chassis containing a controller and four plug-in boards from National Instruments. The controller comprises a PC connected to the Internet which hosts the plug-in boards in the form of two function generators, an oscilloscope and a digital I/O board. The instrument settings are controlled from the host computer, and there are no buttons or control knobs on the generators or the oscilloscope, only connectors. Next to the PXI chassis is a Data Acquisition/Switch Unit that functions as a multi-meter; this is connected to the controller via the GPIB (General Purpose Instrument Bus). The power supply is to the right. The traditional breadboard is replaced by a remotely controlled switch matrix that is large enough to accommodate most of the

FIGURE 2
A LABORATORY SERVER FOR EXPERIMENTS IN CIRCUIT THEORY AND BASIC ELECTRONICS

circuits used in electrical and basic electronic experiments in undergraduate education. The matrix used in the remote laboratory at BTH has five main nodes and ten main branches. The main nodes are denoted A, B, C, D and GND. The ground terminals of the function generators and the oscilloscope are connected to GND. Each branch between the main nodes can be composed of either a jumper lead or up to four components with two leads mounted in parallel in sockets on the printed circuit boards as shown in Figure 2. In this way, the instruments and a total of 40 different components can be connected to the main nodes [7].

The window appearing on the client PC screen when a user starts an exercise is shown in Figure 3. The circuit to be formed is defined using the PSpice-compatible net list shown. In PSpice, the ground terminal, GND, is denoted "0." The net list is sent to the server in the closed room to set the switches in the matrix in order to assemble the circuit. Virtual front panels are used to set and connect the instruments. The oscilloscope panel is shown in Figure 4. The server software is written in LabVIEW 6.1 (Laboratory Virtual Instrument Engineering Workbench), and the client software is written in Visual Basic 6 [9].

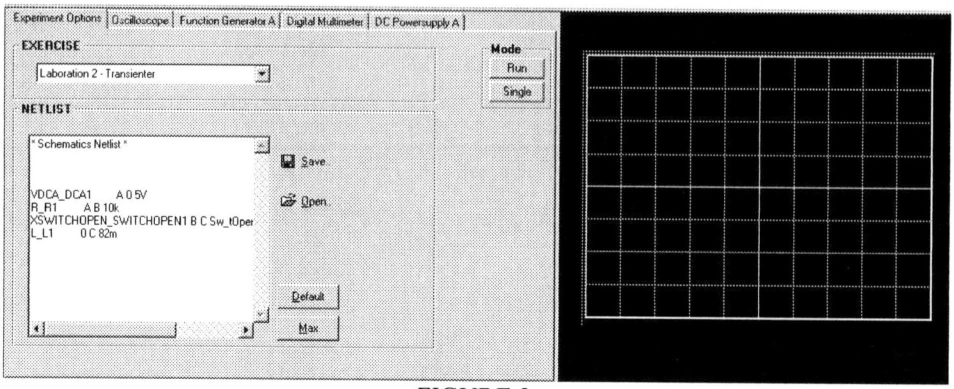


FIGURE 3
THE EXPERIMENT OPTIONS PANEL

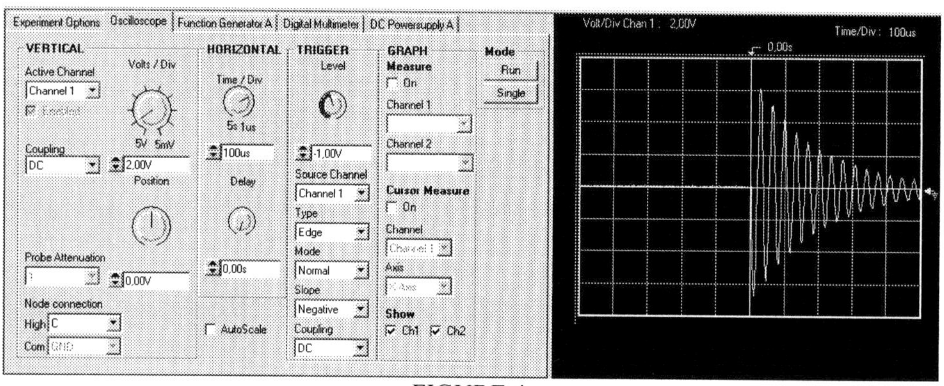


FIGURE 4
THE VIRTUAL FRONT PANEL OF THE OSCILLOSCOPE

## REMOTE EXPERIMENTS

The normal procedure for performing a single experiment in the remote laboratory is as follows:

- The student starts the client and selects the desired exercise. The experiment options panel will be displayed on the screen. S/he can choose an experiment and, for example, use a default net list representing the circuit in the instruction manual. The student can delete components or sources in the list. S/he connects one or more instruments to desired test points. Finally, the student presses the "Single" button to send the list, the instrument settings, and the test probe connections to the server.
- When the user request is dequeued on the server side, a virtual instructor compares the list received with a checklist belonging to the experiment selected in order to ascertain that the desired circuit is safe. Where this is the case, the circuit will be formed, the instruments set, the test probes connected, and the voltage applied. The outcome will then be sent to the client computer. In all other cases an error message will be returned.

- If there is no error message, the student reads the result on the virtual front panels of the instruments and evaluates it. If the result is acceptable, s/he will record it in the laboratory report. Where the result is not acceptable, troubleshooting must be carried out - if necessary, with the support of a real human instructor when available.

The reason why the list, all the instrument settings, and the test probe connections are sent to the server in one message only when the user presses the "Single" button is to ensure that the hardware is engaged for only a short time. In this way, a number of student teams can share the same server. The main parts of the server response time are divided up as follows:

- The server decodes the message received.
- The virtual instructor checks the desired circuit.
- Some switches are opened; others are closed to turn the last circuit into the desired circuit.
- The sources in the desired circuit are activated. Transients will now appear. A delay is entered here to allow these to disappear before the oscilloscope is armed or the multi-meter is read. The delay is currently fixed at 5 ms.
- If the oscilloscope is used the time needed to sample a signal depends on the time/division setting.
- When data from the instruments are available the outcome is converted into a message to be returned to the client PC.

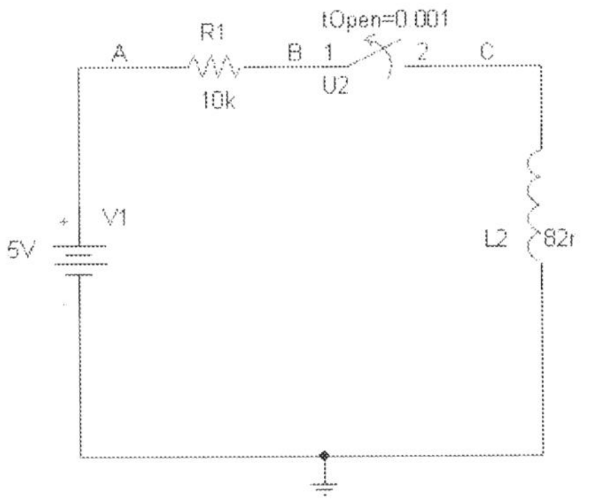


FIGURE 5
A TRANSIENT EXPERIMENT

An example of a desired circuit drawn using PSpice symbols is shown in Figure 5. One well-known circuit of this kind where you make use of energy stored in an inductor after the power source has been disconnected is the classic breaker-point ignition system for an automobile. Students are expected to test what happens when a DC current through the inductor is broken. First the server forms the circuit and activates the power source to generate the current to be broken. When this has taken place it is necessary to wait at least five time constants, i.e. 40 μs in this case, for the current to increase. However, the oscilloscope is armed only when the fixed delay (5 ms) expires and the switch is later opened. The outcome of the experiment is shown in Figure 4. The time setting of the oscilloscope is 100 μs/division. It is possible to achieve a server response time below 1 s with eight simultaneous users.

## The Traditional Versus the Remote Laboratory

Traditional experiments are roughly compared with remote ones and to simulations in Table 1. One can anticipate that computer-based instruments with virtual front panels will gradually replace desktop ones. It thus makes no difference if you handle instruments in local laboratories or by remote control. Nevertheless, students need a great deal of hands-on practice. At BTH students gain this experience in practical projects, which provide the right methodology to cope with real-world problems such as EMC issues, finding bad connections etc.

| | **Simulated Experiments** | **Traditional Hardware Experiments** | **Remote Hardware Experiments** |
|---|---|---|---|
| Distance learning students | Yes | No | Yes |
| Campus students | Yes | Yes | Yes |
| Occasional experiments | Anywhere Anytime | Lab sessions only | Anywhere Anytime |
| Required equipment | PC with a simulator program | Provided by the university | PC with a 56 kbit/s Internet connection |
| Verify physical laws | No | Yes | Yes |
| Solve problems | Convenient | Time-consuming and a limited number of components available | Convenient but only a few components |
| Gain instrument experience | Possible | Much | Much |
| Manual handling of components and test probes | No | Yes | No |
| Easy to provide several models of each instrument | Should be | No | Yes |
| University expenses | Low | High | Low |

TABLE 1
SIMULATIONS AND EXPERIMENTS – A COMPARISON

The limited size of the switch matrix in terms of nodes and sockets for components is acceptable because only small circuits containing few components and nodes are required in courses in circuit analysis and basic electronics. In the circuit analysis course at BTH there are four laboratory sessions, and all the components required for these sessions can be accommodated in the matrix. However, circuits assembled using the switch matrix need more cabling than that required by an experienced person using a conventional breadboard. The former circuits can perhaps be compared to those made by the average student. The extra wires and relays limit the bandwidth somewhat; the oscilloscope time base is currently restricted to 1 μs/division. The teacher can choose an appropriate time scale for the experiments by selecting

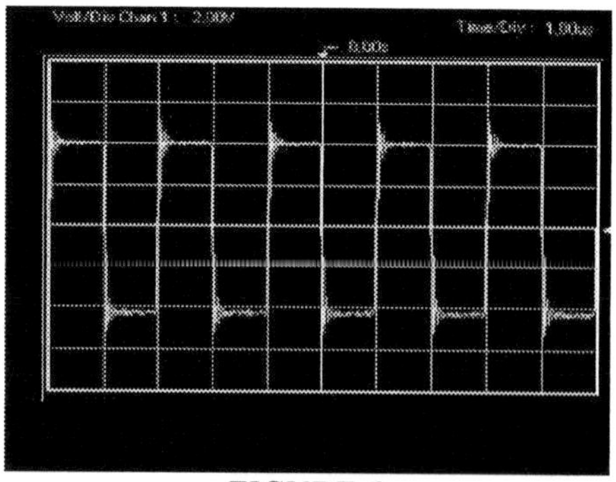

FIGURE 6
A SQUARE WAVE SIGNAL FROM THE FUNCTION GENERATOR DISPLAYED IN MAXIMUM TIME RESOLUTION ON THE OSCILLOSCOPE

proper values for the components to be used by the students. Time settings on the oscilloscope in circuit theory and electronics experiments normally range between 1 μs/division and 10 ms/division. Thus, a single shot is usually 10 μs to 100 ms because the time axis is 10 divisions on most oscilloscopes. A square wave signal from the function generator is shown in maximum time resolution in Figure 6. Both laboratories allow low frequency experiments only to be performed.

It is possible to replace the computer-based instruments and the other hardware by a simulator program in the experiment server. All data required for a simulation is sent from the client PC or is already available on the server. If people are present in the closed laboratory when a remote client is using the lab server the only thing they will notice is the clattering of the relays. One may ask if a microphone should be put in the vicinity of the server so that the clatter can be transmitted to the client PCs to convince the user that s/he is performing a real experiment. Perhaps, but it is also possible to simulate the sound of the relays. You might say that the user interface for a simulator looks different. It would probably be hard to sell a simulator program with an oscilloscope-type display as the only output.

## CONCLUSIONS

In a remote laboratory students can conduct physical experiments from the comfort of their own home. It is possible to emulate traditional laboratories. The size of the switch matrix in terms of the number of nodes is less than that of a conventional breadboard but sufficient for experiments in, for example, circuit analysis. If the traditional laboratory were equipped with computer-based instruments there would be no difference in the method for making instrument settings. The server response time is added to the duration of an experiment but the delay is less than 1 s if eight simultaneous users are active, and every experiment requires less than 100 ms; selecting adequate time constants can easily satisfy this. The main trade off for the flexibility in time and space is the fact that it is impossible for the user to touch the physical setup, i.e. the degrees of freedom are reduced. However, less resources are required for the remote laboratory whereas the number of simultaneous users is the same.

The students use a text-based interface to control the switch matrix in the current version of the laboratory. An important item is thus missing. A new experiment server where the user interface is replaced by a virtual breadboard was released in March 2004. The virtual breadboard, which controls the switch matrix, is a photograph of a traditional breadboard displayed on the client PC screen. Photographs of the components mounted in the switch matrix are displayed in a window below the breadboard. To form a circuit, students use the mouse pointer to move the components to the breadboard and connect them. The virtual breadboard is the last piece to fall in place. This new server will be used and evaluated in a circuit analysis course for distance learning students soon after the release.

The current version of the laboratory can be used by anyone with an Internet connection whenever it is not being used in regular education. The client software, which uses the TCP/IP port 5000, is an unsigned ActiveX control embedded in an html page. The address of the laboratory home page is: http://www.its.bth.se/distancelab/english/. The address of the new experiment server is: http://distanslabserver.its.bth.se/.

## REFERENCES


1. Foss, B.A., Malvig, K.E., Eikaas, T.I., "Remote Experimentation – New Content in Distance Learning," Proceedings of the 2001 International Conference on Engineering Education (ICEE-2001), Oslo, Norway, August 6 – 10, 2001.

2. Benson, H., *University Physics,* John Wiley & Sons Inc, 1991, pp 41 - 42.

3. Gillet, D., Crisalle, O.D., Latchman, H.A., "Web-Based Experimentation Integration in Engineering Curricula: Experience in Deploying Resources Distributed Among Universities," System Theory, Proceedings of the Thirty-Fourth Southeastern Symposium on, 2002.

4. Strandman, J.O., Berntzen, R., Fjeldly, T.A., Ytterdal, T., Shur, M.S., "LAB-on-WEB: Performing Device Characterization via Internet Using Modern Web Technology," Proceedings of IEEE Caracas Conf. on Devices, Circuits and Systems, ICCDCS 2002, Aruba, IEEE Catalog No 02TH8611C, April 2002, pp. 1022.1 – 6.

5. Gustavsson, I., "A Remote Laboratory for Electrical Experiments," Proceedings of the 2002 ASEE Annual Conference, Montreal, Canada, June 16 - 19, 2002.

6. Gustavsson, I., "Laboratory Experiments in Distance Learning," Proceedings of the International Conference on Engineering Education (ICEE-2001), Oslo, Norway, August 6 – 10, 2001.

7. Gustavsson, I., "User-Defined Electrical Experiments in a Remote Laboratory," Proceedings of the 2003 ASEE Annual Conference, Nashville, USA, June 22 - 25, 2003.

8. Johnson D., Johnson J., Hilburn J., Scott P., Electric Circuit Analysis, Third edition, Prentice Hall International, Inc., 1997.

9. Naghedolfeizi, M., "Survey of LabVIEW Technologies for Building Web/Internet-Enabled Experimental Setups," Proceedings of the 2002 ASEE Annual Conference, Montreal, Canada, June 16 - 19, 2002.


**Ingvar Gustavsson** was born in Karlskrona, Sweden in 1943. He was awarded the MSEE and Dr. Sc. degrees from the Royal Institute of Technology (KTH), Stockholm in 1967 and 1974, respectively. In 1982, he was appointed Head of the Instrumentation Laboratory. Together with another research scientist he founded a private company providing automatic inspection systems for industrial customers in 1983. In 1994 he returned to the academic world to take up his current position as Associate Professor of Electronics and Measurement Technology at Blekinge Institute of Technology (BTH), Sweden. His research interests focus on industrial electronics and distance learning.

# Chapter 17

# eMerge: A European Educational Network for Dissemination of Online Laboratory Experiments

RUBEN CABELLO[1], IVAN GONZALEZ[1], FRANCISCO GOMEZ-ARRIBAS[1], JAVIER MARTINEZ[1], MICHEL BILLAUD[2], THOMAS ZIMMER[2], DIDIER GEOFFROY[2], HANS EFFINGER[3], WILHELM SEIFERT[3], REINHOLD S. JAEGER[4], TOR A. FJELDLY[5], KJELL JEPPSON[6], HERMAN MANN[7], NIKOLAOS ASIMOPOULOS[8] and ZOLTAN GERMAN-SALLO[9]

*[1]Lab. Microelectrónica, EPS, Universidad Autonoma, 28049 Madrid, Spain. [2]Université Bordeaux, 351 Cours de la Libération, 33405 Talence Cedex, France. [3]Fachbereich Elektrotechnik, Fachhochschule Muenster, Stegerwaldstr. 39, 48565 Steinfurt, Germany. [4]Center for educational research, University of Landau, Friedrich-Ebert-Str. 1. Landau, Germany. [5]UniK - University Graduate Center, P.O. Box 70, Instituttveien 25, N-2027 Kjeller, Norway. [6]Chalmers University of Technology, Department of Microelectronics ED, S-412 96 Göteborg, Sweden. [7]Czech Technical University, CTU, Zikova Street 4, CZ-166 35 Prague 6, Czech Republic. [8]Tech. Education Institute of West Makedonia, Kozani, Greece. [9]Petru Maior University, Tirgu Mures, Romania. E-mail: javier.martinez@ii.uam.es*

*This paper presents the development of the eMerge project, an innovative and advanced educational network infrastructure that will permit the dissemination of online laboratory experiments to support engineering and science education. Partners from nine different educational institutions are involved in this project, which was performed under the framework of the European Community Socrates – Minerva program. The current work is based on previous experiences in which prototypes of virtual laboratories were created and successfully demonstrated; a primary objective of the current project is to extend these technologies beyond the participating institutions, making the services widely available to European students. To this end, the project employs web technologies and computer controlled instrumentation to allow students to access the remote laboratories. To simplify access to the laboratories, system requirements for clients are minimal: simply a computer with Internet access and a web browser.*

*This project also emphasizes the creation of a variety laboratory experiments and supporting course material and educational practices. From a pedagogical point of view, it is important to present students with a lab-like view of the experiments, so the client interface was designed such that the student interacts with instrument-like images and receives experimental data in standard formats such as graphs and tables. In many cases, the experiments are accompanied by online simulation tools to allow further in-depth investigation and comparison. In other cases, we provide the ability to reconfigure the experiments online, for example by means of switching matrices.*

*The project employs advanced server systems that provide the laboratory services in a reliable and secure fashion. Innovative solutions concerning the networking of partner laboratories, server technologies, access control, security, queuing, scalable user interfaces, are being developed using standard web technologies such as applets, SOAP, ASP, COM+, and .NET.*

## INTRODUCTION

Laboratory experiments are a vital part of engineering and science education that have so far been considered impractical for distance learning. In addition, many institutions cannot afford to buy the expensive instrumentation needed to provide their students with a satisfactory range of modern engineering and science experiments. Recently, however, advances in Internet and web technologies and computer-controlled instrumentation permit network-based techniques to be used to provide remote laboratory access [1-3] to advanced instrumentation for students throughout Europe in a cost-effective manner.

The goal of the eMerge [4] project is to develop an innovative and advanced educational network infrastructure that will permit the dissemination of online laboratory experiments to support engineering and science education. This project has been performed in the framework of European Community Socrates – Minerva program. The Minerva projects aim to promote understanding among teachers, learners, decision-makers and the public at large of the implications of Open Distance Learning (ODL) and Information and Communication Technology (ICT) for education. Partners from nine different educational institutions are involved in this project. The actual work is based on previous experiences like Retwine [5], the Lab-on-Web [6] and the Socrates RichODL [7] projects, in which prototypes of virtual laboratories were created and successfully demonstrated. By using web technologies and computer-controlled instrumentation, students could access to these remote laboratories. The main objective of the current project is to extend these technologies out of the individual institutions, making the services widely available to European students. In the project, we emphasize the creation of a variety laboratory experiments, and the development of supporting course material and educational practices.

To keep the scope of the project manageable, and since all of the partners have a solid background in education and research in electronics or computer science, we are limiting the range of experiments to the field of microelectronics. The technology developed on this project will be quite generic, however, and will be open for additional experiments in a wide range of other disciplines within engineering and science.

The remainder of this paper is organized as follows: the following section describes the distributed network architecture of the eMerge project based on a main server and several local servers; the next section presents the services offered by each server, with examples of educational activities that can be performed by students; and the final section concludes with a summary.

## NETWORK ARCHITECTURE

The eMerge network has been designed with a distributed architecture. Servers local to the different institutions involved in the project offer control of the laboratory experiments and the course material. Each partner develops and manages a range of educational content and shares it with the rest of the group. One central server is dedicated to providing information about the features of the project and to establishing the conditions to give students access to the contents of the local servers. The central server is called the main server.

### Local Servers

For a remote experiment, a schematic of the instrumentation pool in one of the institutions' laboratories is shown in Figure 1. A group of instruments is connected to the sample or device to be measured. The measurement instruments must be equipped with the GPIB IEEE-488 interface (General Purpose Interface Bus) for remote control capability. In the laboratory, the instruments are controlled by a server application executed from a workstation connected to the Ethernet LAN and labeled as the instrumentation server. The local web server is responsible for performing all types of interaction with the instrumentation server regarding the measurement process and for collecting the measured data. The local web server also provides information, documentation and tutorials about how to choose and perform the laboratory experiments. Although the configuration shown in Figure 1 represents all laboratories, some variation is possible due to the specific GPIB-LAN [8] device used or the need for additional servers to perform simulation exercises.

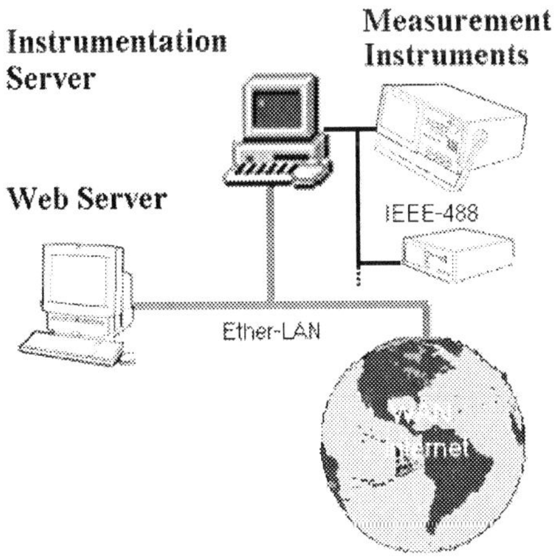


FIGURE 1
SCHEMATIC OF THE INSTRUMENTATION POOL

It is important to note that the equipment used by the eMerge project is shared with other research projects. Because the equipment is shared, individual laboratory experiments are sometimes unavailable. An experiment may be unavailable if the equipment it requires is in use by another eMerge task or experiment. For example the electrical output characteristics of a bipolar transistor cannot be measured simultaneously with the electrical characteristics of a field effect transistor because both devices use the same equipment. For these reasons, the laboratories offer the ability to perform an experiment during a limited period of time that can be several hours a day, several days a month, etc.

Additionally, when necessary the period during which an experiment is available can be sub-divided into reservable time slots. The reservation system is necessary for long-running measurements or for experiments involving consecutive measurements when the current value depends on previously measured values. In these cases, users must book a time slot to ensure they have exclusive access to the equipment for some period of time. The duration of the time slot naturally depends of the type of experiment being performed, and should be established by the partner responsible for the experiment. In some experimental work, queuing the different queries can control the concurrent access by several users to the equipment. In these cases, the reservation process is unnecessary.

This distributed architecture, where local servers at the institution responsible for the experiment manage control of the laboratory experiment and help content, has several advantages. First, both the web server and the instrumentation server are located on the same LAN. This allows the rapid communications resulting from error-free packets that are very important in such real-time applications. Moreover, modifying and testing a laboratory experiment can be performed far easier when the servers are located on the same LAN. Similarly, new experiments can be tested and added more quickly to the eMerge project.

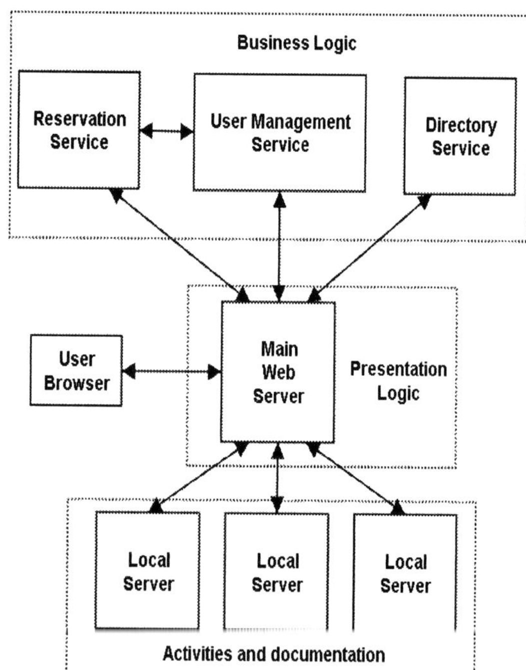


FIGURE 2
NETWORK ARCHITECTURE

### The Main Server

The eMerge network architecture includes a main server whose goal is to offer several services for users and partners of the project. The services offered to users are: registering for access to experiments; viewing educational materials; viewing a schedule of laboratory experiment availability; and booking reservations for the experiments that require them. The services offered to partners are: adding new educational content, including help information; and scheduling the time periods when the content can be accessed.

The main server is a web server that communicates with several services, and is the primary website for all eMerge project users. After a user selects an activity, the main server checks the user's access rights, then redirects them to a page on the local server responsible of the selected activity.

A diagram of the services included in the main server is shown in Figure 2. The main web server is the backbone of the structure. This server stores general information on eMerge objectives and the institutions involved in the project. Instructions on how to access the educational content is also presented to the user. Three services are available to the web server: the directory service, the user management service and the booking service.

The user management service maintains information about the registered users. Four user types are defined with different access privileges. These are guest, registered, local administrator and main administrator users.

- The guest user has access to general information about the project and the institutions. A list of the educational content with brief descriptions is also available for this type of user. Some simulated examples are presented in order to give the guest a general overview of the activities that can be performed.
- Registered users have access to all of the educational material of the project, restricted only in the experiments they can perform and the reservations of time slots for experiments that require them. Registered users can also modify their user profile on the system. To become a registered user, a guest must fill out a web form with the user's academic and personal information.
- Local administrators can modify and add new educational content-related to the local server. These users can insert time periods during which experiments are available and establish the duration of reservation the time slots. Information useful for the directory service should be added by this type of user.
- The main administrator user has all privileges of the other users and can also add or modify the general information stored in the main web server. New local administrator accounts can only be created by the main administrator.

The directory service stores information related to educational content that can be accessed by registered users. This information includes the activity name, activity type (which can be a laboratory experiment, a simulation exercise or a course), the time periods during which the activity is available and keywords associated with the activity. The directory service offers a list of activities a user can perform along with the additional information described above.

An example of a web page generated from the data supplied by the directory service is shown in Figure 3. This page can be accessed by clicking the "Portal" link in [4]. The user can search activities by activity type, available time period or a specific keyword. In addition to the data visible to users, the directory service also maintains data such as the URL of the page on the web local server where the activity resides and the URL where help information and a tutorial about the content are located. Local administrators maintain this information.

The booking service maintains information about the availability of each activity provided by the eMerge project. For each activity, it divides a day into time slots whose duration is established by the appropriate local administrator. Before a user can invoke an activity, he must first reserve a time slot. For each activity, the booking service lists free

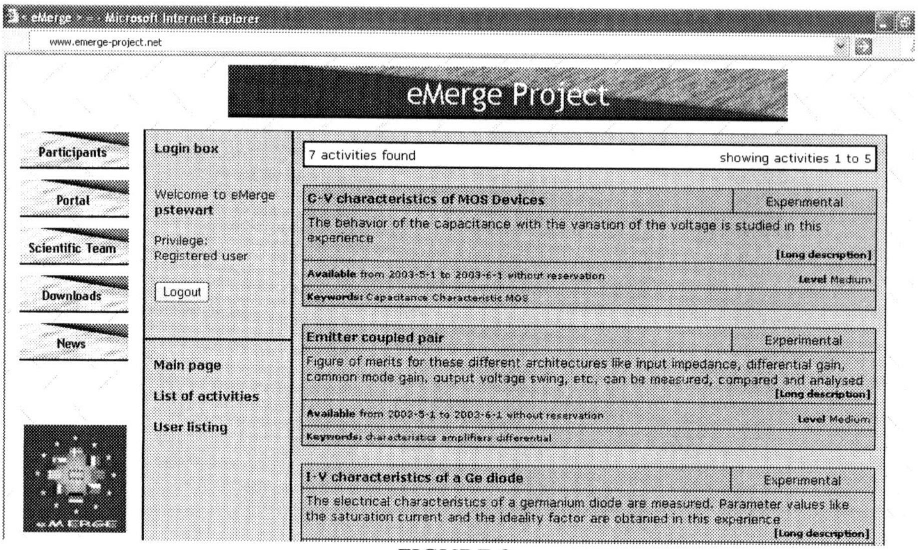


FIGURE 3
LIST OF ACTIVITIES PROVIDED BY THE DIRECTORY SERVICE

time slots for a desired date. The user selects a time slot from this list to ensure exclusive access to the measuring device.

From the list of activities obtained from the directory service, a user may choose to start an activity, which redirects the user to the corresponding URL on the local web server. Previous to the redirect, several checks are performed: Is the user registered? Is the activity available at this time? Does this activity need a time slot reservation? If so, does this user have a reservation at this time? All checks are performed by querying the directory, user management and booking services. If all checks succeed, the redirection to the local server proceeds.

Communication between main web server and the services is performed using Simple Object Access Protocol (SOAP) [9] with HTTP as the transport. HTTP is a protocol based on TCP, which is widely used in request-response model service scenarios. SOAP is a lightweight protocol for distributed computing. It uses XML encoding and a variety of protocols as the transport, HTTP being the most common. A SOAP message consists of an envelope, which contains one or more messages. SOAP messaging is commonly used to perform remote function calls in a Remote Procedure Call (RPC) style. SOAP has the advantage of being an internationally accepted standard, so external applications could easily interact with our services. Because the web services standard [10] uses SOAP for messaging, the system can be easily integrated with web services provided by external entities; and because SOAP is a widely accepted standard, libraries are available for most popular languages such as C++, Python, or PHP. Additionally, common platforms such as Java2 Enterprise Edition (J2EE) and the .NET framework include SOAP support in their standard class libraries.

The web server acts as a SOAP client and the services offer methods for remote invocation. The web server sends a SOAP request to a service, which maps it to a local service method, and returns the result of the local method in a SOAP response. The web server's PHP pages present the data returned from the services so that the web server controls the presentation and user interaction while the services encapsulate the business logic.

## EDUCATIONAL ACTIVITIES DEVELOPED

The different partners involved in the project have created a variety of laboratory experiments and developed supporting course material and educational practices. As we explain later, each partner is responsible for a group of activities offered on their local server. In all cases, special attention has been devoted to creating a lab-like atmosphere on the client side. Visual tools have been used to retrieve images from the original laboratory, thereby making the students feel as if they were in the actual laboratory. Client interfaces have been developed so that the students work with instrument-like images, and receive experimental data in suitable formats such as graphs and tables. The resulting data can be saved by the student on his local computer for further manipulation. The fitting of data and parameter extraction can be performed by the student using his favorite tools. If possible, the experiments are accompanied by online simulation tools for further in-depth investigation and comparison. In other cases, we allow participants to reconfigure the experiments online, for example, the user might use switching matrices for this purpose. Some of the activities developed by four of the partner laboratories are described below.

### An Experimental Basic Electronics Course

The contents of this course are related to the solution of circuits with active and passive devices. The experiment covers the basic topics of an electronics course:

1. The study of resistive and capacitive circuits: Ohm's Law, charge of a capacitor.
2. Diode characterization: Zener diode, Light Emitting Diode (LED).
3. The polarization of Bipolar Junction Transistors (BJT): input and output electrical characteristics.
4. Metal Oxide Semiconductor Field Effect Transistor (MOSFET) characterization.

The main window contains a circuit diagram and several buttons as shown in Figure 4. The object used to represent circuit components is called an *entity*. Each entity defines the virtual equivalent of a single circuit component, such as a voltage source, current source, etc. In the current implementation, only four types of entities are available when defining a circuit. It is useful to distinguish between input entities like voltage and current sources and output entities like ammeters and voltmeters. The main difference is that the former need to be configured by the user before the measurement process. We have defined these four components because a maximum of 4 sources and 4 measurement test units are available in the real instrument. The system was designed with reusability and scalability in mind, so it can be easily extended to implement new components types.

Each entity also contains an entity control panel object [11]. This object manages the related Graphical User Interface (GUI) for the entity. The entity's graphical control panel window is displayed by clicking the entity symbol on the diagram. Typical components of an entity control panel include power-on lights, LCD displays and text entries for component configuration where the user specifies measurement parameters, for example the start and end values of the source in the sweep measurements.

After all input entities in the circuit have been configured by the user, pushing the measure button in the main window starts the measurement process. When the measurement process is done, the resulting data are delivered to each entity control panel and displayed on the virtual LCD during a second time period, as shown in Figure 4.

By pressing the List button on the circuit diagram, the user opens a window containing all the measurement results. Due to Java security limitations, these data cannot be manipulated on the user's computer, so a client-server subsystem was developed. It consists in a web gateway developed with PHP and a SQL database to store the measurement result data. The user presses the Export button to export the data to standard HTML pages on the server. The client can then retrieve these result pages for further analysis with other tools. For standard graphic representation, SmartPlot2D [12] Java classes were included in the system. Several output formats can be viewed by using the different application menus. An example is shown in Figure 5.

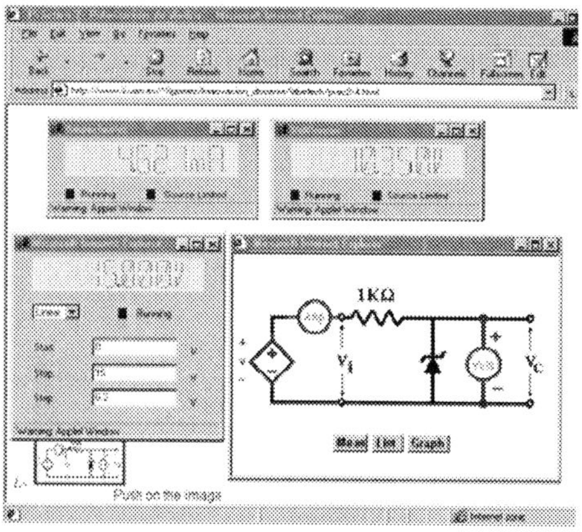


FIGURE 4
APPLICATION WORKSPACE

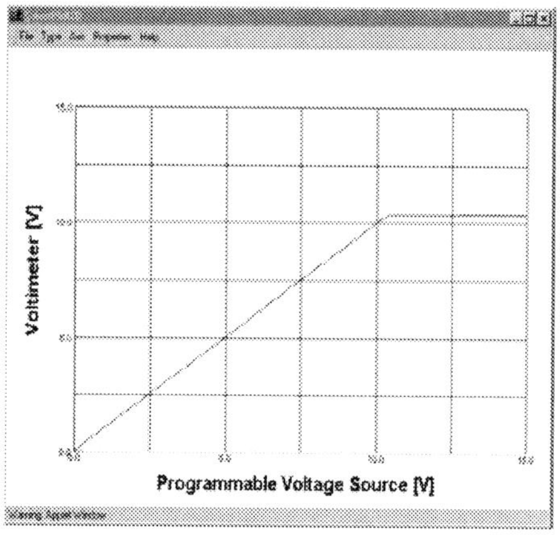


FIGURE 5
MEASUREMENT RESULT PLOT

## LAB-on-WEB

LAB-on-WEB [6], physically located at UniK – the University Graduate Center in Norway, seeks to familiarize the students with the behavior of the building blocks of modern integrated circuits. The lab is also designed for the extraction of device and processing parameters needed for circuit design. The extracted parameters are suitable for use with circuit simulators such as SPICE to predict the behavior of more complicated circuits. This experience is important for gaining a deeper insight into the practical aspects of electronic circuit engineering. Individual discovery activities among students are also strongly encouraged. The system has been successfully tested in senior and graduate level courses in electronic device technology.

LAB-on-WEB relies on modern web and instrument control technologies, including COM+ (Component Object Model with extensions), ASP (Active Server Pages), web services using .NET, and LabVIEW. Advanced capabilities of modern web browsers are utilized, allowing the system to communicate via of XML (eXtensible Markup Language) and SVG (Scalable Vector Graphics).

The current experimental setup of the LAB-on-WEB server/laboratory unit is shown in Figure 6. The server and the instruments are connected via a GPIB bus. The main instruments are an HP 4142B Modular DC Source/Monitor, a high-speed, accurate, and computer controlled DC parametric measurement instrument for characterizing semiconductor devices, and an Agilent 4284A Precision LCR Meter for measuring capacitance/voltage characteristics. In addition, the setup includes an Agilent E5250A Low Leakage Switch unit for remote configuration of experiments, and a Tektronix TDS 3052 500 MHz oscilloscope used for servicing and testing the setup, but also for allowing the measurements of transient events or waveforms. Figure 6 also shows the test chip used in LAB-on-WEB [13]. This chip, which contains diodes, capacitances, MOSFET arrays, inverters and a ring oscillator, was developed as part of a project under the European Commission within the ALFA exchange program between universities in the European Union and Latin America.

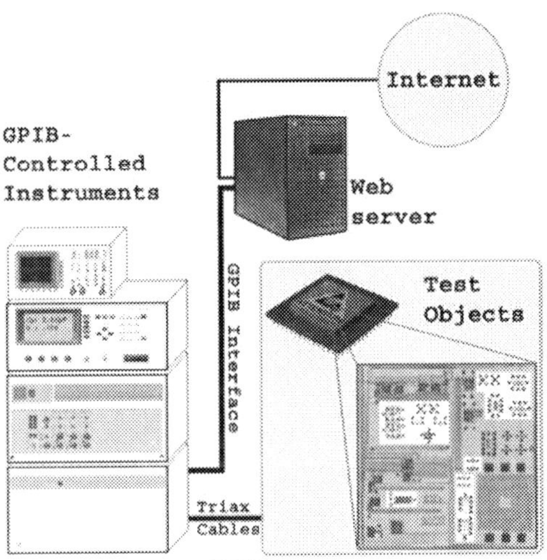


FIGURE 6
LAB-ON-WEB SETUP

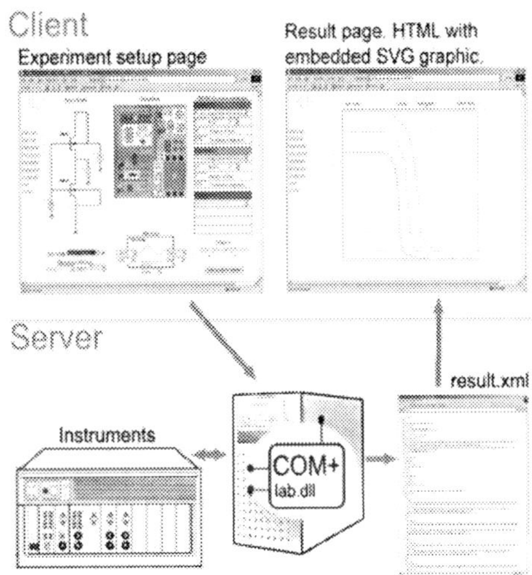


FIGURE 7
LAB-ON-WEB ARCHITECTURE BASED ON A COM+ SOLUTION

In the COM+ mode of LAB-on-WEB, the system operates as indicated in Figure 7. The lab is called from an HTML form in the active client page via an ASP page. Subsequently, this ASP page receives the measurement results from the COM+ component and relays them to the client, for example, in the form of SVG presentations. The layout of the interactive client page and a diagram with measurement results are depicted in the upper part of Figure 7. We note that the experiment may be graphically reconfigured using active areas of the client HTML form.

## eLab

eLab [14] is the name of the remote laboratory located at the Bordeaux University. Through an Internet browser, the client (student, remote training class, or professor) can access the eLab and make real measurements. The activities offered by eLab are dedicated to the electronic field. One of the difficulties in remote laboratory implementations is the connection of the device or circuit to the corresponding measurement instruments. Switching matrices allow us to overcome these roadblocks. In fact, the matrix switch configuration is the most versatile because it can connect multiple inputs (in our case measurement instruments) to multiple outputs (circuits or devices to be tested: DUT). The drawback of switching matrices is the cost and the limitation in the number of connections. For this reason, we developed a special switching matrix dedicated to remote lab applications. Its main advantage is its versatile architecture and low cost. In the actual implementation, more than 100 lab experiments can be switched (connected) for measurements in both the time domain and the frequency domain. Each DUT has 5 connections: the input and output signal, the positive and negative power supplies and the ground connection. These connections (besides the ground) are fed through the switching matrix to the DUT. The selection of the switches to be set is done by logic components decoding an 8-bit word applied by the instrument server through the parallel port.

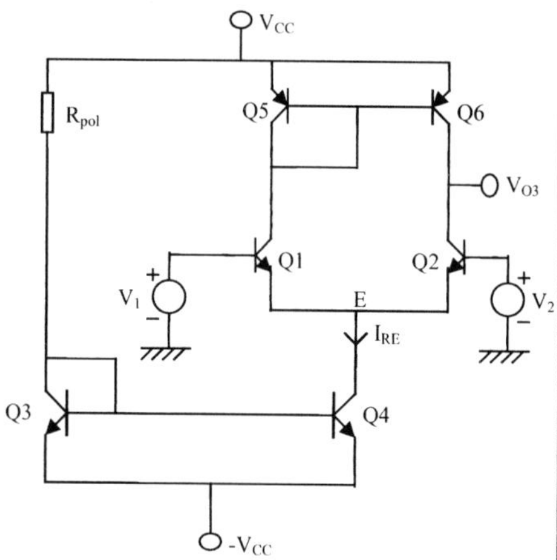


FIGURE 8
EMITTER COUPLED PAIR WITH ACTIVE CURRENT SOURCE AND ACTIVE LOAD

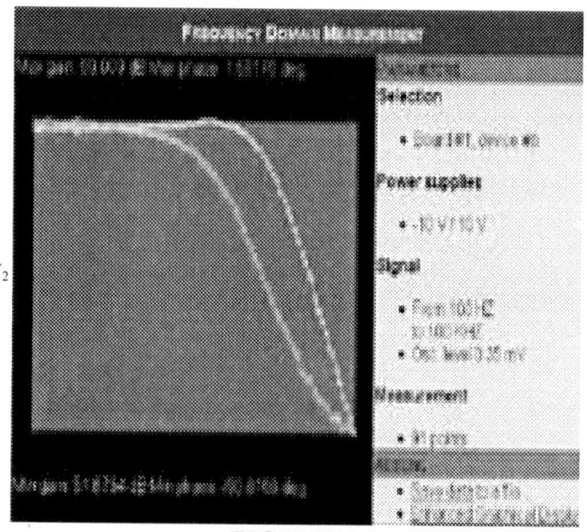


FIGURE 9
MEASUREMENT INSTRUMENT SCREEN SEEN BY THE CAMERA

An example of the activities developed in eLab is the emitter coupled pair experience. Emitter and source-coupled pairs are certainly the most widely used two-transistor subcircuits in analog circuit design. It is therefore of particular interest to investigate their characteristics in lab exercises. Figure 8 shows a configuration for integrated circuit implementation. This configuration and others similar to it have been implemented in eLab [14]. Figure of merits for these different architectures like input impedance, differential gain, common mode gain, output voltage swing, etc, can be measured, compared and analyzed. The measurement data are transferred to the user in the form of ASCII tables, and measurement curves can easily be drawn. Furthermore, a fast investigation is possible using the camera. A standard web camera and a motorized camera are connected to the system for remote system surveillance and creation of a real lab atmosphere. Figure 9 shows a screen capture of a measurement result, in this case the frequency response (gain and phase) of the voltage gain ($V_{O3}$ / ($V_1$-$V_2$)) of the emitter coupled pair (Figure 8).

## iLab

iLab [16,17] is the name of the remote diode characterization laboratory located at Chalmers University of Technology. Currently, iLab is designed for I-V, C-V and I-t characterization of semiconductor diodes. An example of a transient I-t measurement is shown in Figure 10. First, it should be noted that iLab is a cost effective way to opening up the laboratory measurement hall 24 hours a day. Through iLab, expensive measurement equipment is made available to large group of students (~350 Electronic Engineering and Computer Engineering students). Also, through user-friendly graphical interfaces, measurements are made simple. Odd multi-function instrument push-buttons are replaced by clear and simple dialog boxes on the personal computer screen. Because of this, the students' focus can shift from collecting data to analyzing data and fitting it to device models. Our belief is that this new learning tool offers new possibilities to support student learning and to enhance understanding of physical modeling.

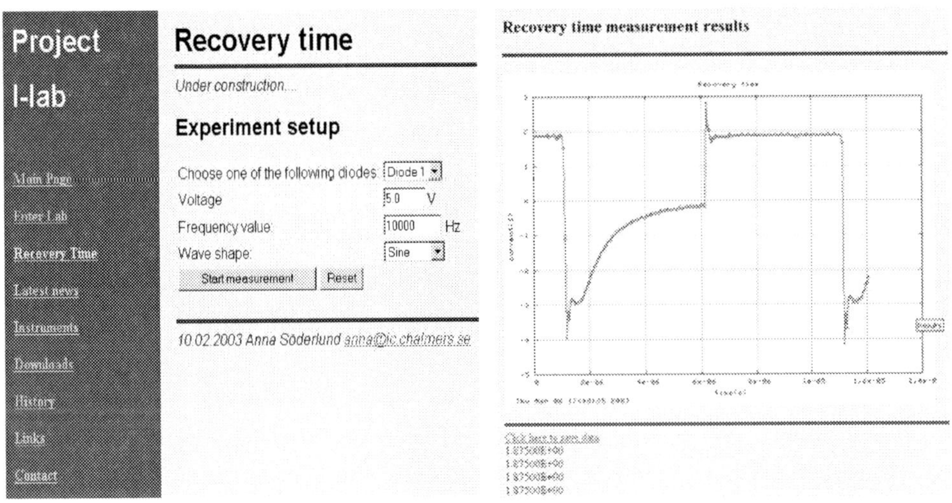


FIGURE 10
EXPERIMENT SETUP AND MEASUREMENT RESULTS FOR THE ILAB DIODE REVERSE RECOVERY TIME (I-T) EXPERIMENT

With this new Internet-based learning tool at hand, we have looked at new methods of teaching that would encourage students to undertake more advanced data analysis now that less time must to be spent on practical details. When used in the second year introductory microelectronic device course, the laboratory task was changed from a closed to an open task. Students were given the freedom to independently organize their own laboratory exercise. They were given two weeks to perform the task. An important factor here, which must be emphasized in future work, is to make available textbooks that support exploratory learning through measurements. Some such book chapters are available, for instance about diodes [18], but much work needs to be done in the area of transistor characterization. Here, most textbooks focus on theoretical explanations using band-diagrams rather than on device characterization and modeling. Finally, after having performed their laboratory exercise, each group of students was given 10 minutes to present and discuss their findings to teachers and peer students. We found this to be a very positive experience with good results, as already discussed elsewhere (see chapter 6 in [13]).

## Summary

In this article, we have presented some of the results obtained prior to and in the early phases of the Socrates – Minerva project eMerge. A network infrastructure linking the collaborating laboratories has been outlined, and the educational activities offered in conjunction with the partner laboratories have been presented. The experiments focus on characterization of semiconductor devices and circuits, and on training for the operation of advanced instruments used for this purpose. The laboratories have been tested as modules in electrical engineering courses in the partner institutions and have been well-received by the students.

## Acknowledgement

This project has been carried out with support from the European Community within the framework of the Socrates program.

## References

1. Berntzen, R., Strandman, J.O., Fjeldly, T.A., and Shur, M.S., "Advanced Solutions for Performing Real Experiments over the Internet," Proceedings of 2001 International Conference on Engineering Education (ICEE- 2001), Oslo, Norway, August 2001, Session 6B1, pp. 21-26.

2. Strandman, J. O., Berntzen, R., Fjeldly, T. A., Ytterdal, T., and Shur, M. S., "LAB-on-WEB: Performing Device Characterization via Internet Using Modern Web Technology," Proceedings of IEEE International Conference on Devices, Circuits and Systems (ICCDCS 2002), Aruba, IEEE Catalog No. 02TH8611C, April 2002, pp. I022.1–6.

3. Fjeldly, T. A, Strandman, J. O., and Berntzen, R., "LAB-on-WEB – A Comprehensive Electronic Device Laboratory on a Chip Accessible via Internet," Proceedings of 2002 International Conference on Engineering Education (ICEE-2002), Manchester, UK, August 2002, Paper No. O337.

4. The eMerge Project, http://www.emerge-project.net

5. Socrates-ODL, Retwine http://www.retwine.net

6. LAB on WEB, http://www.lab-on-Web.com

7. Socrates-ODL RichODL, http://virtual.cvut.cz/odl/

8. Gomez, F.J., Cervera, M., and Martinez, J., "A World Wide Web-Based Architecture for the Implementation of a Virtual Laboratory," Proceedings of the 26th Euromicro Workshop On Multimedia And Telecommunications., Vol. II, Netherlands, Sept 2000, pp. 56-62.

9. SOAP, http://www.w3.org/TR/SOAP/

10. Web Services, http://www.w3.org/2002/ws/

11. Cabello, R., Gonzalez, I., Gomez F.J., and Martinez, J., "A Web Laboratory for a Basic Electronics Course," Proceedings of the World Conference of the WWW and Internet, WebNet 2001, Vol. 1, Orlando, USA, Oct. 2001, pp. 816- 821.

12. SmartPlot2Dlibrary (1998), http://www.et.fh-muenster.de/research/mmedia/software/smartplot

13. Fjeldly, T.A, and Shur, M.S., "Electronics Laboratory Experiments Accessible via Internet," in LAB-on-the-WEB, Running Real Electronics Experiments via the Internet, T. A. Fjeldly and M.S. Shur, eds., John Wiley & Sons, New York, 2003, ISBN: 0-471-41375-5.

14. http://centrevirtuel.creea.u-bordeaux.fr/

15. MIT Microelectronics Weblab http://Weblab.mit.edu/

16. http://www.ic.chalmers.se/ilab/

17. Söderlund, A., Ingvarson, F., Lundgren, P., and Jeppson, K.O., "The Remote Laboratory – A New Complement in Engineering Education," Proceedings of 2002 International Conference on Engineering Education (ICEE-2002), Manchester, UK, 2002.

18. Pierret, P. F., *Semiconductor Device Fundamentals*, Addison-Wesley Publishing Company, Reading, Massachusetts, 1996


**Nikos Asimopoulos** is Head of the Electrical Engineering Department of the Technological Education Institute of West Macedonia in Greece. He received his M.S. and Ph.D. degrees in Electrical and Computer Engineering from Virginia Polytechnic Institute and State University (VPI), Blacksburg, VA, USA. He has been Vice President of Engineering at Image Processing Technologies and a Visiting Assistant Professor at VPI.

**Michel Billaud** is currently Associate Professor in the Department of Informatics at the Technological Institute of the University of Bordeaux. His special interests include multimedia, open and distance learning and the implementation of ICT (Information and Communication Technologies) in practical courses.

**Ruben Cabello** was born in 1979 in Madrid, Spain. He received the degree of Computer Science Engineer in 2003 from the University Autonoma of Madrid. He is currently an Associate Professor in the Computer Science Department of the University Autonoma of Madrid. His research interests include web services, distributed applications and developing of ODL tools.

**Hans Effinger** received his Ph.D. in physics from the University of Marburg, Germany, in 1986. From 1987-1993 he was scientist in the industrial research laboratories at Dornier GmbH and Daimler AG. Since 1993 he has been a Professor of Computer Science and Mathematics at the University of Applied Sciences in Münster and is head of the operating systems laboratory. Main topics of research include operating systems, distributed systems, distributed computing, multimedia, mathematical modeling and simulation.

**Tor A. Fjeldly** is a Professor of Electrical Engineering at UniK – University Graduate Center and Norwegian University of Science and Technology, and an

Adjunct Professor at University of Oslo, Norway. He received his Ph.D. degree in physics from Brown University, and was a Scientist at Max-Planck Institute of Solid-State Physics in Germany, and at the SINTEF research organization in Norway. He has been a Visiting Professor at University of Virginia and Rensselaer Polytechnic Institute, USA. He is a Fellow of IEEE, and a Member of the Academy of Technical Sciences in Norway.

**Didier Geoffroy** was born in 1953 in Epernay, France. He received the M.Sc. degree in applied physics from the University of Orsay, Paris, in 1980. He is currently an Associate Professor in the Department of Electrical and Electronic Engineering at the University of Bordeaux, and head of the pedagogical lab centre in that Department. His special interests include multimedia, open and distance learning and the implementation of ICT (Information and Communication Technologies) in practical courses.

**Ivan Gonzalez** was born in 1978 in Vigo, Spain. He received the degree of Computer Science Engineer in 2000 from the University Autonoma of Madrid. He is currently an Associate Professor in the Computer Science Department of the University Autonoma of Madrid. His research interests include Reconfigurable Hardware, Codesign and implementation of distance learning tools for these areas.

**Francisco J. Gómez-Arribas** was born in 1966 in Segovia, Spain. He received his Ph.D. in physics from UAM. He is currently a Professor of Computer Architecture at University Autonoma of Madrid (UAM). His main research interests concern reconfigurable computing applications in FPGA circuits, embedded systems and the implementation of remote control instrumentation in C.S and E.E. practical courses.

**Prof. Dr. Reinhold S. Jäger** received his Ph.D. in Psychology from the University of Mannheim. Since 1987, he has been Head of the Center for Educational Research (zepf) of the University of Koblenz-Landau, Landau Campus, and a chair holder in Department of Psychology.

**Kjell O. Jeppson** received his Ph.D. in Solid-State Electronics in 1977 from Chalmers University of Technology. He was with Rockwell International in Anaheim, California during the academic year 1973/74. He became a senior lecturer at the School of Electrical Engineering at Chalmers University of Technology in 1978. In 1996 he was appointed Professor in the Department of Microelectronics. His main research interests are within the fields of transistor modeling and parameter extraction, and CMOS VLSI design. He has authored one textbook on semiconductor devices (in Swedish). He was the initiator of Chalmers′ E-96 project for reforming the EE curriculum for undergraduate studies.

**Herman Mann** is an Associate Professor of the Czech Technical University in Prague from which he graduated and later received his Ph.D. and Sc.D. degrees.

Besides his Alma Mater, he taught at universities in Sudan, Canada and Belgium. Currently, he is involved in research projects focused on computer-assisted modeling and simulation of multi-disciplinary engineering systems as well as on development of web-based courses. He is a Life Member of the IEEE and a Member of the IEEE Technical Committee on Computer Aided Control System Design

**Javier Martinez** is a Professor of Computer Networks at the Computer Science Department of University Autonoma of Madrid. He received his Ph.D. in physics from the Madrid University in the field of the Electrical Engineering. He has been a Visiting Scientist at the laboratory Phase CNRS at Strasbourg, France. His main research interests concern web services, distributed applications, embedded systems and the implementation of remote control instrumentation in C.S. and E.E. practical courses.

**Wilhelm Seifert** received the Dipl.-Ing. degree in physics and the Dr.-Ing. degree in electrical engineering from the Technical University of Berlin, Germany, in 1970 and 1975, respectively. From 1975 to 1979, he was with the AEG-TELEFUNKEN GmbH, Warstein, Germany, where he worked on the development of silicon power diodes, thyristors and GTOs. In 1979, he returned to the Institute for Electronic Material at the Technical University of Berlin, where he was the leader of the technology group. Since 1995 he has been a Professor at the Fachhochschule Münster in the Electrical Engineering Department. His research interests include the development of semiconductor devices and microsensoric biosensors.

**Thomas Zimmer** was born in Wollbach, Germany. He received his M.Sc. in physics from the University of Würzburg, Germany, in 1989 and his Ph.D. in electronics from the University of Bordeaux, France, in 1992. Since 2003 he has been a Professor at the University of Bordeaux. His research interests include characterization and modeling of high frequency devices. His education interests concern open and distance learning and the implementation of ICT in practical courses. In particular, he was the coordinator of the European ODL Transnational Cooperation Project "RETWINE: REmoTe Worldwide Instrumentation Network" and is currently the coordinator of the Socrates - Minerva project eMerge: http://www.emerge-project.net/.

# Chapter 18

## Developing an Advanced IT-Based Visualization Classroom for Enhanced Engineering Learning

TARA C. HUTCHINSON and FALKO KUESTER
*University of California, Irvine, CA 92697, USA. E-mail: thutchin@uci.edu, fkuester@uci.edu*

*Interactive learning, critical thinking, creative problem-solving and problem-based learning are all critical elements to enhancing engineering education. Visualization can provide the much-needed computer-assisted design and analysis environment to foster problem-based learning, while Virtual Reality (VR) can provide the environment for hands-on manipulation, stimulating interactive learning in the engineering classroom. At the University of California, Irvine a new interactive learning environment, termed VizClass, is being developed. VizClass incorporates a specially designed lecture room and laboratory integrating both 2- and 3-dimensional spatial learning by coupling a series of interactive projection display boards, also called touch sensitive white boards, and a semi-immersive 3D wall display. This paper will describe the hardware architecture designed for this new environment as well as the initial visualization framework to be integrated with an open source object-oriented general finite element platform. Example numerical analysis results from a representative model building structure are described, illustrating the integration of simple visual paradigms to provide an enhanced understanding of the physical model.*

### INTRODUCTION

Though the field of engineering has changed dramatically in the last 20 years, the study of engineering has changed relatively little. Students continue to passively listen to lectures in chalkboard-based environments that afford them little opportunity for visualization, hands-on manipulation, interaction, or creative design. Indeed, almost all of these skills, which are vital for engineering, are learned through on-the-job practice rather than at universities. A continuation of this trend threatens to weaken the relevancy of engineering and computer science study at the university. Such modes of learning have an especially negative impact on under-represented groups in engineering, such as African Americans, Hispanic Americans, and women, all of who have been shown to especially benefit from interactive forms of instruction [1,2].

Interactive learning, critical thinking, creative problem-solving and problem-based learning are all critical elements to enhancing engineering education. Visualization can provide the much-needed computer-assisted design and analysis environment to foster problem-based learning, while Virtual Reality (VR) can provide the environment for hands-on manipulation, stimulating interactive learning in the engineering classroom. At the University of California, Irvine a new interactive learning environment, termed *VizClass*, is being developed. VizClass incorporates both a specially designed lecture room and laboratory integrating both 2- and 3-dimensional spatial learning by coupling a series of interactive projection display boards, also called touch sensitive white boards, and a semi-immersive 3D wall display. The use of VizClass and the accompanying pedagogical and curricular innovation will allow for the redesign of existing engineering courses, the development of new courses, and the integration of courses and curriculum across engineering disciplines. Verification of VizClass is targeted towards finite element methods applicable to civil and structural engineering problem-solving at the graduate education level. Different pre- and post-processing tools are being developed for an open source finite element software package to assist in smoothly integrating primitive visualization concepts into the numerical analysis process.

This paper will describe the hardware architecture designed for this new environment as well as the initial visualization framework that was integrated with an open source object-oriented general finite element platform. Example numerical analysis results from a representative model building structure are described, illustrating the integration of simple visual paradigms to provide an enhanced understanding of the physical model. First, a brief review of the literature is provided describing environments that have been or are being developed using similar technologies.

## PREVIOUS IT-ENVIRONMENTS

In recent years, a number of information technology (IT) environments have been developed at universities and research laboratories across the United States and internationally. These environments provide infrastructure support for classrooms, office activities, tele-conferencing and meeting spaces. A common thread when comparing these environments is the enhanced visualization and display capabilities provided to support activities held in these spaces. In some cases, multiple tiled 2D display systems are used, while in others 3D active or passive stereo support is provided on high-resolution display systems, allowing for full immersion in the space. Important to this, from a user perspective, is the capability of the environment to provide comfortable human-computer-interaction devices to support user activities. Of particular interest to the work described in this paper are the *eClass*, *iRoom*, and the *Office of the Future* environments.

The *eClass* project, formerly called *Classroom2000*, has been developed by the Future Computing Environments group at the Georgia Institute of Technology ([3,4,5]). Since 1995, *eClass* has supported teaching at Georgia Tech in the areas of Computer Science and Engineering by providing a digital environment for course lectures and discussion sessions. The environment supports playing, recording, and animating (in real-time) lectures. A Java-based client-server system, ZenPad, records and stores lecture video clips, slides, and annotations during each lecture. Lectures collected and compiled by ZenPad provide a detailed record of the discussion, including the original presentation, annotations compiled during the presentation, and a full audio record of the session. *eClass* has a unique 2D display layout for lecturers, primarily including a digital active LiveBoard, and two projection surfaces for passively displaying information. Figure 1(a) shows a photograph of the *eClass* environment and highlights the components of this IT-classroom.

The Interactive Workspace Project, more commonly termed *iRoom* (6, 7]), at Stanford University uses a broad array of interactive devices, such as multiple large 2D and 3D displays and hand held and other portable devices to provide an interactive research working space. A software layer developed by Shankar et al. [8], termed *iROS* (Interactive Room Operating System), is used to provide the underlying middleware architecture integrating the different interaction devices and computational activities together. The *iRoom* project brings together interaction at a broad range of scales and using a variety of interfacing techniques; both wired and wireless devices are integrated into the environment. For example, the user can control multiple displays with a handheld device such as an iPAQ, or remotely activate devices, such as scanners, printers, etc. during working sessions. Figure 1(b) shows a photograph of the *iRoom* environment and the components of this IT-workspace.

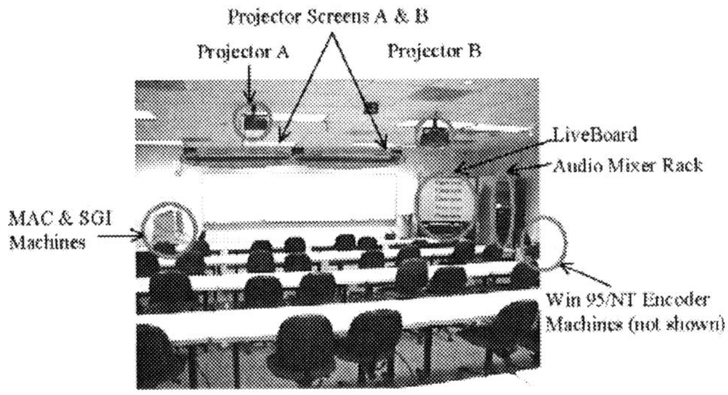


(A) *ECLASS* (AT GEORGIA TECH UNIVERSITY)

(B) *IROOM* (AT STANFORD UNIVERSITY)

FIGURE 1
EXAMPLE IT-ENVIRONMENTS, PHOTOGRAPHS COURTESY OF [5] AND [6], RESPECTIVELY

The *Office of the Future* environment ([8,9]) was developed at University of North Carolina at Chapel Hill to provide an immersive space allowing 3D graphical model display and manipulation for the future office environment. The primary focus of the office of the future project is to capture and display 3D data dynamically and on random surfaces within the space in which it is constructed. To do this, Raskar et al. [10] have replaced ceiling lights with projection systems and carefully controlled structured light within the room. Using advanced computer vision techniques, the authors create a flexible display environment whereby changes in geometry, intensity and resolution variations across multiple display areas are dynamically controlled.

These and other systems are providing state-of-the-art interactive digital workplaces, whether it is for an office, educational, or entirely research environment. Thus far, however, none of these systems has integrated interactive visualization and advanced 3D display components with mathematical solvers. Furthermore, with the exception of the *iRoom* project, previous IT-spaces that provide advanced visual tools have used *either* 2D or a 3D semi-immersive-type display paradigm. Little emphasis has been given to integrating the two space scales into a single environment. In engineering education, mathematics and modeling are critical to developing and solving a physical problem. Engineers first learn the fundamental mathematics defining a boundary value problem and subsequently must develop a physical representation of the problem. In most cases, the scale of the problem is broad; thus, 3D interpretation and spatial correlation is generally important to understanding the physical constraints imposed. Therefore, in the engineering education environment, it is important that students first interpret the 2D information, such as equations, algorithms, etc., and then work with the 3D information, such as geometric models.

## VIZCLASS: AN ENVIRONMENT FOR EXPANDING ENGINEERING THINKING

A new IT-Classroom is being developed at the University of California, Irvine, with the primary objective of providing a completely digital, interactive workspace for research and education in the areas of Computer and Civil Engineering. Application courses include computer graphics, scientific visualization, and numerical methods (finite element methods) in structural engineering. The IT-space serves not only as an educational environment, but it also provides a space for interdisciplinary research centered on simulation and visualization as well as an IT development test bed advancing system-level research and education.

A 65-$m^2$ classroom in the Engineering Gateway building on the UC Irvine campus was allocated for the development of VizClass. The original layout of this room is shown in Figure 2(a). In the original room, approximately 20 students passively listened to white board lectures provided at the working front space of the room. Over the past six months, the room was reconstructed, redesigned and remodeled to provide support for a new advanced IT-hardware layout incorporating a range of 2D and 3D displays, driving computational systems, and interactive devices. The layout of VizClass, showing the 2D active display area and the 3D stereo working area is shown in Figure 2(b). The following sections describe the advanced display systems, interactive devices, and driving computational systems designed for VizClass.

### Advanced Display Systems

The 2D display component consists of three interactive 1.8-meter diagonal (72-inch) SmartBoards™ connected to three computers with DVI output. The SmartBoards™ are compactly designed to fit within a thin 60 cm deep space, by backwards-projecting images onto an angled mirror and then forward projecting them onto the optical screens. A single NEC 1060 2500 ANSI Lumens projector with a short throw lens is used per screen to provide an on-screen resolution of 1024x768 pixels. The front of the room layout of 2D displays provides a large laterally tiled screen for users. The digital SmartBoards™ have multiple input/output sources - with the primary source being an optical on-screen input providing active interaction for annotating and controlling input on the displays. Each SmartBoard™ is also equipped with a digital pen-tray, providing digitally color-coded pens, supporting '*digital ink*,' a display surface eraser, and a wireless keyboard with a trackball mouse. These interface devices support input scenarios via direct application of the *digital ink* from touch events, or query or data access interpreted by mouse or keyboard events. Figure 3(a) shows a photograph of the 2D interactive SmartBoards™ installed in VizClass. For interested readers, the cost of these digital display systems, in 2003 U.S. dollars, was $18,000. Cost can be significantly reduced if the slimline frame and/or rear projection are not required. For comparison, a typical whiteboard or chalkboard can range in cost from $500-1000.

(A) PRIOR TO VIZCLASS DEVELOPMENT

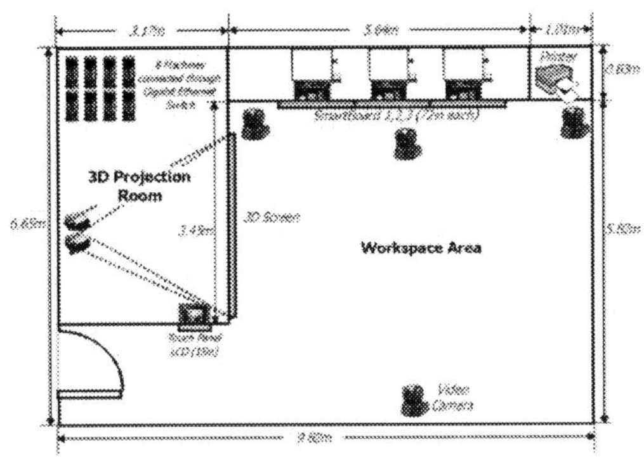


(B) SCHEMATIC ILLUSTRATION OF VIZCLASS LAYOUT

FIGURE 2
PRE- AND POST-VIZCLASS DETAILS

The 3D display component is designed with two projectors vertically stacked and mounted at floor level, projecting from the rear onto a large 3-meter (10-foot) diagonal screen. The projectors are digital light projectors (DLP) with a brightness of 3000 ANSI Lumens and resolution of 1280x1024. Each projector is equipped with polarizing filters and together the two projectors are used to generate the stereoscopic image. A single PC with an nVidia™ Quadro FX-1000 graphics card using quad-buffering is designated to provide the stereo image pairs and a CYVIZ™ 3D stereo converter is subsequently used to assign the left and right eye images to the respective projector. Users can wear simple polarized glasses to engage in a 3D immersive presentation.

In the workspace area (lecture and discussion space), a 3D surround sound and motion tracking system are planned for capturing lectures and discussions in all modalities. Video capture is accomplished by using a single motion tracking camera at the rear of the room and multiple passive, fixed-viewing cameras arrayed towards the audience. Sound input and output (via multiple microphones) requires a hardware echo-cancellation system and multiple sound cards to receive and send spatial sound information into the workspace area. Properly synchronized video and audio acquisition is required to allow for archiving course material and later playback.

### Device Control

To control all of these devices, the 2D/3D displays, the touch-sensitive LCD display, cameras and sound system, an 8-node PC cluster connected via a Gigabit Ethernet switch is provided in the 3D projection room. System software supporting the environment is primarily being developed in C/C++ and Python and integrated with wxWindows and wxPython, respectively, for GUI development with an operating system specific look and feel. Figure 3(b) shows the PC cluster and working space for development of the software interface layer. The PCs have been actively named based on their function, e.g.VizSmart1, VizSmart2, and VizSmart3, each independently controlling the SmartBoards™, Viz3D controlling the 3D stereo display, VizAudio controlling the sound system, etc. A single PC, VizFEA, is dedicated for numerical analysis solution processing.

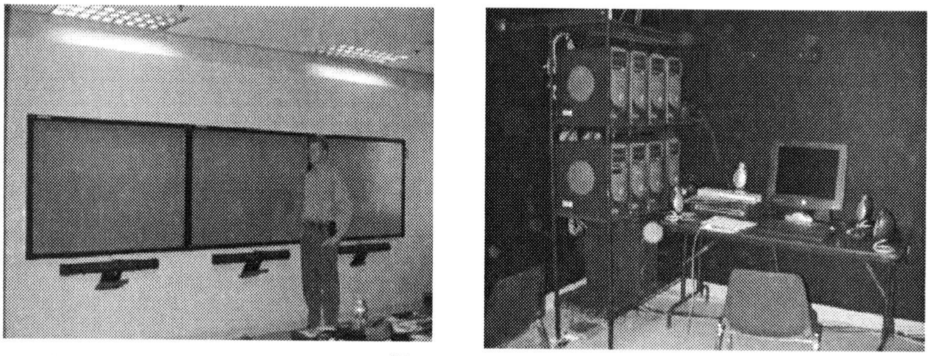

(A) ARRAY OF THREE-2D SMARTBOARDS™ (B) PC CLUSTERS IN VIZCLASS

FIGURE 3
EQUIPMENT INSTALLED IN VIZCLASS

The synchronization of the multiple input/output devices is controlled through a keyboard-video-mouse (KVM) matrix switch also located in the 3D projection room. The KVM switch supports high-resolution video display with no degradation of the signal. Ultimately, additional interactive devices will be integrated to respond to both the 2D and 3D display systems. A prototype of this idea is shown in Figure 4, where the user, to manipulate a model, wears datagloves.

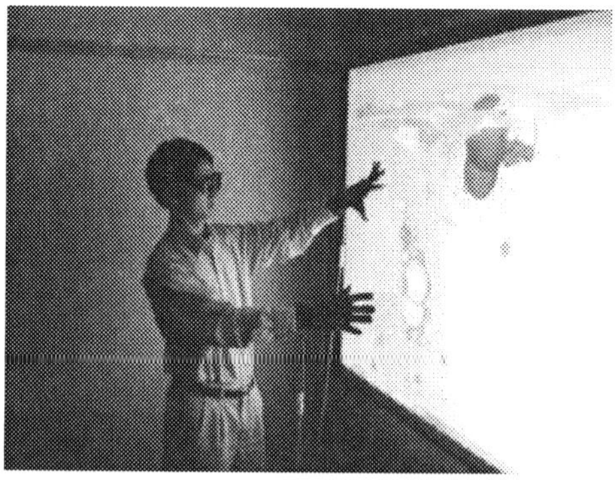

FIGURE 4
EXAMPLE OF INTERACTING WITH THE 2D ACTIVE

Access and control to multiple devices integrated into VizClass is made possible through an interface software layer running on a LCD touch-sensitive display, mounted at eye-level at the entrance to the environment. Presenters will be able to load their digital data via the VizClass webserver prior to lectures and subsequently activate this data onto the 2D and/or 3D display systems via the touch-sensitive display.

Digital material may include slides, web contents, video clips, sound files, numerical simulations, animations or interactive visualization. This may be further augmented with hand written notes, annotations, the instructor's voice and gestures tying information together as well as instructor-student and student-student interaction. The flexibility of this design also lends itself to distance learning scenarios.

## VISUALIZING NUMERICAL MODELS IN VIZCLASS

Finite element analysis (FEA) is one form of numerical modeling commonly used in civil, mechanical and aerospace engineering. In FEA, the problem domain is spatially discretized into known primitive elements (1D, 2D, or 3D elements) and nodes with given boundary and loading constraints. In classic FEA environments, engineers usually spend 70% of their time working on 3D shape modeling and mesh generation – the pre-processing phase, 20% on analysis and review of results – the post-processing phase, and only 10% on running the FEA solver. Thus, enhancing the visual representation of the numerical model, especially during the pre-processing model development stage, will increase productivity, while providing increased time for analysis interpretation. Most current commercial FEA software, such as ANSYS, ADINA, and MSC-Nastran, have good 2D graphical user interfaces for the pre- and post-processing steps; however, such commercial packages have yet to achieve stereoscopic visualization and full interaction during the visual previewing stages.

Virtual Reality (VR) technology is premised upon creating a real-time synthetic environment, believable to users as being as near to the physical world as possible. Human-computer interaction within VR is a critical component to realizing this true immersion and grasping near-physical world reality. Wrapping VR technology around FEA can assist in immersing the analyst in the model, driving a clearer picture of the design during both the pre- and post-processing phases. Such an approach is particularly appealing since the cost of computational systems and stereoscopic devices has seen a reduction over recent years. A brief review of the literature indicates that there have been various application-specific VR-FEA integrations.

### Literature Review - Virtual Reality Techniques Applied to Finite Element Analysis

Research has been conducted to integrate VR technologies with FEA, resulting in both commercial and proprietary software packages. In general, previous approaches have focused on first providing a wrapping layer for the FE analysis package with VR engaged with either pre- or post-processed data and subsequently, attempting to enhance the level of immersion within VR by providing new interaction devices for event triggering and model manipulation. Interaction is accomplished using, for example, data gloves, pens, and head-mounted-displays. Additional devices can help enhance the feeling of immersion within the scene if the user can grab, lock, and manipulate objects or the scene. In an application by Yeh and Vance [11], for example, the integration of post-processed data from FE analyses within a virtual environment provided a visual space in which the model deformation and stress contours could be projected in stereo view for users. Using a flexible data set, the user could interactively change design parameters using data gloves, without re-running the analysis. In an application by Ryken and Vance [12], virtual reality techniques were applied to study the stress analysis of a tractor lift arm. NURBS-based free form deformation, sensitivity analysis, and collision detection were combined with VR and FEA to create an interactive environment for designers to view and modify the shape of the part, evaluate the resulting stresses, and check for interference of the new part shape and surrounding parts in real-time. Their application provided a particularly novel hardware solution for attempting to fully immerse the user into the model scene by using a surround screen environment, C2, which is similar to a CAVE environment.

Others have developed wrappers for either pre- or post-processing stages of FEA. Liverani et al. [13], for example, developed a semi-immersive real-time 3D-enabled VR software layer for FEA called VRFEA. Interfacing tools that were encoded into VRFEA were fairly flexible. These include stylus and glove devices that provide an interface to the model and allow users to operate on the elements and nodes, create and adjust the mesh, specify boundary conditions and control which results are visualized. The software framework was directly integrated into the popular OpenInventor™ toolkit by SGI. In a similar API-wrapper approach, Lu et al. [14] studied two different API software packages for interfacing VR and FEA: the popular dVISE 3D program and SUN's Java3D package. In their study, it was concluded that dVISE, although powerful when interfacing with VR hardware, is relatively slow when loading objects, mostly due to its file reading approach, which required transforming data from multiple formats. Java3D was selected to maximize performance and speed of the VR-FEA interaction.

## VRML-Based Approaches

The expansion and popularity of the Internet has led to implementation of 3D graphics contents based on the Virtual Reality Modeling Language (VRML). For a discussion of the use of VRML viewer and of VRML as a component modeling language, see Chapter 34 by Mhamdi et al. in this book. VRML is appealing since it is operating system independent, only requires a VRML plug-in and may be run stand-alone in a network browser. FE analysis modules have been integrated with VRML in an effort to provide a VR-interface via the Internet. For example, Del Pino et al. [15] directly integrate a ray-tracing package with the FE model of a 3D acoustic problem. In their approach, a VRML-based wrapping of the FE modeling layer is used to provide model interaction. Similarly, Ranga and Gramall [16] present an application of FEA actively running over the Internet using a VRML-based software layer.

Although not using the FEA-framework, in a related geometric modeling application specific to building structures, Cross et al. [17] developed VRML application tools for visualizing the earthquake retrofitting of structures. The educational module developed by Cross et al. provides a powerful, stand-alone visual learning tool. Similarly, Haque [18] developed VRML-based structural design modules allowing students to become immersed in the 3D Internet-based module.

## Integrating VR with FEA for Engineering Education - OpenSees

The approach taken in this project includes first developing an application wrapper framework for an open-source finite element platform. The platform selected is the Open System for Earthquake Engineering Simulation (OpenSees) [19] being developed by researchers within the Pacific Earthquake Engineering Research Center (PEER). OpenSees was designed as an open source framework for simulating the seismic response in structural and geotechnical systems. The platform independent framework was developed entirely in C++ and provides an intuitive Tcl/Tk scripting front-end. With this front-end, complex structural systems can be defined, covering aspects such as specimen geometry and material properties, boundary conditions, and loading, among others. However, the primary focus for this system is on delivering precise numerical results, while data analysis and interpretation are left to other external post-processing tools.

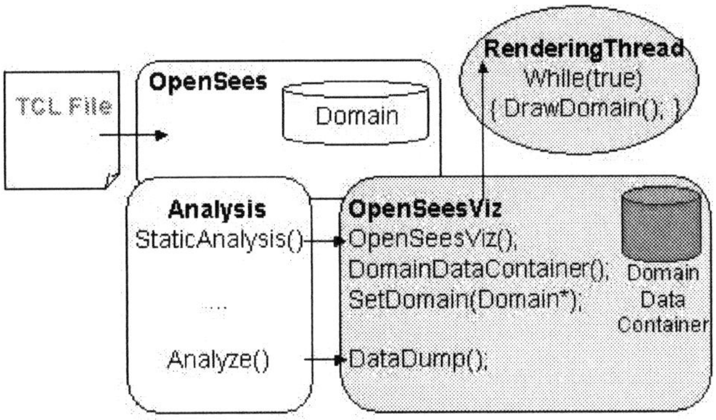


FIGURE 5
SOFTWARE FLOW - VISUAL INTERPRETER FOR OPENSEES

A general library called OpenSeesViZ is being developed as a wrapper for OpenSees based on the general software flow shown in Figure 5. Upon initialization of OpenSees to perform an analysis, OpenSeesViz is initiated. This instance creates a thread that maintains a separate data structure for visualization-specific tasks. OpenSeesViz then displays snapshots of images of the analysis in an OpenGL window as the analysis is performed, i.e. at each user-defined time step. Simple visual paradigms are used to represent common elements used in structural analysis, such as trusses, beams, and solid elements. Nodes are represented using geometric proxies (spheres), and some specialty sections have been implemented to graphically represent physical members, which are simplified numerically by the user.

### Example Structure

An example structure, modeled in OpenSees and visualized during analysis using OpenSeesViz, is shown in Figure 6. Figure 6(a) shows the finite element representation of the 2-story, 3-bay structure. The structure is composed of varying sectioned girders (beams) and columns and constructed of reinforced concrete. Floor-level dead and live loads representative of practical field conditions are applied. The structural model itself is a modified form of an example structure provided with the OpenSees analysis package. Gravity (vertical load) analysis, lateral load analysis, and dynamic (base input) analysis are performed on the structure and the visual deformations of the structure are observed. Figure 6(b) shows an example snapshot of OpenSeesViz during the lateral analysis. The horizontally placed reinforcing bars, which would otherwise be hidden from the user, are articulated by using translucent colors for the outer concrete patches. In such a fashion, the user can understand the extent of the sectional detailing as it occurs along the finite element representation of the girders.

## SUMMARY REMARKS

At the University of California, Irvine a new interactive learning environment, termed *VizClass*, is being developed. VizClass incorporates both a specially designed lecture room and laboratory integrating both 2- and 3-dimensional spatial learning by coupling a series of interactive projection display boards, also called touch sensitive white boards, and a semi-immersive 3D wall display. Verification of VizClass is targeted towards finite element methods applicable to civil and structural engineering problem-solving at the graduate education level. Different pre- and post-processing tools are being developed for an open source finite element software package to assist in smoothly integrating basic visualization concepts into the numerical analysis process. The initial hardware architecture designed for this new environment, as well as the initial visualization framework to be integrated with an open source object-oriented general finite element platform, is described in this paper.

### Future Plans: Integration of VizClass into Course Curriculum and Assessment of Student Learning

VizClass is planned for integration into two specific engineering education areas: (i) Civil Engineering and (ii) Computer Engineering. Integration is planned by holding lectures and discussion sessions in VizClass. Specifically, for structural engineering students, the new approach holds promise for enhancing their ability to understand modeling and analysis via geometric interpretation.

To assess the impacts and merits of VizClass on student learning, pre-VizClass data collection was conducted on three courses taught in a non-digital, chalkboard-based environment. This data will be compared with material collected after these courses have been taught within VizClass. In addition, surveys, student and instructor interviews, videotaping, in-class monitoring of instructor-student interaction and other techniques are being used to assess the impact of VizClass on student learning.

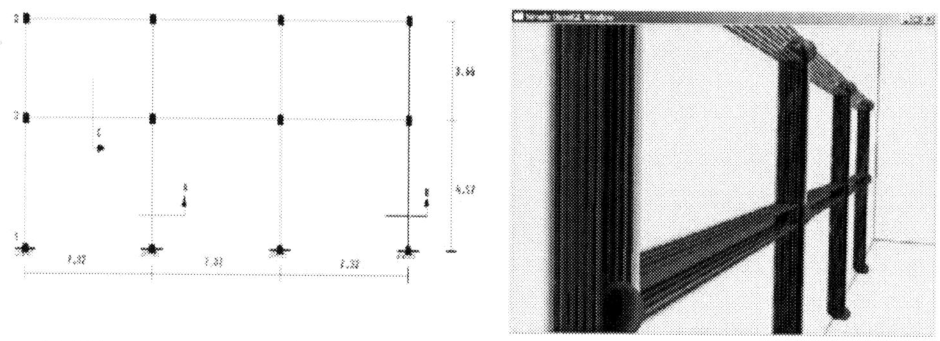

(A) FEA REPRESENTATION OF 3-BAY STRUCTURE

(B) CLOSE-UP VIEW SHOWING REINFORCING STEEL SECTIONS (REPRESENTED AS CYLINDERS)

FIGURE 6

EXAMPLE STRUCTURE SHOWN IN 2D AND IN 3D (WITHIN OPENSEESVIZ)

## ACKNOWLEDGEMENTS

The National Science Foundation, under Grant Number EIA-0203528, supports this research. Support for the hardware infrastructure was provided by the California Institute for Information and Telecommunications (Cal-(IT)$^2$). The above support is greatly appreciated. In addition, the assistance of Mr. Bob Kazanjy, the student researchers of the Structural Engineering Test Hall and the Visualization and Interactive Systems (VIS) group at UC Irvine are greatly appreciated. Mr. Rui Lu and Mr. Sung-Jin Kim are acknowledged for their early contributions and assistance to the hardware installation.

# References


1. AAUW Educational Foundation, Tech-Savvy: Educating Girls in the New Computer Age, Washington, D.C., 2000.

2. Warschauer, M., *Electronic Literacies: Language, Culture, and Power in Online Education,* Lawrence Erlbaum Associates, Mahwah, N.J., 1999.

3. Abowd, G.D., Brotherton, J.A. and Bhalodia, J., "Classroom 2000: A System for Capturing and Accessing Multimedia Classroom Experiences," Proceedings of the Conference on Human Factors in Computing Systems, CHI 98, May 1998.

4. Abowd, G.D., "Classroom 2000: An Experiment with the Instrumentation of a Living Educational Environment," IBM Systems Journal, Vol. 38, No.4, October 1999, pp. 508–530.

5. eClass (Formerly Classroom2000), Future Computing Environments Group, Georgia Institute of Technology, http://www.cc.gatech.edu/fce/eclass/, 2003.

6. iWork, "Stanford Interactive Workspaces Project," Stanford University, Department of Computer Science, http://iwork.stanford.edu/, 2003.

7. Johanson, B., Fox, A. and Winograd, T., "The Interactive Workspaces Project: Experiences with Ubiquitous Computing Rooms," IEEE Pervasive Computing, Vol. 1, No. 2, April-June 2002, pp. 71–78.

8. Ponnekanti, S.R., Johanson, B., Kiciman, E. and Fox, A., "Portability, Extensibility and Robustness in iROS," Proceedings of the IEEE International Conference on Pervasive Computing and Communications (Percom 2003), Dallas-Fort Worth, TX, March 2003, pp. 11–19.

9. UNC, "Office of the Future Project," University of North Carolina at Chapel Hill, Department of Computer Science, 2003, http://www.cs.unc.edu/Research/ootf/

10. Raskar, R., Welch, G., Cutts, M., Lake, A., Stesin, L. and Fuchs, H., "The Office of the Future: A Unified Approach to Image-Based Modeling and Spatially Immersive Displays," Computer Graphics Proceedings, SIGGRAPH, 1998.

11. Yeh, T.-P. and Vance, J.M., "Combining MSC/NASTRAN, Sensitivity Methods, and Virtual Reality to Facilitate Interactive Design," Finite Elements in Analysis and Design, Vol. 26, 1997, pp. 161–169.

12. Ryken, M.J., and Vance, J.M., "Applying Virtual Reality Techniques to the Interactive Stress Analysis of a Tractor Lift Arm," Finite Elements in Analysis and Design, Vol. 35, 2000, pp. 141–155.

13. Liverani, A., Kuester, F., and Hamann, B., "Towards Interactive Finite Element Analysis of Shell Structures in Virtual Reality," Proceedings of the International Conference on Information Visualization, London, England, July 14-16 1999, pp. 340–346.

14. Lu, L., Connell, M. and Tullberg, O., "The Use of Virtual Reality in Interactive Finite Element Analysis by JAVA3D API," Proceedings of the Applied Virtual Reality in Engineering and Construction, AVR II and CONVR, Goteborg, Sweden, October 4-5 2001, pp. 60–68.

15. DelPino, S., Heikkola, E., Pironneau, O. and Toivanen, J., "A Finite Element Method for Virtual Reality Data," Numerical Analysis/Analyse num´erique, Vol. 330, No. 1, 2000, pp. 1107–1110.

16. Ranga, K. and Gramoll, K., "3-D Finite Element Analysis on the Internet using Java and VRML," Proceedings of the American Society of Engineering Education (ASEE) Conference and Exposition, 2000.

17. Cross, B., Smith, S., Kennedy, C. and Arbogast, S., "Virtual Reality Retrofit Demonstrations," Proceedings of the National Conference on Earthquake Engineering, 2002.

18. Haque, M.E., "Web-Based Visualization Techniques for Structural Design Education," Proceedings of the American Society for Engineering Education (ASEE) Annual Conference and Exposition, 2001.

19. PEER, "Open System for Earthquake Simulation," Pacific Earthquake Engineering Research (PEER) Center, University of California, Berkeley, Online documentation at: http://opensees.berkeley.edu/, 2003

**Tara C. Hutchinson** received her M.S. degree in Civil Engineering from the University of Michigan, Ann Arbor in 1995 and her Ph.D. degree from the University of California, Davis in 2001. She is Assistant Professor in structural engineering in the Department of Civil and Environmental Engineering at the University of California, Irvine. Current research activities include both experimental and analytical studies primarily in earthquake engineering and emphasizing seismic soil-structure interaction and performance-based earthquake engineering. Other areas of interest include research in Integration of visualization and IT-based techniques.

**Falko Kuester** received his M.S. degrees in Mechanical Engineering and Computer Science and Engineering in 1994 and 1995, respectively, from the University of Michigan, Ann Arbor. In 2001, he received his Ph.D. from the University of California, Davis and currently is Assistant Professor in the Department of Electrical Engineering and Computer Science at the University of California, Irvine. His research interests include image-based modeling and rendering, tera-scale scientific visualization and virtual reality as well as distributed and remote visualization. Application areas of his research include earthquake engineering, biomedical engineering, rapid prototyping, and command and control.

# Chapter 19

# Online Experimentation with a Spectrum Analyzer in a Signal Processing Virtual Laboratory

KEITH HOLBERT[1], MIHAELA ALBU[2] and FLORIN MIHAI[2]
*[1]Arizona State University, Department of Electrical Engineering, Tempe, Arizona 85287-5706, USA. [2]Universitatea Politehnica Bucureşti, Department of Electrical Engineering, 313 Spl. Independentei, Bucharest, 060042, România. E-mail: holbert@asu.edu, mialbu@pcnet.ro, fmihai@electro.masuri.pub.ro*

*This paper describes a remote experimentation project that is part of a virtual laboratory (VLab). The activity is organized around a spectrum analyzer and a function generator, as real devices mirrored into web nodes. A logically structured replica of the spectrum analyzer is realized on a LabVIEW platform, in a client-server approach. The application also integrates other modules including function generator control, chat, user authentication, and video broadcast of images from the device display. The corresponding server modules are distributed between the VLab nodes in Politehnica University of Bucharest and Arizona State University. Students from worldwide locations and enrolled in different programs with different subject-related knowledge levels can access the spectrum analyzer, while enjoying increased schedule freedom.*

## INTRODUCTION

According to the World Intellectual Property Organization (WIPO), about 10% of the world's population, representing more than 605 million users, were online as of 2002 [1]. WIPO reports that this number is increasing more rapidly than earlier foreseen (the 1999 forecasts for 2002 envisaged only a third of the actual number of Internet users) so that the world online population could reach one billion by 2005. Internet use is directly affecting the means by which education is delivered, particularly in the area of distance learning (DL). Distance education via the Internet is already part of many university programs, and its penetration is increasing. Electronic learning has smoothly transformed itself, in less than a decade, into I-learning (Internet learning). We consider e-learning to be education provided by electronic media such as video segments, computer visualization and, in general, an organized collection of knowledge that can be asynchronously retrieved by the members of a learning community. In contrast, we define I-learning is a subset of e-learning, as long it is using an electronic version of the teaching material. I-learning is also an extension of e-learning, in the sense that the knowledge is distributed (into web nodes) and can be simultaneously accessed; that is, knowledge is *shared* by multiple users who intercommunicate.

A challenge that educators are now facing is that the daily, Internet-based environment is driving a follow-up response from instructors. This is visible in electrical engineering education - a domain where simply using computers and the appropriate software is no longer attractive enough for prospective students. For example, today's students have become accustomed to, and often expect, advanced graphics and visualization that they routinely see in electronic media and motion-picture special effects.

DL is evolving into a standard delivery mode in higher education. The initial emphasis on implementing DL for college courses was directed toward lecture-only courses; however, most engineering subjects require experimental work. In the U.S., Engineering Criterion 3(b) from the Accreditation Board for Engineering and Technology (ABET) requires that "engineering programs must demonstrate that their graduates have an ability to design and conduct experiments, as well as to analyze and interpret data" [2]. Presently, there is a technology gap between non-laboratory and laboratory classes in terms of the incorporation of e-learning. Nevertheless, in the case of electrical engineering instruction, pursuing hands-on experimentation in classical laboratories is still a dominant component. Only a few laboratories are integrated in a remote access mode, usually over the Internet, and most of them are simulation-based. A recently developed subset of DL efforts is that of web-based laboratory experiments.

We believe that a new paradigm of teaching is evolving toward distributed access to an online-available, continuously-changing pool of information. This virtual classroom becomes as real as the classical one [3]. The laboratory, a subset of learning places, is particularly open to confusion when considered virtual: it is the place that has no physical determinant, and thus is called virtual, while the activity performed is tangible. Questions that have arisen from studies on cyber culture robots and artificial intelligence include [4]: How do we process measured signals from a distant place? How do we share observations and interact with the phenomena under study as part of a geographically dispersed team, all while preserving the sense of "being present"?

This paper first examines existing work in establishing remote instrumentation-based (versus simulation) experiments. There are more examples of simulation-based remote laboratories than instances of actual hardware use. Present implementations can be classified into two approaches: (i) delivering experiments via the Internet, while targeted to a single online user, and (ii) broadcasting experiments, where students do not share the same physical location, but rather a virtual one: a multi-user laboratory platform. We then present a virtual laboratory (VLab) that utilizes real devices in an online mode. In addition, we propose that web-based laboratory experiments might be more economical, as noted in [5]. Within this paper, our philosophical approach is conveyed and implementation details are presented for a digital signal processing laboratory experiment that targets electrical engineering students.

## VIRTUAL LABORATORIES

Early virtual laboratories did not employ physical equipment. An early definition of virtual laboratory was "a simulation of a physical laboratory, created for the purpose of providing instruction to students in the use of laboratory facilities" [6]. One of the major advantages provided by simulation is its ability to help trainees develop technical skills. The student uses a simulator to develop skills for manipulating the corresponding real-world system. For the past ten years, apart from providing other services, the World Wide Web has been looked upon as an environment for hosting modeling and simulation applications [7]. There are three major application areas for online simulation [8]: (1) to support the operation of a real-world system, (2) to support online diagnostics, and (3) to augment reality. In the case of augmented reality simulation, real and virtual entities (humans or equipment) as well as the environment can coexist at the same time and, therefore, operations can take place in a richer environment. An online simulation might induce a lack of responsibility toward the activity performed when people are not provided the opportunity to interact with the physical equipment. This is the case with virtual laboratories, where students located at different nodes of the global network share data and software. A solution is to provide online access to real experiments, thus enhancing the role of a virtual laboratory as not only a teaching mode, but also a learning tool.

Some educators feel that virtual laboratories are not as effective as directly manipulating actual equipment. For example, it can be argued that students miss the hands-on experience; however, remotely located hardware is nonetheless superior to simulation in that respect. Often overlooked is the fact that, psychologically speaking, the remote student is shielded against the adverse consequences of misconnecting equipment; furthermore, software protection interlocks that prohibit the student from making such mistakes increase the longevity of the equipment. Nevertheless, this is a major restriction on the student's freedom in adopting a personal path in acquiring knowledge; often, the possibility of committing errors during the experiment is a major motivation in performing high-quality activity. A similar drawback to simulation is the fact that the student does not enter the actual laboratory, so there is less hardware troubleshooting (*e.g.*, loose wiring or connections). The non-simulation approach taken here retains these possibilities but in a restricted form.

To leverage the advantages of the Internet, web-based laboratories require a tradeoff between simulation and actual operation of the laboratory equipment. This solution is particularly suitable when applied to technical universities, where effective instruction of certain engineering topics requires experience with actual equipment, rather than small-scale replicas or simulation. Recently, scheduling flexibility for laboratory classes has become an important issue since many students have extended commitments, especially part-time students. Further, we acknowledge that the dominance of the Internet in the development of information and communication technology has made web-based distributed solutions increasingly attractive. A response addressing these issues is to develop web-based laboratories on an Internet platform from which real (vs. simulation) experiments can be conducted at anytime, without instructor surveillance or guidance. We denote virtual laboratory as being grounded in the definition of laboratory as a place providing opportunity for experimentation. When this place becomes a virtual location on the web, having a distributed nature and a dynamic configuration, it is called a *virtual laboratory*.

Significant effort has been expended into organizing off-campus delivery of lessons using multimedia tools [9]. Laboratories based on simulation techniques have been set-up for local [10] and remote access [6,7]. There are many endeavors reported in the field of remote-experimentation [11-13], which allows more optimal use of facilities and personnel time in engineering education. A subset of the literature reporting various implementations of virtual laboratories is dedicated to the signal processing discipline. A popular approach has been to utilize LabVIEW [14] or Java [15-17] simulations. As illustrations, Siddiqui *et al.* developed a LabVIEW-based simulation of a spectrum analyzer that students use as a pre-lab experiment [18], and Spanias *et al.* developed an object-oriented J-DSP (Java Digital Signal Processing) tool to provide online simulations of modern statistical and spectral analysis algorithms through web browser access [19].

Virtual laboratories based on remote access to physical laboratory equipment have also been reported. The use of LabVIEW™ has been particularly prevalent for such applications. The interested reader is directed to [20] for a comparison of LabVIEW technologies for Internet-enabled applications. Several papers in a special issue of the *International Journal of Engineering Education* in 2000 were dedicated to DL applications of LabVIEW in engineering education [21]. Educators have employed virtual laboratories to utilize various equipment, such as a network analyzer [12], a waveform generator and data acquisition card [16], and a semiconductor parameter analyzer [22]. Still other educators have chosen to create virtual laboratories by combining simulation and physical experiments in an integrated approach [23,24].

Some investigators are delivering experiments via the Internet geared toward a single online user. For example, Ko *et al.* limited access to a spectrum analyzer to one-to-one interaction [11]. Solitary use is also an approach of commercial experiment providers [13]. Another technique is that of allowing simultaneous remote users to share a virtual, multi-user laboratory platform [25]. Implementation options include chat, video, archiving, hardware and software platforms organized in a multi-user environment. The sense of teaming is exploited by networked games, and this similarity with Internet gaming can be seen as an attractive feature to today's students, but can also induce a lower level of responsibility toward the experiment itself.

## A Distributed Signal Processing Laboratory

VLab is a series of laboratories, with VLab experiments being organized into two groups: (1) those dedicated to applications that mainly offer remote access to digital measuring equipment, such as oscilloscopes, multimeters and signal generators, and (2) a group of applications that use data acquisition cards for controlling events and processing signals from measurement systems. This application fits into the first group.

Although computer simulation is very useful when teaching signal processing topics, practical situations often require sophisticated (and implicitly expensive) equipment whose use might be difficult because of insufficient availability. An efficient approach for students to develop a deeper understanding of signal properties and analysis is to combine the following two teaching modes: (1) simulation, carried out with appropriate software such as MATLAB, and (2) online illustration of basic principles in signal processing using remote control of measurement devices, together with an in-depth presentation of the signal conversion and analysis theory.

In our approach, we restrict the local students' control of the equipment to software access (vs. allowing them to physically manipulate the hardware). This leads to two different access modes, depending on the user type: (1) *local users,* who are present in the laboratory during the actual experiment, and (2) *remote users,* who attend the laboratory and access the equipment via the Internet. Both implementations require a multi-user, client-server architecture for the applications and allow complete data management [5,11,26,27]. However, major differences appear when considering the communication speed, limitations in Internet bandwidth and, more critically, the multiple computing platforms.

## Hardware Platform

This signal processing virtual laboratory (VLab) utilizes equipment and software distributed between two universities from which remotely located students can perform experiments. A diagram of the VLab architecture is presented in Figure 1. The software solution is a multi-user, client-server architecture developed in the LabVIEW environment. Users accessing the client applications from different network nodes can interact and perform experiments consisting of data acquisition, signal generation and waveform visualization and/or analysis.

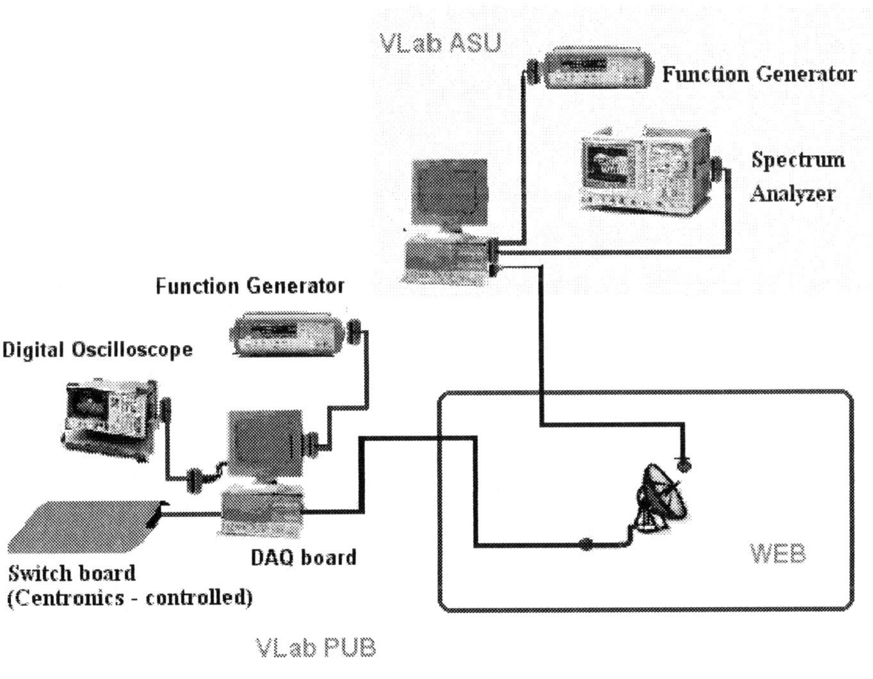


FIGURE 1
DIAGRAM OF SPECIFIC ARCHITECTURE EMPLOYED IN THIS PROJECT (PUB IS POLITEHNICA UNIVERSITY OF BUCHAREST, AND ASU IS ARIZONA STATE UNIVERSITY)

Each remotely accessible device is directly connected to a PC running a specific hardware server application. Each server application is unique to the communication interface and protocol (*e.g.*, serial–RS, USB or wireless; parallel–GPIB or Centronics-based) between the computer and the specific laboratory device. The server alone controls the communication parameters between the server host and the particular laboratory hardware. Changing certain features of the experimental equipment is restricted to the server administrator. Splitting the application server into modules corresponding to various features provides an increased modularity in the software layout of the experiment. Different server applications can be implemented for chat sessions, video broadcasting, user authentication, and online help. These servers may be operating on different machines connected to various web nodes; however, these application modules are normally hosted by computers within the same Intranet that the hardware server occupies (vs. remote Internet nodes).

In Figure 2, the front panel of the function generator client application is shown. Users operate the equipment via a client application that can be formed by multiple client modules. For instance, the application exemplified in Figure 2 might use a client for the chat module, another client for the signal generator module, and a third client for controlling the data acquisition card. The servers corresponding to those applications are running on PCs located in VLab PUB. Another server application enabling user authentication for each experiment is running on another machine in VLab PUB. Figure 3 shows a hardware layout that employs a connecting device that is computer controlled (via a parallel interface) which allows different configurations of the measurement circuit.

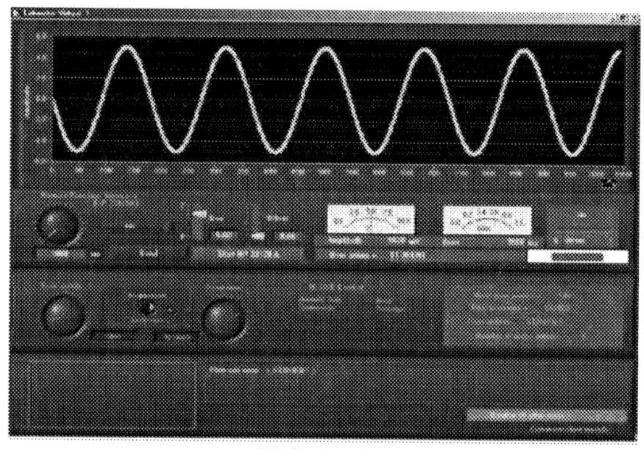

FIGURE 2
EXAMPLE OF A CLIENT MODULE EMPLOYED IN A REMOTE EXPERIMENT USING A SIGNAL GENERATOR AND A DATA ACQUISITION CARD

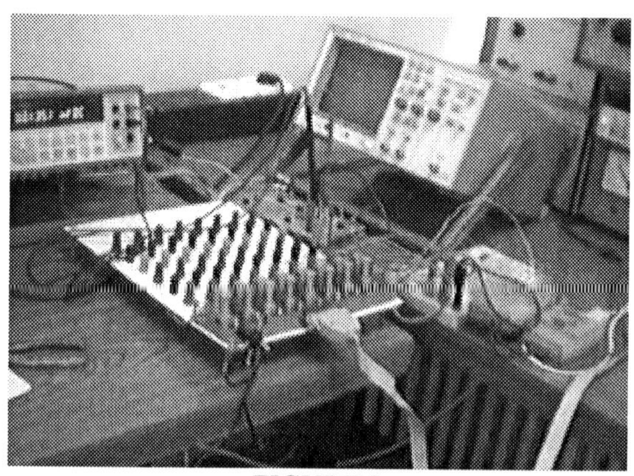

FIGURE 3
EXAMPLE OF LABORATORY LAYOUT USING REMOTE CONTROLLED SIGNAL CONNECTOR

The chat module within the VLab experiments provides a means for users of a given experiment to intercommunicate. This approach stimulates teaming, even when simultaneous remote users are geographically dispersed. Instead of using

Internet instant messaging software, the chat module was developed in LabVIEW to create a private virtual space for the VLab community. The internal chat software allows seamless integration of the messaging into the screen of a particular application. The chat design (both in terms of appearance and use) is intended to mimic software that college students are already familiar with, thereby reducing the learning time for utilizing the experiments. All users of VLab equipment in a specific experiment see the text messages for that experiment, and the chat server generates a log file of the student interaction.

Presently, the default mode is to grant a fixed duration of five minutes to the user who takes control of the experiment, while the other students observe and communicate with each other. Note that an instructor client always has a higher priority in taking control of the hardware. An interesting development for the future is to employ artificial intelligence techniques for natural language processing to assess student interaction. Specifically, semantic data retrieval would be applied in real-time to the log file created by the chat module. Reference [28] describes dynamic profiling of users using navigational patterns. One indicator is word histograms in the chat discussions. We could use the same technique, although not for adapting the client interface, but to dynamically reallocate the time each student must assume control of the equipment.

The maximum number of VLab users is constrained not so much by boundaries created by the hardware-software networking access, but rather in the students' feeling of passivity toward the actual equipment use. VLab has been tested with as many as eight simultaneous users: six located on a network relatively near to the hardware server and two users on distant networks. However, like physical laboratories, as the number of simultaneous users of a single experiment (*i.e.*, a single set of dedicated laboratory equipment) becomes larger than a small group, some students find themselves more passive listeners than active participants.

## Spectrum Analyzer Application

Of particular interest for signal processing courses is the use of spectrum analyzers. They are expensive and are generally not available to all undergraduate students. VLab utilized an Advantest Model R3131A Spectrum Analyzer (SA), with a frequency range of 9 kHz to 3 GHz. It is a suitable instrument for illustrating basic signal analyses for teaching purposes. Specific laboratory experiments include:

- How to use and access the features of the spectrum analyzer
- Fundamentals of sampling and aliasing
- Arbitrary waveform generation and analysis
- Fnding the signal bandwidth for different signal-to-noise ratios
- Demodulation of FM signals
- Spectral analysis of a user-generated waveform and comparison with numerical analysis performed on the simulated signal

Students submit a written laboratory report as is normally done in the case of in-person laboratory experiments. Besides experiments by students, instructors can use this system to illustrate a particular topic during lectures away from the laboratory.

The spectrum analyzer communication with an external processing unit can be either serial (RS232) or parallel (GPIB). Though slower, we choose to use serial communication between the analyzer and the server application host because the client-server communication speed is the actual limiting factor here. Because the spectrum analyzer is located at ASU, both server applications for controlling the instrument and for video broadcasting are running on the same machine at VLab ASU. It is not necessary for both server applications to be housed on the same machine; in fact, the bandwidth availability can be improved if the video server is placed on a separate computer.

Some educators have utilized Java platforms for greater portability and easier access via web browsers. There are also projects [27] developed in the LabVIEW environment that can be executed on a LabVIEW player freely distributed by National Instruments. Figure 4 depicts the user interface of a client developed for performing simple signal analyses on the acquired data. The portability of this application is limited only by LabVIEW, which requires different coding for Windows and Macintosh platforms. The transferability is restricted to equipment that uses the Standard Commands for Programmable Instrumentation (SCPI) driver.

During the initial development of the SA virtual laboratory, the use of web browsers

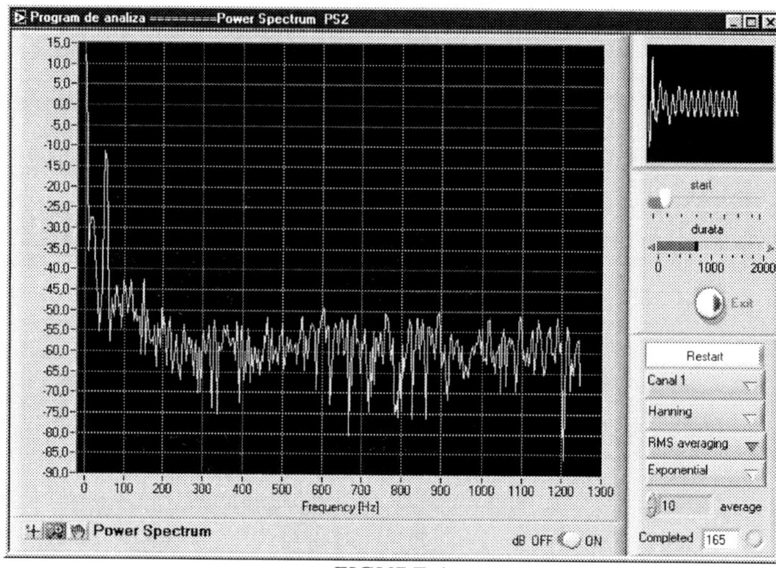


FIGURE 4

EXAMPLE OF A CLIENT MODULE EMPLOYED IN A REMOTE EXPERIMENT CONCERNING BASIC SIGNAL PROCESSING TASKS

running platform-independent Java applets was eliminated from consideration due to prohibitive development time. Instead, LabVIEW Version 6*i* (now 6.1) was chosen. Two LabVIEW implementations using the runtime engine (player) were tested: (1) with the code running on a web browser, and (2) using LabVIEW directly. The web browser implementation was found to have considerable delay time when the user is remotely located on the Internet vs. on an intranet where the hardware server is located. Experience with the application using direct client-server communication in LabVIEW showed an order of magnitude increase in communication speed with image refresh rates of 1 sec even in the case of a 48K modem dial-up connection through a remote Internet connection. A drawback to this option, however, is that the user must be cognizant of the possible need for firewall protection.

In the present solution, the students obtain access to the client application after entering their user name and password. This login is validated by a server application that looks in a database maintained by the administrator. The authentication application and database are located on different computers to enhance security. The master program requires a fixed IP address, and consideration was given to requiring users to have certain fixed IP addresses to increase system security, but this was deemed problematic given that students may access the virtual laboratory via a modem and public Internet provider.

Perhaps the most difficult task in the case of the SA experiment was deciding how to re-organize access to each SA function while balancing the preservation of the actual structure on the equipment front panel and a logical arrangement appropriate for learning to use such complex equipment. We have chosen to design a generic, functionality-based graphic interface versus an exact replica of the actual instrument. This allows us to use the same client interface for other spectrum analyzers. This allows easy upgrading of the software when new hardware is purchased or added. Moreover, this implementation provided an opportunity to re-organize the function buttons in a more straightforward, logically structured arrangement.

Figure 5 shows the client application interface as seen by the student user. The bottom left portion of the front panel provides information on the number of online users and the master user who is presently controlling the experiment. Even in the online team environment, control of the equipment must be conferred to a single *master user*. In the main window (upper portion) of the front panel, a software replica of the spectrum analyzer buttons is supplied together with an information box on the highlighted function. The information box contains Advantest technical data. After selecting a specific function and its parameter(s), the command string is shown in the bottom part of this window. To send this command to the SA, the student clicks the Send RS232C button. If the

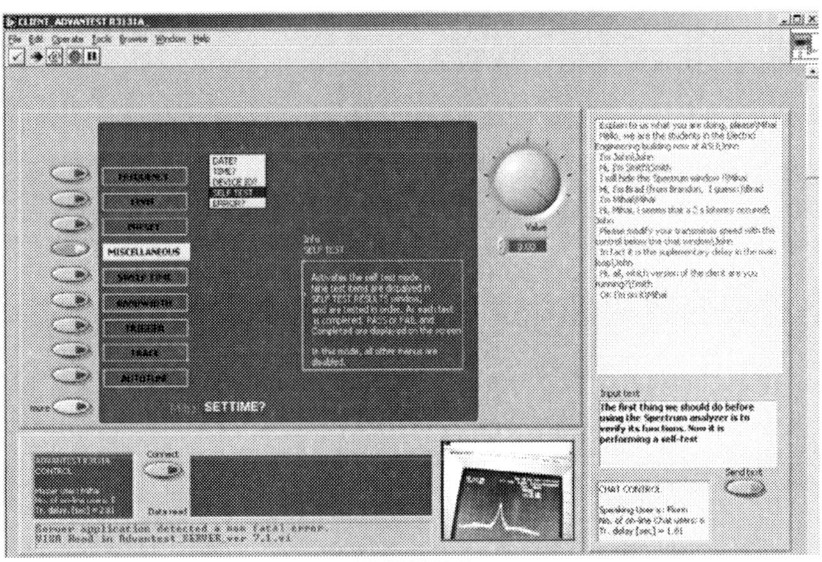


FIGURE 5

ADVANTEST SPECTRUM ANALYZER CLIENT APPLICATION FRONT PANEL. NEW USER (SMITH) PREPARING ANALYZER FOR SELF TEST BEFORE TAKING CONTROL FROM THE PRESENT MASTER USER (MIHAI)

command string ends with a question mark, the SA will respond with an answer string, which is displayed in the window to the right of the Send RS232C button. In the bottom right corner, the video output from a camera monitoring the SA display screen is shown. The video output is an important feature as it not only allows students to visualize the signals and the signal processing results, but it also provides students with a connection to the physical equipment. For those developing their own virtual laboratories, we note that LabVIEW does not have resident video broadcast software; we therefore purchased software from a third-party vendor to accomplish this.

In the classical laboratory environment, students can carry out experiments either chaotically or courteously. In the chaotic setting, students jostle for the right to operate the equipment; in the courteous atmosphere, students share the responsibility for controlling the measurement system. Similar possibilities exist for an online, multi-user laboratory. In the online environment, the chaotic situation could occur when multiple users send competing command strings. To avoid this, we have chosen to utilize the structure of a *master user* whose position alternates among the students. Presently, a student can assume the role of master user by sending the appropriate request string. To eliminate the possibility of students fighting for master status, the server application mandates a minimum period that a master user has control. Conversely, the server also limits the maximum time a student can assume master user status to insure that no student becomes a purely passive listener.

Figure 5 shows the typical front panel on the remote *master user's* computer when the *chat module* is selected. The previous *speaking user* was Florin according to the *Chat Control* information window in the bottom right corner of the panel. The present *chat user* is Smith, who is announcing his intention of requesting a *Self Test* of the spectrum analyzer. The *master user* is Mihai. Smith selects the *Self Test* suboption of the *Miscellaneous* menu before taking control by pressing the *Connect* button.

The ASU engineering college has developed an extensive web-based course assessment tool; however, present security restrictions on that site limit its use to internal students. We therefore have not had significant opportunity to gather anonymous student feedback. Students who otherwise do not have access to a spectrum analyzer consider the remote access rewarding compared with their previous option, *i.e.*, to use numerical simulations. Students who have direct access to a SA deem the tutorial helpful before actually working with the equipment, and they appreciate the availability of the equipment during non-traditional laboratory hours.

Since engineering faculty are constantly being asked to do more, another goal of VLab is that these experiments are carried out without instructor supervision. In addition, the capability of providing access to the laboratory equipment for extended hours also has the potential to reduce costs. However, the time devoted to developing such virtual laboratories may be prohibitive without the benefit of sharing between universities. Cost reduction may be possible because (a) laboratory equipment can be shared between institutions, (b) laboratory instructors need not be present at all times, and (c) equipment utilization can be spread out over larger segments of the week, thereby reducing the number of identical setups required to equip a particular laboratory course. In an ever-changing field such as signal processing, however, the period required to see a return on the initial development time invested may be too long.

## CONCLUSIONS AND FUTURE WORK

One of the main advantages offered by VLab is that students from all over the world can use equipment located in a particular laboratory. Each university that is part of the ring of virtual laboratory users can provide a different subset of experiments. The diversity of signal processing topics taught and realistically performed can be extended in this way.

A separate question arises from [1], which states that the culture of the Internet, which was largely Anglo-American in its early stages, is in transition. It is envisaged that by the end of 2003, non-English speakers will represent one-third of the community of Internet users [1]. It is expected that this tendency will be accommodated with the almost exclusive use of English as a technical language for many topics. Virtual laboratories integrated in international DL programs will also reflect the increased diversity of our postmodern society. Specific topics to be further addressed consist of defining authorship and protecting intellectual property rights in a digital, distributed and online learning environment hosting experiments, apparatus and licensed code.

## ACKNOWLEDGEMENT

The authors gratefully acknowledge the Swiss National Science Foundation (Grant 7RUPJ062223.00/1, within the SCOPES framework) for supporting the VLab PUB activity.

## REFERENCES

1. World Intellectual Property Organization, "Intellectual Property on the Internet: A Survey of Issues," Report WIPO/INT/02, Geneva, December 2002. Available: http://ecommerce.wipo.int

2. Engineering Accreditation Commission, "Criteria for Accrediting Engineering Programs", Accreditation Board for Engineering and Technology, Inc., Baltimore, MD, March 2002.

3. Mihalache, A., "The Virtual and its Virtues. An Interpretation of the Virtual Reality Techniques in Terms of Plato's Cavern Myth (Le virtuel et ses vertus)," *Les Autoroutes de l'Information*, Actes de la XVI-e Biennale de la langue française, Bucarest, 1994.

4. Van Der Veer, G.C., "Mental Models of Computer Systems: Visual Languages in the Mind," *Cognitive Aspects of Visual Languages and Visual Interfaces*, M. J. Tauber, D. E. Mahling, F. Arefi (eds.), Elsevier, 1994.

5. Benetazzo, L., Bertocco, M., Ferraris, F., Ferrero, A., Offelli, C., Parvis, M., and Piuri, V., "A Web-Based Distributed Virtual Education Laboratory," IEEE Trans. Instr. Meas., Vol. 49, No. 2, April 2000, pp. 349-356.

6. Mosterman, P.J., Dorlandt, M.A.M., Campbell, J.O., Burow, C., Bouw, R., Brodersen, A.J., and Bourne, J.R., "Virtual Engineering Laboratories: Design and Experiments," Journal of Engineering Education, Vol. 83, No. 3, July 1994, pp. 279-285.

7. Guru, A., Savory, P., and Williams, R., "A Web-Based Interface for Storing and Executing Simulation Models," Proc. of the 2000 Winter Simulation Conf., Vol. 2, Orlando, FL, Dec. 10-13, 2000, pp. 1810-1814.

8. Oren, T., "Impact of Data on Simulation: From Early Practices to Federated and Agent-Directed Simulation," Proc. of EUROSIM 2001, Shaping Future with Simulation, Delft, NL, June 26-29, 2001.

9. L. Harasim, "A Framework for Online Learning: the Virtual-U," Computer, Vol. 32, No. 9, Sept. 1999, pp. 44-49.

10. Grima-Palop, J.M., and Andres-Teruel, J.M., "Virtual Work Bench for Electronic Instrumentation Teaching," IEEE Trans. on Education, Vol. 43, No. 1, Feb. 2000, pp. 15-18.

11. Ko, C.C., Chen, B.M., Hu, S., Ramakrishnan, V., Cheng, C.D., *et al.*, "A Web-Based Virtual Laboratory on a Frequency Modulation Experiment," IEEE Transactions on Systems, Man, and Cybernetics—Part C: Applications And Reviews, Vol. 31, No. 3, August 2001, pp. 295-303.

12. Joler, M., and Chistodoulou, C.G., "Virtual Laboratory Instruments and Simulations Remotely Controlled via the Internet," IEEE Antennas and Propagation Society International Symposium, Vol. 1, Boston, MA, July 2001, pp. 388-391.

13. VirtuaLabs at TechOnLine, Bedford, MA. Motorola Project: Motorola's DSP56F826EVM Development System with Motorola's SDK2.5; Hitachi Project: H8 "Tiny" (EDK3664F) Hitachi Semiconductor; and Wind River visionClick (PPC8260), January 2003. Available: www.techonline.com.

14. Kostic, M., "Sampling and Aliasing: An Interactive and On-line Virtual Experiment," Proceedings of ASEE Annual Conf., Nashville, TN, June 22-25, 2003.

15. Fortino, G., Grimaldi, D., and Nigro, L., "An Agent Based Measurement Laboratory over Internet," Proceedings of IEEE Systems Readiness Technology Conf. (AUTOTESTCON '99), San Antonio, TX, Aug. 30-Sept. 2, 1999, pp. 61-66.

16. Ferrero, A., Salicone, S., Bonora, C., and Parmigiani, M., "ReMLab: a Java-Based Remote, Didactic Measurement Laboratory," Proceedings of the International Symposium on Virtual and Intelligent Measurement Systems (VIMS2002), Mt. Alyeska, AK, May 19-20, 2002, pp. 117-122.

17. Zhou, G.T., and Lo, H.-J., "Developing Java-Based Virtual Laboratory Tools for an Undergraduate Random Signals and Noise Course," Proceedings of ASEE Annual Conf., Charlotte, NC, June 20-23, 1999.

18. Siddiqui, A.A., Zia, Y.I., Aamir, M., and Haque, S.E., "Virtual Remote Electronic Laboratory," Proceedings of IEEE Students Conf. (ISCON'02), Aug. 16-17, 2002, pp. 94-98.

19. Spanias, A., Urban, S., Constantinou, A., Tampi, M., Clausen, A., Zhang, X., Foutz, J., and Stylianou, G., "Development and Evaluation of a Web-Based Signal and Speech Processing Laboratory for Distance Learning," Proceedings of IEEE Intern. Conf. on Acoustics, Speech, and Signal Processing (ICASSP2000), Vol. 6, Istanbul, June 5-9, 2000, pp. 3534-3537.

20. Naghedolfeizi, M., Arora, S., and Garcia, S., "Survey of LabVIEW Technologies for Building Web/Internet-Enabled Experimental Setups," Proceedings ASEE Annual Conf., Montreal, June 16-19, 2002.

21. Special Issue: Applications of LabVIEW in Engineering Education, International Journal of Engineering Education, Vol. 16, No. 3, 2000.

22. Gomez, F.J., Cervera, M., and Martinez, "A World Wide Web Based Architecture for the Implementation of a Virtual Laboratory," Proceedings of the 26th Euromicro Conference, Vol. 2, Maastricht, NL, Sept. 5-7, 2000, pp. 56-61.

23. Stonick, V.L., "Teaching Signals and Systems Using the Virtual Laboratory Environment in ECE at CMU," IEEE Intern. Conf. on Acoustics, Speech, and Signal Processing (ICASSP-93), Vol. 1, Minneapolis, MN, April 27-30, 1993, pp. 36-39.

24. Seixas, J.M., Bittencourt, M.C., and Jandre, F.C., "A Virtual Laboratory for Teaching and Developing Applications on Digital Filtering," Proceedings of International Conference on Engineering and Technology Education (INTERTECH 2002), Santos, Brazil, March 17-20, 2002.

25. Albu, M., Mihai, F., Buiu, C., and Gillet, D., "Enhancing Virtual Control Laboratories Using Software Agents," Proceedings of Internet Based Control Education Symposium (IBCE 2001), Madrid, December 2001, pp. 47-52.

26. Gillet, D., Latchman, H.A., Salzmann, C., and Crisalle, O.D., "Hands-On Laboratory Experiments in Flexible and Distance Learning," Journal of Engineering Education, Vol. 90, No. 2, April 2001, pp. 187-191.

27. Bertocco, M., Ferraris, F., Offelli, C., and Parvis, M., "A Client-Server Architecture for Distributed Measurement Systems," IEEE Trans. Instr. Meas., Vol. 47, No. 5, October 1998, pp. 1143-1148.

28. Kurhila, J., Miettinen, M., Nokelainen, P., and Tirri, H., "Dynamic Profiling in a Real-Time Collaborative Learning Environment," Proc. of Intelligent Tutoring Systems (ITS2002), Biarritz, France and San Sebastian, Spain, June 2-7, 2002, pp. 239-248.

**Keith Holbert** is an Associate Professor of Electrical Engineering at Arizona State University. He holds the Ph.D. degree in Nuclear Engineering from the University of Tennessee, Knoxville. Dr. Holbert is a registered Professional Engineer, and is also the Associate Chair for Undergraduate Studies in Electrical Engineering at ASU.

**Mihaela Albu** is a Professor of Electrical Engineering at Universitatea Politehnica Bucureşti, România. She holds a Ph.D. degree in Power Systems from the same university. Her research interests include distributed measurement systems, remote experimentation, and power quality.

# Chapter 20

# MIMIC: A Multimedia Course for Teaching the Operation of Communications Equipment

CARMEN BACHILLER, FRANCISCO RAMOS, JORGE GOSÁLBEZ and CARLOS HERNÁNDEZ
*Departamento de Comunicaciones, Universidad Politécnica de Valencia (UPV), Camino de Vera, s/n, 46022 Valencia, Spain. E-mail: mabacmar@dcom.upv.es*

*The paper presents an interactive multimedia course for teaching the operation of communications and audiovisual equipment. The objectives of the course are: to provide a tool for undergraduate students that enables self-learning of laboratory equipment function and operation; to familiarize the students with laboratory equipment in order to improve their understanding of the equipment's operation, use and theoretical basis; to support the students during laboratory classes; and to speed up laboratory classes. The course consists of a set of videos that are indexed and synchronized with oral explanations, supported texts, schemas and graphics, and integrated in a unique visualization framework. In this way the students can browse an entire video or a specific section of the video that is of interest.*

## INTRODUCTION

In a telecommunications or audiovisual communications laboratory, there are several instruments that are of fundamental importance to the job of an engineer. These instruments are used extensively in practical classes for several engineering subjects (courses), such as Signal Theory, Radio-communications or Digital Signal Processing. Nevertheless, the objective of these practical classes is not learning the operation of the instruments, but performing practical experiments to investigate the concepts taught in theory classes. Although there are several subjects in the degree programs that include detailed learning of such instruments, these subjects are unfortunately optional and planned for advanced courses study. Therefore, students who must use laboratory instruments in basic practical classes are usually unfamiliar with the operation of these instruments. Since the use of instruments can be quite complex in some cases, professors must devote a significant portion of class time to explaining their operation and solving problems related to them. This detracts from the real objective of the practical classes, which is the experimentation and investigation of theoretical concepts.

This problem can be alleviated to some degree by providing the students with interactive manuals for the laboratory instruments. The aim of these manuals is to support the explanations of the professors in practical classes and to be a self-learning tool for the students. In this way, students can familiarize themselves with the operation of the instruments even before attending the practical classes, answer their own questions during the execution of experiments, and improve their knowledge of concepts (the theoretical basis for the instruments), which are currently handled in a superficial way due to constraints on time and space in the use of laboratories. The expected benefit of this initiative is an improvement in the quality of practical classes and a better understanding of the operation of communications instruments by the students.

In this paper, an interactive multimedia course for teaching the operation of communications and audiovisual equipment is described. In contrast to previously proposed interactive manuals [1-10] which are mainly based on HTML text, Java-programs, simulators or virtual laboratory environments, this course consists of a set of videos that explain the theoretical fundamentals and operation of several communications and audiovisual instruments. The primary advantage of using interactive videos conducted by the same professors is that students learn the operation of the instruments with real explanations in place of multimedia animations, so we think that it is easier for them to remember the concepts in addition to being much less time-consuming. These videos can be played at the appropriate speed, and the student has the ability to jump forward/backward to the exact point where a specific concept is explained. Furthermore, the interactive videos are complemented by a wealth of easy-to-access information: user's manuals, methodologies, measures, handling precautions for the instruments, etc. In summary, this is a simple and powerful tool for both students and professors involved in the teaching of telecommunications equipment.

## MIMIC PROJECT

The implementation and publication of an interactive multimedia course has been performed in the framework of a Teaching Innovation Project funded by the Polytechnic University of Valencia. The Project is named MIMIC: multimedia course for teaching the operation of communications equipment. The project involves professors and students from the Communications Department at the Polytechnic School of Gandia and the Telecommunications Engineering School of Valencia.

This course helps students learn to operate the instruments used in laboratory lectures (Signal Theory, Video Systems, Video Instrumentation, Introduction to Digital Signal Processing, Radio Communications and Laboratory for Radio Communications) in the three telecommunication engineering degrees and the audiovisual communications degree presently taught at the Polytechnic School of Gandia. The main objectives of the course are:

- To provide the students with an interactive self-learning tool related to the communications instruments.
- To familiarize the students with the communication equipment that an engineer might find in a communications laboratory.

- To improve the students' comprehension of each instrument, its theoretical basis and its functionality.
- To assist in the use of each instrument.
- To energize the lectures in the laboratory.

| Equipment | Maker | Model |
|---|---|---|
| Video-Recorder Trainer | Centro nuevas tecnologías Tomás Perales S.L. | EV-6B |
| Video S-VHS Camera | Panasonic | AG DP-800 |
| Mini-DV Digital Camera | Sony | DCR PC101 |
| TV Signal Generator | Promax | GV-698 |
| TV Signal Generator | Tektronix | TSG-271 |
| Video Measurer | Tektronix | VM 700 A |
| Vectorscope | Tektronix | 1741A |
| TV Trainer | Distesa | ETV 364 |
| TV Monitor | Panasonic | TC-1451RC |
| Radio Communication Analyzer | Marconi | 2965 |
| Spectrum Analyzer | Tektronix | Tek 2711 |
| Spectrum Analyzer | Hewlett-Packard (Agilent) | HP 8590L |
| Analog Oscilloscope | Hameg | HM 303-4 |
| Analog Oscilloscope | Tektronix | TDS 320 |
| Signal Generator | Hewlett-Packard (Agilent) | HP 33120 A |
| Network analyzer | Agilent | 8714ET |
| Satellite RF detector | Promax | MC677 |
| Terrestrial RF detector | Promax | MC360B |
| Digital Oscilloscope | Tektronix | TDS 320 |

TABLE 1
SOME OF THE SELECTED EQUIPMENT

To fulfill these objectives, the course involves explaining the operation of several instruments (Table 1). The format of the course is basically a multimedia presentation framework, accessible via the Internet, where the students are able to surf and select the instrument of interest. For each instrument, a multimedia presentation is shown. This presentation includes video, speech, diagrams and animations that explain the instrument's functionality, theoretical basis, operation and safety considerations. A final aim is to publish the final product on a web site so that students can access it throughout the course, either during the laboratory lectures or while studying the subjects related to the instruments.

The project was implemented in three main stages:

- MATERIAL PROVISION: In this stage, the structure of the interactive course and the scripts for the descriptive videos were made. The videos, online manuals, animations and graphics for the course were also implemented.
- SYNCHRONIZATION: This was a critical stage since the comprehensibility and usability of the course for students depend on the correct ordering and integration of the material. An important effort within the project was dedicated to the organization, editing and structuring of the teaching material.
- SETTING UP and EVALUATION: Finally, the course is set up and evaluated by the students. This will be done in two stages. In the first stage, the interactive course is used in pilot subjects. For the evaluation stage, a survey is designed to gather information from the students about different aspects of the interactive manuals.

## MATERIAL PROVISION

The selection and preparation of the equipment is the first stage of this project. The equipment that is used in practical classes is listed and analyzed in order to determine which pieces will be added to the interactive course to support the laboratory classes.

Some of the selected equipment is listed in Table 1. From the list of selected equipment, the following didactic material is prepared for each piece:

- An index that contains the main items to be explained.
- A video of the standard operation of each piece of equipment.
- Slides and explanation of equipment operation, to be synchronized with the video.
- A transcription of the video voice-over, to be used to create subtitles.

### Video Recording

The aim of the video recording is to allow students to watch and listen to a description of how to use each instrument. A video clip lasting about 10 minutes explains each instrument's main functions: buttons, menus, and examples (Figure 1). In anticipation of publication on a digital platform, these recordings have been made with a mini-DV digital camera. The result is not a high quality video, but sufficient for the purposes of the project. The editing and indexing of the video is done with Adobe Premier 6.0, which is an easy and convenient program since, among other characteristics, it allows nonlinear editing that gives random access to recorded digital videos, and provides easy communication with Mini-DV cameras using the Firewire bus.

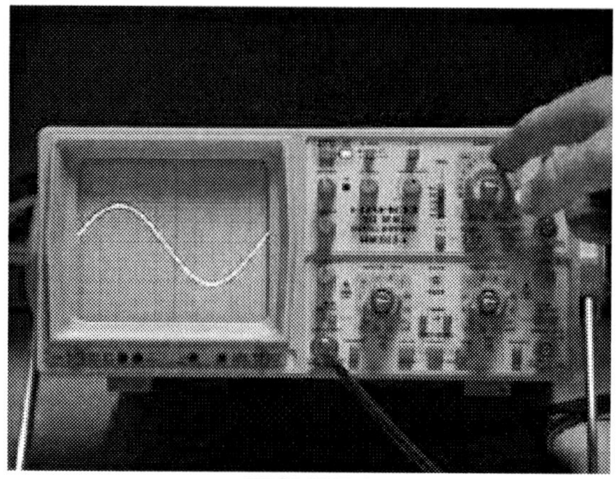

FIGURE 1
ANALOG OSCILLOSCOPE RECORDING

### Slides

The video support slides (Figure 2) are created with Microsoft PowerPoint. The aim of these slides is:

- To reinforce the oral explanation of the video
- To highlight the main features and characteristics of the equipment, especially those that are not conducive to being recorded on video. Graphics and diagrams are included.
- To provide theoretical explanations. In some instances it has been deemed necessary to augment the video with additional explanations in order to ensure that students fully understand the operation of the equipment.

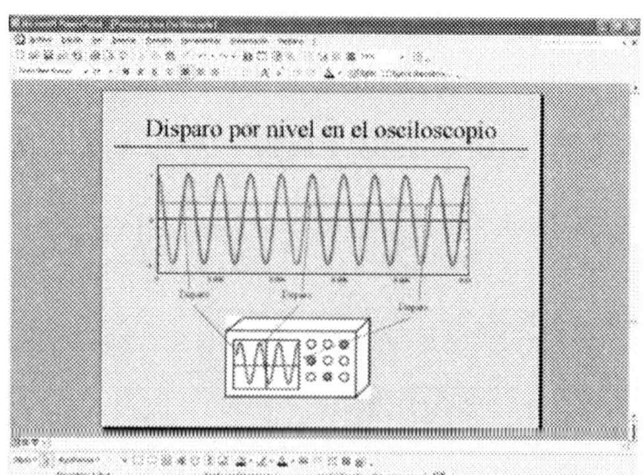


FIGURE 2
EXAMPLE OF THE SLIDE FOR THE EXPLANATION OF THE OSCILLOSCOPE TRIGGER

### Transcription

Finally, the audio is transcribed in order to subtitle the video. Initially, use of voice recognition software was considered, but these programs require personal training for each speaker, so the transcription was done manually.

## SYNCHRONIZATION TOOL

In order to develop the final multimedia course, a tool was needed to integrate all of the didactic material into a single environment. After evaluating several tools, for example Microsoft Media Producer, we chose a synchronization tool [4,5] developed by the Distributed Real Time Applications and Systems Group (SATRD) of the UPV. This tool allows us to specify the exact time when a specific segment of media should be played. It also allows content without temporal references to be synchronized and successfully played, e.g. text and PDF files (intra-media marking). It can develop "multimedia presentations" based on media flows without merging the results of this composition into a single media archive; rather, it references the multimedia archives and stores the starting time for each stream (inter-media marking). This tool was chosen due to its simplicity, the ease of marking and synchronizing of different media (videos, text and slides), and the excellent final results.

## RESULTS

As stated above, the multimedia course, accessible via the web or locally on a PC, consists of a set of videos, indexed and synchronized with oral explanations, supporting texts, diagrams and graphics all integrated into a single visualization framework, which explains the theoretical fundamentals and operation of several communications and audiovisual instruments. First, the videos were mounted and edited using Adobe Premier 6.0. Second, the synchronization tool was used in order to determine the specific timing of support text, diagrams, graphics, etc., associated with a specific video. Figure 3 shows the different windows available in this tool when we specify a given instrument. Finally, the videos were indexed and synchronized with oral explanations, support texts, diagrams and graphics. Each element was put into a specific sequence in time and integrated into the visualization framework, which explains completely the theoretical fundamentals and the operation of the instrument selected by the student. Figure 4 shows the final result of this process.

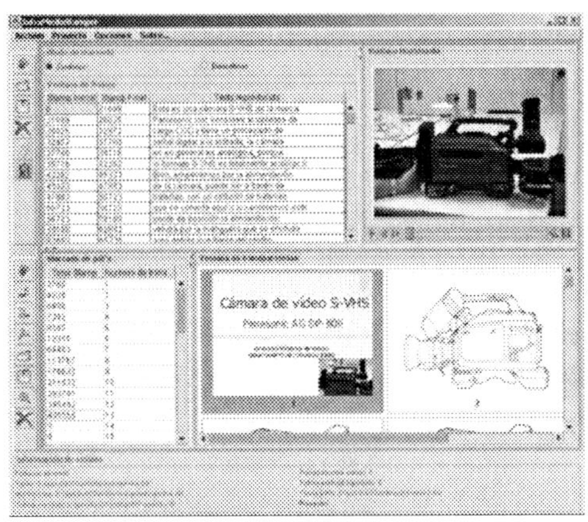

FIGURE 3
USING THE *INTRAMEDIASTAMPER* TOOL, OR TIME STAMPER

## Setting Up the Project

The project was set up in the first term of 2003/2004 as a pilot project in the practical classes for the Signal Theory and Microwaves subjects.

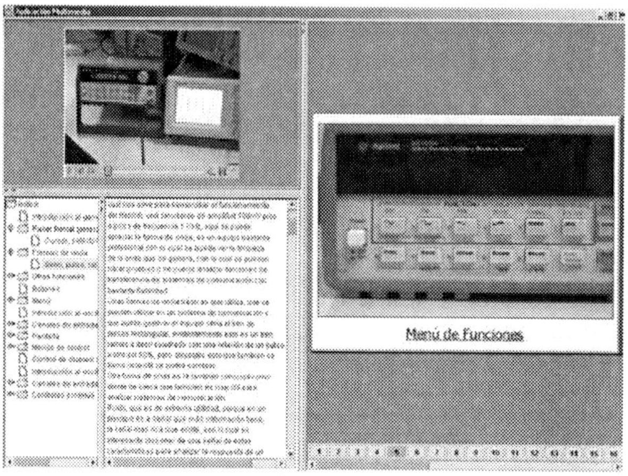

FIGURE 4
VISUALIZATION FRAMEWORK FOR THE INSTRUMENT *HP 33120 A*

Signal Theory is a second year subject, common to Image and Sound Engineering, Telecommunication Systems Engineering and Electronics Engineering, with more than 200 students regularly attending the classes. The instruction is divided into 40 hours of theory and 15 hours of practical classes in the laboratory. There are five sessions of laboratory classes: three of them deal with simulations of analog modulations (lineal and angular) and digital modulations (base band and pass band); the other two are measurements of these modulations by using communications equipment: function generators, digital oscilloscopes and spectrum analyzers. When the students begin the laboratory classes, they do not have previous experience with the instruments. We have, therefore, included the use of the MIMIC course during the first session to introduce the equipment the students will use in the following sessions.

Microwaves is a third year subject in Telecommunication Systems Engineering, with about 70 students regularly attending the classes. The instruction is divided into 35 hours of theory and 15 hours of practical classes in the laboratory. There are six sessions of laboratory: three in analysis and design of microwave circuits using a CAD tool, and three in measurements of microwave circuit parameters using communications equipment: function generators, digital oscilloscopes, spectrum analyzers and network analyzers. The only instrument that the students have not used previously is the network analyzer, so we have included the MIMIC course in one of the laboratory sessions to explain the operation of this instrument.

The students have evaluated the experience positively and have shown their interest in using the MIMIC course at home.

## Experience Gained, Lessons Learned, Challenges and Problems

Since our aim was to create virtual manuals for laboratory instruments, it was very important for us to know all of their characteristics. During the process, the professors needed to create synergies to achieve this objective. It was necessary to maintain a "common" technical language, "common" operational procedures, etc. At this point in the MIMIC project, there were two basic challenges: (1) to decide what we wanted to explain, and (2) to determine how to explain it using a computing tool. Although the students would improve their educational results by using this course, we found that it was not easy to introduce new methods of action into the teaching process. In this case, changes were required of both the professors and the students, such as establishing a "periodic cycle" for updating the manuals, videos, slides, etc.

Setting up the project consisted mainly of giving students access to the interactive laboratory instrument manuals that they would use during the practical sessions. This material is now available in the laboratory, but due to the interest the students have expressed, we plan to publish it through additional methods.

Due to the large volume of multimedia material included in each interactive manual, the best mechanism for distributing the material is by means of a DVD that includes the manuals classified by thematic areas. In this way, the majority of students could play the interactive manuals at home before attending the laboratory. About 90% of the students of Telecommunication Engineering have a PC at home. The publication of the MIMIC course on the web is another possibility we are currently studying jointly with the Distributed Real Time Applications and Systems Group of UPV. This option presents technological and economic challenges since the publication and delivery of the content of the MIMIC course in the web environment requires high performance servers.

The professors of the relevant subjects are beginning to recommend that students learn to use the necessary laboratory instruments before attending each practical session. Since the interactive manuals explain the use of each instrument in a general way, students who view the manual are prepared to use the instrument in all laboratory sessions in which it is used. This avoids the situation where a student learns only the commands required for a specific laboratory session. MIMIC gives students a general overview of the instruments' operation and application.

## EVALUATION

At the time of publication of this paper, the assessment of the MIMIC course for several different laboratory classes was in progress. For this purpose, we have designed a survey instrument to gather information from the students about different aspects of the interactive manuals such as utility, usability, clarity, methodology and personal view. Some of the questions included in this survey are:

- Do you feel more comfortable attending the laboratory sessions having previously learned to use the instruments?
- Do you think that the time of the laboratory session is more efficiently used through this method?
- Please evaluate from 1 (less adequate) to 5 (more adequate) the following aspects of the interactive manuals: content, clarity, usability, extension, methodology, etc.
- For each interactive manual, do you have a general understanding of the instrument or do you know how to use it only for a specific application?

The survey sample has included the students attending the practical classes of Signal Theory and Microwaves. Additionally, we expect to extend this study to students and professors in other subjects not directly involved in this project, and to determine their opinion of the multimedia tool and the possibility of including it the teaching-learning process of those subjects. An interesting study would be the comparison of the results of practical classes in which the students have previously studied the interactive manuals of the laboratory instruments versus conventional classes in which the professor takes time at the beginning of the session to explain the use of the necessary equipment. After compiling and discussing these results, we will summarize the main conclusions in a follow-up to this study.

## CONCLUSIONS

Familiarization with communication and audiovisual equipment is part of the education of telecommunication engineers. Nevertheless, due to time or space limitations in laboratories, students may encounter difficulties learning to use the instruments during classes. This problem can be alleviated to some degree by the creation and publication of online interactive multimedia manuals for the different instruments. The MIMIC Project has selected the instruments, prepared the multimedia material (video records, slides and transcriptions), selected the video editing tool and the synchronization tool for the presentations, and integrated all the material in a multimedia framework, which was accessible via the web or locally on a PC. We have set up the online manuals for two of the subjects of the Telecommunication Engineering program, and are conducting a survey to assess the opinion of the students about the utility, usability and other aspects of the interactive manuals developed in this project.

## ACKNOWLEDGEMENTS

The authors want to acknowledge the work performed in the MIMIC project by the professors: Dr. Santiago Cogollos Borrás, Prof. Amparo Girona Coma, Prof. Pablo Soto Pacheco and Prof. Vicent Miquel Rodrigo Peñarrocha. We also want to thank the Distributed Real Time Applications and Systems Group (SATRD) of the UPV for their support in providing the synchronization tool for the implementation of the course, in particular to Dr. Juan Carlos Guerri Cebollada and Juan José Cermeño Portet.

## REFERENCES

1. Sears, A.L., and Watkins, S.E., "A Multimedia Manual on the World Wide Web for Telecommunications Equipment," IEEE Transactions on Education, Vol. 39, No. 3, August 1996, pp. 342-348.

2. Koh, J-J., Kwon, S-D., Kim, B-U., Nam, G-M., Jang, Y-H., Lee, W-J., and Lee, D-Y., "Implementation of an Interactive Electronic Technical Manual Based on Web-Multimedia Technology," Proceedings of 4th Korea-Russia International Symposium on Science and Technology, Ulsan, South Korea, Vol. 2, July 2000, pp. 21-24.

3. Bachiller, C., Esteban, H., Cogollos, S., San Blas, A., and Boria, V.E., "Teaching of Wave Propagation Phenomena Using MATLAB GUIs at the Universidad Politecnica of Valencia," IEEE Antennas and Propagation Magazine, Vol. 45, No. 1, February 2003, pp. 140-143.

4. Chan, W.L., and So, A.T.P., "A Cost Effective Interactive Multimedia System for Electrical Undergraduate Laboratory Sessions," Proceedings of IEEE First International Conference on Multi-Media Engineering Education, Melbourne, Australia, Jul. 1994, pp. 219 –224.

5. Hassapis, G., "An Interactive Electronic Book Approach for Teaching Computer Implementation of Industrial Control Systems," IEEE Transactions on Education, Vol. 46, No. 1, February 2003, pp. 177-184.

6. Drofenik, U., Kolar, J.W., Van Duijsen, P.J., and Bauer, P., "New Web-Based Interactive e-Learning in Power Electronics and Electrical Machines," Industry Applications Conference, October 2001, Chicago, USA.

7. Khalig, M.A., "Interactive Multimedia Courseware for Integrated Circuit Fabrication Courses," Proceedings of 31st Annual Frontiers in Education Conference, October 2001, Reno, USA, Vol. 3, pp. S1C-1-4.

8. Cvetkovic, S.R., and Cohen, B., "Intelligent Tutoring System for Signals and Systems Course Based on the cT Programming Environment," IEE Colloquium on Intelligent Tutoring Systems, London, UK, November 1990, pp. 5/1-5/4.

9. Antao, B.A.A., Brodersen, A.J., Bourne, J.R., Cantwell, J.R., "Building Intelligent Tutorial Systems for Teaching Simulation in Engineering Education," IEEE Transactions on Education, Vol. 35, No. 1, February 1992, pp. 50-56.

10. Anderson, B.L., Pelz, L.J., Ringel, S.A., Clymer, B.D., and Collins, S.A., "Photonics Laboratory with Emphasis on Technical Diversity," IEEE Transactions on Education, Vol. 41, August 1998, pp. 194-202.

11. Pajares, A., Guerri, J.C., Belda, A., Cermeño, J.J., Palau, C., and Esteve, M., "JMFMoD: A new System for Media on Demand Presentations," EUROMICRO 2002, Dortmund, Germany, 2002.

12. Belda, A., Pajares, A., Guerri, J.C., Palau, C., and Esteve, M., "JMFVoD: Design and Implementation in Java of a Media Streaming on Demand System", EUROMEDIA 2002, Módena, Italy, 2002.

**Carmen Bachiller** was born in Castellón, Spain, in 1972. She received the M.Sc. degree in Communications Engineering from Polytechnic University of Valencia in 1996. She worked from 1997 to 2001 for the company ETRA I+D, S.A as a Project Engineer in research and development for automatic traffic control, public transport management and public information systems using telecommunications technology. In 2001 she joined the Communication Department of Polytechnic University of Valencia as an Associate Professor. She is secretary of the IMCO2 Informatics Multimedia Communications and Computers Research Institute of the Polytechnic University of Valencia. She teaches signal and systems theory and microwaves. She has participated in several teaching innovation projects such as the development of GUIs for teaching electromagnetic phenomena and the MIMIC project. She is now working on her Ph.D. thesis on electromagnetism and radio frequency circuits.

**Francisco Ramos** was born in Valencia, Spain, on April 2, 1974. He received the M.Sc. and Ph.D. degrees in Telecommunication Engineering from the Polytechnic University of Valencia in 1997 and 2000, respectively. Since 1997, he has been with the Communications Department at the same university, where he is now a Professor. He has participated in several national and European research projects in areas such as optical access networks, broadband wireless systems and optical networking. Prof. Ramos has co-authored more than 50 papers in international journals and conferences, and he has acted as reviewer for the IEEE and Taylor and Francis Publishers. His research interests include nonlinear fiber optics, compensation of intensity noise and nonlinear distortion due to the optical fiber, optical-phase conjugation, microwave and millimeter-wave optical systems, broadband access networks and high-speed optical networks. He is the recipient of the Prize of the Telecommunication Engineering Association of Spain for his study on the application of optical nonlinear effects in microwave photonics.

**Jorge Gosálbez** was born in Valencia, Spain, in 1975. He joined the Communications Department in 1999 and received the M.Sc. degree in Telecommunications Engineering from Polytechnic University of Valencia in 2000. At the present time, he works as teacher in this department and is involved in ultrasound signal processing for a non-destructive cement evaluation project. He has participated in the BRITE Programme of the European Community. He has also published several papers and conference communications in the statistical signal processing area.

**Carlos Hernández** was born in Sancti-Spiritus, Cuba, on July 16th, 1966. He received his M.Sc. degree in Telecommunications Engineering from the Polytechnic University of La Havana, Cuba, in 1989 and his Ph.D. degrees in Telecommunication Engineering from the Polytechnic University of Valencia, Spain, in 1998. From 1998 to 2001 he was a Professor in a Computing Department at a private university. Since 2001 he has been with the Communications Department at the Polytechnic University of Valencia, where he is now an Associate Professor. Prof. Hernández works with the IMCO2 Informatics Multimedia Communications and Computers Research Institute of the Polytechnic University of Valencia, teaching Audiovisual Technologies and Digital Signal Processors. His research interests mainly include digital signal and image processing and the application of the Information Technologies to the control of natural protected areas. Prof. Hernández also works on projects related to improving educational methods at the University.

# Chapter 21

# A New Approach on Teaching Circuit and Electronic System Design

JOAQUÍN CERDÁ BOLUDA, MARCOS MARTÍNEZ PEIRÓ, MIGUEL ÁNGEL LARREA TORRES, RAFAEL GADEA GIRONÉS and RICARDO JOSÉ COLOM PALERO
*Polytechnic University of Valencia, Camino de Vera, s/n, 46022 Valencia, Spain. E-mail: joacerbo@eln.upv.es*

*The study of Programmable Logic Devices (PLDs) is one of the more accessible branches of microelectronics, given the conceptual simplicity and relative ease with which implementation resources can be found that enable fairly large projects to be undertaken. The Circuit and Electronic System Design course – offered as part of the telecommunication engineering study plan at the Polytechnic University of Valencia – teaches digital design methods based on PLDs. This implies an understanding of structures and resources, as well as design methods based on Hardware Description Languages (HDLs). Given the broad and essentially practical nature of the course, we decided to develop new materials to aid active classroom teaching. This includes material for self-teaching so that the student can acquire practical design skills when working away from the classroom. A procedure has been designed for student evaluation that is based on moderately difficult practical designs that have been developed using design tools and logical synthesis. This methodology provides added motivation for students as they find themselves tackling real problems associated with digital design. Evaluation is structured around three methods: completely specified designs, partially specified designs, and hardware-oriented physical implementation.*

## INTRODUCTION

The teaching of digital electronics has been revolutionised over the past decade. The development and introduction of Programmable Logic Devices (PLDs) and Electronic Design Automation (EDA) tools have enabled students to productively design moderately complex logical systems and then implement and test them in-circuit – all at very little cost. This was unfeasible using traditional approaches such as Discrete Normalised Components or classic VLSI techniques. As a result, these techniques are being progressively relegated in current study plans.

It is now difficult to establish what should be the minimum level of knowledge for an electronic engineer. Our objective should not be simply teaching students to become simple users of a given EDA. Students should understand the application and fundamentals of a system, as well as being able to compare one EDA with another. In this way, students will be better able to handle the inevitable migration between systems. This implies a multi-discipline training (electronic and manufacturing technologies, computer science, logic design, etc.) and, above all, the development of an intellectual maturity and initiative in the analysis and synthesis of logical systems. It is this intellectual approach that constitutes real know-how.

Ideally, the desired approach to teaching electronic system design is to spend a large number of hours in the laboratory. But this is difficult when the number of students grows [1]. In this sense, the challenge is to find an efficient method of teaching our subject [2]. Here, we describe some of the solutions that the authors have found through their experience in teaching the subject of Electronic Circuit and Systems Design [3].

## Teaching Approach

Electronic Circuit and Systems Design is part of the third year of the Telecommunications Engineering degree curriculum taught at the Higher Technical School of Telecommunications Engineering at the Polytechnic University of Valencia. This is an obligatory core subject with 60 hours of classwork divided equally between lectures and laboratory sessions. The total number of students is about 300, divided into 4 theoretical classes. The practical sessions are performed in a laboratory in student groups of about 20 students. The course teaches general concepts that enable students to acquire a global perspective of microelectronics as well as specific concepts that prepare students for future courses in the subject [3].

The syllabus covers the following objectives:

- **Introduction to the technology and manufacturing of integrated circuits (IC) -** This topic offers a global, as well as specific, vision of integrated circuits (ASIC). The technologies and economies of flat process manufacturing are presented.
- **Design techniques -** This topic covers the main approaches and strategies for complex design, methodologies of design, and the main design tools. We emphasis those tools that use Hardware Description Languages (HDLs). These tools enable the student to use synthesizers to move from a description domain that is closely linked with functioning behaviour to a structural domain. Choosing the correct tool is not easy and various factors must be taken into account, including:
    - Ease of use. It is important to choose a tool that does not require long boring hours of instruction and that offers efficient online documentation.
    - The student must be able to install and use the tool on his or her own computer without restrictions. Student computers may be old and slow.
    - The tool must offer many possibilities with various HDLs, e.g. generic, configuration, VHDL and VERILOG, FSM detection, test-bench generation, etc.
    - It must enable students to perform complex tasks with economical programmable devices, and include low-cost development boards.

  - The tool should speedily compile simple designs as this quickens the pace of practice sessions.
  - It should offer a project structure that is easy to export.
  - It must include diagram output, simulator, time analysis, and a floorplan editor. These must be simple to use, yet comprehensive. The tool must be fairly representative of other tools that perform similar tasks and that the student may encounter in the future.
  - The synthesiser should allow the student to see what is being synthesised. A picture the student can understand is worth more than a thousand words – logic gates and LUTs are the most interesting options.
  - Programming should be straightforward, using either serial or parallel ports.
  - The insertion of macros should be intuitive and should not require changes in the normal design flow. These macros should be parametrical, but useable by VHDL and Verilog without tricks or shortcuts for simulation and synthesis.
  - The tool must interface easily with HDL simulators and synthesisers from other manufacturers.
- **Design technological support** - This topic involves the study of ASIC programmables (CPLD and FPGA). It is important to select devices used by students in previous courses (SPLD). A designer can exploit the best features of design in terms of area, speed, and consumption, only by having adequate background knowledge. The choice of CPLD and FPGA programmable devices over ASIC semi-custom masked programs (gate-arrays, standard cells) is justified by cost. The various design tools and kits, as well as the physical devices make a more ambitious approach impractical. It must be remembered that both design flows are increasingly similar in terms of methodologies, verification systems, design inputs, RTL description styles, and synthesisers. We have verified that our students easily adapt to working with masked program ASIC devices. The main target is to give a complete perspective of the System-on-Chip design methodology [4].
- **Design Criteria** - This topic covers testing the reliability of a design. It is necessary for students to have a good understanding of the timing hazards that can arise in combined and sequential circuits and to establish margins for activation and positive retention. Students will have received a digital education centred on reading transition, operation, and truth tables. They must begin to understand that design is more demanding when implementing on a programmable physical device or PCB.
- **Design Structuring and Specification** - This topic helps students learn how to tackle fairly complex designs. Students learn that systems are organised so that part of the system handles the data, and another part of the system controls when, and under what conditions, the data should be handled (Datapath and Controlpath, respectively). Students also incrementally learn to respect form and clearly specify digital systems. It is here that we find the very reason for the existence of hardware description languages: to ensure that information can be easily exchanged between teams of engineers. The same aim should be clear when specifying – and in this area, the ASMD (Algorithmic State Machine Diagrams) seems to be the best option.

FIGURE 1
OPENING SCREEN OF THE MULTIMEDIA APPLICATION

## COMPLEMENTARY MATERIALS

With the aim of centralising the contents and increasing the level of motivation, the course teachers have prepared support material for developing the theoretical concepts and undertaking designs with programmable devices [5].

### Material for Theoretical Classes

Theoretical classes are given in lecture rooms using video projectors. All of the principal themes within the subject are presented in PowerPoint presentations, which are shown in the lecture room and made available to students in a compilation that includes additional comments and explanations [6].

In addition to the PowerPoint slides, several works have related the benefits of inluding media resources in the material for a subject (see, for example, [7] and [8]), so a multimedia application [9] that encompasses all of the theoretical subject content (see Figure 1) has been developed. This application was written with the support of Polytechnic University of Valencia Educational Innovation Project 11034.

The application took two years to complete. Three students worked on it as a final degree project, following the directions of the teachers. We selected Macromedia Director 8 for the implementation. The images, graphic presentation and animations, which were especially important, were developed in design programs like Adobe Photoshop or 3D Studio MAX.

This application includes all the material presented in the theoretical classes, complete with comments, drawings, graphics, and animations (see Figure 2). Special features in the multimedia tool allowed us to include a self-assessment system that enables students to evaluate their level of comprehension (see Figure 3) [10].

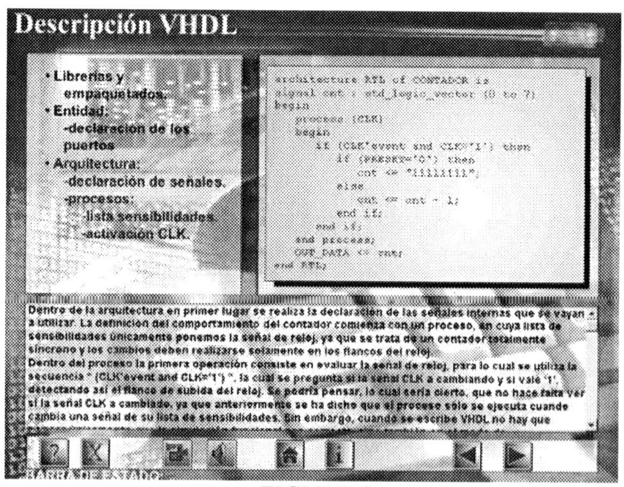


FIGURE 2
PHOTOGRAPH SHOWS THE PRESENTATION OF ONE OF THE THEMES IN THE MULTIMEDIA APPLICATION

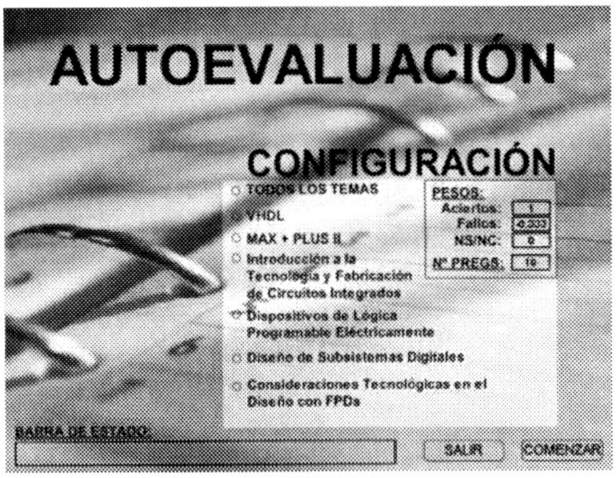


FIGURE 3
PHOTOGRAPH SHOWS THE OPENING OF THE CONFIGURATION PAGE FOR THE SELF-EVALUATION SYSTEM

### Material for Practical Classes

It is clear that EDA tools increase the productivity of a designer and enable more complex logical circuits to be tackled. However, they require consideration of issues related to timing as well as function.

The student version of Altera MAX+PLUS II was selected as the most suitable EDA. It enables highly abstract designs to be introduced using graphic descriptions, diagrams and text as a sub-set of VHDL for synthesis. The software is used with the Altera UPI educational board and, where possible, designs are verified in-circuit in the laboratory [11].

A prepared text [12] uses demonstrations and design proposals to explain the use of the tool and programmable devices. In addition, an expansion module has been designed for the Altera board MEPUA1 (see Figure 4). This board connects the FLEX10k expansion buses with the Altera UPI board. MEPUA1 expansion modules are also explained in the text.

Student evaluation is difficult because the large number of students makes individual tutorials impossible, and the complex concept of 'system' is difficult to digest in the limited time available to students before their examination. The approach and evaluation of examinations requires serious reflection and study by teachers [13].

As part of the preparation for examinations, a new publication [14] has been prepared which includes 43 problems and solutions taken from previous practical examinations. These exercises emphasise a concept that has been especially stressed in recent courses – structured blocks that cooperate in time and space. This book of exercises includes a CD with solutions to exercises for MAX+PLUS II, as well as the student version of the program.

## EVALUATION METHODOLOGY

The methodology includes three methods of evaluation. We explain the basic characteristics of each method as well as the theoretical objectives and advantages and disadvantages. The methods are based respectively on designs that are completely specified (top-down method); partially specified (bottom-up method); and physical implementation (hardware oriented method). We describe the three methods and draw various conclusions. Examples can be found at [14].

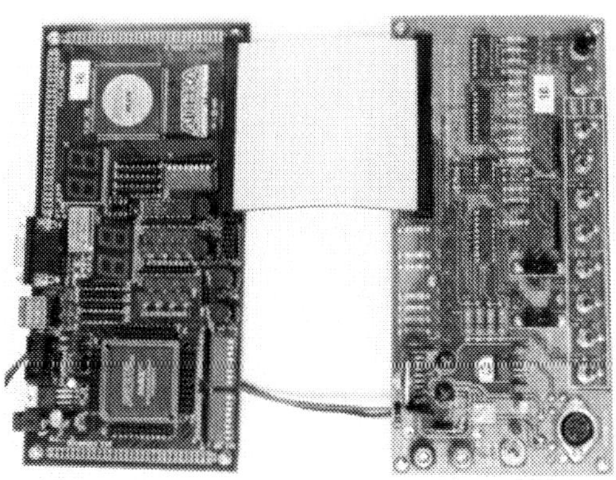

FIGURE 4
PHOTOGRAPH SHOWS THE BOARDS USED IN THE PRACTICAL SESSIONS. THE UPI ALTERA BOARD (LEFT) IS USED TOGETHER WITH THE MEPUA1 BOARD (RIGHT)

### Bottom-Up Methodology: Partially Specified Design (PSD)

The methodology proposed in this section is a shortened specification of a complex digital system [15]. Initially, the details of a design are implemented (*bottom*). The specification is then increased as well as the resolution complexity (*up*) until the final design is reached.

An example of the PSD method can be seen with the design of an FIR filter with given coefficients. In attempting these designs, students should be motivated to explore other subject areas that are important in the field of telecommunications, such as signal filtering, and to draw on knowledge acquired from their studies on signal digitisation, which also forms a part of the degree course. The educational advantage is that students will better understand the concepts of sampling, impulse, and response in an LTI system, as well as the limitations of working and activity frequencies.

### Top-Down Methodology: Completely Specified Design (CSD)

The top-down methodology starts from a complete specification of the system. It is usually a new system, so the student can comment on the specifications. This type of evaluation is termed a 'learning exam'. Part of the design is given as data, and the student can be offered simulated results to verify. As an example, we offer a self-timed parallel multiplier that was not presented during theoretical classes and is partially offered as data.

### Physical Methodology: Hardware-Oriented Design (HOD)

This methodology uses an initially simple design as an image of the complete digital system is not necessary. It is implemented on two boards as described in the previous section. The student must use tools to synthesize the design, then copy it to a programmable device and verify the simulation and correct working of the system. In this way, self-evaluation is carried out in the laboratory. An example would be the step-by-step design of a motor control system.

| Type | Advantages | Disadvantages |
|---|---|---|
| PSD | Simple Complete system specification. Various design techniques (VHDL, schematic...). Wide work spectrum (hierarchy, activity, frequency, timing...). | No physical programming. Lack of depth in specification and time design requirements. |
| CSD | Learning in practical sessions. Complete specifications. In-depth vision of digital system. Various design techniques. Wide work spectrum. | Difficult student comprehension. No physical programming. |
| HOD | Simple specification. PLD and physical card programming. | Poor system vision. |

TABLE 3

ADVANTAGES AND DISADVANTAGES OF THE THREE METHODS OF EVALUATION

### Advantages and Disadvantages of Evaluation Methods

Table 1 shows the main characteristics of each of the three established methods. While each model has advantages and disadvantages, the proposed methodology contains the majority of the theoretical points covered in the subject. The advantages of this methodology can be seen in later courses when students demonstrate their ability to understand microelectronic circuits, as well as the design and modelisation of advanced digital systems. Students interviewed in polls positively rate the teaching mix of theoretical and practical sessions. This high level of satisfaction is difficult to achieve in these types of courses.

## SUMMARY AND CONCLUSIONS

The *Circuit and Electronic System Design* course, offered as part of the telecommunication engineering study plan at the Polytechnic University of Valencia, teaches digital design methods based on PLDs. This implies an understanding of structures and resources, as well as design methods based on hardware description languages (HDLs).

Some materials were developed to aid active classroom teaching and self-teaching so that the student can acquire practical design skills when working away from the classroom. The first of these materials is a complete set of PowerPoint slides used in class and available to students on the Internet. As well as the slides, a multimedia tutorial has been developed that explains the fundamental concepts of the course in a dynamic, simple, and attractive format. The tutorial includes a self-assessment section. Two manuals have been prepared as a third resource for students. They offer greater depth and provide examples of design solutions.

For student evaluation, we have designed an active and practical methodology based on the production of certain designs. These designs can be separated into three categories depending on their level of specification.

The teaching result of this effort has been clearly satisfactory. Students finishing the course have acquired a good foundation and are better able to tackle the subsequent courses. Furthermore, students show an increasing interest in the subject. These conclusions were drawn from student polls. Results of the last four years are given in Figure 5.

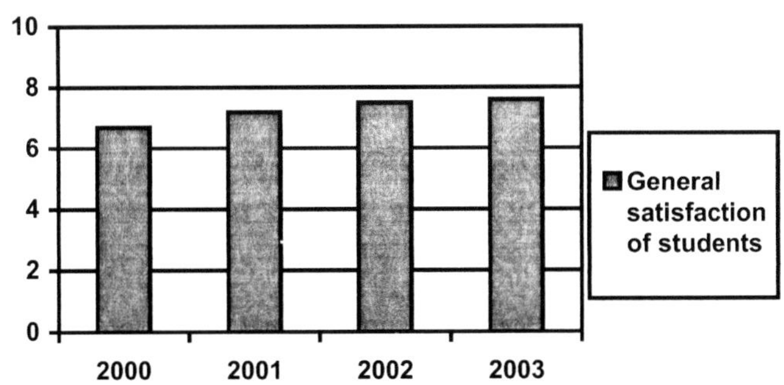


FIGURE 5
GENRAL SATISFACTION WITH THE SUBJECT BY STUDENTS GIVEN IN AN OPINION POLL

## Acknowledgements

Development of the multimedia application was supported by Polytechnic University of Valencia Educational Innovation Project 11034. The authors thank R. García for their laboratory expertise.

## References

1. Artigas, J.I., and Falcó, J., "Evaluación del aprendizaje de los alumnos en grupos numerosos: una propuesta concreta" IV Congreso de Tecnologías Aplicadas a la Enseñanza de la Electrónica, 2000, pp. 59-62.
2. Han, J.T., "Ideas to Make Engineering Education More Efficient" Proceedings of 2000 International Conference on Engineering Education (ICEE-2000), Taipei, Taiwan, 2000.
3. Gadea, R., Colom, R.J., and Larrea, M.A., "Diseño de circuitos y sistemas electrónicos" IV Congreso de Tecnologías Aplicadas a la Enseñanza de la Electrónica, 2000, pp. 561-564.
4. Mensch, W.D., and Silage, D.A., "System-on-Chip Design Methodology in Engineering Education," Proceedings of 2000 International Conference on Engineering Education (ICEE-2000), Taipei, Taiwan, 2000.
5. Cavin, R.K., Joyner, W.H., and Wiggins, V.C., "A Semiconductor Industry Perspective on Future Directions in ECE Education," IEEE Transactions on Education, Vol. 46, No. 4, 2003.
6. Larrea, M.A., Gadea, R., Colom, R., Martínez, J.M., and Cerdá, J., "Diseño de Circuitos y Sistemas Electrónicos. Transparencias," Reprografía Universidad Politénica de Valencia 2002, pp 1-430.
7. Alonso, A., Ferrero, M., Redondo, C., Alegre, E., and Rodríguez,, F.J., "Estructuración Gráfica de la Electrónica Mediante Sistemas Multimedia" II Congreso de Tecnologías Aplicadas a la Enseñanza de la Electrónica, pp. 58-63, 1996.
8. Shiga, A.A., and Pegollo, C.A.G., "The Media as a Complement of Engineering Teaching" Proceedings of 2000 International Conference on Engineering Education (ICEE-2000), Taipei, Taiwan, 2000.
9. Cerdá, J., Colom, R.J., Larrea, M.A., Gadea, R., Martínez, M., "Desarrollo de un Tutorial Multimedia como Apoyo a la Docencia en VHDL," V Congreso de Tecnologías Aplicadas a la Enseñanza de la Electrónica, 2002, pp. 311-314.
10. Rosado, L., and Herreros, J.R., "Hiperdocumentos en la enseñanza de la electrónica: tutoriales, unidades didácticas, test de evaluación y páginas WEB," IV Congreso de Tecnologías Aplicadas a la Enseñanza de la Electrónica, 2000, pp. 249-252.
11. Hamblen, J.O., and Furman, M.D., *Rapid Prototyping of Digital Systems, A tutorial Approach*, Kluwer Aademic Publishers, 2000.
12. Larrea, M.A., Gadea, R., and Colom, R.J., "Diseño Práctico con FPGAs," Editorial UPV, 2001.
13. Martínez, M., Larrea, M.A., Colom, R.J., Gadea, R., and Cerdá, J., "Metodología Docente para la Evaluación de Sistemas Digitales mediante Software de Síntesis Lógica," V Congreso de Tecnologías Aplicadas a la Enseñanza de la Electrónica, 2002, pp. 153-156.

14. Larrea, M.A., Gadea, R., Colom, R.J., Martínez, M., and Cerdá, J., "Ejercicios Prácticos con Lógica Programable," Editorial UPV, 2003.

15. Cappuccino, G., Cocorullo, G., Corsonello, P., and Perri, S., "Educational Design of High-Performance Arithmetic Circuits on FPGA," IEEE Transactions on Education, Vol. 42, No. 4, 1999.

**Joaquín Cerdá Boluda** received his M.Sc. degree in Telecommunication Engineering from the Polytechnic University of Valencia in 1999. He is currently a Ph.D. student in Telecommunication Engineering, and an M.Sc. student in Physics. Since 2000, he has been in the Department of Electronic Engineering at Polytechnic University of Valencia working in digital design, programmable devices, and hardware description languages. His main research interests are fundamental physics, microlectronics, and cellular technology.

**Marcos Antonio Martínez Peiró** was born in Oliva, Spain. He received the M.Sc. and Ph.D. degrees from the Universidad Politécnica de Valencia, Spain, in 1993 and 2000, respectively. Since 1993, he has been a lecturer in the Department of Electronics at the Universidad Politécnica de Valencia. During the first half of 1999, he was a research student with the Department of Electrical Engineering, Linköping University, Sweden. He is Titular Professor in the Telecommunications Engineering School of the Universidad Politécnica de Valencia, Spain. His areas of research interest include VLSI signal processing, digital filter design and custom digital signal processing for video applications.

**Miguel Ángel Larrea Torres** received his B.S. degree in Electrical Industrial Engineering from the Universidad Politécnica de Valencia (UPV) in 1984 and M.Sc. in 1987. Since 1988, he has been a lecturer at UPV teaching VLSI Design Skills Techniques at the newly-founded E.T.S.I de Telecomunicación. His primary research interests are microelectronics and the related CAD.

**Rafael Gadea Gironés** received his B.S. and Ph.D. degrees in Industrial Engineering from the Polytechnic University of Valencia in 1990 and 2000, respectively. Since 1992, he has been in the Department of Electronic Engineering at Polytechnic University of Valencia working on digital design, programmable devices, and hardware description languages. His main research interests are microlectronics, neural networks, image compression and recently cellular automata.

**Ricardo José Colom Palero** was born in Cullera, Spain. He received the M.Sc. and Ph.D. degrees from the Universidad Politécnica de Valencia, Spain, in 1993 and 2001, respectively. Since 1993 he has been a lecturer of the Department of Electronics at the Universidad Politécnica de Valencia. Currently, he is Titular Professor in the Telecommunications Engineering School of the Universidad Politécnica de Valencia, Spain. His areas of research interest include VLSI signal processing, design of FPGA-based systems and custom digital signal processing for audio and video applications.

# Chapter 22

# A Modular Educational Robotic Toolbox to Support University Teaching Efforts in Engineering

UWE GERECKE, BERNARDO WAGNER and PATRICK HOHMANN

*Learning Lab Lower Saxony (L3S), University of Hannover, Expo Plaze 1, 30539 Hanover, Germany. E-mail: gerecke@learninglab.de*

*Robots offer an excellent basis for teaching a number of different engineering disciplines and cross-disciplinary fields. To implement courses and projects, universities need robotic platforms that are flexible and modular so they can be easily customized to the requirements of different subjects. This paper introduces the concepts of our Modular Educational Robotic Toolbox (MoRob) project, which aims to design and develop an advanced mobile robot system for teaching undergraduate and graduate students as well as for Ph.D. research. At the core of our system is the Scalable Processing Box (SPB), a standardized processing platform for easy experimentation with any kind of robotic platform. Key characteristics are scalable performance, modular setup to facilitate tailoring to individual courses, flexible interfaces and easy configuration. Furthermore, MoRob aims to provide a standard set of control modules and teaching units, with a particular focus on project-based learning. Our MoRob Toolbox, therefore, not only supplies an educational robotic platform and improves student learning, it also supports teaching by providing reusable and sustaining material.*

## INTRODUCTION

Robotics concerns the construction of artifacts that couple perception to action to allow a system to carry out purposeful tasks for humans [1,2]. In recent years, research activity in the field of mobile robots has gained considerable momentum and, with its growing success, both the scientific community and industry are paying attention. Autonomous robots constitute a clear emerging market for a large variety of potential applications in a range comprising service robots, entertainment robots, as well as robots for hazardous environments or disaster areas.

Going along with this development, universities are offering an increasing number of courses in robotics, and research students are carrying out more and more dissertation projects in this area. It is an interdisciplinary subject that involves mechanical, electrical, control, and computer engineering. A key component is in integrating these disciplines in order to construct autonomous systems. Therefore, mobile robotics offers an excellent basis for teaching a number of different engineering disciplines and their integration into systems.

Robots also provide an excellent platform for demonstration of basic engineering problems in education. Practical exercises in student projects help to develop skills such as creativity, teamwork, designing and problem-solving. A great advantage of using robots as demonstration tools is that abstract concepts can be turned into real-world problems and solutions [3]. Student learning motivation is improved when real-world objects are included. This is the case in robotics projects as students can see, touch and play with their project [4]. Experiences with courses utilizing robots have been reported, e.g., on software engineering projects [5], genetic programming [6], artificial intelligence [7], and data structures [8]. Using robots, progress can be achieved in gaining both theoretical knowledge and practical knowledge. Ahlgren [9] identified 17 fields in which considerable progress could be made while working on a robot contest, including physics, mathematics, systems design and teamwork practice. Robots not only appeal to children [10], but also are very popular and stimulating for university students [11]. Furthermore, they can be used as a motivational factor to attract high-school students to take engineering courses [12]. Contests like RoboCup or the fire fighting contest [13], as well as robot building projects offer interesting tasks that challenge students to put theoretical knowledge into practice.

## MODULAR EDUCATIONAL ROBOTIC TOOLBOX

In order to use robots in teaching, practical projects and research, universities need platforms which are flexible and modular, yet powerful, so they can be easily customized to the requirements of individual subjects. No existing platform which can be purchased "off-the-shelf" meets all these requirements. A well-designed educational kit available with associated learning material is provided by Lego Mindstorms; however, this has been designed for use in schools and is inappropriate for advanced classes at universities, such as those in automatic control, real-time systems, or even in robotics itself. While robots provide an excellent basis for teaching a variety of topics, a notorious problem is the lack of a suitable platform for such an effort. Often lecturers have to resort to "primitive" platforms, or construct their own systems. Using existing platforms has the disadvantage of too many limitations to really illustrate the key concepts of a particular course. On the other hand, the effort to set up a reasonable system by the lecturers on their own requires numerous resources for implementation and maintenance over time.

The goal of our MoRob (Modular Educational Robotic Toolbox) project is to overcome this problem and provide a suitable platform together with a standard set of control modules and teaching units. This is aimed at teaching undergraduate and graduate students as well as for Ph.D. research. The MoRob project is embedded into the interdisciplinary research in educational technology at the Learning Lab Lower Saxony (L3S) in Hanover, Germany. The studies are carried out within the international framework of the Wallenberg Global Learning Network (WGLN). Project partners are the Centre for Autonomous Systems, KTH Stockholm, and the Robotics Laboratory at Stanford University.

This paper focuses on three important aspects of our project dealt with at the L3S. At the core of our system is the Scalable Processing Box (SPB), a standardized processing unit for easy experimentation with any kind of robotic platform. Key characteristics of this platform are standardization, scalable performance, modular setup to facilitate tailoring to courses in different disciplines, flexible interfaces and robustness. In the following section, we report on applying robots in project-based learning classes and integrating robotics into our engineering curricula. Following this, we focus on educational evaluation of the project. Within the project, we are establishing a complete framework for using robots in education. A complete account of all aspects is not feasible within the scope of this paper; however, before concluding the paper, we provide a brief sketch of our other workpackages within MoRob.

## Scalable Processing Box

A key part of our research is the development of a Scalable Processing Box (SPB), a standardized processing unit for easy experimentation with any kind of robotic platform. It is targeted at users ranging from students taking their first steps in robotics to researchers using the SPB as a base system for their scientific work. The range of applications running on a SPB depends on the user. Small entry-level projects for undergraduates, as well as complex low-level programming for research projects can be adapted on the SPB.

To fit the requirements of educational robotics, the box has been designed to meet the following requirements: simple handling, easy configuration, flexible interfaces, robustness from a hardware and software point of view, real-time capability and user-oriented documentation and examples. The design of the box follows an open source approach. Such approaches have been shown to be an important issue for Lego robots [14], and are becoming increasingly popular for higher-level robot platforms [15,16]. There are several processing platforms established already in the embedded and robotics world. On the embedded side, there are some small distributions available (e.g. Linux VR [17] or ElinOS [18]), but often these distributions are too specialized for a particular platform, or they are commercial solutions, reducing the flexibility needed in an educational environment. On the robotics side, there is, for example, the Orocos [16] system, which is a CORBA-inspired approach of a large selection of open source software for robotics-related control tasks. As it is based on standard Linux desktop distributions, it is difficult for Orocos to meet embedded demands such as economical use of all system resources. The SPB, on the other hand, is designed as a Linux-based open-source processing platform to fit embedded as well as robotic requirements.

The SPB provides a real-time framework for simple usage of interfaces and algorithms. It is divided into two major parts, a runtime environment (SPB-REnv) and a development environment (SPB-DEnv). SPB-REnv comes with everything needed to run programs. To use the SPB-DEnv, only a common desktop Linux system is required. The basis of the SPB-REnv is an embedded Linux, extended by the Linux Real-Time Application Interface RTAI [19], real-time support for common devices (e.g. CAN bus, Ethernet, serial, etc.) and real-time networking. For the UNIX look-and-feel, a standard set of UNIX commands (e.g., ls, ifconfig, cp, su, etc.) is realized by the tools BusyBox

[20] and Tinylogin [21]. These tools are widely used in the embedded Linux community and they combine tiny versions of many common UNIX utilities into two single small executables. The system provides a suitable UNIX administration background covering standard file and network management as well as user handling without requiring much memory on the boot media. To have a simple and secure remote access to and from a SPB-REnv, a secure shell client and server are included (SSH). For remote diagnostics and interactive presentation of results, a simple HTTP server is added to the basic system (Monkey HTTP Daemon [22]).

The Installation of the SPB-REnv is very simple. All programs and a suitable kernel are available as pre-compiled packages (in the first stage for i386 platforms). To get a system up and running, there is a set of interactive installation scripts to select, configure and install the packages needed for the required environment. To handle these scripts, only simple UNIX user knowledge is necessary. The installation will not take longer than 15 minutes. For expert users and other people who need to optimize their system, all source code for the programs, RTAI and the kernel are available. Therefore, the system can be set up and enhanced step-by-step by the user. As it is a normal Linux system it is also possible to install any other suitable software needed for special tasks on the SPB-REnv (e.g. debugging tools). The SPB-REnv is bootable from any suitable type of media (hard disk, Compact Flash card Zip, network-boot, etc.). It is of great advantage to use removable media like CF-cards, which makes it simple to switch between development (SPB-DEnv) and execution of the software in a different place (SPB-REnv). Furthermore, in a scenario with multiple user groups working on a single hardware (e.g. one robotic platform and several groups of students) every user group can have its own experimental environment on a removable data carrier to avoid interference with other students' software development processes. The same idea is valid for a research scenario with different approaches. It is possible to have one media for each project without unsolicited interaction between the solutions.

The core real-time support of the SPB is handled by RTAI. With this improvement of a standard Linux kernel, the SPB is a proper real-time system. On top of the basic real-time structures a set of tools, application interfaces and Linux scripts supplement the SPB in handling real-time processing and networking structures [23] (see Figure 1 for a schematic overview of the architecture). For development of programs within a SPB structure, the SPB-DEnv is equipped with everything needed to work in a standard Linux environment using the GNU C-compiler (gcc). A set of code examples and documentation simplify software development in a robotic context with real-time requirements using the SPB. For fundamental development tasks on RTAI structures and Linux core level, all required kernel header files are included. Therefore, this environment allows the

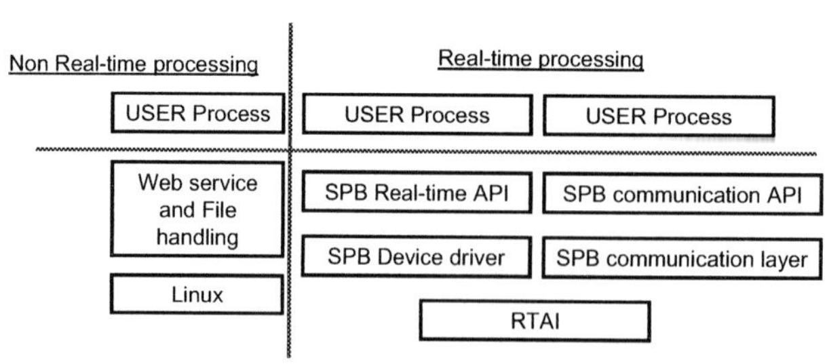


FIGURE 1
ARCHITECTURE OF THE SCALABLE PROCESSING BOX

development of RTAI kernel modules without the need to use the entire SPB infrastructure. This can be used, for example, for further expansion of special hardware drivers. For high-level development, a set of APIs is available to handle the mechanisms provided by the SPB, for example real-time driver access, real-time networking, prepared algorithms for basic data processing, non real-time systems maintenance and debugging.

In the first stage, the SPB is designed to run on an Intel platform. In following stages it is planned to port it to several embedded systems (e.g. PPC or Geode). This will be promoted by the use of Linux/RTAI in the core of the system and free publication of the SPB under an open source license system. To increase the flexibility and modularity of the system, SPB runtime and development environments will be provided for all systems supported by Linux/RTAI. This gives the user a choice in specifying requirements such as low power consumption, high processing power or physical size.

Armed with the capabilities of distributed real-time data processing, a well-structured software interface, and a simple setup procedure, the Scalable Processing Box is designed to become a flexible and powerful processing standard for education and research in the field of robotics. Due to the modular and structured design, it is possible to rearrange existing SPBs similar to Lego building blocks, creating different possible scenarios. With Linux/RTAI as a base and an open source approach of the SPB packages, a broad community can use the SPB and continue developing further components, such as device drivers or other real-time network components. Providing an operating framework based on Linux/RTAI, further developments can follow standard specifications for this environment[3], or existing drivers can be used. As the embedded context is limiting the use of ISA or PCI extension cards, network controlled sensors and actuators preferably should be used for interactions with the robot.

More information about the technical details of our SPB were presented in [24], while current developments including documentation, downloads and sample implementations can be found on our website[4].

## PROJECT-BASED ROBOTICS COURSES

Within MoRob, we promote problem-based learning, which increases the effectiveness of learning by having students perform projects through the learning-by-doing approach. This approach also helps to develop engineering skills such as creativity, teamwork, designing and problem-solving. Robots offer an ideal tool to challenge students with such problem-based tasks. Our initial focus is on the development of three courses at different levels: a Lego Mindstorms project for teaching introductory courses to undergraduates; a lab project with an ER1 platform by Evolution Robotics (see Figure 2) for advanced undergraduate teaching; and a series of robotics projects throughout an M.Sc. course in three consecutive semesters. Development of the first two courses has been completed, while the third is under development.

### Lego Mindstorms Project

The Lego project "Autonomous Service Robots" was carried out by groups of 5th semester students pursuing the B.Sc. degree in "Applied Computer Science" as part of a hardware laboratory class. Within this class, after following a number of set exercises,

[3] A lot of information is provided, e.g., on http://www.rtai.org

[4] http://www.learninglab.de/spb

FIGURE 2
EVOLUTION ROBOTICS ER1 PLATFORM

students have to complete a "mini-project" over a 4-week period. Here, the Lego project is one of 4 possible projects among which students can choose. Although the Lego system is not suitable for use in advanced engineering courses, it is ideal for introducing students to problem-based exercises, to working in teams, and to applying previously acquired programming skills in a practical problem. The Lego project can accept the largest number of students for projects within this lab. Yet, it was the first project to have its available spaces filled up, indicating the popularity of the subject. Equipped with the LeJos Java environment, the students' task was to solve a problem using the Lego Mindstorms kit. Tasks for the students are different every year. In 2003 the application scenario was a series of burning oil wells that was generating intense heat, giving rise to such high temperatures that firefighters could not get close. The task, therefore, was to build a fire-fighting robot to extinguish the fires, which were simulated by a number of burning candles placed in an unenclosed area on the floor (for a similar task, see [9]). Within the groups, the students had to solve the task through self-organization with little interference by the tutor. There were 4 groups in the project, all successfully achieving the goal. See Figure 3 for one of the robots[5]. This course was also used for initial evaluation (see a following section).

**Lab Project for Advanced Undergraduate Teaching**

A new lab project for advanced teaching was developed and carried out in the summer semester of 2003. Here, students of Communication Science had to solve a localization task using information based on signal strength from Bluetooth access points. The application scenario was a "smart house" where users can communicate with appliances and mobile service robots from a PDA. For the robots, in order to navigate according to the user's requests, and to find out where the user is, localization is needed to identify the position. A number of undergraduate student teams carried out this project-based task as a lab session within a degree course. The robots here were not used to teach the principles of robotics, but as a vehicle for implementing a practical task in Communication Science. Students benefited from the hands-on experience which also increased their interest in problem-based learning.

[5] A video of the final presentation of the students can be found via the link: www.learninglab.de/morob

FIGURE 3
FIRE-FIGHTING LEGO ROBOT BUILT BY STUDENTS

## M.Sc. Program with Robotics Projects

A new M.Sc. program in Systems Design began in autumn 2003 at the University of Hanover. From the second year on, it will incorporate a series of robotics projects as an essential part of the degree. It will take the concept of robotics in a project-based scenario further than in introductory courses, using it in a far more sophisticated setting. Here, robotics will be integrated as a core topic for the understanding of systems analysis and design.

The curriculum of this course requires the students to undertake a series of three projects, building on each other over the first three semesters of the course. Students work in teams with 4 students on each team. In Project I, the task of system analysis and modeling is carried out. The students use MATLAB/Simulink to create a simulation model of an existing system with all its software and electrical and mechanical hardware components, with static and dynamic properties. Project II takes the students to the design phase. Here, the aim is to create a prototype as a virtual model, using the same tools from Project I (MATLAB). Finally, Project III is concerned with integration. Here, MATLAB is not used; instead, the system is implemented. The focus is on software development and electrical sensors development, leading to the mechanical assembly of the system. The assembly takes into account all of the concerns of the real world, including noise and real-time constraints. A key necessity is multi-disciplinary interaction.

The concept of the course is to be independent of a particular robot platform. We have been developing our own platform (see below), which together with the SPB will be the first choice for the project work. However, it could also be a Pioneer robot (see Figure 4) with our SPB, the Evolution robot with our SPB, or any other available robot which allows the mounting of the SPB. The platform-independence of the SPB is important as it allows flexible usage. Furthermore, as the software of the box is run from a CF-card, it is easy to organize different groups working on the same platform while developing software on independent workstations. Therefore, all resources are exploited to great advantage.

As we are using high-level platforms, students have the capability to tackle design tasks that are much more complex and sophisticated than those possible with Lego. This also implies that, by applying our approach, a much broader range of engineering concepts can be taught. The M.Sc. program also offers an excellent testbed for pedagogical and didactic evaluation of our project. The building blocks and technology can be tested during a complete course sequence of four semesters. See [25] for more details about this course.

FIGURE 4
THE PIONEER 2AT MOBILE ROBOT

## EVALUATION OF EDUCATIONAL IMPACT

A particular focus of our project is the evaluation of the learning process and educational impact so as to determine the suitability of robotics and the MoRob system for use in engineering education. To achieve this, we plan to employ a summative evaluation to determine program effectiveness. Evaluation of our concepts and the use of robotics in teaching will be carried out at all three partner institutions of MoRob. The questions of interest include the following: Does MoRob and project-based learning improve student learning in robotics and other courses? Is our system better able to integrate research and scientific methods in teaching? What is the best way to integrate the project, and problem-based learning, with courses? To answer these questions, we plan to conduct comparative evaluation with students from the following control groups: B.Sc., M.Sc. and Ph.D. We will focus on the following elements in the students' learning process: motivation, creativity, knowledge gain, theoretical concepts used, precision, quality of results, the way students cope with technical challenges, and their understanding and use of scientific and engineering methods.

Evaluation testbeds will be the new M.Sc. program on Systems Design at Hanover; a new international M.Sc. Robotics course at KTH; and the course Robotics 225 at Stanford. The main part of the evaluation of the MoRob system will be carried out once the new courses are established. However, during the first year, a questionnaire was used on existing courses at the three institutes to examine aspects of the current state of robotics education. The evaluation approach with an accompanying questionnaire was designed in Hanover with the help of expertise from our experienced evaluation team. The same evaluation material was used on three different courses at Hanover, Stanford and KTH, respectively, to acquire information about course contents and style across the three participating departments. For this first phase of evaluation within MoRob, the mini-project within our hardware lab was one of the courses to be examined. At the end of the course, students had to fill out a questionnaire. First insights can be reported after an initial look at this part of the evaluation with our Lego participants: within the problem-based learning task, students could bring in their own ideas to a large extent and were able to acquire new knowledge on their own. The tutor only had to give few directions and students could organize their work freely. Teamwork and communication skills were felt to be among the most important things learned during this project. Many students felt that this was the best or most interesting project work they had ever done. Overall, the students judged the Lego platform and its environment as positive and appropriate. However, there were a large number of individual comments highlighting the shortcomings of this platform, including imprecise sensors, limited software, small memory, problems with multi-threading and arithmetic. This confirmed that the Lego platform reaches its limitations very easily (programming power, sensors, scalability, mechanical robustness, etc.), that it is inappropriate for other than simple tasks, and that we need a more sophisticated robot platform. Students were highly motivated on this course and usually achieved more than was expected. For example, the quality of presentations was higher than expected, showing that the students put in a lot of effort to show their work. One group even had the first prototype ready between the time when the kits were handed out and the time of the first official group meeting, when the project was meant to be kicked off. Similar observations have been made in other Lego projects as well, e.g. [4].

## OTHER MOROB WORKPACKAGES

Our experiences with different commercially available robot platforms in education showed that each had certain advantages, and a number of shortcomings as well. For example, the modular approach of Lego and its kit approach are perfectly suited for educational applications. However, its technical limitations restrict the use to certain courses only. While the ER1 is more powerful due to its use of a laptop with a robot development environment under MS-Windows, it is still very limited with respect to sensing facilities and flexibility of the software. The use of aluminum profiles to build robot bases for the ER1 offers enough flexibility to build a variety of different robots. Pioneer 2AT robots are state-of-the-art mobile platforms widely used in research, and it can be equipped with a large variety of sensors and actuators, but these are not designed as a kit and changing the hardware configuration is not always trivial.

To get a more suitable robot platform for education, we designed a new kit that combines the advantages of these 3 systems. With modular components, flexible platforms tailored to teaching and research can be built. The mechanical setup is easily customizable to the requirements of the users (just like Lego, except much more robust). To equip the robot individually, the kit includes a selection of components, e.g. servo drives and sensor modules like odometry, laser scanner, ultrasound, inertial sensors, etc. Individual components (aluminum profiles, sensors, actuators, controllers) can be complemented to obtain reasonable kits that can be used in a variety of courses and lab exercises. A picture of the new kit can be found in Figure 5.

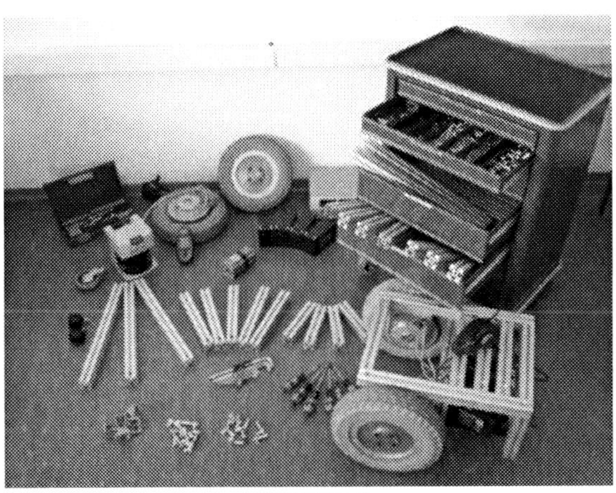

FIGURE 5
MOROB KIT TO BUILD ROBOT PLATFORMS

Based on the basic infrastructure of the kit and SPB, we need a comprehensive library of toolboxes to use the system in lectures and exercises in a variety of different subjects. In robotics, for example, we need standard components for navigation, object recognition, planning, etc. to demonstrate the key concepts of our courses. These toolboxes are higher level software modules, which can also be used in research settings at the Ph.D. level. MoRob allows the design of a number of such demonstrator modules that can easily be exchanged on the robot depending on the required robot tasks. As a sample module, we are developing a demonstrator for localization experiments, e.g. using a laser scanner for simultaneous map building and localization (2D/3D). With this demonstrator we can work with students in our lab, evaluate the pedagogical merit, and test our software architecture and the whole system.

To supplement the platform and its application in a teaching environment, one of the aims of MoRob is the provision of a variety of course materials to cater to different courses within the robotics and engineering curriculum. Therefore, the system will be accompanied by teaching units, which can be easily used and adapted to a variety of courses. This will comprise a selection of different topics, offering a choice of teaching units and exercises in different subjects. It will comprise different approaches at different levels and integrate complementary expertise from different institutions. Individual modules or whole courses from the robotics curriculum can be drawn from this model.

To enable other institutions to access the resources created within MoRob, these resources have to be made available on the Internet. Therefore, we are creating an Internet repository for sharing, that is, a www-portal for course material and technical documentation. This way, everything developed can be shared. Furthermore, a detailed documentation should ensure an easy re-use of all material. The repository will consist of the following: lecture notes, slide presentations, exercises, SPB, robot control software and toolboxes, sample code, construction examples, documentation, and curriculum.

## CONCLUSIONS

Robots are a perfectly suitable tool to support education in many different disciplines. They can be applied in teaching, demonstration, exercises and lab classes and in research projects at universities. In this paper, we have presented the concepts of our MoRob project, which seeks to provide the infrastructure to support the teaching efforts when using such a tool. The Scalable Processing Box at the core of our system is a powerful and flexible processing platform for easy experimentation with any kind of robotic platform. It provides a framework for simple usage of standardized interfaces and algorithms when working with robots. The different subjects involved in our design tasks clearly underline the interdisciplinary nature of robotics. Our new system will not only allow for easy construction of robot systems and their use in robotics courses, but it will also be applicable as a vehicle for experimentation in courses on other subjects like programming, signal processing, control theory and artificial intelligence. Therefore, a standardized teaching environment will emerge which will be easy to use in a variety of applications.

A standard toolkit for teaching with robots can serve as a standard platform with which students can become well acquainted during their studies. Initially they can use the platform for exercises in programming. At a later stage, with sensors on the platform, it can be used for exercises in signal processing. In advanced control applications, the students can design a trajectory-following algorithm or a new PID controller for improved traction. At a later stage in their studies, students can assemble a new system for a particular exercise or competition. The basic robot modules which we are developing are flexible and powerful enough to be used at all levels from introductory courses to advanced research projects. In Ph.D. research, students benefit from a system which is flexible and versatile so it can be adapted to their specific research needs. MoRob and the SPB make it possible to combine quickly the modules that the students need with only minimal customization. Therefore, Ph.D. students can concentrate on their actual research tasks and questions, rather than having to deal with creating their research tools.

Through this project, we have developed an educational robotic platform that improves student learning and increases the attractiveness of learning through problem-based and experimental learning in student projects ("learning by doing"). The platform supports teaching through reusable materials. In addition, the resultant educational resources can be shared across a number of institutions so that the materials and exercises can be utilized by several instructors. This way the investment in new educational material becomes worthwhile and the community of educators can combine efforts and share experience.

## ACKNOWLEDGEMENTS

This project has been supported with funding from the German Ministry of Research and Education (BMBF). We would also like to acknowledge the cooperation from the Institute for Systems Engineering, University of Hanover.

## REFERENCES

1. Arkin, R.C., *Behavior-Based Robotics*, MIT Press, London, 1998.
2. Murphy, R.R., *Introduction to AI Robotics,* MIT Press, Cambridge, 2000.
3. Kumar, D. and Meeden, L., "A Robot Laboratory for Teaching Artificial Intelligence," SIGCSE Bulletin, Volume 30, Number 1, 1998, pp.341-344.
4. Garcia, M.A. and Patterson-McNeill, H., "Learn How to Develop Software Using the Toy Lego Mindstorms," Proceedings of Frontiers in Education Conference, Boston, MA, 2002.
5. Gustafson, D., "Using Robotics to Teach Software Engineering," Proceedings of Frontiers in Education Conference, Tempe, Arizona, 1998.
6. Bianchi, R.A.C. and La Neve, A., "Studying Electrical and Computer Engineering through the Construction of Robotic Teams and Systems," Proceedings of 2002 International Conference on Engineering Education (ICEE-2002), Manchester, UK, 2002.
7. Paul, C., Hafner, V. and Bongard, J.C., "Teaching New Artificial Intelligence using Constructionist Edutainment Robots," Workshop on Edutainment Robots, Sankt Augustin, Germany, 2000.
8. Dannelly, R.S., "Use of a Mobile Robot in a Data Structures Course," The Journal of Computing in Small Colleges, Vol. 15, No. 3, 2000, pp. 85-90.
9. Ahlgren, D.J., "An International View of Robotics as an Educational Medium," Proceedings of 2002 International Conference on Engineering Education (ICEE-2002), Manchester, UK, 2002.
10. Druin, A. and Hendler, J., *Robots for Kids - Exploring New Technologies for Learning*, Morgan Kaufmann, 2000.

11. Wang, E.L., "Teaching Freshmen Design, Creativity, and Programming with Legos and Labview," Proceedings of Frontiers in Education Conference, Reno, Nevada, 2001.

12. Oppliger, D.E., "University - Pre College Interaction through FIRST Robotics Competition" Proceedings of 2001 International Conference on Engineering Education (ICEE-2001), Oslo/Bergen, Norway, 2001.

13. Verner, I. and Ahlgren, D., "Robot Contests Framework for International Cooperation in Engineering Education," Chapter 24, Aung, W. et al. (eds.), *Innovations 2003: World Innovations in Engineering Education and Research*, Int. Network for Eng. Ed. and Res. (iNEER), Arlington, VA, USA, 2003, pp. 237-245.

14. O'Hara, K.J. and Kay, J.S., "Investigating Open Source Software and Educational Robotics," Consortium for Computing Sciences in Colleges Eastern Regional Conference, Bloomsberg, PA, 2002. Also to appear in The Journal of Computing Sciences in Colleges.

15. CARMEN: Carnegie Mellon Robot Navigation Toolkit, http://www-2.cs.cmu.edu/~carmen/

16. Orocos: Open Robot Control Software, http://www.orocos.org

17. Linux VR, http://sourceforge.net/projects/linux-vr/

18. ElinOS, http://www.elinos.com

19. Mantegazza, P., et al., "RTAI: Real-Time Application Interface," Linux Journal [Online], April 2000, http://www.linuxjournal.com/article.php?sid=3838

20. BUSYBOX - The Swiss Army Knife of Embedded Linux, http://www.busybox.net/

21. TinyLogin - The world's smallest login/passwd/getty/etc., http://tinylogin.busybox.net/

22. Monkey HTTP Daemon, http://monkeyd.sourceforge.net/

23. Kiszka, J., Hagge, N., Hohmann, P. and Wagner, B., "RTnet - Eine Open-Source-Lösung zur Echtzeitdaten-kommunikation über Ethernet," Proceedings of Telematik, Siegen, Germany (2003).

24. Hohmann, P., Gerecke, U. and Wagner, B., "A Scalable Processing Box for Systems Engineering Teaching with Robotics," Proceedings of International Conference on Systems Engineering, Coventry, UK, 2003.

25. Gerecke, U., Wagner, B. and Hohmann, P., "Educational Robotics in a Systems Design Masters Program," 3rd IEEE International Conference on Advanced Learning Technologies, Athens, Greece, 2003.

**Uwe Gerecke** graduated in 1996 with a M.Sc. in Artificial Intelligence with Engineering Applications from the University of Wales, College of Cardiff (UWCC), UK. In 2002 he received his Ph.D. from the University of Sheffield, UK, following research in the areas of Robotics and Neural Networks. In 2002 he was a visiting researcher in the Mobile Robotics Group at the Department of Cybernetics, Czech Technical University of Prague (CTU). He moved on to become a Research Scientist at Learning Lab Lower Saxony (L3S), University of Hanover, Germany. He now works in the area of Educational Robotics as the project leader of the international WGLN (Wallenberg Global Learning Network) project MoRob (Modular Educational Robotic Toolbox).

**Bernardo Wagner** received a M.Sc. (1984) and a Ph.D. (1989) degree in Electrical Engineering from the University of Stuttgart, Germany. From 1985 to 1988 he was head of the Reverse Engineering Group of GPP in Oberhaching/Munich. In 1991 he was appointed Professor of Computer Science at the University of Applied Sciences in Ulm, Germany. Dr. Wagner has been full professor at the University of Hanover since March 1997. He is a founding member of the Computer Science department and in charge of the Real Time Systems Group. Moreover, he is director of the Center for Technical Didactics at the University of Hannover, a member of the Center for Mechatronics (MZH) in Hanover, and a member of the directorate of the Learning Lab Lower Saxony (L3S).

**Patrick Hohmann** received his Diploma in Computer Science from the University of Hildesheim/Germany in 1999. Following 3 years developing embedded car multimedia systems for Bosch/Blaupunkt, he joined the Educational Robotics team of the Learning Lab Lower Saxony (L3S) in 1999. His Ph.D. research project deals with human machine interfaces for service robots in a deterministic real time context.

# Chapter 23

## Cooperative Learning in a Digital Electronics Course at EPSC-UPC

FRANCESC JOSEP SÁNCHEZ ROBERT
*Technical School of Castelldefels (EPSC), Technical University of Catalonia (UPC), Canal Olímpic, s/n, Castelldefels, 08860, Spain.*
*E-mail: francesc.josep.sanchez@upc.es*

*We present the experience of introducing the cooperative learning (CL) method in a course on Digital Electronics. This course is compulsory and is one of the subjects that are included in the selective phase (the first two semesters) of the Bachelor's degree in Telecommunications Engineering at the Technical School of Castelldefels (EPSC), a school within the Technical University of Catalonia (UPC). This paper explains the basics of the cooperative learning (CL) method used, the context and syllabus of the subject in which CL was applied, and the process that has been applied for restructuring the course, which until recently was taught in the traditional passive lecturing way. We include the results of assessment by students.*

### INTRODUCTION

The cooperative learning (CL) method differs substantially from the classical, passive lecturing approach in which students merely take notes during the class while watching and listening to the professor. The latter approach represents the old paradigm of teaching, as Johnson, Johnson and Smith [1] indicate, while the former represents a new paradigm of teaching that can produce a great improvement in the learning process. Many authors have demonstrated the feasibility of applying the CL concepts at a technical or engineering school, e.g. [2,3].

Cooperative learning derives its strength by forming base groups in which students are put together for the whole semester for the purpose of learning collectively as a group. In the classical method of lecturing, which is predominantly in use at UPC, a student learns in isolation, in competition with his or her colleagues. The cooperative approach represents a real change in the way teaching and learning take place at the school. At EPSC, other professors have carried out several new experiments with students at the second- and third-year levels of the bachelor degree, and problem- or project-based learning (PBL) has been applied at the second-cycle or master's level in Telecommunication Engineering [4]. Our colleagues' experience as well as the experiences reported in the literature that covers applications of PBL to upper-level graduate courses [5] show that it is highly desirable to instruct students to work cooperatively as early as possible.

Furthermore, the new trends in engineering education that include (in addition to technical subjects) increased emphasis on communication skills, teamwork, group problem-solving skills and preparation for lifelong learning are leading to changes in approaches for teaching and learning around the world, e.g. [6,7].

Therefore, we decided to begin utilizing cooperative learning techniques with our new students during the second semester (1B) in the bachelor's program. The work was conducted with full support from EPSC and the Grup d'Interès en l'Aprenentatge Cooperatiu (GIAC) [8]. The objective is to redesign the way the subject of Digital Electronics is taught with the aim of improving the quality of learning and transferring the experience to other subjects at the selective phase, including the courses of the first semester (1A).

It was recognized at the outset that it is impossible to achieve the whole potential of the new method in a single semester. One has to proceed step-by-step, by first reorganising the subject towards the new paradigm, and then trying to inculcate the five elements of cooperative learning: mutual interdependence, individual accountability, face-to-face interaction, interpersonal and small group skills, and group processing. Our goal was implementation through evolutionary changes adopted in successive semesters, as the instructors themselves and the students gain experience.

This paper seeks to share the experience gained and the results of implementation in this process.

## Digital Electronics Context and Syllabus

Figure 1 shows the curriculum for the selective phase of the studies for the bachelor's degree in Telecommunications Engineering, showing that Digital Electronics is offered during semester 1B of the first year. The course is worth 6 credits: 4.5 credits of theory (3 hours per week) and 1.5 credits (1 hour per week) of application. No specific laboratory classes are scheduled, but some laboratory work on digital circuits is included in Electronics Laboratory during the same semester. Figure 2 shows the relationship between Digital Electronics and other subjects, which include compulsory and optional topics in Telecommunications Systems and Telematics. It is intended that by the time students finish their bachelor's degree, they will acquire enough know-

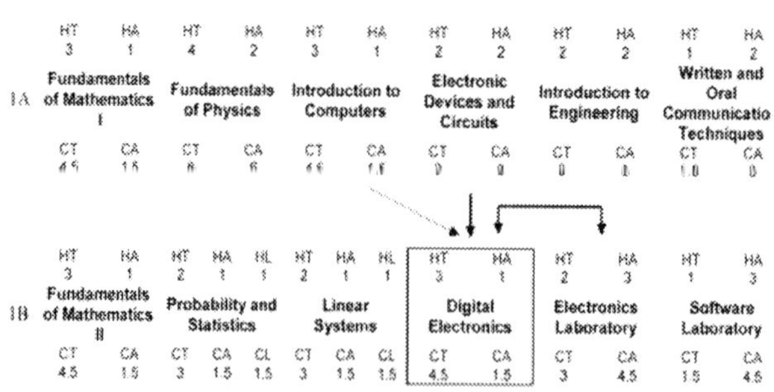


FIGURE 1

SUBJECTS OF THE SELECTIVE PHASE CONSISTING OF SEMESTERS 1A AND 1B OF THE FIRST COURSE

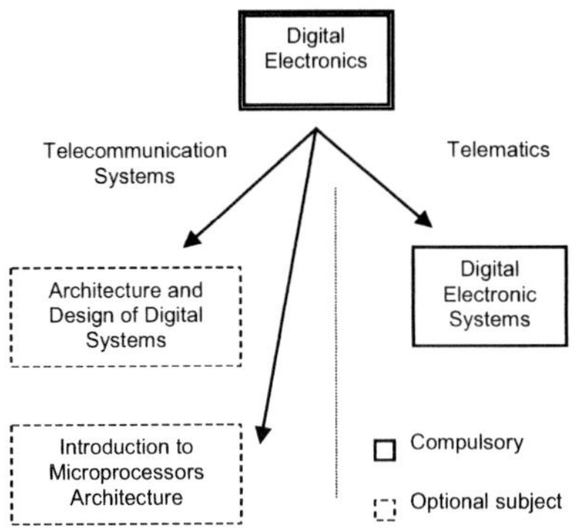


FIGURE 2
DIGITAL ELECTRONICS AND OTHER RELATED SUBJECTS IN CHARGE OF THE DEPARTMENT OF ELECTRONIC ENGINEERING

ledge to design digital systems with certain degrees of complexity. As indicated in Figure 3, the course has been divided into two parts of identical length: Combinational Systems for the first seven weeks and Sequential Systems for the last seven weeks. Additionally, some course materials have been included that complement the core program and that introduce the subject matter to be taught the following years, including lectures about microprocessors, Very High Speed Integrated Circuit Hardware Description Language (VHDL), Programmable Logic Devices (PLD), and software for computer aided design of digital systems. To introduce a more practical orientation to the subject, the design of an application project (AP)—a practical exercise including real integrated circuits—is also scheduled during the semester.

## IMPLEMENTATION OF THE COOPERATIVE LEARNING METHOD

The systematic implementation of the cooperative learning method requires major changes in academic activities and an initial preparation for the teachers far beyond that needed for the traditional method. It is compulsory to notify students that they will have to work an average of 8 hours a week. That is, they will have to invest double the time shown in Figure 1 for the course. Concerning teachers, it is recommended that they start to prepare the documentation for the course several weeks before the beginning of the semester, and to consult colleagues who have used the method before. Each base group is comprised of three students and they will stay together for the whole semester.

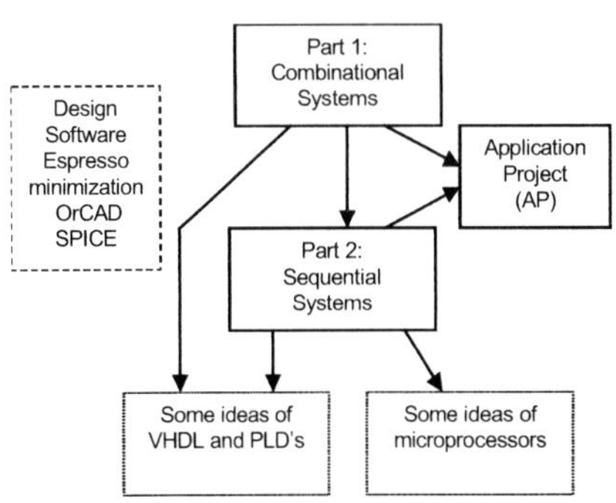


FIGURE 3
COURSE SYLLABUS

The academic activities and the weekly sessions employed for cooperative learning are listed in Figure 4. As noted therein, there is theory to be learned, exercises to be attempted, problems to be solved, an application project to be designed, a portfolio to be prepared, and several examinations to be scheduled to assess the students' progress. The regular sessions in the classroom with the teacher (4 hours per week) will consist of a lecture session (L), conducted by the professor, of 1.5 hours, a teamwork session (TWA) of 1.5 hours, and a teamwork session (TWB) of 1 hour, which the students develop themselves. In addition, the students will have to conduct by themselves an extra teamwork session (TWC) of 2 hours minimum in the study classroom, the library or even their own home. Therefore, if the lecture is considered individual work, each student spends between

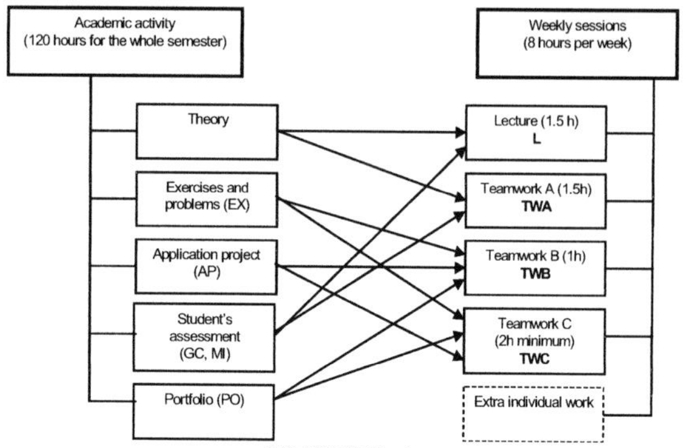


FIGURE 4
WEEKLY SCHEDULE FOR ACADEMIC ACTIVITIES AND THE SESSIONS WHERE THEY TAKE PLACE

minimum of 56% and a maximum of 82% of the time working cooperatively with his or her classmates from the base group.

Theory and student assessment are developed in both sessions L and TWA, while problem-solving, exercises, portfolio making, and the application project are carried out in cooperative sessions TWB and TWC. The teacher can monitor the group activity during regular sessions and provide feedback to each team, thereby monitoring the progress of every student in the classroom. Figure 5 depicts how the cooperative environment is working in sessions TWA and TWB, illustrating positive interaction between team members and between the team and the teacher.

FIGURE 5
PICTURE OF THE COOPERATIVE LEARNING IN ACTION IN THE CLASSROOM

The portfolio consists of the following: the table of contents; exercises and control tests proposed during the course; class notes and additional documents processed by the team; a register of the work done in every TWC session; and the memory of the application project and its oral presentation. An index of terms or glossary of the subject can also be included. The portfolio is a way for organising and classifying the academic activities during the semester and represents, without any doubt, something new as compared to the traditional teaching approach where no one takes into account the students' study habits or how each student structures the notes.

Before the beginning of the semester, it is essential that the instructors prepare carefully all the documentation and tasks for the students and for themselves, including:

- A detailed course syllabus including learning objectives, bibliographical references and a guide to describe how students will be assessed. Additionally, a document is needed to explain how the subject is organized and the basics of the cooperative learning method.

- The weekly working schedule with indications of what will be taught in every session (there are about 45 sessions in the semester), which exercise will be proposed and which exercise will be picked up for correction.
- Instructions on how to use the digital campus – the Intranet -- and even the web site where most of the digital pfd-formatted documents will be stored.
- Reservations of special classrooms for teamwork sessions.
- Spreadsheets for entering assessment results for individuals and groups, and other information needed to track the students' progress. Even the weekly time each team devotes to the subject can be recorded and graphically plotted.
- A document that lists the assessment standards, including requirements for passing the course.
- A set of exercises of 'minimum knowledge' for practice as a baseline for solving more complex design problems during the semester.
- Class notes to stimulate a learning of quality, including transparencies, concept maps, etc.
- A list of titles of application projects and their template document to be selected by each team.
- A document for explaining the professor's grading criteria of exercises and examinations.

In addition, documents should be prepared for self-assessment and self-improvement by the team for its activities and its portfolio. New faculty members may request copies of the above documents in the Catalan language to help simplify the task of applying the CL method to other engineering courses. To reduce the possibility of failure, it is very important for faculty members not to have to begin from scratch, but to ask for help from someone who has demonstrated the feasibility of the method. Also, it is one thing to try new teaching methods on senior students or in optional subjects, and another to try them on overloaded freshmen where a failure in the methodology can have a significant impact.

## ASSESSMENT SYSTEM

Assessment of students in the classical lecturing method simply requires putting marks on a pair of exams and determining the mean grade. Assessment in the cooperative learning method, however, is more complex because all the academic activities of the students contribute to the final grade. Most of the activities and tests are done cooperatively

$$Q = (EX + GC) \cdot 20\% + MI \cdot 30\% + (AP + PO + SUB) \cdot 10\% \qquad (1)$$

producing the same grade for the three members of the team. Equation (1) shows the various components that lead to the final grade at the end of the semester. Here, EX stands for the component grade on exercises, of which 12 are carried out and graded on a weekly basis; GC stands for the marks for group controls, arising from an average of 2 or 3 tests conducted each by a different member of the team; MI represents the results of the minimum knowledge tests, consisting of 9 individual tests dealing with basic concepts for subjects that everyone must pass in order to get a passing grade in the course; AP stands for the marks for the application project, involving a design project and an oral presentation; PO is the grade for the portfolio, where all the activities are recorded; and, finally, SUB is the subjective grade from the instructor. The concept of minimum knowledge is an original idea that has proven to be very effective for sustaining continuous concentration over the whole semester. The importance of the application project cannot be overemphasized, as the development and presentation of a well-structured application project early in the career of a student strongly reinforces the quality of learning, giving, among other impacts, significance and justification to many concepts included in the syllabus. Of course, this approach is central to the PBL-based learning approach.

Student marks are updated and posted in the digital campus every fifteen days allowing continuous evaluation of the progression of each team. These marks are not the final ones, but serve as an interim reference. Furthermore, as shown in Figure 6, each team is asked to record weekly the study time spent on the course, so that even from the firsts weeks of the course, it is possible to detect problematic situations inside the teams, such as students not completing the expected 8 hours of work per week. The current approach represents a major step toward satisfying key features of the European Credit Transfer System (ECTS), which has been developed with support by the European Commission [9].

We have demonstrated that successfully passing Digital Electronics is not an easy task for students. As seen in Figure 6, most of the groups follow the theoretical line of the weekly workload, thereby guaranteeing success. The final score is heavily influenced by personal attitude and disposition towards learning. An instructor can obtain disparate performance results even among similar classes of 40 students each [10]. Better interaction and trust between team members and the whole class may prevent individual failures. Our experience for several semesters indicates that it is not easy for students to

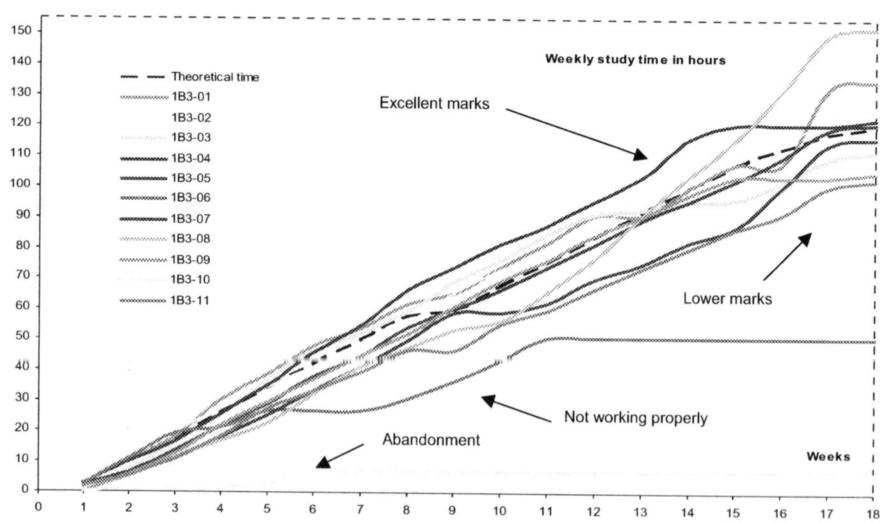


FIGURE 6

RECORDING STUDY TIME FOR EACH TEAM FOR THE COMPLETE SEMESTER WILL HELP MONITOR THE ACTIVITY OF EACH TEAM AND WILL FACILITATE ADAPTATION OF EUROPEAN CREDIT TRANSFER SYSTEM. WORKLOAD OF EACH TEAM IS STRONGLY CORRELATED WITH THE FINAL GRADES.

get involved in cooperative learning. Consequently, training the students on social skills and small group dynamics is required for solving and preventing situations where, for example, students refuse to work together. In some ways, instructors must work at once as teachers, facilitators and tutors.

## DISCUSSION AND SUMMARY

The initial results show that CL is a viable alternative to the classical way of lecturing at our university. Figure 7 shows the high passing rates (around 82%) that have been achieved since the introduction of the new methodology in semester 02/03-1. Our conclusions are as follows: (a) CL can be successfully applied to a technical field with freshmen students; (b) the main syllabus of a classical course on digital systems has been developed, covering even demonstration lessons for introducing advanced concepts for the following semesters (see Figure 3); (c) students can sustain out-of-class base group work and maintain a high level of workload for 15 continuous weeks; (d) most of them can take an active role for the whole semester, learning new skills for constructive team work, technical discussions, structured oral presentations, quality reporting, etc.

Every semester, input from a large questionnaire adapted from the Students' Evaluation of Educational Quality (SEEQ) is processed. The acceptance of the new cooperative learning method grows as the number of students who know it increases. As seen in Figure 8, students also recommend the adaptation of CL in other courses, since they perceive clearly that they have learned valuable and interesting concepts and are able to obtain higher marks.

However, the new method has a serious drawback: that of imposing a considerable initial workload for the instructors -- estimated to be more than 2 hours for each hour of lecturing. Even after the experiment has been set up, the major changes have been in place, and the documentation has been developed over the course of the first few semesters, CL still requires more time than the traditional lecturing method, essentially due to the large amount of material to be regularly graded and delivered back to the students (roughly 1 hour for every hour of lecturing). Nevertheless, without any doubt, restructuring Digital Electronics for CL has been a worthwhile experience. The new method represents a commitment to excellence as it has had extraordinary impact over the quality of learning, and it must be continued and expanded to other courses and even to the whole degree program. Many of the features and positive actions taken that are worthy of future adoption include self-evaluation of exercises and portfolios between

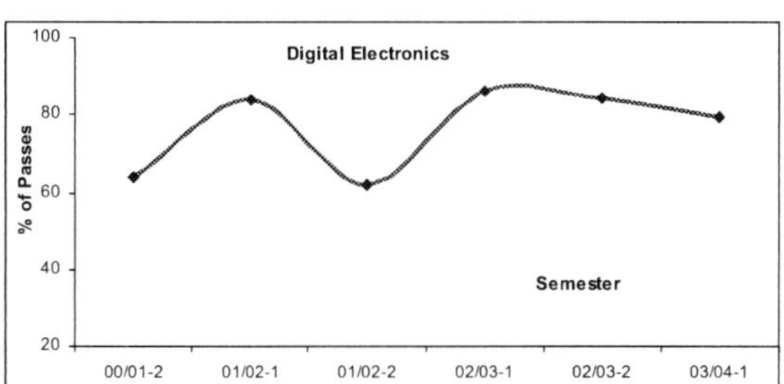


FIGURE 7
PERCENTAGE OF STUDENTS WHO PASSED DIGITAL ELECTRONICS IN THE PREVIOUS SIX SEMESTERS

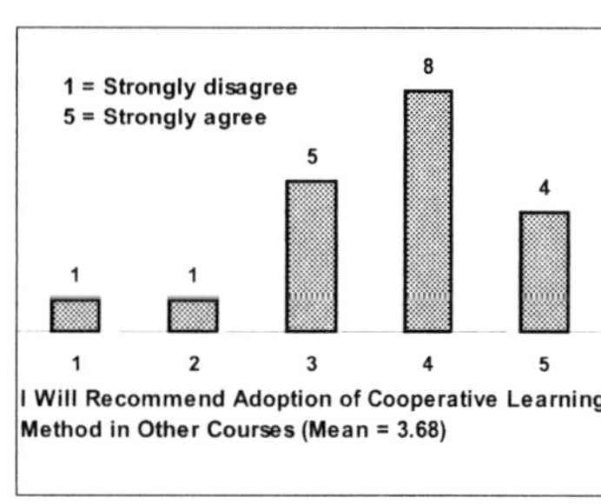


FIGURE 8
SAMPLE FEEDBACK RESPONSES FROM A CLASS OF 27 STUDENTS (NUMBER ABOVE EACH BAR IS NUMBER OF RESPONSES)

teams; preparation of lesson and other class materials within the CL context; introduction of generic sessions to explain the basics of the new method to the education community; and substitution of lecture sessions by teamwork sessions.

## ACKNOWLEDGEMENTS

This work was sponsored in part by the projects number 86MQD2002 and 105MQD2003 of the Ministry of Universities, Research and the Information Society (DURSI) and the Agency for Administration of University and Research Grants (AGAUR) of the Generalitat, the Government of Catalonia. This work was sponsored in part by the Technical School of Castelldefels (EPSC) of the Technical University of Catalonia (UPC). The author will also acknowledge the support received from the Grup d'Interès en l'Aprenentatge Cooperatiu (GIAC) of the Institute of Education Sciences ICE (UPC).

## REFERENCES

1. Johnson, D. W., Johnson, R. T., and Smith, K. A., *Active Learning: Cooperation in the College Classroom,* Interaction Book Company, Edina, MN, 1991
2. Jacobson, D., Davi, J., and Licklider, B., "Ten Myths of Cooperative Learning in Engineering Education," IEEE Frontiers in Education Conference, 1998, pp. 790-794.
3. Watson, A.K., "Utilization of Active and Cooperative Learning in EE Courses: Three Classes and the Results," IEEE Frontiers in Education Conference, 1995.
4. Alcover, J., Ruiz, S., and Valero, M., "Evaluación de la Implantación del PBL en la EPSC," XI Congreso Universitario de Innovación Educativa en las Enseñanzas Técnicas (CUIEET), Vilanova i la Geltrú, Spain, 2003.
5. Watkins, S.E., and Hall, R.H., "Complex Problem-Solving using Structured Collaboration," Chapter 29, *Innovations 2003: World Innovations in Engineering Education and Research,* iNEER, Aung, W., et al. (ed.), Int. Network for Engineering Education and Research (iNEER), 2003, pp. 285-296.
6. Curriculum Development Guidelines (online): http://www.career-space.com/downloads
7. Bradbeer, R., "An Introductory Course in Electronic Engineering using a Collaborative PBL approach in a Studio Environment," Proceedings of the 2003 International Conference on Engineering Education (ICEE-2003), Valencia, Spain, 2003.
8. Institut de Ciències de l'Educació, UPC, Grup d'Interès en l'Aprenentatge Cooperatiu (GIAC) (online): http://giac.upc.es
9. European Credit Transfer System, ECTS (online): http://europa.eu.int/comm/education/programmes/socrates/ects_en.html
10. Sánchez, F. J., Casanella, R., Fernández, I., "Introducción Sistemática del Aprendizaje Cooperativo en la Asignatura Electrónica Digital de la EPSC," XI Congreso Universitario de Innovación Educativa en las Enseñanzas Técnicas (CUIEET), Vilanova i la Geltrú, 2003.

**Francesc J. Sánchez Robert** received the Technical Engineering in Telecommunications degree (B. Eng.) from the Technical University of Catalonia and the Electronic Engineer (M. Eng.) degree from the University of Barcelona, Spain, in 1987 and 1997, respectively. He is a college lecturer of Telecommunication Engineering at the Technical School of Castelldefels (EPSC) of the Technical University of Catalonia (UPC), Barcelona. At present, he teaches courses in the area of digital electronics and is working towards the Ph.D. in electronic engineering at the same university. His research interests lie in the area of EMC measurements and ESD detection.

# Chapter 24

# ABET's Engineering Criteria 2000: Our Efforts in a Nutshell

ANAND D. SHARMA[1], ISADORE T. DAVIS[2], RAMON E. VASQUEZ[3] and LUENY MORELL[4]
*[1]University of Puerto Rico, Mayaguez, PR, USA. [2]LSI-PSI-EPUSP, Raytheon Missile Systems, Tucson, AZ, USA. [3]University of Puerto Rico, Mayaguez, PR, USA. [4]Hewlett Packard, Palo Alto, CA, USA.*
*E-mail: sharma@centennialpr.net*

*Ever since the approval of the new Accreditation Board for Engineering and Technology (ABET) Engineering Criteria (EC 2000) by the Board of Directors of ABET on November 2, 1996, and its mandatory application as of Fall 2001, educational institutions across the United States have had to assess and evaluate their undergraduate engineering programs from a different perspective. That is because ABET EC 2000 has been designed to employ the philosophy and practice of continuous quality improvement to engineering programs. Engineering Criteria 2000 calls for each engineering program to identify its constituencies and, based upon their feedback, to formulate the services that each program will provide. The first step is the development of program educational objectives – statements describing expected achievements of graduates in the early years of their careers after graduation as a result of their educational preparation. Usually, educational objectives are expectations of graduates' performance after they have left the school and been in the workforce for three to five years. Subsequently, the second step requires the more specific definition of program outcomes – skills, knowledge and behavior that would be expected of students at the time of their graduation. All this requires considerable time and effort, not only to clearly define the program educational objectives and program outcomes, but also to develop proper and continuous assessment methods and tools, the documentation, the processes, and the necessary culture and philosophy changes that would be introduced as a consequence of the cyclical processes. We, at the University of Puerto Rico at Mayaguez, recognized the significance of EC 2000 early on and undertook planning and organizational steps fully four years prior to the actual site visit. All six of our undergraduate programs were evaluated during November 2002.*

## OBJECTIVE

This paper outlines the planning and implementation processes/steps that we undertook to educate the general populace, the industry-sponsored EC 2000 mock visits we conducted, and the processes and respective offices with dedicated staff and working committees we established in order to not only prepare for the ABET EC 2000 site visit, but also to develop an organizational structure which will sustain the process beyond the EC 2000 site visit for continuous quality enhancement. Furthermore, the intent of this paper is to share the experience with other institutions that may also be in the early stages of preparation. Various other independent accounts of such experiences can also be found in literature, such as the Ohio State University [1] and Drexel University [2] experiences.

## INTRODUCTION

The University of Puerto Rico at Mayaguez (UPRM) is one of the 11 campuses of the University of Puerto Rico System. The UPR System is a public institution, which was created by the Puerto Rico Legislative Assembly on March 12, 1903. It collectively enrolls about 67,000 students. The Mayaguez Campus (UPRM) is a land grant institution that began in 1911 with the College of Agricultural Sciences. Subsequently, other colleges were added as follows: College of Engineering (1913), College of Arts & Sciences (1943), and the College of Business (1970).

The student body consists of about 11,000 undergraduate and 970 graduate students. The College of Engineering has an undergraduate enrollment of 4458 students, of which 36 percent are females, which is one of the highest in engineering among U.S. institutions. This enrollment resulted in UPRM ranking 15$^{th}$ nationally in terms of the number of bachelors' degrees awarded (695) during 2000-2001 [3].

The strategic plan of the College of Engineering was approved by its faculty on October 13, 1998. The vision and mission statements, which are an integral part of the strategic plan, are well in consonance and they subscribe to preparing the "best professionals in engineering" and providing a "strong education in engineering." This commitment to excellence is reflected in the philosophy of our college "to provide a firm educational foundation [4]." Undergraduate education is our strength. While emphasis on research and graduate education with newer doctoral programs is gaining increased attention, the fact remains that graduate degrees have consistently accounted for less than ten percent of the total number of degrees conferred.

The College of Engineering comprises six academic units or departments, which are: Electrical & Computer Engineering, Mechanical Engineering, Industrial Engineering, Chemical Engineering, Civil Engineering & Surveying, and General Engineering. These departments collectively offer seven bachelor's degree programs, of which six are in engineering with a separate program in surveying. All of the six bachelor's degree programs in Electrical, Computer, Mechanical, Industrial, Civil, and Chemical Engineering are accredited by the Accreditation Board for Engineering and Technology (ABET).

The bachelor's degree programs at UPRM are of five-year duration as opposed to four years at most U.S. institutions. This provides both breadth and depth, along with ample opportunities for summer internships, undergraduate research, exchange programs, and a strong cooperative education program in partnership with industry. Approximately 27 percent of all undergraduate engineering students avail themselves of this cooperative education industry experience with the bulk of the students, about 45 percent, coming from the Mechanical and Chemical Engineering programs. The programs underwent a re-accreditation site visit during November 2002 as per ABET's new Engineering Criteria 2000.

## FOCUS OF ABET'S EC 2000

ABET's periodic review of engineering programs is well accepted as a form of program assessment and quality assurance. Since its enactment, Engineering Criteria 2000 have been well publicized in various sources and, effective Fall 2001, all programs up for accreditation review are being evaluated for compliance against these criteria. These essentially consist of eight criteria with a goal of continuous program improvement as opposed to the earlier focus on rigid quantitative inputs [5]. These criteria encompass: (1) Students, (2) Program Educational Objectives, (3) Program Outcomes and Assessment, (4) Professional Component, (5) Faculty, (6) Facilities, (7) Institutional Support and Financial Resources, and (8) Specific Program Criteria.

George D. Peterson, ABET's Executive Director, states that at the core of these criteria "is an outcomes assessment component that requires each engineering program seeking accreditation or re-accreditation to establish its own internal assessment process, which in turn, will be assessed by ABET [6]." M. Dayne Aldridge and Larry D. Benefield point out that it is not sufficient to merely demonstrate the achievement of educational objectives (Criterion 2) and program outcomes (Criterion 3), but additionally, "a commitment to continuous improvement and the stability to continue its achievement record over the next six years [7]. Frank G. Splitt succinctly remarked, "Engineering education reform presents a formidable challenge, given academe's interest in preservation of the status quo [8]."

The entire review process can be summarized into the following key steps [9]:

a. The Institution requests an evaluation visit.
b. The Engineering Accreditation Commission selects the Team Chair.
c. The Team Chair contacts the Dean of Engineering to select dates for the campus visit, and to determine the status of self-study materials.
d. The Institution provides all applicable Program Self-Study reports to ABET Headquarters.
e. The Team Chair assembles a team by selecting Program Evaluators.
f. The Institution provides all applicable Program Self-Study reports and other applicable materials to the Team Chair, the Program Evaluators, and the Headquarters.
g. The Team visits the campus.
h. The Team conducts an exit interview with university officials and issues a draft statement at the time of departure.
i. The Engineering Accreditation Commission revises the draft statement, if necessary, and takes final action.

## OUR EARLY EFFORTS & CHALLENGES

The leadership of the College of Engineering recognized the importance of EC 2000 as far back as November 1998, fully four years ahead of ABET's next review visit. The upcoming review would be unlike earlier traditional accreditation-related efforts, where the crux of the effort was put into producing a self-study document and demonstrating that the minimum curricular requirements were met or exceeded. Preparation could not be relegated to the last minute. It was a whole new approach that required early action and planning, and with which few of us were well conversant.

The administrative leadership of the College encouraged a team consisting of department heads to participate in the 2nd Working Symposium on Best Assessment Processes in Engineering Education at the Rose Hulman Institute of Technology in Terre Haute, Indiana. This was the start of our efforts, which was soon followed by the formal establishment of a Faculty ABET Committee with a lead coordinator within the College of Engineering. As John W. Meredith corroborated much later, "The most important element in conducting a successful EC 2000 implementation is commitment at the highest level [10]." Some of the early challenges we faced were: simply getting to understand and digest the implications of EC 2000; team composition and stability; the holding of regular meetings to identify constituents; and discussion and decision on possible assessment methods and tools.

It was quite evident early on that Criterion 2 and Criterion 3 in particular would be the most demanding. Criterion 2 calls for a clear establishment of Program Educational Objectives, with input from the key stakeholders or constituents. An assessment mechanism is required as an embedded self-improvement process (the first loop) for the program educational objectives.

Criterion 3 calls for the definition of Program Outcomes that should, as a minimum, embrace the eleven (a-k) outcomes listed under the criterion, along with proper assessment methods, which would constitute the second loop of the self-improvement process. Criterion 3 could be viewed as a subset that had to map on to Criterion 2, which, up the ladder, was also required to satisfy the mission and vision of the College of Engineering, and eventually, that of the institution as a whole.

Course syllabi were restructured to incorporate applicable (a-k) outcomes. All six programs were required to develop not only their own strategic plans but also, within these, their specific program educational objectives with input from their constituents, and program outcomes. Consequently, this led to the establishment of each program's ABET sub-committee and the scheduling of numerous working retreats by each department. Each department chose its coordinator, who in turn, became member of the Faculty ABET Committee at the college level. Some of the committee members also participated at different stages in the ABET Program Evaluator Training workshops.

## Assessment Methods & Tools

Given the new accreditation paradigm that every engineering program establish an assessment process and document results, George D. Peterson was quite correct in his statement "No one expects that the outcomes assessment component of Engineering Criteria 2000 will be easy to implement. Establishing measurable objectives and evaluating their outcomes are sophisticated activities with which most engineering educators have had little or no experience [6]." Thomas Angelo, former director of the Assessment Forum of the American Association of Higher Education (AAHE), defines assessment as "an ongoing process aimed at understanding and improving student learning [6]." An analogy to this departure from the so called "comfort zone" was provided by Lyle Feisel, past president of ASEE, who compared it to the five stages of grief in Elizabeth Kubler-Ross' book entitled *On Death and Dying* [11]. These five stages were identified as: denial, anger, bargaining, depression, and acceptance.

The Faculty ABET Committee organized a series of one-day workshops in mid-year 2001 that led to the development of assessment tools and strategies package. This was adopted for common use by all programs with each one at liberty to modify or be selective about the recommended methods or tools. The package contained an outcomes assessment matrix, an assessment strategies matrix, and various custom-designed assessment forms for integrating ethics, oral and written reports, teamwork, peer evaluation, course/project evaluations, exit survey, alumni survey, employer survey, and internships. Felder and Brent have also reported on a strategy for integrating program-level and course-level activities to fulfill the ABET criteria [12].

## PRINCIPAL DRIVERS FOR CHANGE

Peggy L. Maki, Director of Assessment, AAHE, stated, "All too frequently higher education institutions view the commitment to assessing their students' learning and development as a periodic activity – most often driven by an impending accreditation visit [13]." While this would generally be the case with external drivers such as industry, or ABET, or other institutional accreditors such as the Middle States Commission on Higher Education, we, at the University of Puerto Rico at Mayaguez, had had some experience related to assessment in earlier educational projects, such as MEEP (Manufacturing Engineering Education Partnership) Learning Factory, and PaSCoR (Partnership for Spatial and Computational Research) [14]. Ideally, such should really be the case: universities should be driven by institutional curiosity, an internal motivator, versus attempting to comply, an external motivator [13]. In order to institutionalize this assessment process as part of the various courses, the College of Engineering established a physical office called System for the Evaluation of Education (SEED) in mid-year 2001.

## THE SEED OFFICE

With the goal of developing assessing strategies for the undergraduate engineering programs, the principal goals of the SEED office (Figure 1) were to:

- Establish and facilitate a strategy for continuous evaluation of engineering programs and assessment of student learning outcomes.
- Coordinate with engineering departments and accreditation committees (ABET EC2000 and the Middle States Association) the College of Engineering's activities regarding accreditation processes, including their implementation strategies.
- Assess outcomes of the College of Engineering's Strategic Plan.
- Become the College of Engineering's repository of assessment strategies, assessment instruments, and assessment results and reports.
- Coordinate professional development activities concerning evaluation and assessment.
- Disseminate assessment results to stakeholders and decision-makers for their information and decision-making.

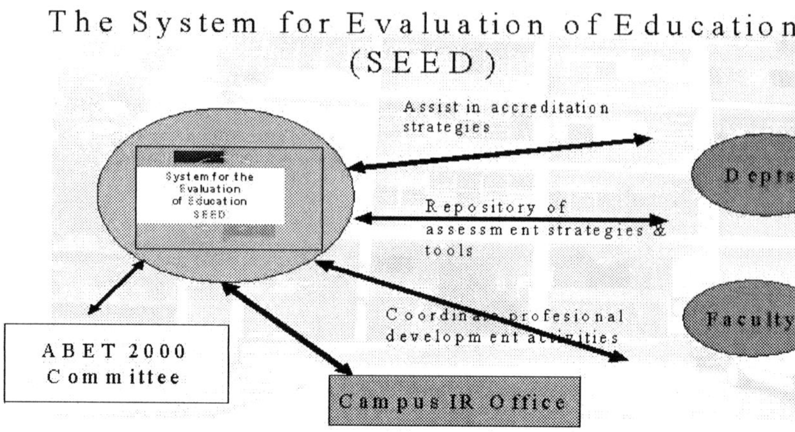


FIGURE 1
SEED OFFICE

The SEED Office counts on the services of a coordinator on a half-time basis, a person in-charge of database management and web page maintenance, and a full-time secretary. As an extension to this concept, similar offices were created in each of the six departments with names such as Continuous Improvement Center, The Curriculum Renewal Plan, and Center for Academic Research.

## AWARENESS CAMPAIGN

A series of workshops and retreats were organized, with guidance and support from industry and other U.S. universities, on the definition of outcomes and the development of outcomes-based course syllabi, the development and redefinition of mission statements at individual program as well as faculty level, the mapping of outcomes to program educational objectives, and the implications of ABET's EC 2000 criteria – not only to the faculty and staff in the College of Engineering, but also much later to the faculty and staff of the entire campus as part of a much wider awareness outreach. A number of one-day workshops were also organized on the integration of ethics across the curricula, which were well distributed throughout the period. There were also workshops on assessment and student mentoring which were given by invited faculty members representing the NSF's SUCCEED and Foundation coalitions.

## INDUSTRY SUPPORT

Criterion 4 (Professional Component) and Criterion 5 (Faculty) allude to interactions with industrial and professional practitioners as well as employers of students. Robert M. Laurenson from ASME stated, "A positive result of EC 2000 was the involvement of the program Advisory Boards. These groups have provided a very useful resource to the programs in establishing educational objectives and defining associated measurements of student outcomes [15]."

We, at the University of Puerto Rico at Mayaguez, sought out industry partnership very early in the process of preparing for the re-accreditation site visit in November 2002. For example, early in the process, Hewlett Packard donated the server used to collect all data regarding the College's outcomes assessment and strategic plans. A college-wide ABET EC 2000 Retreat led by Raytheon engineers and quality improvement personnel, and co-sponsored by Microsoft, was organized in November 2000 to assist each program to define their Program Educational Objectives (Criterion 2) and Program Outcomes (Criterion 3). This retreat workshop enabled each program to develop or re-define their mission statement, to develop outcomes-based course syllabi, and to map the outcomes to program educational objectives. This retreat led to an ABET EC 2000 mock visit sponsored by Raytheon Missile Systems, Microsoft, Hewlett Packard and Boeing in January 2002 with team members representing both academia and industry. The objectives of the mock visit were to review the laboratory facilities, conduct interviews with faculty and students, evaluate the first drafts of the individual self-study reports, and to offer candid comments and recommendations to incorporate assessment and continuous quality improvements within the programs.

## EARLY LESSONS LEARNED

The results from the mock visit were an eye-opener for many of us, both the faculty and the administration, and provided vital external feedback on our status. In essence, it provided the impetus towards redesigning the course syllabi, incorporating the applicable outcomes of Criterion (3), and in general becoming more sensitive towards the new criteria. There were severe flaws in the self-study report drafts.

The visit was soon followed by the formation of the first Industrial Advisory Board (IAB) of the College of Engineering in June 2002, although each of the departments had individually been interacting formally or informally with industry representatives. The college saw the need to form the IAB to receive direct feedback from their senior-level industry constituents. Holly correctly pointed out that "EC 2000 has made us much more attentive to the advice and observations of industrial advisory boards [16]."

As an outcome of this IAB meeting, it was recommended that we organize a second ABET mock visit in September 2002, fully two months prior to the actual ABET site visit. The second mock visit team was similarly composed of members representing both academia and industry, like the team in January 2002. The team was provided, prior to their arrival on campus, copies of the final self-study reports for all six programs that would be submitted to ABET by the end of June 2002. This mock visit was sponsored by a much wider group of industries, such as Raytheon Missile Systems, Hamilton Sundstrand, Abbott, Microsoft, Boeing, Merck Sharp & Dohme, and Eli Lilly. During this time, we fared significantly better in all aspects of our preparation, and the mock visit served to provide the confidence for the real evaluation visit in November 2002.

## CONCLUSION

Though the detailed comments with regard to each of the programs are not meant to be publicly listed, all six programs, which were evaluated for the first time as per the new EC 2000 criteria, were accredited by the Engineering Accreditation Commission of the Accreditation Board for Engineering and Technology. However, the comment that "the institution's systematic and innovative effort to introduce the culture of outcomes-based assessment to the College of Engineering community is especially noteworthy" is an indicator of the successful process.

## Documentation and Continuation of Activities

Throughout the course of the preparatory efforts during the span of last four years, leading up to the ABET site visit, proper documentation was maintained. This included all reports as well as the minutes of all meetings conducted by the Faculty ABET Team which can be found in the website http://www.abet.uprm.edu. The website can also be accessed from the College of Engineering website http://ing.uprm.edu.

The SEED Office under the College of Engineering is expected to continue providing support to ensure the smooth functioning of similar offices in each department. This will require conducting assessments on a regular basis, utilizing common assessment practices and methods. Mulrine briefly summarizes the various accreditation visit experiences, and at the same time stresses on the need for the post-visit continuous quality improvement phase [17].

In our own particular case, the experience gained from this effort is already being applied towards UPRM's institutional re-accreditation efforts for a decennial visit from the Middle States Commission on Higher Education (MSCHE) in 2005. The new team, using its gathered experience with ABET's EC 2000, has not only formalized some institutional student learning outcomes from the newly-developed plan for the assessment of student learning, but also developed questionnaires for campus-wide surveys with the hope of institutionalizing a continuous improvement process. These efforts are well documented under http://www.uprm.edu/msa.

## References


1. Soundarajan, Neelam, "Preparing for Accreditation Under EC 2000: An Experience Report," Journal of Engineering Education, Vol. 91, No. 1, January 2002, pp. 117-123.
2. Fromm, Eli, "The Changing Engineering Educational Paradigm," Journal of Engineering Education, Vol. 92, No. 2, April 2003, pp. 113-121.
3. American Society of Engineering Education, "Databytes," PRISM, September 2002, p. 16.
4. University of Puerto Rico at Mayaguez, "College of Engineering: Aims and Objectives," Bulletin of Information – Undergraduate Studies, 2002-2003, p. 260.
5. Engineering Accreditation Commission, "Criteria for Accrediting Engineering Programs," Accreditation Board for Engineering and Technology, Inc., Baltimore, Maryland, 2001-2002 Visits, http://www.abet.org.
6. Peterson, G., "A Bold New Change Agent," *How Do You Measure Success*, ASEE Professional Books, 1998, pp. 6-7.
7. Aldridge, M.D., and Benefield, L.D., "Assessing a Specific Program," *How Do You Measure Success*, ASEE Professional Books, 1998, pp. 33.
8. Splitt, Frank G., "Systemic Engineering Education Reform – A Grand Challenge," The Bent of Tau Beta Pi, Spring 2003.
9. Engineering Accreditation Commission, "Manual of Evaluation Process," Accreditation Board for Engineering and Technology, Inc., Baltimore, Maryland, 2001-2002 Visits, pp.2-3, http://www.abet.org.

10. Meredith, J.W., "Preparing for an ABET Visit: The Basic Elements," THE INTERFACE, April 2002, No. 1.

11. Feisel, L., "Accepting the Challenge," *How Do You Measure Success*, ASEE Professional Books, 1998, pp. 65.

12. Felder, Richard M., and Brent, R., "Designing and Teaching Courses to Satisfy the ABET Engineering Criteria," Journal of Engineering Education, Vol. 92, No. 1, January 2003, pp. 7-25.

13. Maki, P.L., "Developing an Assessment Plan to Learn about Student Learning," Journal of Academic Librarianship, January 2002.

14. Buxeda, Rosa, Jimenez L., and Morell, L., "Transforming an Engineering Course to Enhance Student Learning," Proceedings of 2001 International Conference on Engineering Education (ICEE-2001), Oslo/Bergen, Norway, August 2001.

15. Laurenson, R. M., "Initial Assessment of the Impact of ABET/EC2000 Implementation Using Mechanical Engineering Programs as the Pilot Study Group," A Project of the ASME Committee on Engineering Accreditation, ASME, 2001, NSF Grant DUE-0086486.

16. Holly, Forrest M., "ABET's EC 2000 and Engineering Practice," *Licensure EXCHANGE,* National Council of Examiners for Engineering and Surveying, Clemson, South Carolina, December 2002.

17. Mulrine, Anna, "The ABET Visits Were a Huge Success at Your School But Now that They're Over, How Do You Put Continuous Improvement into Practice?" PRISM, ASEE, February 2003, pp. 39-40.

**Anand D. Sharma** is a Professor in the General Engineering Department and Special Assistant to the Dean of Engineering at the University of Puerto Rico at Mayaguez. He was involved with ABET accreditation-related efforts as a team member from November 1998 until July 2000, and later, as the Coordinator of the Faculty ABET Team from July 2001 until June 2002. He now leads the campus team for a re-accreditation visit from the Middle States Commission on Higher Education.

**Isadore T. Davis** is Manager of Engineering University Relations at Raytheon Missile Systems in Tucson, Arizona. He is a member of the ABET Industry Advisory Council and Vice Chair (2002-03) and Chair-elect (2003–05) of ASEE's Corporate Member Council. He led the effort to develop Raytheon's ABET EC 2000 and TC 2000 Assistance Programs; and actively led and sponsored ABET EC 2000 retreats and mock visits. He also serves as Interim Chairman of the UPRM College of Engineering Industrial Advisory Board (IAB).

**Ramon E. Vasquez** is a Professor in the Electrical & Computer Engineering Department and Dean of the College of Engineering at the University of Puerto Rico at Mayaguez. He has been involved with and actively supportive of all ABET accreditation-related efforts, including the formation of the SEED Office.

**Lueny Morell** is Director of University Relations, Latin America with Hewlett Packard in Palo Alto, California, and also a certified ABET reviewer. She also holds a rank of Professor in the Chemical Engineering Department at the University of Puerto Rico at Mayagüez. She has been actively involved with ABET accreditation-related efforts and curriculum development, and also led the Faculty ABET Team at the University of Puerto Rico at Mayaguez until July 2001.

# Chapter 25

# Transformation of the Teaching Methodology in a Laboratory Course of the Industrial Biotechnology Curriculum in Response to Student Profile

LORENZO SALICETI-PIAZZA, ROSA BUXEDA and JUAN LÓPEZ-GARRIGA
*University of Puerto Rico, Mayagüez Campus, P.O. Box 9010, Mayagüez, Puerto Rico 00681-9010. E-mail: saliceti@uprm.edu*

*Biotechnology is a discipline with a large impact on the economic development of many countries. As a result, academia has developed curricula that respond to new professional profiles. In response to the economic development of the island of Puerto Rico, the University of Puerto Rico at Mayagüez offers an undergraduate degree in Industrial Biotechnology. The program's interdisciplinary curriculum exposes students to diverse skills in the fields of science and engineering. One of the integrating courses in the curriculum is the Unit Operations III Laboratory. In this lab course, students develop practical experience in bioprocessing, with emphasis on applications to industrial biotechnology. A pilot study was devised to revise the course's teaching methodology and learning assessment in response to the professional profile of this discipline. Richard Felder's model of learning styles was used to determine the learning profile of the student population. The data indicated a predominance of visual learners vs. verbal learners, but no difference among the active/reflective, sensorial/intuitive, and sequential/global learning pairs. This information was used to design activities that corresponded to the population's diversity of styles. The course revision incorporated development of writing, oral presentation, and teamwork skills. For each skill, assessment tools were designed to establish the pattern of students' development throughout the course. The learning experience with regards to knowledge and skills, as well as the impact of the teaching methodology, was documented by means of student portfolios. The assessment data were used to provide feedback to the students and to improve the course's teaching/learning process.*

## INTRODUCTION

College education nowadays brings new challenges which constitute a new educational paradigm to course instructors in all areas of knowledge. This educational paradigm is characterized by new trends in active learning, integrated development of mathematics and science concepts within context, close interaction with industry, appropriate use of information technology, and faculty functioning as mentors and coaches instead of dispensers of information. In essence, the educational paradigm has shifted from a traditional instructional model to an innovative learning model where the student, not the instructor, is the center of the paradigm, and the success of the curriculum is measured by outcomes-based criteria established by the academic program constituents. This new educational paradigm has been catalyzed by several driving forces, which include: institutional and departmental advisory boards of science and engineering programs, professional societies, private foundations and government granting institutions (NSF, for instance), accreditation agencies (ABET, Middle States Association), educational research, and government legislation on state colleges and universities. The science, mathematics, engineering and technology (SMET) curricula are particularly influenced by these driving forces, which bring forth the challenges of innovation in learning methodologies [1].

This paper describes a pilot study on how the teaching methodology of a chemical engineering laboratory course was transformed in order to respond to the new educational paradigm and a new student profile. The course title is InQu 4004-Unit Operations Laboratory III, offered by the Department of Chemical Engineering to students enrolled in the Industrial Biotechnology Program of the University of Puerto Rico at Mayagüez. This course is also open to chemical engineering majors as an elective. The course load is for one-credit hour, which corresponds to a total of 45 contact hours in a semester, 3 hours per week in a 15-week period.

In previous offerings, the standard course teaching methodology was essentially traditional in style. At the beginning of the semester, students were divided in groups of three and were assigned up to four experiments related to biotechnology unit operations, such as fermentation, ultrafiltration, ion exchange, evaporation, and crystallization. Each group was given a short laboratory handout and all available experiments were alternated for each group, so at the end of the semester all groups would have finished all experiments. If additional information was necessary beyond lab handouts, students were encouraged to search for literature in the library or on the Internet. After each experiment was completed, a detailed written report was to be submitted by each group and graded by the instructor. In general, there was no link between the different experiments, and grades were based on laboratory work performance and evaluations of the written reports.

In the Fall 2002 semester, the course teaching and learning assessment methodologies were transformed in response to the professional profile of the industrial biotechnology discipline. Richard Felder's model of learning styles [2] was used to determine the learning profile of the student population. The data indicated a predominance of visual learners vs. verbal learners, but no difference among the active/reflective, sensorial/intuitive, and sequential/global learning pairs. This information was used to design course activities that corresponded to the population's diversities of styles [3]. The course revision incorporated the development of writing, oral presentation, and teamwork skills [4]. For each skill, assessment tools were designed to establish the pattern of students' development throughout the course. The learning experience with regards to knowledge and skills, as well as the impact of the teaching methodology was documented by means of student portfolios [5]. The assessment data were used to provide feedback to the students and to optimize the course's teaching/learning process. In this paper, each of the teaching methodologies and teaching tools used for the course transformation are described below.

## STUDENT PROFILE

One of the main goals in the new educational paradigm is to convert the incoming student, with a certain level of knowledge and skills, into a professional with improved knowledge and skills. These skills include the capability to work in teams, a full concern for ethical issues, a global awareness, an understanding of diversity, and the ability to communicate properly and effectively in both written and oral channels. The curriculum and its courses should provide the conversion tools to achieve this transformation. In terms of the unit operations lab, the instructor specified that the student's profile after passing the course would consist of the following:

- Knowledge of fermentation and recombinant product recovery and purification;
- Technical skills in the standard operating procedures (SOP) of the lab fermentor and the liquid chromatography column;
- Improvement of the soft skills of teamwork, oral and written communication.

An important feature of the course transformation was integrating the laboratory experiences planned for the course. Previously, each course lab experiment was not necessarily linked with the other experiments. Now, a current research topic by a faculty member from the Department of Chemistry was used to show at a bench scale the bioprocessing steps for production of a recombinant polypeptide, from the stage of production by fermentation to the stage of product recovery and purification by means of cell rupture and liquid chromatography. The product of the lab experiments is a recombinant hemoglobin (rHb) that can be naturally obtained from a sea clam. The sea clam rHb gene has been incorporated into a genetically transformed *Escherichia coli* strain. The rHb has a potential commercial value if used as a biosensor in environmental monitoring applications. This level of integration of a current research project into the undergraduate experience should greatly enhance the industrial biotechnology student's understanding of a real bioprocess production sequence.

## PLANNING TO ENHANCE LEARNING

The plan implemented to transform and enhance student learning in the Unit Operations III laboratory followed four main strategies [6]:

1. Establish course instructional objectives:
    a. Using action verbs.
    b. Developing soft skills.
    c. Complying with the ABET Engineering Criteria 2000 (ABET 2000) [7].
2. Communicate course innovations to the students by providing a syllabus that included objectives, learning activities, and assessment strategies.
3. Design activities using student's learning styles to accomplish course objectives [3].
4. Identify and apply assessment tools to evaluate student performance.

5. Prepare portfolios [5].

### Strategies 1 & 2: Establish Course Instructional Objectives & Communicate Course Innovations to Students

The first step in determining instructional objectives was to document the plan in a well-written syllabus. The educational objectives in terms of knowledge, technical skills and soft skills goals were tabulated, and for each one a teaching/learning strategy and an outcomes assessment tool were defined. The final plan is shown in Table 1. Teaching/learning strategies included workshops, mini-lectures, peer teaching, preparation of SOPs, lab work, oral and written presentations, review of videotaped oral presentations, and written reports [8]. Assessment tools included grading of written work, peer, self-, and instructor evaluations, technical skills self-assessment, teamwork self- and peer evaluations, and portfolio preparation [8].

| | **Learning Strategies** | **Assessment Tools** |
|---|---|---|
| Understand the principles of bioprocessing. | Mini-lectures and peer teaching, literature search workshop. | Instructor, peer and self-evaluations. |
| Operate fermentation & chromatography equipment. | Development of SOPs, equipment demonstrations & hands-on production of purified rHb protein. | Instructor grading of SOPs, written reports. Technical skills self-assessment. |
| Collect and analyze experimental data. | Mini-lectures, experimental work. | Instructor assessment of technical skills. |
| Present oral & written reports of experimental results. | Preparation of oral and written reports, cooperative learning, videotape oral presentations. | Instructor, peer and self-evaluations of oral presentations. Instructor grading of reports. |
| Develop teamwork skills. | Assignment of groups. | Teamwork peer and self-evaluations. |

TABLE 1

COURSE SYLLABUS, DOCUMENTING THE PLAN TO ENHANCE LEARNING

The instructional objectives were stated using operational verbs that reflected the level of student cognitive development following Bloom's taxonomy levels of knowledge, comprehension, application, analysis, synthesis and evaluation. In addition, the objectives were structured following the ABET 2000 a-k competencies [7], in order to direct the course results in an outcomes-based fashion. The resulting instructional objectives and ABET 2000 criteria stated for the course are shown in Table 2 below.

| Instructional Objectives | ABET 2000 a-k Competencies |
|---|---|
| Know, comprehend, apply, and analyze general principles and techniques of bioprocess engineering, particularly in the unit operations of fermentation and protein purification. | Competencies a, b, d, e, j, k. |
| Evaluate and analyze experimental data. | Competencies c, e, i, k. |
| Evaluate and apply principles and techniques to troubleshoot and solve problems in a critical, creative and effective manner. | Competencies i, j. |
| Search, evaluate, comprehend, and apply current scientific literature related to the topics covered in the course to solve current issues. | Competencies i, j. |
| Present current issues related to biotechnology. | Competency j. |
| Work in an interdisciplinary team to solve problems in bioprocess engineering. | Competency d. |
| Develop written reports and make oral presentations on the assigned laboratory subjects as part of a team. | Competencies d, g. |

TABLE 2
INSTRUCTIONAL OBJECTIVES AND ABET 2000 ACCREDITATION CRITERIA FOR THE UNIT OPERATIONS III LABORATORY

### Strategy 3: Activities Designed According to Student Learning Styles

The Unit Operations Laboratory course is taken by fourth-year students in the Industrial Biotechnology Program. Richard Felder's model of learning styles [2] was used to determine the learning profile of the student population. Course activities would be planned according to the group's learning styles profile. By determining how the student population's learning styles are distributed, the instructor of the course would design learning activities that match the students' learning preferences so that the instructional objectives of the course would be achieved. This strategy empowers students to learn.

The course enrollment consisted of seven industrial biotechnology senior students, no chemical engineering students were enrolled in this course. The enrollment consisted of four female students (57%) and three male students (43%). The results of Felder's test of learning styles provided the following profile:

- Active learners: 3; reflective learners: 4.
- Sensorial learners: 4; intuitive learners: 3.
- Visual learners: 6; verbal learners: 1.
- Sequential learners: 3; global learners: 4.

The data indicated a predominance of visual learners vs. verbal learners, but no bias among the active/reflective, sensorial/intuitive, and sequential/global learning pairs.

The selected teaching/learning activities to achieve the instructional objectives were:

- Mini-lectures.
- Laboratory safety awareness.

- Learning styles, teamwork & literature search workshops.
- Oral presentations.
- Laboratory notebook.
- Laboratory equipment inspection.
- Development of SOPs.
- Laboratory experiments.
- Incorporation of ongoing research into the undergraduate laboratory to connect the sequence of unit operations.
- Written reports.
- Teamwork.
- Portfolios.

The following activities were selected in order to adapt to the student's learning styles profile. Note that only in the category of visual vs. verbal learners was there a predominance of one type of learner. Activities to cover all of the categories of learners were implemented, therefore, and fewer verbal learner activities were included as compared to the rest of the activities.

- Sensorial:
  - Laboratory equipment inspection, demonstrations, hands-on fermentation and purification of rHb protein.
- Intuitive and global:
  - Mini-lectures, objectives, integrated experiences in UO using a well-defined project, literature search.
- Visual:
  - Laboratory equipment inspection, demonstrations, process diagrams & unit operations equipment photos.
- Verbal:
  - Mini-lectures, literature search of assigned unit operations topics.
- Active:
  - Team formation, team rules and team evaluation.
  - Laboratory practices.
  - Cooperative learning.
- Reflective:
  - Laboratory notebook.
  - Written reports.
  - Portfolios.
- Sequential:
  - Development of SOPs.
  - Laboratory experiences introduced in a logical manner as they would in a real industrial setting.

For the development of teamwork skills, a workshop was provided to the class in which various aspects of teamwork were discussed, including the importance of working in teams, awareness of differences in learning styles and their relevance in team building, conflict resolution, evaluation of team performance, and design of teamwork assessment tools. Other important issues such as cooperative learning and assessment of teamwork from the perspective of individual and peer standpoints were discussed. A certificate of participation in this workshop was issued to each student.

**Strategy 4: Identification of Assessment Tools to Evaluate Student Performance**

Various assessment tools were designed and applied to evaluate the student's performance in the Unit Operations III laboratory. These were evaluations of oral presentations by the instructor, peers and the students themselves; technical skills self-assessments in fermentation and chromatography; teamwork evaluations; portfolios and final evaluations by the course instructor.

In the assessment of oral presentations, the assessment criteria were incorporated in an evaluation form that was filled in by peers and the instructor [8]. Oral presentations were videotaped and digital copies were given to students for self-evaluation. The criteria included the following items: organization, clarity of expression, knowledge of subject, effective use of audiovisuals, presentation of objectives, adequate voice tone, selection of relevant aspects of the topic, management of time, overall quality, management of questions, adequate appearance and eye contact, and effective achievement of objectives.

Each group had the opportunity to present twice during the semester. In the first oral presentation, the most common shortcomings were either too much reading from notes or the screen resulting in poor visual contact with the audience, and interference with the projected slides on the screen. Both peers and the instructor commented on these common errors and in the final oral presentations the group showed very good eye contact, avoiding reading from screen or notes, a proper use of audiovisual aids, and with high confidence and rapport with the audience. The improvement in oral presentation skills was a note-worthy outcome of this course.

The assessment of technical skills was classified into two key areas of the lab experiments: fermentation and liquid chromatography. Table 3 shows the technical skills that were identified as key skills for each area. Each student had to report his level of confidence in performing each of the technical skills shown in the table using the following scale: 0 - no confidence at all, 1 - very low confidence, 2 - some confidence, 3 - medium confidence, 4 - high confidence and 5 - very high confidence. Students were requested to perform a self-assessment before and after each of the experiments. The instructor also provided his assessment of the students' final mastery of technical skills. In their portfolios, students plotted the data for the three skills and wrote reflections on their technical skills outcomes. In general, students showed an increased level of confidence after the experiments were completed. Figures 1 and 2 show an assessment plot for one of the students, chosen at random. Improvement in technical skills confidence levels is evident from the figure.

For the assessment of teamwork skills, each group rated the work of peers as well as performed a self-assessment. The criteria consisted of the following: commitment to assigned task, performance of task within a timeline, quality of individual contribution, individual accountability (responsibility), interpersonal skills (ability to work with people), management of conflict resolution, active participation, and individual performance in comparison to other team members. Figure 3 shows the teamwork assessments of one of the groups. The students were more severe in their self-assessments than their peers. This tendency was consistent not only in the students' teamwork assessments, but also in other assessments. It is not clear that this indicates improvement in teamwork skills, but such generosity in peer evaluations does point to an affinity among group members.

| Fermentation | Liquid Chromatography |
|---|---|
| pH electrode calibration | Flow of moving phase |
| Dissolved oxygen calibration | Installation of pre-column |
| Use of peristaltic pumps | Installation of column |

TABLE 3
ASSESSMENT ON LEVELS OF CONFIDENCE IN TECHNICAL SKILLS
SCALE OF CONFIDENCE LEVELS: 1-NONE, 2-VERY LOW, 3-SOME, 4-HIGH, AND 5-VERY HIGH

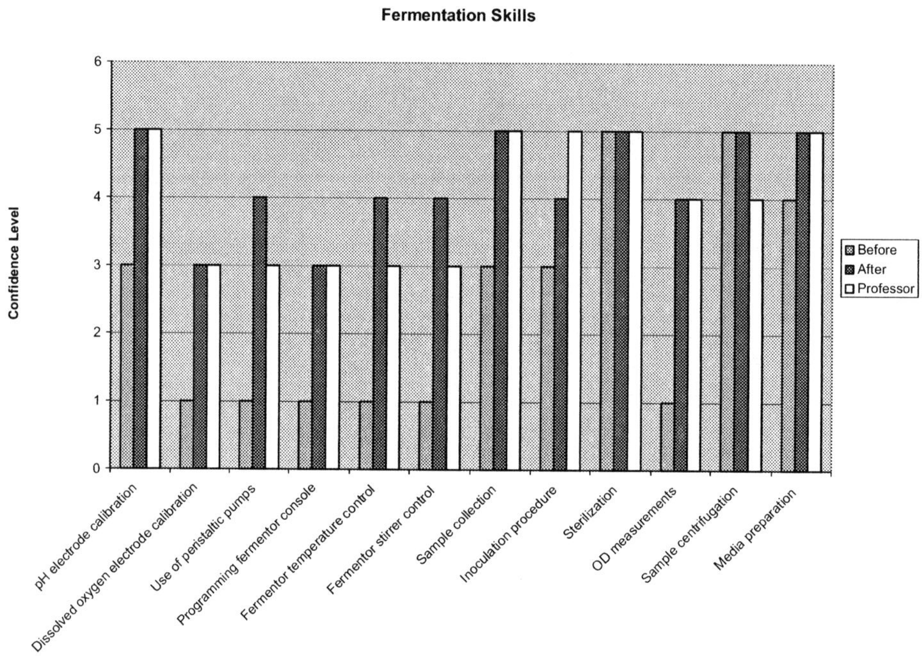


FIGURE 1
SELF-ASSESSMENT OF A RANDOM STUDENT IN FERMENTATION TECHNICAL SKILLS BEFORE AND AFTER LABORATORY EXPERIMENT, COMPARED WITH ASSESSMENT BY INSTRUCTOR

### Strategy 5: Student Portfolios

Portfolios were required from each student and were collected at various dates during the semester. A portfolio is a systematic, integrated and organized collection of the student's work documenting acquired knowledge and skills [5]. As such, it is a useful assessment strategy. The portfolios for this course were required to contain all of the following material:

- Presentation page
- Index

- Learning styles profile
  - Individual
  - Team
- Oral presentation skills
- Written presentation skills
  - Development of SOPs
  - Written experiment reports
- Teamwork skills
  - Self- and peer assessments
- Technical skills
  - Fermentation
  - Purification
- Reflections

Students were also required to reflect on what skills they developed after performing each laboratory experiment (fermentation and chromatography) and how the activity helped them become more effective as a student, a professional, a member of society. In general, students felt that they acquired better technical laboratory skills, deeper knowledge of the field of bioprocessing, and that their acquired knowledge would help them in becoming better contributors in their professional and community lives.

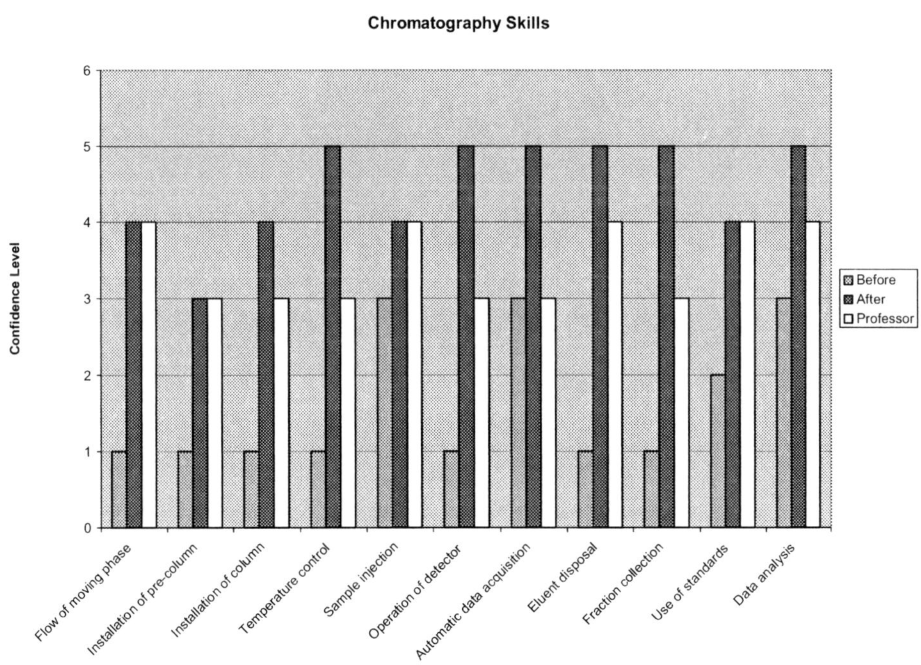


FIGURE 2

SELF-ASSESSMENT OF A RANDOM STUDENT IN LIQUID CHROMATOGRAPHY TECHNICAL SKILLS BEFORE AND AFTER LABORATORY EXPERIMENT, COMPARED WITH ASSESSMENT BY INSTRUCTOR

## Faculty Teaching Evaluation

A final evaluation of the instructor's teaching was distributed to the students. The evaluation criteria consisted of seven items: organization, overall quality, clarity, knowledge of subject, adequacy of materials, acquired ability to use new information, and overall understanding of the subject materials. A scale of 1-5 was used, with 5 being the highest rating and 1 the lowest. The resulting data is graphed in Figure 3 below. With the exception of the organization criteria, which received an overall 3, all other criteria were evaluated with a 4. More care should be put into the organization of the class materials and the logistics of experimental runs.

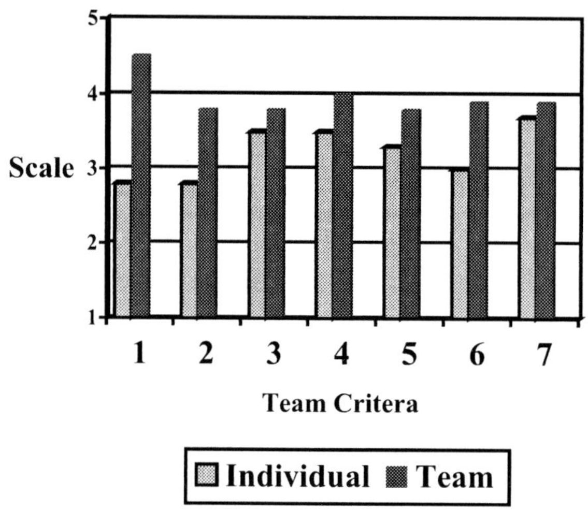


FIGURE 3
STUDENT'S SELF- AND PEER-ASSESSMENTS ON TEAMWORK TEAM CRITERIA:
1-ORGANIZATION, 2-OVERALL QUALITY, 3-CLARITY OF EXPOSURE, 4-KNOWLEDGE OF SUBJECT, 5-ADEQUACY OF MATERIALS, 6-ACQUIRED ABILITY TO USE NEW INFORMATION, 7-OVERALL UNDERSTANDING OF SUBJECT MATERIALS

Based on the student's comments on the last day of class, where a 10-minute conversation was held in the classroom, students' level of satisfaction with the course was very high. Their only complaint was that the paperwork for assessments and portfolio building was excessive.

Future offerings of the course will integrate several of the instructional tools applied, and a more comprehensive series of lab experiences will be incorporated due to availability of new bioprocessing equipment.

## Conclusions

The teaching methodology of a laboratory course can be transformed in response to student learning profiles to enable better comprehension of the course material by the students while hopefully simultaneously instilling in them lifelong learning skills.

Five general strategies were used to transform the course learning methodology:

- Establishment of well-defined course objectives.

- Communication of course objectives and innovations to students through a well-designed course outline.
- Design of adequate activities reflecting student learning profiles.
- Identification of student assessment tools to properly evaluate performance.
- Requirement for student portfolios to document all learning and skill development activities.

In summary, the results from this pilot study in transforming a laboratory course indicate that student attitudes about course contents can be improved by using student learning style tests to guide design and incorporation of learning experiences that adjust to those learning profiles. The use of a portfolio to document the individual students' learning and development of technical and soft skills is useful in the assessment of course objectives and to achieve a prescribed professional profile in any curriculum.

## ACKNOWLEDGEMENTS

The authors would like to thank Ms. Ruth Pietri and Ms. Evys Collazo, from Dr. Juan López-Garriga's research team, for their excellent work as teaching assistants to the Fall 2002 Unit Operations III Laboratory.

## REFERENCES

1. Felder, R.M., "Reaching the Second Tier- Learning and Teaching Styles in College Science Education," Journal of College Science Teaching, Vol. 23, No. 5, 1993.
2. Felder, R.M. and Silverman, L.K., "Learning and Teaching Styles in Engineering Education," Engineering Education, Vol. 78, No. 7, 1988.
3. Buxeda, R. and Moore, D., "Using Learning Styles Data to Design a Microbiology Course," Journal of College Science Teaching, Vol. 29, No. 3, 2000.
4. Morell, L., Buxeda, R., and Vélez-Arocho, J.I., "Helping Faculty Developing an Outcomes-Based Course," Proceedings of 3rd IEEE International Conference on Information Technology Based Higher Education and Training, 2002.
5. Buxeda, R. and Moore, D., "Expanding a Learner-Centered Environment using Constructivist Portfolio," Journal of Microbiology, Vol. 2, No. 1, 2001.
6. Buxeda, R., Jiménez, L., and Morell, L., "Transforming an Engineering Course to Enhance Student Learning," Proceedings of 2001 International Conference on Engineering Education, ICEE-2001, Oslo, Norway, 2001.
7. Engineering Accreditation Commission, Accreditation Board for Engineering and Technology, "Criteria for Accrediting Programs in Engineering in the Unites States," ABET, Baltimore, 1996-1997, Education, p. 7, May 2001.
8. Buxeda, R., Vásquez, R., Vélez-Arocho, J. I., and Morell, L., "Summer Station: Initiating the Undergraduate Research Experience in RS/GIS," IGARSS Proceedings, 2000.
9. Cutright, M., "What are Research Universities Doing for First-Year Students," About Campus, September-October, 2002.

10. Felder, R. M., "Meet your Students: Stan and Nathan," Chemical Engineering Education, Vol. 24, No. 1, 1989.

11. Felder, R.M., "Meet your Students: Susan and Glenda," Chemical Engineering Education, Vol. 24, No. 1, 1989.

**Dr. Lorenzo Saliceti-Piazza** is Associate Professor in the Department of Chemical Engineering, University of Puerto Rico, Mayagüez Campus. His teaching interests are in the area of biochemical and bioprocess engineering, mass and energy balances, and unit operations laboratories. He is currently the Associate Coordinator of the Industrial Biotechnology Program at the University of Puerto Rico, Mayagüez Campus.

**Dr. Rosa Buxeda** is Professor in the Department of Biology, University of Puerto Rico, Mayagüez Campus. Her teaching interests are in the area of industrial microbiology. She has developed a strong program in education research, having successfully participated in educational programs in the Department of Biology and in the School of Engineering through grants from the Howard Hughes Foundation and NASA. She is currently the Coordinator of the Industrial Biotechnology Program at the University of Puerto Rico, Mayagüez Campus.

**Dr. Juan López-Garriga** is Professor in the Department of Chemistry, University of Puerto Rico, Mayagüez Campus. His teaching interests are in the area of inorganic biochemistry. He has developed Science on Wheels, a successful K-12 educational program. He is also a very well recognized researcher, having secured funding from NIH to establish COBRE II, a research program in protein characterization.

# Chapter 26

# Online Technology for Enhancing Student Learning in a Microcontroller Course

S. M. AZIZ
*University of South Australia, School of Electrical & Information Engineering, Mawson Lakes, SA 5095, Australia. E-mail: mahfuz.aziz@unisa.edu.au*

*In addressing student disengagement with on-campus teaching and learning activities, the author undertook a project in order to stimulate and maintain student interest in a first year engineering course on microcontrollers. Online technology has been used in order to provide students with flexible access to self-paced interactive study materials in the form of online quizzes. These quizzes were also aimed at providing students with an alternative learning style and a mechanism to self-assess their progress. In addition, the online quizzes were expected to aid flexible delivery and internationalisation of programs. This paper examines the impacts of these online quizzes on student engagement and learning, and seeks to obtain student feedback on the suitability of this medium for enhancing students' overall learning experience.*

## INTRODUCTION

Student disengagement in on-campus teaching and learning activities has a profound impact on students' learning outcomes and progression [1]. Declining interest among Australia's brightest high school graduates to study science and engineering make these disciplines more vulnerable than others. Stimulating and retaining students' interest during their first year at university is therefore very important [1-2].

While teaching a large first year engineering class at the University of South Australia, I experienced the problem of increasing student absence/disengagement as the semester progressed. Student attendance in the lecture classes often declined to almost half approximately half way through the semester. On the other hand many students who were regularly attending classes were found to be unable to keep up with the progression of the course despite various active learning strategies [3] implemented in the class. A large proportion of these students were reluctant to spend time at home studying printed material alone for enhancing their understanding of the material covered in the class. This presented a critical teaching problem. The course I taught was one of the foundation courses in engineering offered in the very first semester for students entering the university. Thus, it was seen as a course to stimulate students' motivation to complete a four-year engineering program of study. I felt that augmenting the on-campus teaching and learning activities with additional activities would be useful for stimulating students' interest, and for enhancing their learning experience in the very first year at university. In the context of declining government funding in the higher education sector [4,5] additional learning activities not creating additional demand for face-to-face interaction would be very useful.

All of the above represented a critical opportunity to engage first year students with online technologies as the tools to engage them in a stimulating learning environment. Students who have difficulty attending campus activities due to work commitments [2] could benefit from the flexibility of online learning. Others could supplement and reinforce the learning that occurs in the classroom using self-paced interactive online learning resources. Appropriate online learning activities could also provide students with the opportunity to self-assess their progress in the course [6].

The requirement to comply with rigorous quality assurance processes in Australian universities [7] demand increasing amount of staff time. Academics report of being overwhelmed by administrative work resulting from accountability requirements [8-10]. Since online interactive learning activities do not require regular face-to-face interaction with academic staff, it could be a useful way of engaging students and providing them with more choices without creating additional demand for interaction with staff. This is not to say that the development and maintenance of online learning resources do not involve efforts and time on the part of the academic staff. However, most universities now have well-developed and supported online learning environments. Once an online learning resource is developed for a course, it can then be reused for future deliveries of the course without having to go through the laborious development process on each occasion. Surely academics will endeavour to make enhancements and additions as appropriate (don't they always do), and modifications may be required from time to time with changes in the context and circumstances in which the course is offered. This is applicable to print-based learning resources as well. The idea of using online technologies makes sense as it aims to utilise already-existing learning environments to provide students with more choices to enhance their learning.

The University of South Australia has been committed [11] to the use of technology for flexible delivery of programs so that students can exercise choice and self-management in the way they approach their studies and feel supported. Over the last decade the University has moved in a systematic way to make significant progress toward flexible delivery of academic programs [12]. There is both a practical and logical link between student-centred approaches, a focus on Graduate Qualities, and flexible delivery [13]. It is recognised that flexible delivery is facilitated by technological developments, most notably online technologies. The concept accords with the following definition:

- Provision of learning resources and the application of technologies to create, store and distribute course content, enrich communication, and provide support and services to enable more effective management of learning by the learner [14].

The University is also determined to be an international institution by recruitment of students from other countries with the option of studying either in Australia or remaining at home. It is envisaged that by 2005 40% of the University's total student body will be from overseas and that all students will take part of their program online [12]. For these reasons achieving greater interactivity within the online learning environment has been one of the priorities for the University. Developing interactive online learning resources is therefore totally in tune with the University's future teaching and learning framework.

I decided to develop online quizzes for the microcontroller part of the course *Principles of Computer Systems*. This paper presents the processes involved in developing the quizzes and analyses the student response. It attempts to find answers to some key questions, such as whether students engage with the online quizzes, whether the flexibility (time and place) make them attractive, whether students find the quizzes an interesting way to learn, whether they really benefit from engaging with the quizzes, and whether the quizzes are useful for deeper learning in an engineering course.

## DEVELOPMENT OF THE ONLINE QUIZZES

I used University of South Australia's online environment called 'UniSAnet' [15] to develop the online quizzes. Every course offered by the university has a default web page on UniSAnet. I implemented learning resources for *Principles of Computer Systems* on its web page. These included course information, study guide, lecture notes, tutorial and practical laboratory handouts. I had progressively developed and enhanced these resources over a two-year period in order to give students flexible access to course resources. However, this was merely a repository of course materials. The majority of the students printed the materials off the web. In contrast, the online quizzes were aimed to be truly interactive learning resources.

UniSAnet has a *quiz wizard* that facilitates the creation of multiple choice, true/false and fill-in-the-blank questionnaires. I consulted the *Online Adviser* at our campus and discussed the suitability of various question types for my course. I developed some sample quizzes and tried them in order to ensure that the system was functioning without much undesirable distractions and problems.

## TYPE OF QUESTIONS

Learning microcontrollers requires students to develop in depth (technical) knowledge on the topics covered. A *deep learning* approach [16] is essential to grasping the underlying concepts so that they can be utilised to design, develop and implement simple microcontroller-based systems. This being a foundation course for first year engineering students of all streams in our school, students' performance in future courses in this area would depend on their level of achievement in this course. It is therefore essential to use an appropriate set of questionnaires that would test their understanding of various topics at a deeper level. This meant that I had to design questions that would require students to think and apply the relevant concepts to solve problems before they could answer the questions. The true/false type of questions was obviously not suitable for a deep learning approach. I considered the multiple-choice and fill-in-the-blank formats better suited for the deep learning approach.

In the UniSAnet quiz wizard, there are two different ways to provide feedback:

- Question feedback
- Answer feedback

Although the ways the two feedback mechanisms are implemented in UniSAnet makes it difficult to differentiate between the two, they are both useful to assist students in understanding the concepts associated with a topic. In both cases feedback can be provided for both right and wrong answers.

## PROGRESSIVE PUBLICATION OF QUIZZES

Before publishing each quiz on the course web site, it was important to cover the relevant topics in the lecture and tutorial classes. As I taught only half of the course during the second half of the semester, I had about six weeks to get the whole thing done. I decided that I would prepare three quizzes and release them progressively at the beginning of weeks 9, 11 and 13 (week 13 being the last teaching week). As I published each quiz I sent an e-mail to the students alerting them to the availability of the newly released quiz. In addition, I informed the students about it in the lecture.

## COST

As stated earlier, I used University of South Australia's online learning environment 'UniSAnet'. It already had a web page allocated for the course and a wizard accessible from the course web page for creation of online quizzes. Therefore, it wasn't necessary to acquire or develop any software for the task. From that point of view no cost was incurred for creation of the online environment itself. However, there is definitely a cost associated with the development, maintenance and upgrade of the university's online environment, and an estimate of that is beyond the scope of this paper. There is a direct cost, though, for the development of online quizzes for any course. This is the time required for the development of the quiz questionnaire, feedback and testing. I see this as a non-recurring cost, because I would only have to make refinements, additions etc. in future years requiring much less time compared to the original development time. In total I spent 33 hours for developing, testing and releasing all the three quizzes. This included the time I spent in familiarising myself with the online quiz wizard and reading the relevant documentation.

## STUDENT FEEDBACK AND EVALUATION

I sought feedback from the students on a regular basis during my face-to-face interactions with them. The majority of the students expressed strong interest in the online quizzes and said that they were helpful for learning. After the first quiz was released one student sent an e-mail saying:

- I do not know how much positive feedback you get from students, but I thought I would send you a quick e-mail after doing the quiz to say they are really helpful in learning and applying the material that is being learnt.

I was deeply encouraged by the spontaneous feedback and indeed by all the feedback I had received. The Likert scale questionnaire (using a scale of 1 to 5) shown in Table 1 was used for evaluation of the online quizzes by the students. The questions were designed in consultation with academics engaged in teaching and research in the area of education and flexible delivery in order to ensure that they address the key issues.

**Likert scale questionnaire:**

- I like to do online quizzes frequently as a way of testing my learning in the course.
- The online quizzes helped me to focus my learning on important topics in this course.
- The quizzes were a more interesting way of learning than simply attending lectures, tutorials and reading the text.
- The feedbacks provided were appropriate and adequate.
- The quiz questions required me to think and apply the associated concepts to find an answer.
- I could not answer many of the quiz questions without clearly understanding the underlying concepts.
- I revisited each quiz a number of times until I was convinced that I clearly understood the underlying concepts.
- The online quizzes helped me to judge my strengths and weaknesses in the subject.
- I liked the online quizzes because I could do them anytime and anywhere.
- I liked the online quizzes because I could do them as many times as I wanted.

**Other questions:**

- Approximately how many times did you attempt each quiz on average?
- Would you like to see more online quizzes in future courses?

TABLE 1
QUESTIONNAIRE FOR STUDENT EVALUATION

The evaluation was done anonymously so that students could freely describe their experience and express opinions. Out of 140 students in the course, 70 participated in the evaluation. Twenty-three percent of the students said that they did not attempt any quiz. While half of them did not give any reason, the other half said that they did not have time to do the quizzes due to work commitments and assignments from other courses. In the remainder of this section, I will analyse the results of evaluation by the 77% of the respondents who attempted the quizzes.

Figure 1 shows student responses on the questions of whether the online quizzes were useful for testing their learning and for focusing on important topics. At least 70% of the students felt that way on both questions. Only 2% disagreed on the testing aspect while 6% disagreed on the focusing aspect; others were neutral. Clearly the majority of the students who engaged with the quizzes found them useful for testing their knowledge and for focusing on important topics in the course.

On the question of whether the quizzes were a more interesting way of learning than simply attending lectures, tutorials and reading the text, 58% agreed, 11% disagreed while 31% were neutral. This provides a reasonable indication that a large number of students found their engagement with the quizzes to be an interesting way to learn. On the question of feedback provided, 71% thought that they were appropriate and adequate, 7% disagreed while 22% were neutral. This shows that the feedback I had created with the questions was useful for most students. Nevertheless there is always room for improvement and I am certain that I shall be able to make them even more meaningful.

Seventy-two percent of the students said that the quiz questions required them to think and apply the associated concepts to find an answer, 2% disagreed while 26% were neutral. Sixty-seven percent of the students thought that they could not answer many of the quiz questions without clearly understanding the underlying concepts, no one expressed disagreement, 31% were neutral, while 2% said that they were unable to answer the question. These figures are indications that the students had to grasp the underlying concepts in order to feel comfortable with the questions and answer them correctly.

Only 37% of the students said that they revisited each quiz multiple times until they clearly understood the underlying concepts. In answer to the question on how many times they attempted each quiz, 50% said twice, 11% said three times and 4% said more than three times. These lead to a total of 65%. This means that although many students (65%) attempted the quizzes multiple times, only 37% attempted them enough number of times until they understood the concepts. It was reassuring to me that 65% students attempted the quizzes more than once as it shows that they engaged with the activity in order to improve their understanding of the topics.

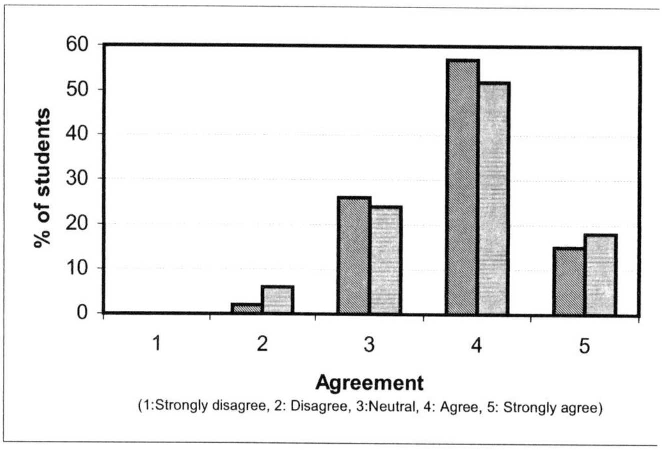


FIGURE 1
STUDENT RESPONSE ON ONLINE QUIZZES AS A WAY OF TESTING AND FOCUSSING LEARNING

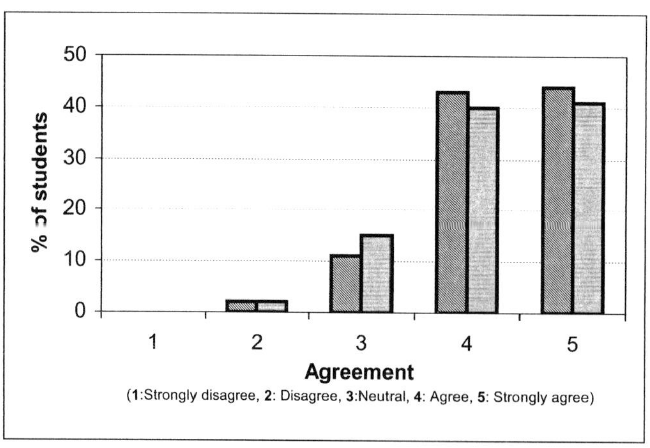


FIGURE 2
STUDENT RESPONSE ON FLEXIBILITY OF ONLINE QUIZZES

Seventy-two percent of the students said that the quizzes helped them judge their strengths and weaknesses in the course. This is a good outcome as long as the students act to enhance their knowledge by building on their strengths and addressing the weaknesses identified.

An overwhelming majority of the students

who attempted the quizzes (at least 80%) said that they enjoyed the flexibility offered by online quizzes as they could do them anytime and anywhere, and because there was no restriction on the number of times they could attempt them. This is depicted graphically in Figure 2. Would students like to see online quizzes in future courses? An overwhelming 93% said 'yes'. These are surely indications that the students found the quizzes useful and an interesting way to learn, and benefited from the activity. There is much debate, however, about the effectiveness of 'online instruction' when compared with traditional classroom [17]. It is argued that the benefits of face-to-face teaching cannot be reproduced using online technology alone. I agree that good 'instructional methods' are fundamental to the effectiveness of any teaching and learning strategy/activity [17]. Nevertheless, online technology can assist in enhancing that effectiveness through appropriately designed learning activities.

## STUDENT COMMENTS

Many students commented on *why they would like to see online quizzes in future courses.* I present some of the interesting comments below:

- They are good self-testing tools. I think they should be available right from the start of the course.
- Because of the instant feedback requiring no appointments/meetings. The animosity (sic) is a bonus for those not willing to see someone for help.
- Because it gives students a chance to check their knowledge of individual sections of the course, rather than testing on a large scale (exam). This allows students to have better focus on what they need to concentrate on.
- I found that many questions were difficult to do and I had to go through the materials, and then redo the quiz. This confirmed that I did not understand the concepts well at the first round.

Many students commented on various aspects of the online quizzes, ranging from user friendliness to the activity not being assessable as a *summative* [16] one. Following are some of the comments on the user friendliness of the system:

- I would have liked to attempt a question (that I got wrong) again after studying the text without going through the whole quiz.

The student is referring to the fact that if a student wants to revisit a particular question in another session then he/she has to go through the entire quiz all over again. In fact most quizzes are likely to be generated randomly from a repository of a large number of questions. A student may not necessarily come across exactly the same set of questions in his/her subsequent attempts of a quiz. The idea is that a particular quiz will test the student's knowledge of specific topics. Therefore, a well-designed online quiz is expected to test the student's understanding of the relevant concepts irrespective of the exact set of questions generated in a certain session. However, a facility for revisiting the questions a student gets wrong in a quiz immediately after taking the quiz would be useful. This way, the time required to browse through questions answered correctly can be avoided. Another student said:

- The use of test fields made answering difficult. For example, I answered "IMM" for "Immediate Addressing", and got the answer wrong. May be pull down menus.

This was in response to one of the fill-in-the-blank questions. The system looks for an exact match to decide whether the answer given is right or wrong. It does not allow the quiz developer to specify multiple alternate answers. Pull down menus could be a solution, but the software needs to have provision for such facility. Another student commented:

- Quizzes are overall a good way of testing our knowledge. However, I think the questions should tell the student what type of input is needed.

This student is also referring to the "fill-in-the-blank" questions, which expect an exact match for the correct answer. The suggestion made by the student makes sense. Some explanation and perhaps examples of the type of inputs allowed could be incorporated.

Students' comments on making the online quizzes assessable are given below:

- I reckon it should be a policy that every subject has online quiz with gradable marks, e.g., 5%.
- Too optional. Needs to be more compulsory.
- Perhaps 5% of final mark would encourage more students to do the quiz.
- There could be an incentive for those who try the quizzes.

Clearly students would want to be rewarded for the efforts they put into the activity. As Harasim et al. stated, "most students will not take part in online activities ... unless there is some form of reward for their participation. It is grading which helps establish the worth and therefore the importance of the activity. Where there is little reward the student will treat the activity as unimportant" [18]. I did not allocate marks because the online quiz system in UniSAnet does not incorporate automatic scoring facility. In the absence of automatic scoring facility, it is difficult to make an online task assessable. I see value in incorporating such a facility into the online quiz system. It will make the online quizzes more attractive for both academics and students. Interestingly enough, a large number of the students in my course (77% of the respondents) attempted the online quizzes, which was really very encouraging.

## CONCLUSIONS

The majority of the students who attempted the online quizzes judged them to be an effective way of learning in a highly technical course on microcontrollers. Online technology can be used to supplement face-to-face teaching with a view to stimulate and maintain student interests, especially for first year undergraduates. Online quizzes provide students with more choice for their learning, an alternative learning style, opportunity for self-assessment, and above all, a level of flexibility that students appreciate. Therefore online quizzes can be very useful for engaging students in the context of increasing disengagements in campus activities. However, students like to see the online activities as summative ones. To facilitate this, it would be useful to incorporate automatic scoring features in the online system. The students provided useful feedback on many other aspects of the online quizzes. It would be worth pursuing some of these in order to improve the students' motivation and learning outcomes. Suitable online activities would facilitate the flexible delivery of programs to offshore international students. This would definitely be in tune with the increasing focus on flexible delivery and internationalisation of programs for universities around the world.

# REFERENCES


1. McInnis, C., "Signs of Disengagement: The Changing Undergraduate Experience in Australian Universities," Inaugural Professorial Lecture, Faculty of Education, University of Melbourne, August 2001, http://www.cshe.unimelb.edu.au/downloads/InaugLec23_8_01.pdf, retrieved on 13 April, 2004.

2. McInnis, C., James, R. and Hartley, R., "Trends in the First Year Experience," Department of Education, Training and Youth Affairs (DETYA), Australian Government, 00/6, July 2000, http://www.dest.gov.au/highered/eippubs2000.htm#00_6.

3. Carbone, E., "Active Learning in a Large Class", *Teaching Large Classes: Tools and Strategies*, Sage Publications, Thousand Oaks, 1998, pp. 47-56.

4. Group of Eight, "Research and Innovation: Australia's Future," Group of Eight - Australia's Leading Universities, December 2000, http://www.go8.edu.au/policy/papers/2000/1220.pdf, retrieved on 13 April 2004.

5. OECD, "Education at a Glance: OECD Indicators 2001 Edition," June 2001 [Online]. Available: http://www1.oecd.org/els/education/.

6. Laurillard, D., "Rethinking University Teaching: A Framework for Effective Use of Educational Technology," Routledge, London, 1993.

7. Skilbeck, M. and Connell, H., "Quality Assurance and Accreditation in Australian Higher Education: A National Seminar on Future Arrangements," DETYA, 2000 [Online]. Available: http://www.detya.gov.au/highered/otherpub.htm#2000.

8. Illing, D., "Driving your Scholar," The Australian Higher Education, 5 June 2002, p. 21.

9. Lloyd, P., "Crisis on Campus," The Advertiser Insight, Adelaide, 1 April 2000.

10. Madden, J., "**Risk of Stress for Uni Staff,**" The Australian Higher Education, 4 July 2002.

11. University of South Australia, "Future Learning Environments for the University of South Australia: A report," June 1994. Available from the author of this paper.

12. King, B., "Introducing the Teaching and Learning Strategy 2010 – 2015," Presentation at the Contextual Awareness Seminar, 17 May 2002. Available from the author of this paper.

13. University of South Australia, "Teaching and Learning Framework 2010," University of South Australia Discussion Paper. Available from the author of this paper.

14. King, B., "Distance Education in Australia," Harry, K., (ed.), *Higher Education through Open and Distance Learning, World review of Distance Education and Open Learning,* Vol. 1, The Commonwealth of Learning, London and New York, 1999, p. 271.

15. University of South Australia, "UniSAnet: The Online Learning Environment of the University of South Australia," http://www.unisanet.unisa.edu.au/unisanet/, 2003.

16. Biggs, J., *Teaching for Quality Learning at University,* The Society for Research in Higher Education and Open University Press, Buckingham, 1999.

17. Ramage, T. R., "The No Significant Difference Phenomenon: A Literature Review," E-JIST Vol. 5, No. 1, 2001, pp. 1-7 [Online]. Available: http://www.usq.edu.au/electpub/e-jist/docs/old/full_papers_5.htm.

18. Harasim L., Hiltz, S. R., Teles, L. & Turoff, M., *Learning Networks: A Field Guide to Teaching and Learning Online,* MIT Press, Cambridge, MA, 1995.

**S. M. Aziz** graduated in Electrical and Electronic Engineering in 1984 and received a Master's degree in 1986. He also received a Ph.D. degree in Electronic Engineering from the University of Kent (UK) and a Graduate Certificate in Higher Education from Queensland University of Technology. He is currently the Program Director of the Computer Systems Engineering discipline in the School of Electrical and Information Engineering at the University of South Australia. He has been engaged in university teaching and research for nineteen years. During this period he has been involved in many industry-sponsored projects. He is the recipient of Commonwealth Academic Staff Scholarship (UK) and certificate of excellent achievement in Network/Internet System Development (Japan). In 1996, he was a visiting scholar at the University of Texas, Austin, Texas, USA. His research interests include VLSI Design, testability, hardware modeling and co-simulation, and engineering education. He is a senior member of the IEEE.

# Chapter 27

# Towards Integrating Graduate Research and Education with "Internal Research Internships": Experiences and Assessment

JORDAN BERG, MARK HOLTZ, YELIN SU, ROMMEL BUNUAN, JENNIFER WILHELM, TIM DALLAS, RICHARD GALE, LAUREN GOLLAHON, SHUBHRA GANGOPADHYAY and HENRYK TEMKIN
*Texas Tech University, Lubbock, Texas, 79409, USA. E-mail: Jordan.Berg@ttu.edu*

*We describe our efforts towards integrating graduate-level research and education in a highly interdisciplinary laboratory environment. The educational innovation we wish to evaluate is the "Internal Research Internship," in which a graduate student with an established research project is placed in a host laboratory for a semester or more to work in a distinct, but complementary, area. The testbed for our study is the third course of a three-semester sequence in microsystem design and fabrication. Participants in this class lead teams of less experienced students to accomplish a difficult design challenge. Each of five teams is responsible for a critical component of a sophisticated device; the five leaders are responsible for coordinating and integrating the separate efforts. Partial success should be attainable through interdisciplinary communication and teamwork. Complete success represents work at the level of the current state-of-the-art. In this study we focus on interactions between the team leaders. In particular, a cell biology graduate student was brought into a microfabrication laboratory specifically to work with a physicist and an electrical engineer who were leading teams developing a DNA amplification chamber and a target DNA pre-filter. Initially awkward, inefficient and unproductive, the relationship between these students gradually became fluid, respectful and ultimately extremely fruitful. Unanticipated benefits also sprang from the biologist's interactions with other team leaders, and with the graduate education student assigned to assess the effectiveness of the project.*

## INTRODUCTION

We describe integrating graduate-level research and education in a highly interdisciplinary microsystems laboratory environment. The framework we use is based on our extensive experience with industrial internships. By applying the industrial template to internship-like positions within university research labs, we hope to incorporate many of the educational benefits of industrial experience into the traditional graduate research model.

Our eventual goal is to develop the "internal internship" approach for broad application in the sciences and engineering for advancing interdisciplinary research. The American Association for the Advancement of Science [1] and the National Research Council [2] advocate collaboration among students from multiple disciplines for fostering innovative ideas. As a developmental testbed, we have incorporated this approach into the final two courses of our graduate three-course microsystems curriculum. In these two courses [3,4], "Topics in Microsystems" and "Advanced Topics in Microsystems," three levels of participants—each playing a role from the industrial milieu—interact to accomplish a significant research milestone. The faculty members teaching the course are the first level, representing a client with specific technology needs.

The faculty—from a variety of science and engineering departments, plus a specialist in science education—define the functionality of the device to be constructed, and a set of design constraints such as size, cost, power consumption, etc. Students in the Advanced Topics course are the second level, representing technical management. They are chosen for the course either because they have broad knowledge and significant experience in the microsystems field, or because they possess deep expertise in a specific area central to the target device. These students work as a team to design a device responsive to the requirements of the Clients. We call students from the Advanced Topics course the "Team Leaders." This refers to their role directing the small teams of one or two students each from the Topics class that make up the third level. Students in the Topics class have a working knowledge of microfabrication techniques, and are either graduate students or advanced undergraduates from science or engineering backgrounds. These students represent practicing engineers, and it is their responsibility to carry out the actual device fabrication and testing. We call them the "Team Members." Component design is entirely handled by the Team Members, with integration aspects overseen by the Team Leaders working as a group. The Clients receive weekly briefings from the Team Leaders where feedback is given and potential problems are discussed. Formal presentations are made approximately monthly by the collected memberships of both courses.

This report focuses on the experience of the students at the second level, that is, on the Team Leaders. Because they are all graduate students with significant research experience, because they are placed in a team setting with a strongly interdisciplinary flavor, and because of the high degree of peer interaction that their duties require, their environment is very close to what we plan to achieve in our internal internship project. Previous implementations of this course structure [3] had Teams competing to develop similar technologies. Here, we combine the Team Leaders into one unit, with a common goal, thereby emphasizing team-building aspects. The result is a far richer learning environment for the Advanced Topics students.

| Team | Title | Majors (* denotes team leader) |
|---|---|---|
| 1 | Mixing, Injecting, Sample Preparation | ME*, EE |
| 2 | DNA Sequence Pre-selection | EE*, EE, EE |
| 3 | On-chip PCR | Physics*, ME |
| 4 | Microfluidic Design/Waveguide | ME*, ME |
| 5 | Optics, Illumination, Detection | EE*, Engineering Physics, EE |

TABLE I

TEAM TITLES AND THE ACADEMIC MAJOR OF THE TEAM LEADERS

The device they are asked to produce is a sensor capable of determining whether a small sample of DNA contains a specified base sequence, for example one characteristic of a specific organism. The device requires a number of steps, including sample preparation, screening for the target base sequence, amplification through polymerase chain reaction (PCR), and signal processing. Exact details are left to the student design teams.

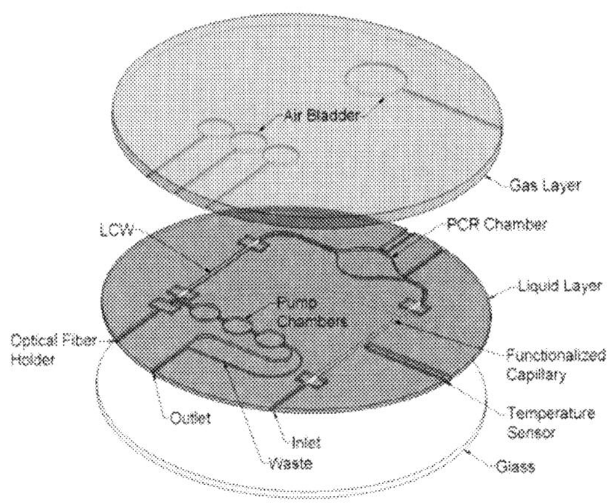


FIGURE 1

SCHEMATIC CUTAWAY OF THE DNA SENSOR CHIP FULLY DESIGNED BY THE STUDENT GROUPS IN THE "ADVANCED TOPICS" COURSE

Each Team Leader was initially asked to prepare an introductory 50 minute lecture in his or her topic area. This was to include a review of the current literature, and a brief discussion of design challenges with some potential solutions. This broad study provided the starting point for each group's efforts. All functions, plus the associated fluid-handling infrastructure, are to be incorporated on a single chip. This is a challenging project, involving several current research areas from the laboratories of the affiliated course faculty. The groups and academic areas of the teams are listed in Table I. Although we attempted to utilize students' skills so that they would have their best chance at succeeding with the device goals, we were also interested in allowing them to expand into areas beyond their comfort zones.

Research success is defined by the degree of functionality of the resulting device. Partial research success is defined by the development of key processes, the design of important device components, etc. Key non-technical goals for this group include development of team leadership skills, and experience with peer interaction in an interdisciplinary environment. Success in these areas is assessed through interviews with the students, journals that they keep, and observation as they interact with each other, with their Team Members, and with the faculty Clients.

## DESCRIPTION OF THE DNA-BASED SENSOR

The functional design of the device is shown schematically in Figure 1. The device is made of two patterned layers of polydimethylsiloxane (PDMS) on a structural glass wafer. The first area encountered by the sample is the capillary, which is functionalized to preferentially immobilize target DNA [5]. The purpose of the capillary is to filter out, through capture, the desired DNA for amplification and detection. The capillary is placed into the chip at the time of application, allowing user customization of the DNA selection. Similarly, the PCR primers are injected at the time of use to maintain flexibility. PCR amplification [6] is done on-chip, requiring heaters and temperature control. Amplified product is transferred to the liquid core waveguide (LCW) [7]. The LCW is made of Teflon® AF, which has an index of refraction of 1.29. This is lower than the index of water. When the LCW is filled with a water-based solution, fluorescence generated within a solid cone will undergo total internal reflection. Light emerging from the end of the LCW is captured by an optical fiber (not shown in Figure 1) and passed to a photodetector. An expanded view of the LCW is shown in Figure 2.

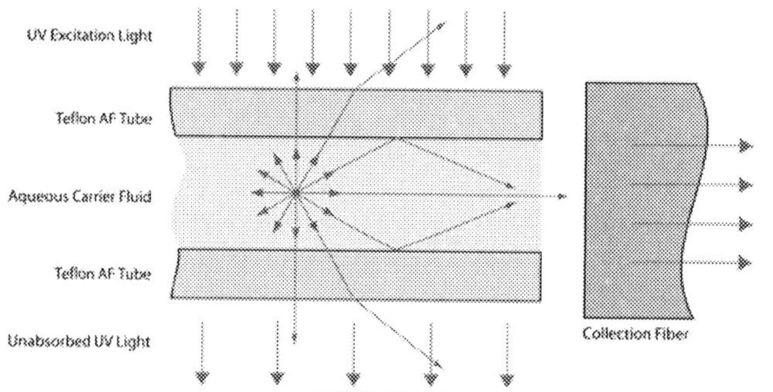


FIGURE 2
SCHEMATIC OF THE LCW FUNCTION. EMITTED FLUORESCENT SIGNAL (GREEN) AT A SUFFICIENT SMALL ANGLE OF INCIDENCE UNDERGOES TOTAL INTERNAL REFLECTANCE AND IS CAPTURED BY THE COLLECTION OPTICS.

On-chip peristaltic pumps [8] were also designed into the system in order to transfer sample and other fluids between the functional areas. An interesting aspect of this device is that the students designed it to function based on a single peristaltic pump. This minimizes footprint but requires a clear understanding of the flow sequences to be used. Their implemented design is clever, but requires further refinement. The sequence of Figure 3 demonstrates the pumps in operation moving a fluorescent test fluid on a prototype device. Figure 4 shows how the functionalized capillary and LCW were sealed into the device using injected PDMS "dams." Fluorescent fluid can be seen moving through the top tube without leaking around the outside of the capillary.

Illumination of the channel is through a series of light emitting diodes (LEDs). Because LEDs are now commercially available in the blue and near-ultraviolet range, their compactness, low power consumption, and low operating temperature make them ideal for incorporation into fluorescence-based sensors.

Each group was instructed to produce an initial design, address integration issues with the other groups, and present their overall design to the Clients for approval and input. Each group was expected to create a testbed for their component so that they could evaluate its effectiveness and troubleshoot it without incurring delays waiting for other groups to complete their tasks. When all components were working, the final device integration was to be accomplished.

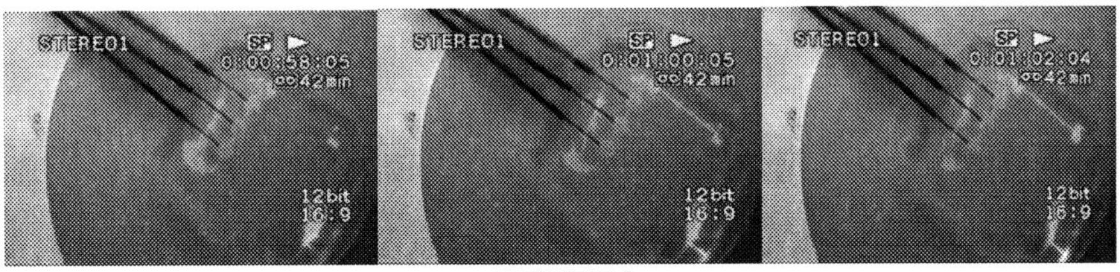


FIGURE 3

PERISTALTIC PUMPING OF FLUORESCENT TEST FLUID. THE RECTANGULAR CHAMBER ON THE RIGHT OF THE LAYOUT IS BEING FILLED. THE TUBES SUPPLY PRESSURIZED AIR TO CYCLICALLY COMPRESS THE THREE PUMP CHAMBERS.

## CASE STUDY

As a case study, we discuss the role of the sole Biology graduate student, J—, in the internal internship. Rather than lead a particular team, J— was expected to act as a Research Intern to the leaders of Teams 2 and 3 (see Table I), combining their fabrication knowledge with his biological expertise to produce PCR and DNA capture schemes, and to spend minimal time with the other Teams. As it turned out, the level of biology required made J— the research mentor to the Leaders of Teams 2 and 3, while interacting significantly with the others. We focus here on his experience because of this comprehensive role and because his experiences were both Internal Internship *host* and *guest*.

In his role as mentor, J— had contact with two groups. The first group was responsible for modifications to a glass capillary for selective binding of target DNA. This is a critical step. The subsequent PCR process preferentially amplifies the target DNA. However, PCR is sufficiently robust that a sample containing large populations of untargeted DNA will also produce non-specific amplification giving a false signal. Therefore, a filtration step is needed to minimize contamination. The filtration is accomplished by first binding an oligonucleotide to the inner walls of the capillary. Samples containing target and unwanted DNA are brought into the capillary, where the target preferentially binds. Following a rinse step to purge unbound "waste" material, the bound species are released by heating the capillary to 95 °C and transferred to the PCR chamber. The capillary part of the design includes glass surface modification, fluid exchange, temperature control, and standard (i.e., off-chip) PCR amplification to determine success of the filtration step.

The team responsible for designing and testing the on-chip PCR system was the second group mentored by J—. The PCR team designed the chamber along with external heating and cooling, each placed in contact with the chip but not in immediate contact with the fluid. This way, heater and cooler were part of the permanent infrastructure. The PCR process requires cycling the temperature between 65 and 95°C, while supplying the amino acid building blocks and primers. The design requires temperature measurement and control, learning standard PCR methods to determine if tests were successful, and interfacing with the microfluidics group for sample transfer.

The design and implementation of these components is clearly interdisciplinary. Before transferring the approaches to the teams, J— first needed to fully research, understand, and experiment with these techniques himself. Because he had to respond to the design direction of his guests in order to act as an effective mentor, he became

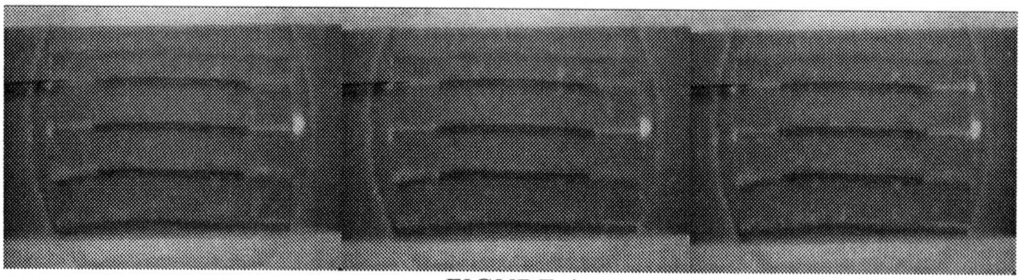

FIGURE 4

CHANNEL SEALING SYSTEM. THE DEVICE SHOWN CONTAINS THREE TEST STRUCTURES, OF WHICH ONLY THE TOPMOST IS BEING USED. IN THE THREE-PICTURE SEQUENCE SHOWN, THE DEVICE IS INITIALLY EMPTY, THEN, FLUID STARTS TO FLOW FROM LEFT TO RIGHT. THE EMBEDDED CAPILLARY IS ROUND WHILE THE CHANNEL IS SQUARE, BUT NO FLUID LEAKS AROUND THE EXTERIOR OF THE CAPILLARY.

familiar with methods not central to his own research. Thus J— gained new capabilities in his *own* field as a consequence of taking this interdisciplinary course! His mentoring of the engineering and physics students began by identifying archival articles at the appropriate level. This was not simple, since J— is accustomed to working with fellow biology students, with a firm grasp of the concepts, language, and methods commensurate with years in the field. In contrast, the engineering and physics students had only introductory freshman-level courses in their undergraduate curricula, and years had passed since that general exposure. With some perspective, both mentor and guests learned the value of interdisciplinary communication through this exercise. Through in-class group presentations and during the practical walk through at the end of the semester, we determined that the teams had learned a substantial amount about the biochemistry of their project components. We view them as conversant in these topics, enabling the team members to conduct interdisciplinary research. This qualifies the internal internship approach as successful.

J— also played the role of an intern in this project when working with the microfluidics and optics groups. From the microfluidics group, J— learned about fluid properties at the microscale. From his own reflective journal, it was clear that he did not appreciate in advance the difficulties of such aspects as mixing. Here, the small characteristic length scale gives small Reynolds number and prevents turbulent mixing. Therefore, absent other steps to enhance advective transport, mixing is driven by the relatively slow process of diffusion. This topic is important for the PCR process, where amplification steps require the intermixing of sample and the nutrient/primer solution. J— learned more about this through discussions with and demonstrations by the pump group.

The optics group worked on all aspects of fluorescence detection. The fluorescent tags, incorporated into the target DNA during PCR, are illuminated using LEDs. The resonant excitation wavelengths and emission spectrum are characteristic of the fluorescent tags, and fluorescence imaging is a standard method in biological science. J— went from being a user of this method to having a detailed understanding of the fluorescence process. The detection of small sample volumes, necessary to obtain a result in the minimum number of PCR cycles (processing time), requires sophisticated detection schemes. In the current context, a LCW was used based on commercial Teflon® AF tubing. Therefore, in order for J— to fully appreciate the fluorescence detection in a microfluidic setting he learned about waveguides and basic optics, minimum detection levels, and methods for testing them. This broadened his experience and prepared him to communicate with engineering and physics students on a much more viable level.

It was very clear to us by the end of the course that the bi-directional flow of capabilities was functioning. For a course in design and implementation of biological sensors, we feel that it was essential to have a biologist strongly involved in the course. In fact, it would be preferable to have at least two to better distribute the workload. It is also important, of course, to see how individuals other than J— function in this setting. While we feel that the experience is generalizable, J— did have some qualities, most notably military training, that may have made him better prepared to handle it. At the same time, we feel that the other five students involved in this process also became clearly better equipped to collaborate effectively in interdisciplinary research. We intend to observe the careers of each of these students to see if they use their interdisciplinary communication skills, and whether they impact their research efforts.

## Role of Assessment

This project placed the graduate students in a situated learning environment [9,10], where they worked on interdisciplinary projects requiring the expertise and talents of all participants. An assessment team of students in Science and Engineering Education evaluated the effectiveness of the structure. They focused on the types of learning and knowing that occur in the peer-to-peer setting. The main reason for this was to extrapolate that experience onto the similar relationship between host and guest doctoral students in an Internal Internship. Thus far, evidence has been provided that supports a connected knowing being established where these six students' knowledge is being facilitated by and co-constructed with the other graduate students in the class.

The effectiveness of the Internal Internship approach was assessed through observations of classroom and group meetings, interviews, and reflective journals. To ensure that the latter were kept up to date, they were made part of the course grade. Students were assured that the interviews and reflective journals would be kept confidential, and that the instructors responsible for the course grades would have no access to the journals outside of sanitized feedback. At the time of this writing, all assessment completed is formative. Evidence in the form of journal entries, survey responses, and videotaped and transcribed class meetings display facilitated learning.

To accomplish the assessment tasks, two education doctoral students participated in the program. Neither student had a science background. They were not enrolled in the course, but were allowed complete access. Class sessions were audio recorded, reviewed, and transcribed. Journals were reviewed on a bi-weekly basis. Several interviews were conducted. A very important finding of this approach was that the education students learned what students were thinking in a way that grade-determining professors could not expect to. Observations included: (1) Group communication was markedly improving. (2) Based on one student's recommendation, meeting notes would include action items to be distributed class wide for better integration. This helps in defining roles and responsibilities. (3) Peer instruction on technical issues was proceeding in the group leader meetings and in the individual task meetings. (4) Students policed each other for understanding and followed up with further discussions to clarify misconceptions. (5) The reflective journals provide us with evidence of improved communication skills. We observe that this improvement over the semester was instrumental in developing the cooperative technical skills needed to make progress on the device.

There were also troubling observations: (1) Students were frustrated that their peers weren't versed in their area, or weren't keeping up with reading they assigned. Students resolved these issues both on their own, with advice from faculty, and with the help of faculty. It was also necessary to remind them and ourselves that specialization happens relatively early in most undergraduate curricula. (2) Several students were likewise frustrated that faculty could not provide hard information on some questions, but recommended further experimentation to find an answer. This shows that students did not fully appreciate that the course was research-oriented—that faculty are also interested in finding the answers to difficult questions. This is in stark contrast to standard courses in which well-known methods and problems are primarily addressed. This distinction will be better drawn in the future.

Interactions between faculty members, while not the main focus of this article, are worthy of brief mention. The members from Mechanical Engineering, Physics, and Electrical Engineering are part of an interdisciplinary center, and frequently collaborate on research projects. This study simply became part of their regular interaction. The Biology faculty also worked well with the team, though not as seamlessly. We believe this is because her graduate student was highly involved, on a daily basis, with the other students and faculty. The result was a healthy interchange of ideas and expertise at the faculty level—a validation of the Internal Internship concept. The Education faculty representative did not interact widely with the others, possibly because her assessment role called for objective detachment, or possibly because her students were not as involved with the others on a daily basis.

However, we find clear evidence that the education doctoral students learned from this experience. This learning included concepts from both biology and engineering, and we were gratified that everybody was benefiting from the experience. However, a keen observation was expressed in a report, in which it was stated:

> *"I begin to see the beauty of interdisciplinary learning. [...] I was really amazed by the lively interaction these team leaders demonstrate especially during those times when they have to discuss the pros and cons of a particular action. Since everyone in the group comes from different fields with varied expertise, neither one is an authority. Everyone is both a teacher and a student as each of them has to share their own knowledge in thinking, planning, designing, testing, implementing, and redesigning the project whenever there is a need to do so. Nobody knows everything and everybody has to share something. [...] I'm saying this because I saw it myself how students in physics and in biology would work together to address for example the problem of DNA sample preparation and bonding in the device which the group is working on. In another occasion, I also witnessed how the Engineering students would have to struggle to come up with a design (i.e. liquid-core waveguide) to detect the presence of DNA. [...], as a doctoral student in Education I can say that my exposure to this kind of learning environment has significantly broadened my view on instructional techniques; that the benefits of learning are better maximized when the teacher does not monopolize the teaching of the subject; that the benefits of learning are enhanced when synergy and creativity among the students are fostered, where their curiosity is best accommodated by a right amount of tolerance."*

This observation, made by a student previously uninitiated to the value of interdisciplinary research and education, provides strong, independent verification of the importance of further developing the Internal Research Internship concept.

## SUMMARY

We have explored the use of Internal Research Internships by placing a biology graduate student in a peer-mentoring learning environment in a team-based microsystem design and fabrication laboratory course. The interaction between this student and others, including graduate students in physics, electrical and mechanical engineering, and education, led to a rich and productive educational experience. In addition to becoming more familiar with technical skills from other disciplines, the students learned to communicate with researchers of different backgrounds. They became more comfortable in their own field, through the struggle to make their knowledge accessible to others. While this is a limited sample, it is extremely positive and encouraging, and we see ample evidence to support pursuing the concept.

## ACKNOWLEDGEMENT

This work supported in part by National Science Foundation Grant EEC-0087902. Invaluable technical assistance was provided by Dr. A. Datta.

## REFERENCES

1. American Association for the Advancement of Science, *Benchmarks for Science Literacy*, London, Oxford, 1993.
2. National Research Council, *National Science Education Standards*, Washington, Nat. Acad. Press, 1996.
3. Dallas, T., Holtz, M., Berg, J., and Gangopadhyay, S., "A Three-Course Sequence in the Engineering of Fluorescence-Based Micro Total Analytical Systems", Proc. 2003 International Conference on Engineering Education, ICEE-2003, Valencia, Spain, 2003.
4. Dallas, T., Berg, J.M., Holtz, M., Gangopadhyay, S., and Temkin, H., "Developing Internal Internships for a Microsystems Engineering Curriculum", Chap. 32, Aung, W. et al. (eds.), *Innovations 2003: World Innovations in Engineering Education and Research*, Int. Network for Eng. Ed. and Res. (iNEER), Arlington, VA, 2003, pp. 319-330.
5. Matsuura, S., Kurita, H., Nakano, M., Komatsu, J., Takashima, K., Katsura, S., and Mizuno, A., "One-End Immobilization of Individual DNA Molecules on a Functional Hydrophobic Glass Surface," J. Biomolec. Struct. & Dynamics, Vol. 20, 2002, pp. 429-435.
6. Dieffenbach, C.W., and Dvksler, G.S., *PCR Primer: A Laboratory Manual*, Cold Spring Harbor Laboratory Press, Woodbury, NY, 1995.
7. Dasgupta, P.K., Genfa, Z., Li, J., Boring, C.B., Jambunathan, S., and Al-Horr, R., "Luminescence Detection with a Liquid Core Waveguide," Anal. Chem., Vol. 71, 1999, pp. 1400-1407.

8. Berg, J.M, Anderson, R.C., Anaya, M., Lalouh, B., Holtz, M., and Dallas, T., "A Two-Stage Discrete Peristaltic Micropump," Sensors and Actuators, Vol. A 104, No. 1, 2003, pp. 6-10.

9. Lave, J. and Wenger, E., *Situated Learning: Legitimate Peripheral Participation,* Cambridge University Press, New York, 1991.

10. Lave, J., *Cognition in Practice,* Cambridge University Press, Cambridge, UK, 1988.

**Jordan M. Berg** received his B.S.E. and M.S.E. in Mechanical and Aerospace Engineering from Princeton University in 1981 and 1984. He received his Ph.D. in Mechanical Engineering, and his M.S. in Mathematics from Drexel University in 1992. He has held postdoctoral appointments at the USAF Wright Laboratory in Dayton, OH, and the Institute for Mathematics and Its Applications in Minneapolis, MN. Since 1996 he has been with the Mechanical Engineering Department and Nano Tech Center of Texas Tech University, where his research interests are control and analysis of dynamic systems, MEMS and microfluidics.

**Mark Holtz** received his B.S. degree in Physics from Bradley University in 1980, and his Ph.D. in Physics from Virginia Polytechnic Institute in 1987. Following this, he was a Fellow at the Max Planck Institute in Stuttgart working in high-pressure physics. From 1989 through 1991 he was a Research Associate at Michigan State University. Since 1991 he has been in the Department of Physics at Texas Tech University working in semiconductors, optical spectroscopy, and recently in microfluidics.

**Yelin Su** is a doctoral student in curriculum and instruction at Texas Tech University. She received her Master's degree in sociology from Morehead University in Kentucky (2003) and Bachelor's degree in Law at Sichuan Normal University, P.R. China (1999).

**Rommel Lazo Bunuan** is a doctoral student in the College of Education, Texas Tech University with a major in Curriculum and Instruction. Prior to his entry in the doctoral program, he taught Economics to undergraduate students at the University of Santo Tomas in Manila, Philippines, for ten years. He also holds a master's degree in Economics.

**Jennifer Wilhelm** received her B.S. in Physics in 1988 from Bowling Green State University, her M.S. in Physics in 1991 from Michigan State University, and her Ph.D. in Math/Science Education in 2002 from University of Texas at Austin. Currently, she is an Assistant Professor of Math/Science Education at Texas Tech University. Her main research interests involve interdisciplinary project-based curriculum and instruction.

**Tim Dallas** received his B.A. degree in Physics from the University of Chicago in 1991 and his M.S. and Ph.D. degrees in Physics from Texas Tech University in 1993 and 1996, respectively. Following a year in the semiconductor equipment industry he was a post-doctoral researcher at The University of

Texas, Austin. Since 1999, he has been an Assistant Professor of Electrical and Computer Engineering at Texas Tech University and the Associate Director of the Nano Tech Center. Research interests include MEMS and microanalysis systems.

**Richard Gale** received his A.B. degree from the University of California at Berkeley (1979) and his M.S. (1981) and Ph.D. (1984) degrees from Lehigh University. He was a member of technical staff at Texas Instruments for 17 years, retiring in the spring of 2001. Currently he is a Professor of Electrical and Computer Engineering at Texas Tech University. His research interests include optical information processing, microelectromechanical systems, optical communications, and high brightness LEDs.

**Lauren S. Gollahon** received her B.A. in Marine Biology from Barrington College (1981), her M.S. in Zoology (1986) and her Ph.D. in Veterinary Anatomy (1990) from Texas A&M University. She held postdoctoral appointments at UT MD Anderson Cancer Center and UT-Southwestern Medical Center. Since 1997, she has been in the Dept. of Biological Sciences at Texas Tech University. Her research interests include investigating differences in cell specific regulation of cancer and exploring avenues of utilizing MEMS and nanotechnology in the identification, diagnosis and prognosis of cancer.

**Shubhra Gangopadhyay** received her B. Sc. and M. Sc. in Physics in 1975 and 1977, respectively, from Jabalpur University, India, and her Ph.D. in Physics in 1982 from the Indian Institute of Technology, Kharagpur, India. From 1985 to 2003, she was a Professor of Physics at Texas Tech University. In 2003, she became a LaPierre Chair Professor of Electrical and Computer Engineering at the University of Missouri-Columbia. Her main research interests are microelectronics and microsensors.

**Henryk Temkin** received his B.S. in Physics in 1969 from the U. Libre De Bruxelles, his M.S. in Physics from Yeshiva University in 1971, and his Ph.D. in Physics in 1975 from Stevens Institute of Technology. He was a member of the technical staff at Bell Laboratories for 15 years. Currently, he holds the Maddox Chair of Electrical and Computer Engineering at Texas Tech University and is the Director of the Nano Tech Center. His research areas include optoelectronics and MEMS for telecommunication.

# Chapter 28

# Evaluation of a Web-Based Learning Environment for Hands-On Experimentation

ANH-VU NGUYEN-NGOC, DENIS GILLET and STEPHANE SIRE
*Swiss Federal Institute of Technology (EPFL), School of Engineering, Ecublens, CH-1015 Lausanne, Switzerland. E-mail: denis.gillet@epfl.ch*

*The evaluation of learning environments depends highly on the evaluation instruments used and on the position of the evaluation in the design cycle. In this article, we present a use-case report of a summative evaluation of a web-based learning environment for engineering education that allows small groups of students to conduct hands-on experimentation sessions with local and remote access to laboratory devices. This use-case illustrates an evaluation methodology based on a choice of usability engineering methods. In particular, we define specific metrics that take into account students' interactions with some of the components of the environment. The metrics defined can easily be extended to any e-learning project involving production and sharing of artifacts in a shared workspace. The methodology is illustrated with the results of a preliminary evaluation with 30 students.*

## INTRODUCTION

The role of evaluation in software development has been well established. It is of the utmost importance in usability engineering [1]. Over the last few years, many interactive systems have benefited from user evaluation, including e-learning applications [2-5]. However, it is not obvious as to how to evaluate educational software. It requires the transposition of the concepts of acceptability, utility and usability to a domain that is at the crossroads of multiple fields such as interactive systems design, collaborative work and learning theories.

This contribution focuses on evaluation in the field of web-based environments that support hands-on experimentation through remote manipulation of physical laboratory devices and/or computer simulation tools, and particularly on the deployment of the

Cockpit environment which is now in its third year of iterative design in the framework of the eMersion project [6]. This project aims at developing flexible engineering education [7] in automatic control, biomechanics and fluid mechanics at the Swiss Federal Institute of Technology in Lausanne (EPFL). More than 100 students are now using the latest version of the prototype.

This article presents the last round of evaluation that we conducted during the winter 2002 semester with 30 students. This design cycle has been the occasion to introduce an electronic laboratory journal in the web-based environment fully connected with the other software components for an automatic importing and exporting of various types of data. The main focus of the article is more on the evaluation methodology by itself than on a complete analysis of its results. In particular we illustrate how the laboratory journal, which is a special case of a shared workspace, gives opportunities to develop new evaluation measures that could serve in other contexts.

## FLEXIBLE, PARTICIPATIVE AND COLLABORATIVE LEARNING ENVIRONMENTS

Web-based environments are usually introduced to let students choose between different learning modalities; that's why we call such a paradigm 'flexible learning'. The modalities vary according to the presence of teaching assistants (face-to-face condition) or not (individual condition), and according to students' location. The location can be a laboratory room with full access to physical devices (local condition), a computer room on the campus (remote-campus condition), and a student's home or any other place (remote-home condition). In the last two conditions students remotely access the physical laboratory devices and/or computer simulation tools.

Hands-on teaching involves multiple interactions between students and their environment; that's why we call such a scheme 'participative learning'. The interactions correspond to trials-and-errors performed while following an experimental protocol. In the eMersion web-based environment, trials and errors are mediated with specialized Java applets that control physical laboratory devices and computer simulation tools.

Finally, students perform hands-on experimentation in small groups (2 students); that's why we call such a setting 'collaborative learning'. They have to collaborate for allocating tasks, taking notes and in some cases editing a report.

The goal of eMersion is to foster advancement in engineering education everywhere (local, remote-campus, remote-home) and, whenever possible, to give students the flexibility to build their own time schedule by mixing face-to-face and individual conditions. The underlying vision is that as more and more experiments are becoming available from different institutions, it will be possible to share these experiments and the expertise of teaching assistants to manage remote students. The second goal of the project is to increase students' participation and collaboration in learning.

We are evaluating the feasibility of these goals by introducing the following learning conditions in some hands-on experimentation activities proposed at EPFL and followed by volunteer students:

1. Hands-on experimentation is organized around regular planned face-to-face sessions with teaching assistants, and students are free to attend. They can also choose to work in computer room available on the campus or at home, with or without their teammates. In the final evaluation of students, their time of presence is not taken into account.

2. User interfaces are designed with continuity of interaction in mind [8]. For that purpose we have built a web-based environment where software components directly import and export data chunks to and from an electronic laboratory journal that we call the eJournal.
3. The eJournal is designed as a persistent centralized repository, and it is visible to students as a shared workspace, comparable in a way to other workspaces such as BSCW [9]. One eJournal is dedicated to one group. We call the data chunks in the eJournal 'fragments'. In fact, these are documents stored in the eJournal. Fragments represent different kinds of data (e.g. snapshots, parameters, experimental results, etc.), and can be annotated and shared between groups and/or assistants. The eJournal also provides the possibility of copying and/or moving fragments from one eJournal to another.

Following the previous conditions, face-to-face learning with teaching assistants takes place either in a laboratory room or in a computer room (thus corresponding to the local and campus-remote conditions). By extension we call it 'face-to-face learning modalities'. All the other conditions (such as individual work at home or on the campus with other students but without teaching assistants) are called 'flexible learning modalities'.

## EVALUATION OBJECTIVES

The first objective of the evaluation is to verify that the web-based environment does not disturb students in conducting flexible hands-on experimentation as compared to the traditional approach. This is the 'acceptance goal'. We even think that this goal needs to be first met in face-to-face modalities before experimenting with flexible modalities. The reason is that if students do not accept the environment in face-to-face conditions they will be less likely to use it in remote conditions.

The second objective of the evaluation is to check if the import and export data functionalities are being used and to which extent. This is a way to validate the design for continuity of interaction principle driving the eJournal design. As a side effect we can consider also that the more the import and export functionalities are used, the more trials-and-errors are performed by students. We can also consider that evidences of use of those functionalities would imply that the environment does not prevent students from actuating the physical devices or their surrogates in remote environments. This is the 'participation goal'.

The third objective of the evaluation is to determine if students choose some flexible learning modalities. The quantification and the determination of the nature of the work done under flexible conditions is also part of that goal. This is the 'flexibility goal'.

The fourth objective of the evaluation is to assess the contribution of fragment sharing and annotation mechanisms built into the eJournal for intra-group and inter-group collaboration. As a side effect we can consider that, the fact that students and/or teaching assistants directly annotate the fragments stored in the eJournal validates the design principle of anchoring communication on these fragments, as well as their use as shared artifacts for getting contextualized assistance. Students can also collaborate with one another by exchanging the fragments. This is the 'collaboration goal'.

For each of the evaluation objectives above, we also plan to better understand by which factors they could be influenced (hence to determine 'how to better fit a web-based environment for hands-on activities', 'how to encourage flexible work', 'how to increase student participation' and 'how to develop group work').

## Evaluation Methodology and Metrics

We have adopted a usability engineering approach to evaluation. This is preferred to a didactical evaluation because, as stated in the introduction, this evaluation also takes place in an iterative design process with the purpose of improving the design of user interfaces.

The web-based environment prototype has been deployed in a real working environment before evaluation. Thus our approach is more summative than formative and we have been constrained to the evaluation instruments that can be applied directly to the field of work. It was not possible for instance to setup user-tests or controlled experiments with volunteer students because the project is part of the students' regular curriculum; that is, we are careful not to create too much perturbation on the students' learning process. However, the option of calling back some students for participating in task-based evaluations of specific interface features is still open. It is also possible to practice such formative evaluations with neutral students that are not enrolled in the laboratory courses.

The evaluation has been based on three evaluation instruments: a questionnaire; a log and content analysis of the eJournal; and interviews with students. These instruments allow the definition of specific metrics that meet the evaluation objectives.

### Questionnaire

The questionnaire has been designed with three purposes in mind: first, to assess the acceptation of the web-based hands-on environment; second, to gather information about students that could help to identify some factors influencing the acceptation; and, third, to collect information for bootstrapping the interviews with students.

The evaluation of acceptance is based on a user-interface satisfaction questionnaire: the Computer System Usability Questionnaire (CSUQ) [10,11]. This is a Lickert scale questionnaire made of 19 usability-related assertions to which the respondent has to agree or disagree on a seven points scale, ranging from 1 (strongly disagree) to 7 (strongly agree). It also comprises open fields for listing the three most positive/negative comments, and open fields for general comments. The CSUQ quantifies satisfaction as a score between 1 and 7 (where 7 represents highest satisfaction).

We have extended the CSUQ questionnaire with 7 factual questions (first name, surname, sex, electronic courses taken, familiarity with computers, questions about the student's own computer if s/he has any, and perceived eJournal use). Some of these questions such as familiarity with computers (ranging from 1, very low, to 7, very high) and the perceived eJournal use (ranging from 1 to 5 as compared to the 4 other main components of the environment) have been included to determine if there is an influence on students' satisfaction. The other factual questions such as those dealing with a student's identity have been included to personalize log analysis and to use the questionnaire as a resource during interviews.

### Log and Content Analysis

The second evaluation instrument relies on the analysis of the content of the database that holds the eJournal. This database contains fragments, annotations and some logs of students' activities such as when and which annotations they viewed. For every fragment the database keeps records of its creation date, origin (either it was directly imported from the web-based environment or uploaded from a local computer) and type (snapshot, mathematical script, experimental result, etc.). Deleted fragments are not removed from the database but simply marked as deleted. This permits the analysis of the evolution of fragment creation over time.

Fragments are categorized according to their origin, their date and their creation time. Fragments that originate from components of the web-based environment and which are directly imported to the eJournal are called **intra-fragments**. In contrast, fragments that are uploaded from local user's computer are called **extra-fragments**. Fragments that are created during face-to-face learning modalities are called **f2f-fragments**, while fragments created during flexible learning modalities are called **flexible-fragments**.

The previous definitions of fragment categories allow the characterization of two features concerning the use of the prototype. The first feature deals with a student's work that takes place within the web-based environment as compared to work that occurs outside. We measure it as an **intra-fragment-ratio**; that is, the number of intra-fragments divided by the total number of fragments. This measure reflects the utility of the environment for performing hands-on tasks. The second feature is linked to the importance of face-to-face learning modalities as compared to flexible learning modalities. It is quantified as the **flexible-fragment-ratio**; that is, the number of flexible-fragments divided by the total number of fragments.

The metrics linked with fragments creation are absolute values that characterize one pattern of a student's work (for instance 'highly flexible with the use of external applications'). It is also interesting to look at the evolution over time of these measures to determine some potential evolutions in the work patterns. The time evolution can be examined at different granularities such as days, or weeks. Some changes in the work patterns could be attributed to some learning effect in the use of the prototype and/or to evolutions in a student's learning conditions such as the proximity of periods of examination.

### Interviews

The last evaluation instrument is an individual interview with students. These interviews are non-directive. Their purpose is to let students remember if they could or could not complete their tasks and to explain why. The interviewer can use critical incident analysis style of questions, for instance by asking students to illustrate the most negative answers given at the satisfaction questionnaire with an example. We have also used a paper-mockup of the user interface to help a student remember specific situations.

Interviews have been recorded in audio and fully transcribed before analysis. This analysis consists in coding the most significant extracts of a student's speech into different categories. We have chosen 7 categories afterwards (positive comments, negative comments, bugs, limitations, methodological remarks, missing explanations).

## RESULTS

There were 32 students enrolled in the automatic control course from the 25th October 2002 to the 7th February 2003. They were divided into 16 groups of two students each, numbered in sequence from A1 to A8, and from B1 to B8. However, because group B6 gave up the course, only 30 students used the prototype as part of their hands-on laboratory assignment. Apart from a presentation and an introduction session, each group had to complete 7 modules and a final practical 'lab' examination. For each module the groups had to prepare first a 'pre-lab' and to submit a report to the teaching assistants. Then they were given access to the 'lab', i.e. they received local and remote access to the physical devices for experimentation. For both the 'pre-lab' and the 'lab' part of their work, students could choose between face-to-face and flexible learning modalities. The teaching assistants were available during planned face-to-face sessions on Thursday and Friday afternoons in the laboratory room. The students were free to make progress at their own pace; however, an indicative calendar was suggested to the groups.

Among the 30 students in automatic control, 16 of them also took part to biomechanical hands-on practice based on computer simulation tools integrated into the same web-based environment. They had to complete 5 modules from the 25th of October to the 7th of February. For each of these modules, they had first a 45 minutes face-to-face session with teaching assistants in a computer room for introducing the exercise. A correction session followed the week after in a non-computer room. Students had to submit answers to the exercise before the correction session; meanwhile, they had full access to the web-based environment.

| Group | Satisfaction<br>Mean (Standard Deviation) | eJournal usage<br>User ranking (Usage category) |
|---|---|---|
| ? | 4.35 (1.17) | 3 (LOW) |
| ? | 4.50 (1.33) | 3 (LOW) |
| ? | 2.79 (1.13) | 1 (HIGH) |
| A1 | 4.45 (1.3) | 3 (LOW) |
| A2 | 3.68 (1.05)<br>4.21 (1.26) | 3 (LOW)<br>3 (LOW) |
| A3 | 2.89 (1.3)<br>4.89 (1.12) | 4 (LOW)<br>2 (HIGH) |
| A4 | 4.84 (0.97)<br>4.16 (0.81) | 2 (HIGH)<br>3 (LOW) |
| A5 | 5.68 (1.15) | 1 (HIGH) |
| A6 | 4.47 (1.74) | 2 (HIGH) |
| A7 | | |
| A8 | | |
| B1 | 3.53 (1.07)<br>5.33 (1.13) | 1 (HIGH)<br>1 (HIGH) |
| B2 | 4.59 (1.06)<br>4.16 (1.68) | 1 (HIGH)<br>3 (LOW) |
| B3 | 2.88 (0.76)<br>4.00 (1.06) | 3 (LOW)<br>2 (HIGH) |
| B4 | 5.05 (1.31)<br>3.79 (1.8) | 3 (LOW)<br>5 (LOW) |
| B5 | | |
| B7 | | |
| B8 | 4.94 (0.8)<br>4.53 (1.26) | 3 (LOW)<br>3 (LOW) |
| Mean | 4.26 (0.75) | |

TABLE 1

SATISFACTION AND EJOURNAL USAGE

## Global Results

Thirty questionnaires were distributed during the last planned face-to-face session in automatic control. Twenty-two of them were returned among which 3 were anonymous. On the 22 students that returned the questionnaire (73%), 6 were also involved in the biomechanical hands-on. The questionnaire also asked if students would volunteer for an interview. Only 4 students answered positively, 1 of them being also involved in biomechanics. The results of the questionnaire are displayed in Table 1.

These results only apply to the use of the environment for the hands-on experimentation sessions in automatic control. The first three lines of the table correspond to the 3 anonymous questionnaires that could not be associated with groups. The environment ranks positively in terms of satisfaction as 72% of the students (16 out of 22) give a mean satisfaction score superior to the neutral point of the scale (4). Question 9 ('The online course gives error messages that clearly tell me how to fix problems') received the worst ranking with a mean satisfaction of 2.65 (S.D. = 1.15) showing that the usability of the environment could greatly be improved by adjusting the help and error messages.

The responses to all the questions are combined together and the overall frequency for each point of the scale are plotted in Figure 1. The plot suggests the existence of two sub-groups in our population of students as it looks like the superposition of two Gaussians. To confirm that idea we have tried to check if some of the factual questions could help to define these sub-groups. Familiarity is not an issue as only 4 students out of 22 declared they didn't feel familiar with computers. The type of operating systems was not an issue either as nearly all students were accustomed to the same operating system (19 out of 22 are Windows users).

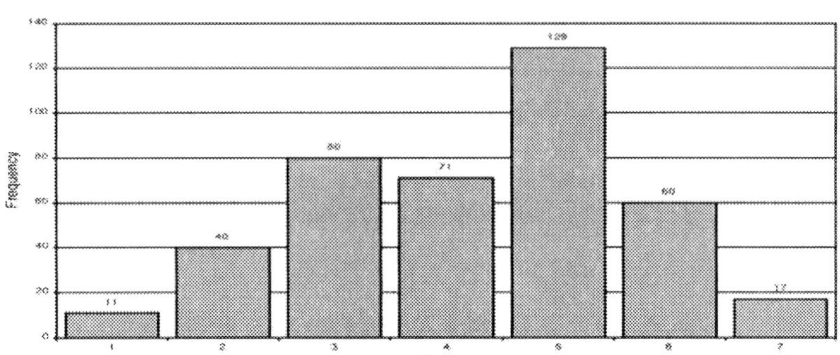


FIGURE 1

OVERALL SATISFACTION RATINGS ALL QUESTIONS MIXED

The last explanation we could check was linked to the eJournal perceived usage. By grouping students according to their ranking of the eJournal usage, we have categorized to 2 sub-groups. A group of high eJournal users (who put the eJournal in 1st or 2nd position in the question: 'attribute a rank to each component of the environment reflecting the time you think you have spent with it') and a group of low eJournal users (who put it in 3rd, 4th or 5th position). Thus we obtained 9 high eJournal users and 13 low eJournal users. One could see the category in Table 1. Both groups differ in their mean satisfaction with high eJournal users ranking it at 4.46 (S.D. = 0.85) and low eJournal users at 4.13 (S.D. = 0.64). That would tend to show that the eJournal had a positive effect on a student's satisfaction. However there are not enough students to check if this result is statistically meaningful.

A total of 1409 fragments have been put in all the eJournals (the 3 fragments put by the group B6 were not taken into account). Figure 2 shows the number of fragments created by all groups. This gives a mean of 93 fragments per group or, if we consider the 7 modules and the pre-lab, about 13 fragments for each module.

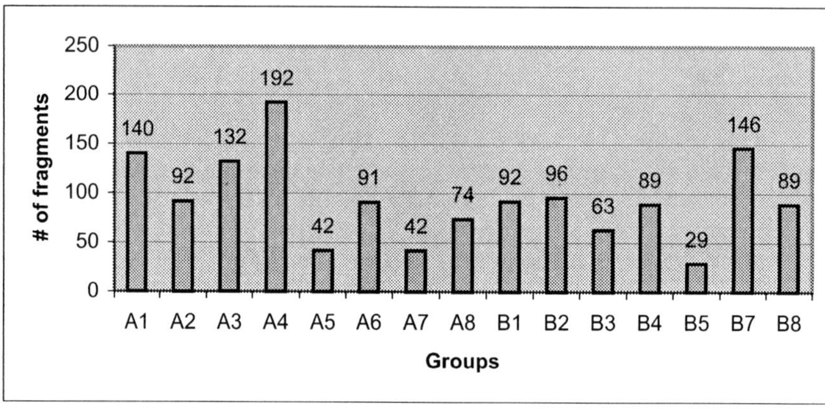


FIGURE 2
NUMBER OF FRAGMENTS CREATED BY ALL GROUPS

Figure 3 shows the percentage of intra-fragments and flexible fragments for all groups. On average, 76.67% of fragments were produced within the environment. During flexible learning modalities 26.29% of the fragments were created. Thus, with a lot of precautions we can interpret that percentage as a measure of the utility of the environment for performing the tasks asked of the students.

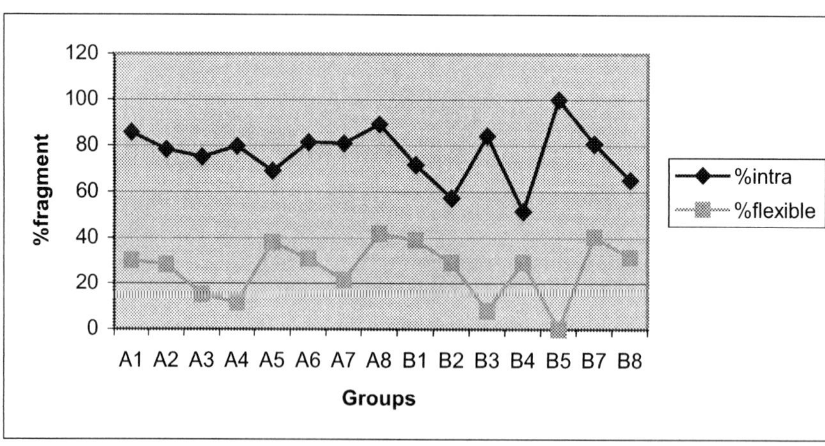


FIGURE 3
PERCENTAGE OF INTRA-FRAGMENTS AND FLEXIBLE FRAGMENTS OF ALL GROUPS

The log analysis also shows that the annotation functionalities of the environment haven't been used. Only 11 comments have been produced by 8 different groups. Thus we didn't analyze the logged annotations.

## Evolution of Fragment Creation over Time

Figure 4 plots the daily evolution of the creation of fragments. It shows a regular pattern with two consecutive peaks every week. These peaks correspond to the planned face-to-face sessions on Thursday and Friday, with an interruption for Christmas holidays. The fragments added in the eJournal outside of these face-to-face sessions represent 26.29% of all fragments. It's significant to see that every week some students added fragments one or two days before the face-to-face sessions.

The number of fragments created on a weekly basis shows a peak the second week and then a decrease until Christmas holidays, with a peak just the week before the holidays. Then it increases again during the last 4 weeks in January until the practical 'lab' examination. The hourly evolution shows that most of the fragments have been created between 2 pm and 6 pm (thus during face-to-face sessions). However it's interesting to notice that one of the groups created 3 fragments around midnight and 5 more fragments at 3 am on another day.

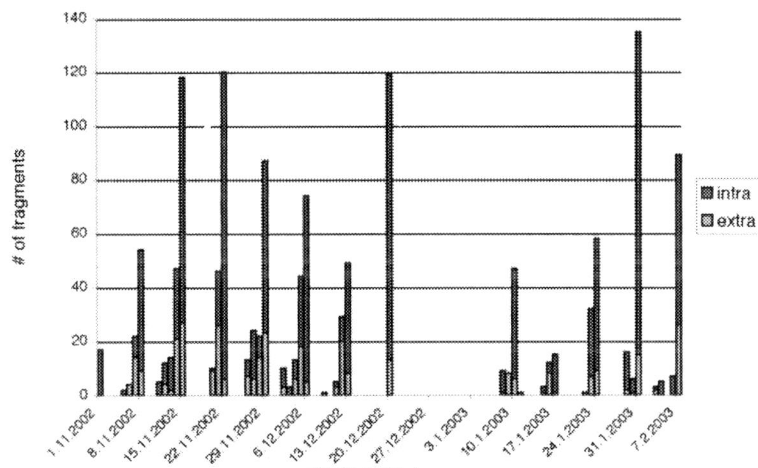


FIGURE 4

EVOLUTION OF FRAGMENT CREATION OVER TIME

## Characterization of work

The proportion of extra-fragment or flexible-fragments didn't change much over time, thus suggesting the students didn't change their working pattern. We have examined the type of fragments created during flexible work to determine the type of work students were performing without the presence of assistants. The graph of Figure 5 shows that 47% of these fragments are text files that most likely correspond to the writing of a report for the 'pre-lab' activities. The 37% for image files and 4% for mathematical scripts suggest that students did use other software tools and imported their results, to share images for writing a report for instance, or to execute the scripts in the mathematical console provided with the environment.

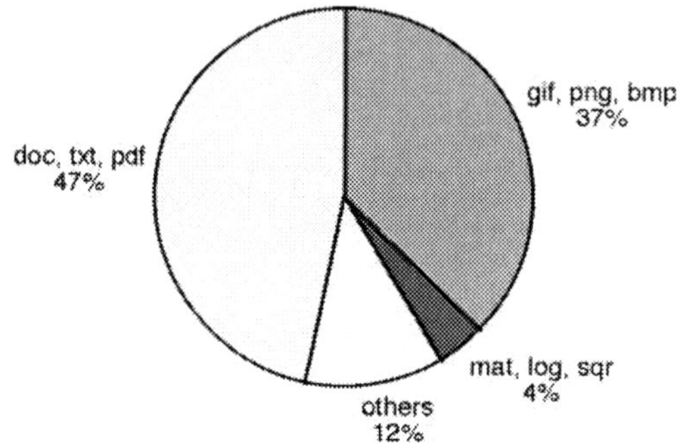


FIGURE 5
EXTERNAL FRAGMENT TYPES (339 EXTERNAL FRAGMENTS)

### Interviews

We recorded a total of 176 minutes of individual interviews with four volunteer students. The interviews were animated by the research assistants (also the developers of the eJournal). Each student received a small gift at the end of the interview for his participation and, we hope, encouragement toward participation in our future studies. The analysis of the interviews gave us 3 positive comments, 11 negative comments, 7 descriptions of possible bugs, 16 comments describing the limitations of the environment, 12 suggestions of improvements, 35 remarks about students' methodology and 4 false ideas students had about the behavior of the environment. Though the number of students was small, the remarks provided us with quite useful information for further developments and studies.

## DISCUSSION

The results above have helped us achieve most of our evaluation objectives. We find the results satisfactory concerning the 'acceptance goal' as the questionnaire shows that most of the students were satisfied. This was confirmed by the interviews, and we are able to identify 1 real bug and write 16 recommendations concerning the user-interface and some missing or unclear basic functionalities from the analysis of the interviews. However as our questionnaire was not anonymous, we question if it had an influence on students' answers. The mean satisfaction is also less than 1 unit above the neutral scale point, thus suggesting much room for improvements.

The 'participation goal' is also reached as all the groups created a significant number of fragments. As a corollary, we think that the 'participation goal' contributes to the 'acceptance goal' as evidences of the use of the prototype such as the creation of fragments shows that it has been accepted. However to be more precise this would require that we estimate what is the standard number of trials and errors necessary to complete a module and hence the fragments. It would be interesting to compare this with traditional hands-on experimentation without the computer environment, to assess more precisely the effect of computer on participation. The interviews also gave us interesting clues about the 23.33% of extra-fragments not covered by the environment. Some of them were created with MATLAB that was preferred as a mathematical tool instead of the mathematical console provided.

The 'flexibility goal' is also reached as the 26.29% of flexible-fragments created shows that some flexible learning modalities have been chosen, and this is independent from our estimation that the students had enough time to perform all the work in face-to-face modalities. However, interviews with students, and Figure 5, show that most of the flexible work was done on the 'pre-lab' part of the hands-on activities, which dealt with mathematical modeling work and didn't involve the direct manipulation of the physical devices.

Finally, the 'collaboration goal' seems to have been the least satisfied. Once again the interviews provide some explanations. First, all the students had a printed version of the experimental protocol where, if they preferred, they can take notes on paper. Second, as lots of the fragments were created under face-to-face conditions in which students could directly talk to each other about these fragments. Third, many annotations required the editing of equations which is not supported at all in our annotation system. Some students also told us that they used the names of the fragments to get some clues about their content and thus didn't need to annotate it. However, this failure to show evidence of collaborative work through the annotations contained in the environment does not mean there was no computer-mediated collaboration at all. In fact we haven't logged enough information to determine if students did look at each other's fragments and if they used the 'copy/move fragments' functionality provided with the eJournal. We plan to do this in future evaluations and to analyze these logs with social network analysis instruments to determine if some computer-mediated patterns of group work emerge [12,13].

Some factors may explain why participation and flexibility had a greater success in our environment than collaboration. This is related to the degree to which certain students' behavior has been 'scripted' either in the tools or in the instructions given to the students. For participation, the export and import functionalities were easily accessible from the software components of the environment. These functionalities provided the students with a continuous interaction mechanism, which improved significantly the hands-on work of students when compared to the discontinuous interaction mechanism in the earlier versions of the prototype [8]. For flexibility, we think that the division of the hands-on into 'pre-lab' and 'lab', where the 'pre-lab' contained all the mathematical and theoretical work that didn't require direct manipulation of the physical device, favored flexibility. This suggests that we could find a similar way, either technical or instructional, to also 'script' collaborative behavior if necessary.

## CONCLUSION

Many dimensions of e-learning environments are subject to an evaluation. For each of these dimensions, one has to choose a methodology and to select the proper evaluation instruments. In this paper, we have presented a use-case study based on a web-based environment for carrying out flexible hands-on experimentation. As the goal of this environment is to supplement traditional laboratory experimentation carried out in the laboratory, we had to check first that the new environment is accepted by the students. As the vision underlying this project is to develop participative, flexible and collaborative learning, we also wanted to measure to which extent we had introduced these dimensions into the pedagogical scenario of hands-on learning. The choice of some classical usability engineering methods have proved useful when combined with clear measures of students' activities based on the use of a laboratory journal for collecting and sharing experimental data. We hope that this contribution has illustrated the challenges involved in evaluation of a web-based education environment. We think this is a necessary step towards the development of standard practices for evaluating e-learning applications. Finally, the results provide some useful indications on how to sustain flexible learning deployment and acceptance.

## ACKNOWLEDGEMENTS

The eMersion project is funded by the Board of the Swiss Federal Institutes of Technology in the framework of its New Learning Technologies program. All the work described in this article would not have been possible without the support of the eMersion staff: Frédéric Geoffroy, Yassine Rekik, Christophe Salzmann, Gamaliel Amaudruz and Karim Zeramdini.

## REFERENCES

1. Nielsen, J., *Usability Engineering*, Morgan Kaufmann Publishers Inc, San Francisco, 1993.
2. Appelt, W., Mambrey, P., "Experiences with the BSCW Shared Workspace System as the Backbone of a Virtual Learning Environment for Students," Proceedings of the World Conference on Educational Multimedia, Hypermedia and Telecommunications (ED-MEDIA 99), Seattle, USA, June 1999, pp. 1710-1715.
3. Brush, B., Bargeron, D., Grudin, J., Borning, A., Gupta, A., "Supporting Interaction Outside of Class: Anchored Discussion vs. Discussion," Proceedings of Computer Support for Collaborative Learning (CSCL 2002), http://newmedia.colorado.edu/cscl/, 2002, visited on January 2003.
4. Crozat, S., Hu, O., Trigano, P., "A Method for Evaluating Multimedia Learning Software," IEEE International Conference on Multimedia Computing and Systems, Vol. 1, Florence, Italy, June 1999, pp. 714-719.
5. Zaphiris, P., Zacharias, G., "Usability Evaluation of an On-Line Modern Greek Language Course," Proceedings of the 8th Panhellenic Conference on Informatics, Nicosia, Cyprus, November 8-10, 2001.
6. Gillet, D., Fakas, G., "eMersion: a New Paradigm for Web-based Training in Engineering Education," Proceedings of 2001 International Conference on Engineering Education, ICEE-2001, Oslo/Bergen, Norway, August 6-10, 2001, pp. 10-14.
7. Gillet, D., "Towards Flexible Learning in Engineering Education," Chap. 11, Aung, W. et al. (eds.), Innovations 2003: World Innovations in Engineering Education and Research, Int. Network for Eng. Ed. and Res. (iNEER), Arlington, VA, 2003, pp. 95-102.
8. Nguyen-Ngoc, A. V., Gillet, D., Rekik, Y., Sire, S. "Sustaining the continuity of interaction in Web-based experimentation for engineering education," Proceedings of the International Conference on Computer-Aided Learning in Engineering Education (CALIE 2004), Grenoble, France, February 16-18, 2004.
9. Horstmann, T., Bentley, R., "Distributed Authoring on the Web with the BSCW Shared Workspace System," ACM Standards View, Vol. 5, No 1, March 1997, pp. 9-16.

10. James, L. R., "IBM Computer Usability Satisfaction Questionnaires: Psychometric Evaluation and Instructions for Use," International Journal of Human-Computer Interaction, Vol. 7, No 1, 1995, pp. 57-78.

11. Perlman, G., "Web-Based User Interface Evaluation with Questionnaires," http://www.acm.org/~perlman/question.html, visited on November 2002.

12. Scott, J., Social Network Analysis: A Handbook, 2nd ed., Sage, London, U.K, 2000.

13. Martínez, A., Dimitriadis, Y., Rubia, B., Gómez, E., Garrachón, I., Marcos, J. A., "Studying Social Aspects of Computer-Supported Collaboration with a Mixed Evaluation Approach," Proceedings of Computer Support for Collaborative Learning (CSCL 2002), http://newmedia.colorado.edu/cscl/, 2002, visited on January 2003.

**Denis Gillet** is MER (Associate Professor) at the Swiss Federal Institute of Technology in Lausanne (EPFL). He received the Ph.D. degree in Control Systems in 1995 from EPFL. His research interests include optimization of dynamic systems, new learning technologies, augmented reality for sustainable interaction, and real-time Internet services. Dr. Gillet received the 2001 iNEER (International Network for Engineering Education and Research) Recognition Award for Innovations and Accomplishments in Distance and Flexible Learning Methodologies for Engineering Education.

**Stéphane Sire** is a research engineer at IntuiLab, a French company that is inventing a new user interface prototyping environment for multi-modal interaction. He earned an engineering degree in Computer Science in 1993 from the ENSIMAG, Grenoble, France and a Ph.D. in 2000 from the Université Toulouse 1, France. In 1993 he was a summer intern at ALTERA, CA, USA. During the last three years he has been a post-doc in several European institutes: the EPFL in Lausanne, Switzerland, the CWI in Amsterdam, Holland and the University of Fribourg, Switzerland. He has participated in different projects on document engineering, e-learning environments and Human-Computer Interaction.

**Anh-Vu Nguyen-Ngoc** earned his B.Sc. Degree in Computer Science from the Vietnam National University, Ho Chi Minh City in 1997. Mr. Nguyen-Ngoc received an award from the Rector for his outstanding academic results at the University. In 1999, he was awarded a two-year Swiss federal scholarship to carry out research activities and to take part in Pre-Doctoral School, Swiss Federal Institute of Technology in Lausanne (EPFL). In July 2001, he obtained a Pre-Doctoral School Degree in Communication Systems, EPFL. Since August 2001, he has been a research assistant and a Ph.D. candidate at the School of Engineering, EPFL. His research interests include collaboration/interaction systems, Computer Supported Cooperative Work (CSCW), Human-Computer Interaction (HCI), and web engineering.

# Chapter 29

# Synchronous Tele-Tutorial Support in a Remote Laboratory for Process Control

ANDREAS BÖHNE, NILS FALTIN and BERNARDO WAGNER
*Learning Lab Lower Saxony (L3S), University of Hannover, Expo Plaza 1, 30539 Hanover, Germany. E-mail: boehne@learninglab.de*

*Traditional laboratories in engineering education are supervised by a tutor. Students can get feedback on their progress and hints when in need of help from the tutor. Often, the tutor will grade the laboratory work. This support is also needed when students work with a remotely controlled laboratory. We believe that live audio-video is a natural way for tutors and students to communicate over a distance. The tutor should be able to see the students' computer screen to understand the work situation. We have built a remote laboratory to teach remote maintenance and real-time Java programming of embedded systems. It supports synchronous audio-video communication with the tutor. Application sharing is provided by an external tool. The laboratory is used by students in electrical engineering and computer science. The educational concept is problem-based and self-directed learning. Students work with the experiment in groups of two and write a program that controls the deflection unit of a laser system and generates a picture. They observe the generated picture and the laboratory devices through a user-controlled web cam. We want to make students self-directed in their learning so that they get maximum responsibility and control over their learning process. The ability to learn in a self-directed mode is an important factor for success in lifelong learning. In a comparative evaluation, we investigate how well self-directed learning works in our remote laboratory setting. We have developed two instructional designs for the evaluation. In the self-directed setting, the student group sets its own learning goals after some predetermined tasks. The tutor acts as e-coach or e-moderator. In the teacher-directed setting, the tasks are given and the tutor acts as e-instructor. We have evaluated student motivation, task success, communication behavior and the consulting effort of the tutor in comparing these two settings. Remote assistance has proven to work very well, with application sharing and audio transmission being the most important communication media. The tutor support style does not have a statistically significant influence on learning.*

## INTRODUCTION

A major goal in engineering education is that students acquire problem-solving and creativity strategies so that they become capable of constructing technical systems. Such strategies can be learned by working on small problems and construction assignments in a problem-based learning environment. Laboratories are a typical example of a problem-based learning setting. They permit the application and testing of theoretical knowledge in practical learning situations.

Reinmann-Rothmeier and Mandl propose the following guidelines for problem-based learning environments, which they derived from the learning theories of anchored instruction, cognitive flexibility and cognitive apprenticeship [1]:

- **Situated and authentic learning**. Letting students solve realistic problems will enable them to apply their knowledge in practical situations. Authentic problems will motivate students.
- **Learning from multiple perspectives and contexts**. Solving several related problems which all require the same basic skill will make it more likely that the skill can be applied in a different practical situation.
- **Learning in a social context.** Learning and problem-solving should be done in student groups. If students have experts as models they can acquire skills, knowledge and orientation.
- **Learning with instructional support**. Complex problem-solving tasks without instructional support will often demand too much from the students and will lead to ineffective learning. Many students need instructions and help for solving problems. The learning environment shall provide knowledge for solving the problem.

Because instructional support is an important element, especially in a problem-based learning setting, remote laboratories should provide such support for students. In typical local laboratories, students get assistance from a local tutor. We think that in a remote laboratory, a tele-tutor, who can communicate by synchronous communications tools with the students, should be an element of instructional support.

## INSTRUCTIONAL SUPPORT IN REMOTE LABORATORIES

The main pedagogical function of a tutor in a local laboratory is to give information and assistance in questions concerning technical problems. He also gives feedback to motivate students, assesses the learning progress of the learners, organizes the learning process, and sets time limits.

Other tasks of a tutor in a local laboratory can be the technical maintenance of the laboratory equipment and the organization of the laboratory. He sometimes develops the laboratory tasks and the documentation for the experiment.

We think that these functions of a tutor in a local laboratory also need to be provided in a remote laboratory. This requires a new design for instructional support because in a remote laboratory students and the experiment are not in one room. Three different solutions for this problem are possible:

**Local tutor:** A tutor, who is in the same room as the students, supports the students. Only the experimental equipment is remote. The advantage of this solution is that the communication and collaboration between the tutor and the students needs no extra technical equipment or bandwidth. However, money must be appropriated for the personnel costs of the tutor and, in the case of laboratory sharing between institutions, tutors must be trained. A distributed scenario, where students from different locations work with a remote experiment, is not possible with this solution because no communication tools are provided.

**No synchronous tutor:** One possible solution for the instructional support in a remote laboratory could be to decrease consulting with the tutor. The students receive no support by a tutor during the laboratory session in this solution. The web-based laboratory environment must fulfill most of the above-mentioned functions of instructional support. The documentation of the laboratory must be very comprehensive and understandable so that students are able to find answers and solutions for their questions and problems on the web pages for the laboratory. An important advantage is that the students can work on the laboratory tasks when the tutor is not available. The costs are comparatively low, because no personnel budget is needed. However, a remote laboratory without any tutor is very unrealistic. The laboratory equipment must be maintained sometimes. Asynchronous support via e-mail must be provided for helping students when they cannot find answers to their questions in the laboratory documentation. The disadvantage of asynchronous support is that the students must accept a time delay for getting an answer to their questions. If the delay is too long, they could lose their motivation [2].

**Remote tutor:** A tutor supports the students over a distance via synchronous and asynchronous communication tools. Synchronous tools like videoconferencing and application sharing allow flexible and fast help for the students. However, synchronous communication restricts the students' time flexibility. Additionally, bandwidth is necessary and technical equipment must be integrated in the learning environment. Therefore, alternating asynchronous and synchronous phases can help to combine the advantages of these two communication forms. The synchronous support can be used when the students work with the laboratory equipment. Problems that arise can then be solved immediately using videoconferencing, chat and application sharing. The amount of interactivity can be reduced during the preparation and postprocessing of the laboratory session. A web-based laboratory environment with synchronous and asynchronous communication tools makes it possible for several students from different locations to work together on the tasks of a remote laboratory.

In the remote experiment, described below, the learners can contact a remote tutor. Videoconferencing, text chat and application sharing are provided in the web-based laboratory environment.

## THE REMOTE EXPERIMENT

We have developed a remote experiment for picture generation by laser deflection. In this experiment, students write embedded Java programs to generate laser pictures and animations.

The laser system of the experiment consists of a green 3mW laser beam, which is deflected by the two mirrors of a galvanometer scanner in the X- and Y- directions (Figures 1 and 2). The scanner can move the beam to 30,000 positions per second. The deflection is controlled by two analogue scanner inputs, which are in turn driven by DA converters controlled by an I/O card in an embedded system.

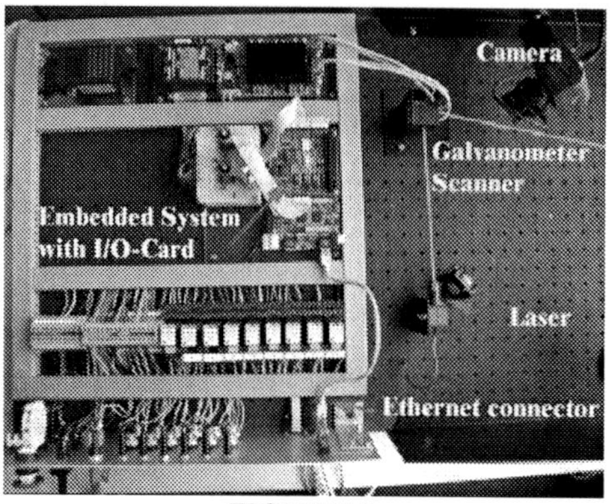


FIGURE 3
LASER EXPERIMENT COMPONENTS

The galvanometer scanner and the laser can be turned on and off by the digital outputs on the I/O card. An analogue input on the I/O card measures the voltage signal of a photosensor that is placed on the canvas. When the laser beam hits this sensor, the voltage output of the sensor increases. A window discriminator allows two threshold values to be set so three different voltage areas can be distinguished by the digital inputs on the I/O card.

The embedded system is a PC 104 single board computer, designed to meet the needs of embedded applications requiring real-time control with web-enabled capability. A Java-based real-time operating system from Esmertec called "JBED" runs on the embedded system. When the board is booting, it loads a boot file from the laboratory server on which the experiment runs. This file contains Java classes that are important for running the system. For example, the IO-Manager allows comfortable control of the I/O card. And the DynamicClassHandler Java servlet allows dynamic loading of Java bytecode from the laboratory server as needed. Once on the embedded system, the bytecode is compiled into native machine code by the JBED software. This ahead-of-time compiling ("AOT Compiling") improves the run time performance of Java programs.

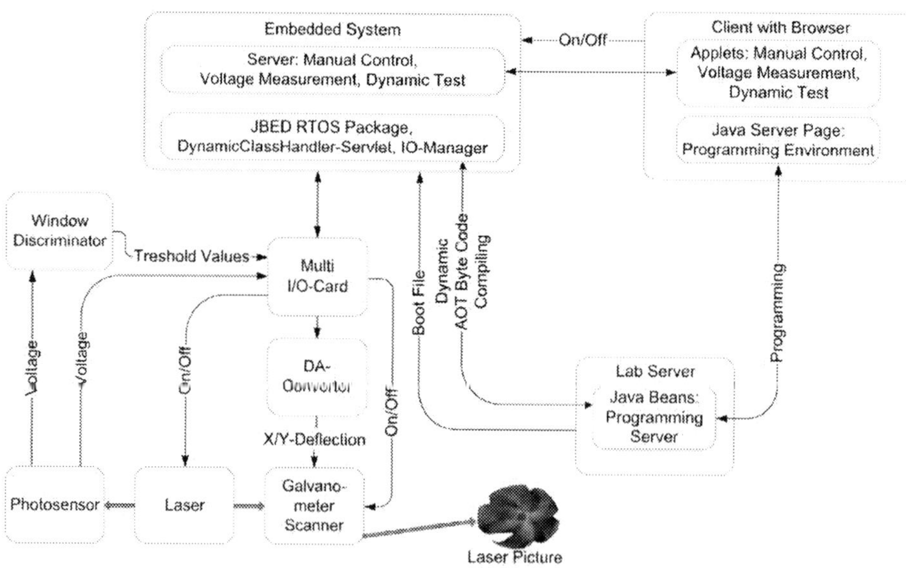


FIGURE 4
BLOCK DIAGRAM OF THE LASER EXPERIMENT

The embedded system uses a real-time earliest deadline first scheduler. When a new task is installed, CPU time and deadline must be provided. Jobs without real-time requirements can be performed with standard Java threads.

Java Programs can be developed in a web-based programming environment that is implemented using Java Server Pages (JSP). The Java server page communicates with a programming server on the laboratory server developed with Java Beans, which in turn communicates with the DynamicClassHandler Servlet.

So, Java programs developed in a browser anywhere in the world can be executed with real-time capabilities on the embedded system to control the I/O card and to generate laser pictures.

The web-based programming environment (Figure 3, middle section) allows editing of Java program files in the editor window. Files can be loaded and saved on the laboratory server with the buttons and input fields in the upper left corner of the editing section. The upper right corner of the programming environment allows Java programs to be compiled and run. Java files are compiled on the laboratory server (Compile button) and compiler messages are displayed in a separate window. To start a Java method, the user asks for a list of Java methods in the file (Inspect button), chooses a method (Select button), enters actual parameter values in a separate window and calls the method (Start button). If the program works correctly, the laser system will generate a picture on the canvas. A separate console output window facilitates the debugging of Java programs.

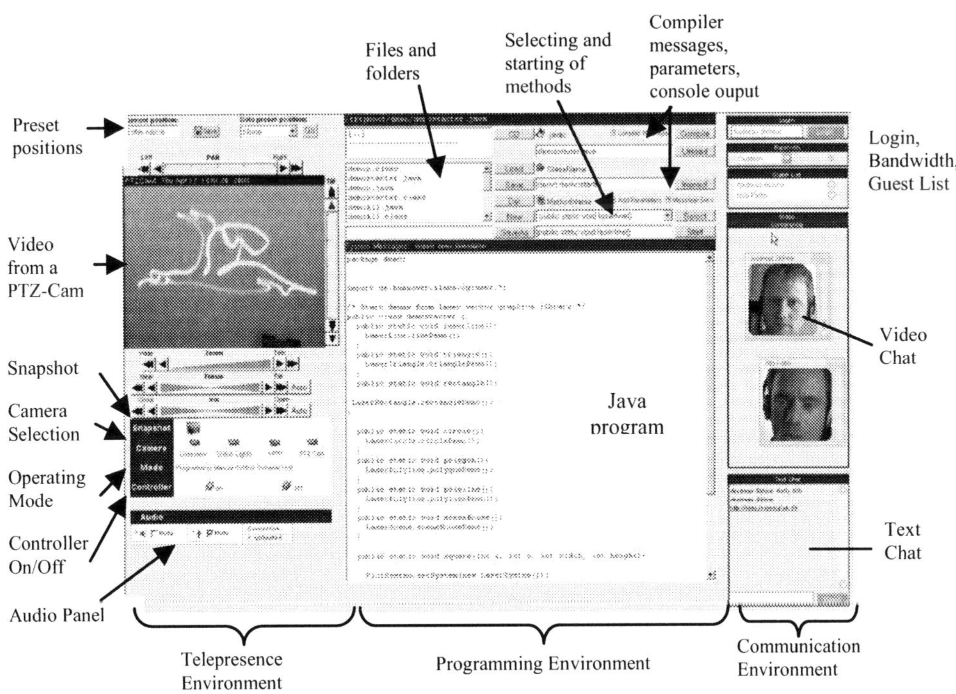


FIGURE 3
WEB-BASED LABORATORY ENVIRONMENT

The left part of the window in Figure 3 provides students with remote access to the laboratory, and is thus a substitute for being physically present in the laboratory (Telepresence). Students can view the output of one of four cameras: three fixed cameras show live video of the canvas, the status lights on the devices, or the moving galvanometer scanner; a fourth camera (PTZ-Cam) allows users to pan, tilt and zoom in on details. A microphone transmits the sound of the operating scanner via an audio server. Users can also switch the embedded system on and off, and take a snapshot of the laser picture that has been created.

The right part of the window in Figure 3 allows different users to communicate with each other over a distance. They can communicate via audio, video and text chat after logging in with a username. Bandwidth usage can be tailored to the bandwidth available. The client side of the communication environment was implemented with Macromedia Flash MX. The communication server is based on Macromedia Flash Communication Server MX. We chose this software solution based on the following criteria:

- Video chat of more than two users is supported.
- Audio quality is acceptable.
- The user need not install a separate communication application. The developed communication application runs in all standard web browsers when the Flash plug-in is installed.
- It allows communication elements to be integrated in a custom-built web page.

Application sharing is provided by an external tool. We are currently testing NetMeeting and VNC (http://realvnc.com) for this purpose. VNC is platform-independent, but only allows complete desktop sharing and cannot be restricted to only a certain application or browser window.

Instead of the "Programming" operating mode shown in Figure 3, the user can select the "Manual Control" or "Dynamic Test" mode. The software components of these modes are implemented by Java applets that communicate with corresponding server applications on the embedded system (see Figure 2). In the "Manual Control" operating mode, the user can deflect the laser beam using two sliders or text input fields in the X- and Y- directions. Furthermore, the current voltage output of the photosensor and the status of the window discriminator are displayed. In the "Dynamic test" mode the users can measure the dynamic characteristic of the laser system.

We have developed a software library for vector-based picture and animation generation in Java. The library eases the development of laser pictures and animations. It supports geometric transformations like translating, scaling and rotating. The library contains classes to create basic shapes like point, line, triangle, rectangle, circle, polyline and polygon. Animations consisting of different frames of laser pictures can be designed. Moreover, animations can be generated with real-time properties.

## THE REMOTE EXPERIMENT IN THE PROCESS CONTROL LABORATORY

The remote experiment "Picture generation by laser deflection" is one of eight experiments in the laboratory for process control at the University of Hannover and is the only one where students are assisted by a remote tutor. Students learn and apply the basics of process control including industrial automation systems, bus systems for process automation, programming of industrial automation systems, hardware-oriented design, and remote maintenance. In the summer term of 2003, nineteen electrical engineering and computer science students in their third year of study took part in this laboratory. Students worked in groups of two or three and performed one of the eight experiments every week. They had to prepare for each experiment by reading documents at home and had to complete a substantial part of each experiment to pass the laboratory.

The laser experiment was organized in two phases: preparation for the experiment and actual performance of the experiment. The preparation for the experiment was supported by asynchronous communication between the tutor and the student. Laboratory documents were provided electronically and student solutions were placed on a web-based collaboration server. We used BSCW (Basic Support for Collaborative Work, http://www.bscw.de) for this purpose. The experiment web site included all necessary information for the lab preparation and the lab tasks for each student group. During the lab preparation, students familiarized themselves with the hardware and software documentation. They completed parts of a Java program for the embedded system. Since we intended to perform an educational evaluation, we wanted to give all students equal working conditions; therefore, we had the student groups come to the University where they had four hours to perform the experiment.

## EVALUATION APPROACH

In our educational research, we wanted to clarify for which students and under what conditions each pedagogical role (e-moderator or e-instructor) is suitable for a remote tutor in a remote laboratory. A tutor can guide the students more or less straight through the problem-solving process in a remote laboratory. If he acts as e-moderator, the students can learn in a self-directed way. The e-moderator stimulates the meta-cognition of the learners and advises them on their choices of learning goals and action goals. If the tutor prefers the role of the e-instructor, he controls the learning process. He acts as knowledge mediator and tries to convey his knowledge to the students.

We investigate this research question in the context of self-directed versus teacher-directed learning in a remote laboratory. Self-directed learning seeks to put the learner as much as possible in control of the learning process. The freedom of choice can refer to different dimensions like learning goals, learning place and time and learning strategy [3]. Especially for learners in continuing education, a self-directed didactic approach in remote laboratories has the advantage that they can choose the learning goals and tasks according to their actual needs. A teacher-directed concept could be more suitable for learners who have less initial knowledge, while a self-directed learning setting may demand too much from them.

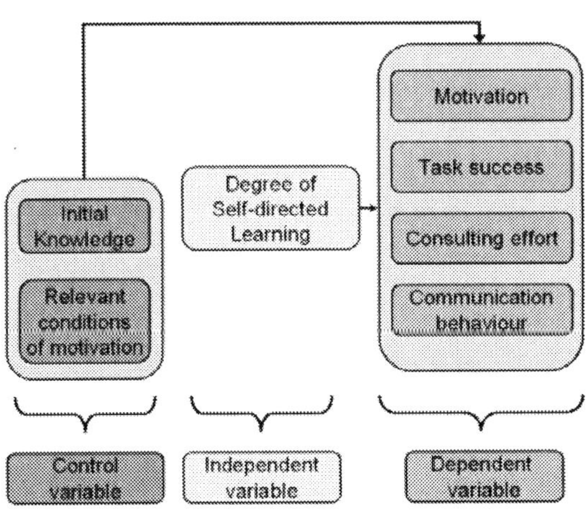

FIGURE 4

VARIABLES OF THE RESEARCH QUESTION

We performed a comparative evaluation study in the laboratory for process control described above. We followed the evaluation methodology of Borz/Döring and distinguished independent, dependent and control variables [4]. Our main research question was: How does the degree of self-control, considering initial knowledge, influence motivation, task success, consulting effort and communication behavior? In addition, we measured conditions for the type of learning motivation as a control variable (Figure 4).

Regarding the **degree of self-directed learning,** we have implemented two different educational settings:

- **Self-directed**. High degree of self-direction on the part of the students in terms of learning goals, tasks, problem-solving process and tutor role. The tutor acts as an e-coach or e-moderator. He supports the students in learning to learn, moderates group discussions and answers the students' questions.
- **Teacher-directed.** Low degree of self-direction in terms of learning goals, tasks, problem-solving process and tutor role. The tutor acts as an e-instructor: He explains the tasks, guides the group in the learning process along the structure of the subject rather than the actual task (i.e. subject-oriented) and answers the questions in a subject-oriented way.

In the teacher-directed setting, students complete parts of a Java program for the embedded system during the laboratory preparation. Students need this program in the practical performing of the laboratory. While performing the experiment, they work on four tasks. First, they get to know the hardware and software by starting different Java methods that already exist to generate pictures. In the second task, they measure the dynamic characteristic of the laser system. The results from this task are helpful for the solution of the third task, in which students develop a real-time Java program for finding the X- and Y-position of the photosensor. The problem-solving process for this task is described in detail, and the tutor gives the students subject-oriented hints. In the fourth task, the students create a laser picture by using the laser graphic library. In this setting, the tutor acts as e-instructor, and might, for example, explain the theory of real-time programming. Students must contact the tutor after finishing certain parts of the tasks.

The self-directed setting gives the students more freedom of choice, as they can set their own learning goals. They have the choice between a focus on real-time programming (finding the coordinates of the photosensor) and a focus on object-oriented programming in Java (generating laser pictures by using the laser graphic library). The laboratory documents are not specific, and describe the tasks only broadly, so students can put the tasks in more concrete terms. The tutor acts as e-moderator, and supports the students in learning to learn. For example, when the students have questions, the tutor gives them hints on where they can find appropriate answers to their questions in the documentation.

We have defined the dependent and control variables in the following way: With **communication behavior** we describe the media that are actually used by the students during the experiment. The **consulting effort** is the time effort of the tutor to support the students. We differentiate between the support for technical questions and for questions concerning the content, learning process and moderation of the collaborative work. The **task success** is evaluated by the tutor. He assesses the results (e.g.: laser picture, Java program) of the group by comparing the results with other groups and with the pre-determined goals. We distinguish six different types of **motivation**: amotivated (referring to a person who lacks motivation), extrinsic, introjected, identified, intrinsic and interested motivation. The **relevant conditions of motivation**, which we measure, are perceived content relevance, instruction quality, support for social-relatedness, support for competence and support for autonomy. This classification of learning motivation types and relevant conditions of motivation is based on the Self Determination Theory (SDT) from Deci, Ryan [5] and the motivation theory from Prenzel et al. [6]. The **initial knowledge** factor is measured with a questionnaire before the experiment, because we consider it to be an important factor in a self-directed learning setting. For example, students who only have a little knowledge of Java programming will have problems with self-directed learning and will need special support from the tutor. Students with experience in programming will be able to work mostly independently.

We use qualitative and quantitative methods for the acquisition of the necessary data. In a pretest, we test initial knowledge relating to the laser experiment software and hardware and Java programming experience. After the laboratory, students answer questions on technical quality and acceptance of the tele-tutorial assistance. For the measurement of the different types of motivation and the relevant conditions of the learning motivation types, we use a validated questionnaire developed by Prenzel [6]. During the lab session the tutor takes notes on the consulting effort and the communication behavior. This is complemented by logging communication for classification after the laboratory session.

## FIRST RESULTS AND CONCLUSIONS

We have performed the evaluation with eight groups of students. A first analysis of the results of our study indicates that the students were content with the remote tele-tutorial support. Most students did not miss a local tutor (M=3.58, 95% confidence interval=+-0.54; scale: 0=strongly disagree, 5=strongly agree). Students and the tutor mainly used videoconference and application sharing for communication and collaboration. Students rated audio chat (M=4.68, ci=+-0.51) and application sharing (M=4.11, ci=+-0.46) as most useful, and video of the tutor (M=1.58, ci=+-0.78) and text chat (M=1.16, ci=+-0.59) as not as important. The tutor rated the video picture of the students as useful in gaining information about the students' working state. Contrary to our expectations, there was no statistically significant relationship between motivation and the task success for students working either in the self-directed setting or in the teacher-directed setting.

Our conclusions from these results are: Remote assistance proved to work very well, with application sharing and audio transmission being the most important communication media. However, in order to answer the question of whether the degree of self-directedness influences the motivation and learning results of students in a remote laboratory with tele-tutorial assistance, further research must be done with a greater number of evaluation participants. New studies should take into account the limited influence of instructional method on learning performance. A meta-analysis of 1763 empirical studies found an average Pearson correlation of 0.17 [7].

In this paper, we are only able to publish the first results of our study. In the near future, we will publish further results of our study and a detailed discussion of these results [8].

## OUTLOOK

In our laboratory, we focus on synchronous communications between a remote tutor and students during the laboratory session. When remote laboratories are used on an international scale, asynchronous communication will be needed. Tutors will assign tasks to students, support them in laboratory preparation and grade laboratory reports using asynchronous communication. Also, organizational questions have to be resolved and business models have to be established for usage of remote laboratories on an international scale. Further research is needed to answer these questions.

## ACKNOWLEDGEMENTS

We would like to thank Christoph Richter from the Learning Lab Lower Saxony evaluation team for his support in designing the educational evaluation. Nils Hagge and Jörg Tuttas from the University of Hanover supported us by developing the programming environment and basic embedded system software.

## REFERENCES

1. Reinmann-Rothmeier, G., and Mandl, H., "Unterrichten und Lernumgebungen gestalten," Chap. 13, Krapp, Weidemann (ed.), Pädagogische Psychologie, Beltz, 2001.
2. Gillet, D., "Towards Flexible Learning in Engineering Education," Chap. 11, Aung, W. et al. (ed), *Innovations 2003: World Innovations in Engineering Education and Research,* Int. Network for Eng. Edu. and Research (iNEER), 2003, pp. 95-102.
3. Böhne, A., Faltin, N., and Wagner, B., "Self-Directed Learning and Tutorial Assistance in a Remote Laboratory," Interactive Computer Aided Learning Conference (ICL), Villach, Austria, September 25-27, 2002.
4. Bortz, J., Döring, N., "Forschungsmethoden und Evaluation für Human und Sozialwissenschaftler," Springer, 2002.
5. Deci, E.L., and Ryan, R.M., *Handbook of Self-Determination Research,* University of Rochester Press, 2002.
6. Prenzel, M., Kristen, A., Dengler, P., Ettle, R., and Beer, T., "Selbstbestimmt motiviertes und interessiertes Lernen in der kaufmännischen Erstausbildung," Zeitschrift für Berufs- und Wirtschaftspädagogik (ZBW), 1996, Vol. 13, pp. 108-127.
7. Fraser, B.J., Walberg, H.J., Welch, W.W., and Hattie, J.A., "Syntheses of Educational Productivity Research," International Journal of Education Research, Vol. 11, 1987, pp. 145-252.
8. Böhne, A., Rütters, K., and Wagner, B., "Evaluation of Tele-Tutorial Support in a Remote Programming Laboratory", to appear in Proceedings of the 2004 American Society for Engineering Education Annual Conference & Exposition, Salt Lake City, 2004.

**Andreas Böhne** received his diploma degree in Technical Education in 2001. He is currently pursuing his Ph.D. studies at the Learning Lab Lower Saxony (L3S), where he is a team member of the I-Labs ("Internet Assisted Laboratories") project. The project develops didactic concepts for online laboratory usage and reusable software and hardware components.

**Nils Faltin** earned a doctoral degree in engineering in 2002 for his dissertation on a new method for structured active learning of computer science algorithms with interactive visualizations. He currently is a postdoc at the Learning Lab Lower Saxony in the I-Labs project mentioned above.

**Bernardo Wagner** has been full Professor at the University of Hannover since March 1997. He leads the Real-Time Systems Group of the Institute for Systems Engineering. Furthermore, he is a member of the directorate of the Learning Lab Lower Saxony (L3S), where he deals with didactic concepts for multimedia and net-based learning environments, especially online laboratories and robotics.

# Chapter 30

# Applying a New Learning Strategy in Thermal and Electrical Engineering Education

EDUARDO MONTERO, CRISTINA ALONSO, MARIA JESUS GONZALEZ, FERNANDO AGUILAR and MONTSERRAT DIEZ
*Universidad de Burgos, Escuela Politécnica Superior, Avenida Cantabria, s/n, 09006 Burgos, Spain. E-mail: emontero@ubu.es*

*Engineering education must emphasise both scientific knowledge and development of organisational skills. Amongst these skills, we can mention communication, teamwork, creative thinking, leadership skills and good technical judgment. The learning and development of these skills is only possible if each student individually engages in the learning process. Especially in engineering topics, laboratory exercises are probably the best area in which to promote the active pursuit by students in acquiring the skills mentioned above. As a contribution to research in higher education in engineering, the present study seeks to analyse and classify the learning strategies that can be developed as a part of laboratory exercises. A strategy that promotes the significant development of scientific knowledge and organisational skills is presented. Eight relevant characteristics of this strategy are described. Two examples of laboratory exercises included in undergraduate courses at the University of Burgos are described: a forced convection heat transfer laboratory exercise in Thermal Engineering, and another dealing with circuit resonance analysis in Electrical Engineering.*

## TRAINING OF ENGINEERS AT A UNIVERSITY

Industrial firms and institutions [1-4] are asking universities to look at the formative aspects of engineering that go beyond scientific and technical knowledge. This refers to the personal capacities engineers must possess to be successful in their various roles. For example, economic-technical roles (in which they study, design, project, calculate, and implement) require innovating, perfecting and adapting; management roles (in which they lead, apply, control, supervise, and delegate) require the ability to set policies; and cultural roles (in which they investigate, learn, train, inform, and develop) require strong interpersonal skills.

Many reviews of engineering and engineering education over the last two decades [5-9] have identified a preponderance of technical and discipline-specific course contents over non-technical and professional practice skills. However, employers value the possession of these professional skills, and the inclusion of these elements in undergraduate curricula is conducive to the development of lifelong learning habits in students. A list of desirable generic skills identified in the literature [10,11] includes, but is not limited to, the following:

- leadership
- teamwork
- creative thinking
- communication
- critical-self awareness
- experiential learning
- resource-based learning
- practical knowledge

The learning and development of these skills is only possible, as happens with knowledge, if it is acquired by the students through self-learning, through situations in which the students play the central role [12-15]. In turn, the teachers must become guides in the process of learning which is not limited only to the subject matter of their own course, but which must be imbibed with a good dose of these skills. We must remember that one cannot teach what one does not know.

Especially in engineering science subjects and in those of a technological nature, laboratories offer an unequalled opportunity for the students to actively participate in the meaningful pursuit of their own learning and to exercise the skills to which we have referred.

There are some studies about how laboratory practice works best and about its effects in engineering and technology education [16-19]. It would seem, therefore, that we could make a major advancement in this field if we were able to make significant improvements in training through laboratory practical work.

In this paper we present a contribution to higher education in technology arising from the experience acquired during the last six years by a group of interdisciplinary university teachers. Different teaching strategies, which can be performed in the laboratory, are studied and a model is adopted which allows one to include the meaningful learning of knowledge and skills. We show two examples in the areas of Thermal Engineering and Electrical Engineering. The difficulties, advantages and inconveniences in their application are dealt with in the final conclusions.

## ALTERNATIVES IN THE PRACTICAL FORMATION OF ENGINEERS

The premise has been established that practical work constitutes an essential component in the education of students in engineering. The first step in our analysis leads us to the need to examine the true situation. To do so we will use the following questions as a guide:

- What are the objectives sought and what is the current practice in laboratory practical work designed for the education of future engineers?
- What perceptions do students come away with through these activities concerning their future professional life?
- Are these activities consistent with the professional role of the engineer and with the society's needs?

- Do they motivate students and create positive attitudes towards the discipline and its education process?

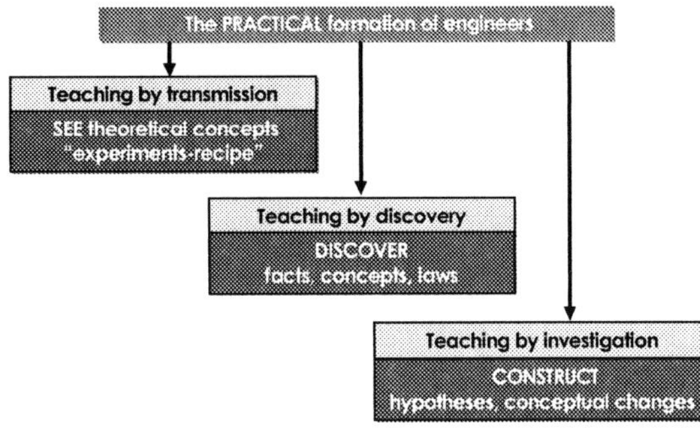


FIGURE 1

THE THREE PARADIGMS OF SCIENCE TEACHING [25]

Studies carried out on laboratory practical work offered in universities reveal rather unsatisfactory results. For example, in many cases, students perform practical work without having a clear idea of just what it is they are doing [20-22] and they are not capable of describing the concepts and phenomena involved in the experiments [23]. Likewise, university teachers, focusing mainly on research work, give little or no consideration to the functions and objectives of the laboratory work they prescribe for their students [24]. Clinging to their preconceptions, the teachers ascribe the failings and scant effectiveness of laboratory work to a lack of preparation, concern or interest on the part of students.

But, are not the normally implicit didactic approaches habitually used in the laboratory also involved? These didactic approaches can be classified into three main types [25] shown in schematic form in Figure 1. The most important characteristics of each of them are summarised in Table 1. The first two types of teaching are normally applied in our laboratory practice. Practical work is frequently used as a means by which to acquire skills in the handling of apparatus, for the learning of specific experimental techniques, and as a way to illustrate or check experimental facts and scientific laws presented earlier by the teacher. This type of work is usually 'guided' by very clear instruction sheets and we believe them to be necessary. Above all in the first stages of learning, the students must learn to measure and to measure well. The experiments also fulfill the aim of reinforcing the knowledge acquired in theory classes.

| Teaching by transmission | Teaching by discovery | Teaching by investigation |
| --- | --- | --- |
| Practical work is used as a means to acquire practical skills in the use and handling of apparatus, to learn specific experimental techniques, and as a way in which to illustrate or test experimentally scientific facts and laws presented earlier by the teacher. The types of work are like "experiment-recipes" which do not include problems to be solved | Practical work consists of activities to discover facts, concepts and laws using scientific processes. This may be through *autonomous discovery,* in which it is the student himself who "discovers" scientific knowledge using scientific processes to do so, and through *guided discovery*, in which the teacher guides the student to a greater or lesser degree. | Through practical work one constructs (not discovers) scientific knowledge from the interaction of our ideas with the ideas of others and with experience. Experiments serve to test ideas or to provoke cognitive conflicts, which help the student to become aware of the weaknesses in his ideas and encourage conceptual *change.* |

TABLE 1

THE MAIN CHARACTERISTICS OF THE BASIC PARADIGMS IN SCIENCE TEACHING [25]

However, we should not stop at this type of practical work. We believe that, if we in the universities are to be able to respond to society's demands concerning aspects other than the purely technical ones, then the work students do in the laboratory must relate to what an engineer does in the exercise of his profession. We believe too that it is in the laboratory more than in the classroom where this is totally feasible - such that students can test their assumptions, work in teams, and make choices. The aim of laboratory practical work is to construct (not discover [25]) scientific knowledge from the interaction of our ideas with those of others, and with real-world experience. Teachers should act not only as transmitters of knowledge and skills, but as guides who facilitate the processes of total learning.

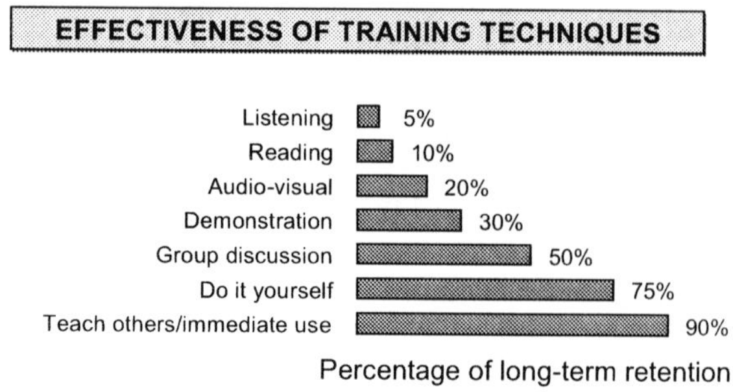


FIGURE 2
EFFECTIVENESS OF TRAINING STRATEGIES

The effectiveness of various training strategies in the long-term retention of knowledge is depicted in Figure 2, and is summed up simply in the old Chinese proverb:

*What you hear, you forget.*
*What you see, you remember.*
*What you do, you learn.*

| Type of Laboratory Work | Characteristics |
|---|---|
| 1. Experiential | Hands-on activities designed to achieve perceptive familiarisation with phenomena occurring in engineering systems |
| 2. Illustrative experiments | Activities to illustrate engineering principles, verify natural laws, or to improve understanding of specific concepts |
| 3. Practical exercises | Activities designed to develop specifically:<br>- Practical skills: measuring, handling apparatus, etc.<br>- Research strategies: repetition of measurements, data treatment, experimental design, control of variables, performance of an experiment, etc.<br>- Communication skills: know how to follow instructions to use a piece of apparatus, communicate results orally and in written format, etc.<br>- Cognitive processes in a scientific context: observation, classification, inference, formulation of hypotheses, interpretation within the framework of theoretical models, application of concepts, etc. |
| 4. Experiments to test hypotheses | Activities established by the students or the teacher for interpretation of phenomena |
| 5. Experiments as investigation | Activities designed to give students the opportunity to work as scientists or technologists in problem-solving. This may be implemented through theoretical or laboratory investigation according to whether it is aimed at solving a theoretical or practical problem |

TABLE 2
TYPES OF PRACTICAL LABORATORY WORK AND THEIR CHARACTERISTICS [23-26]

Accordingly, it seems that the paradigm of "teaching by investigation" is the most appropriate one for achieving our educational objectives. Hence, laboratory work should be designed such that it gives students experience in formulation of questions based on prior knowledge; proposal of possible solutions; testing of these solutions; and testing and discussion of the final solutions.

The orientation of engineering laboratory work (or simply laboratory "practicals") towards this model of "teaching by investigation" is greatly helped if a process is in place that enables teachers to engage in continuous improvement [26-28]. A good initial approach will have a decisive impact on the effectiveness of the process as measured by results obtained versus the effort. Today's laboratory practicals already contain many elements that can be explored for potential improvement.

To evaluate laboratory practicals so as to discover their strengths and weaknesses [23,26], one can use the classification shown in Table 2. The grading from Type 1 to Type 5 designates the degree of approximation to the model of "teaching by investigation."

## The Model of Teaching by Investigation

It is necessary to define the general characteristics of laboratory practicals if they are to be an investigative activity. We believe that the work, in addition to being experimental, must integrate other equally important processes of scientific activity, including aspects that are fundamental to a truly investigative practical session. Summed up in Figure 3, these aspects, suggested in [29], which in no way constitute an algorithm to be followed linearly, are as follows:

1. Present *open-ended problems or situations* of a suitable level of difficulty (corresponding to the *potential state of development* of students) such that students can make the correct decisions.
2. Enable students to develop interest and reflect on the *importance of the situations* proposed, considering possible implications for science, technology and society, etc., and avoid out-of-context, socially neutral study.
3. Foment *qualitative analysis* to understand the situations proposed in light of the knowledge available, the interests of the problem, etc., and formulate operative questions about what is being sought.
4. Propose the *formulation of hypotheses* as a key step in scientific research. Likewise, give these hypotheses a basis and make them operative by deriving testable consequences, giving adequate attention to the control of variables, etc.
5. Give importance to the *elaboration of designs* and of the planning of experiments by the students themselves. Wherever possible encourage the incorporation of current technology into experimental designs in order to obtain a correct vision of contemporary scientific-technical activity.
6. Carry out a *detailed analysis of results* (their physical interpretation, reliability, etc.), in light of available knowledge, of the hypotheses used and of the results of "other researchers. " Pay particular attention to the *conflicts* between results and initial perceptions, thus aiding and paving the way for conceptual changes.
7. Give importance to the preparation of *scientific reports* to document the work that has been accomplished, highlighting the role of communication and discussion in scientific activities.

8. Encourage the *collective dimension of scientific activities* by organising research teams and aiding interaction between each team and the "scientific community" (which is represented in the classroom by the other teams) the body of knowledge already constructed (text books), and other experts (the teacher), etc.

**Characteristics of LEARNING by INVESTIGATION**

**1** Present open-ended problems
**2** Enable students to develop interest
**3** Foment the use of meaningful, qualitative analysis
**4** Propose the formulation of hypotheses, based on student's preconceptions
**5** Give importance to the elaboration of designs and plan of experiments
**6** Carry out a detailed analysis of the results obtained
**7** Give importance to the preparation of scientific reports
**8** Highlight the collective dimension (teamwork, group discussion)

FIGURE 3
CHARACTERISTICS OF LEARNING BY INVESTIGATION (SOURCE: GIL [29])

It must be pointed out that not all practical sessions can fulfill all these requirements, either because of the characteristics of the experimental equipment available, or because some aspects of the discipline may be beyond the students' potential for development. Another recommendation is that changes in practical programmes should be made gradually, through small steps that, following summative assessment, guide the teachers to plan bigger changes [29-32].

## CASE STUDIES

### Case Study 1: Thermal Engineering

We present the example of a laboratory class on heat transmission by convection. We have a single piece of laboratory equipment and we compare the approaches of teaching by transmission which is teacher-centred, and teaching by investigation which is student-centred.

The equipment used is shown schematically in Figure 4. An electrical resistance capable of dissipating up to 100 W when loaded with a continuous current is placed horizontally on the top of a cylindrical, vertical constant section conductor. The air pushed by a centrifugal fan flows transversally over the resistance, thus cooling it. The air speed can be measured with an anemometer situated in the pipe and can be varied within the range between 1 and 7 m/s by choking the airflow at the entrance of the centrifugal fan with the help of a gate adjustable with a screw. The temperature of the airflow can also be measured as can the surface temperature of the resistance using Type K thermocouples. Finally, the electrical potential supplied to the resistance can be varied using a potentiometer, measuring the voltage and current supplied at any given moment.

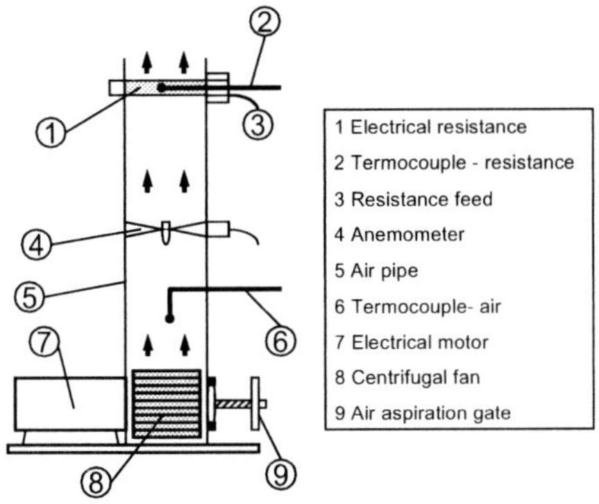


FIGURE 4
SCHEME OF THE EQUIPMENT USED TO STUDY HEAT TRANSMISSION BY CONVECTION

The objective of the practical work is to investigate the heat transmission by forced convection from the surface of the cylinder of the resistance at different air speeds. The aim is to obtain the relation between air speed and surface temperature of the cylinder.

In the teaching by transmission approach, the students will already have, before entering the laboratory, a set of instructions prepared by the teacher for the practical work. These instructions contain a definition of the objective of the practical work, the description of the experimental equipment, a review of the theory of forced convection heat transmission, a detailed explanation of the operation of the equipment, tables for the experimental data and calculated results, and the request for students to state their conclusions. As an option, an estimation of experimental errors may also be requested. The teacher's role in the laboratory typically consists of an oral explanation of how the equipment works, providing safety warnings regarding the handling of the equipment, and providing a reminder for the students to include the results and conclusions in their final report. Assessment is done by analysing the scientific results presented by the students in this final report, including an assessment of their accuracy and precision, graphic representations of the results, and the scientific conclusions reached.

The text of the instructions, the teacher's oral explanation, and the assessment system are designed to support the idea that the practical work is aimed at imparting an understanding of the experimental techniques and the verification of scientific laws previously explained in the classroom. If students are to make the most of the practical work in the laboratory, they must have a good grasp of the law of energy conservation and of the mechanisms of heat transmission. In terms of skills, since the experiment is totally guided by the teacher, all that is required of the student is that he uses the apparatus correctly.

In the teaching by investigation approach, the students will likewise have a set of instructions for the practical work. These instructions contain an explanation of the objective of the practical work, the methodology required to achieve it, the criteria used for assessment, a review of current knowledge in the scientific field being studied, and a description of the specific experimental objectives. The practical work will require the active participation of the students and of the teacher in the role of moderator. The most significant differences between this and the set of instructions described in the previous paragraph for the teaching by transmission approach are as follows:

a. The introduction is brief and focuses on students reviewing what they already know about energy conservation. The examples used are very familiar to them (a TV, a personal computer, etc.)
b. The experimental equipment is not described. Groups are made up of four students each and they are asked to design an experimental setup to determine the cooling capacity of air flow over an electrical heating element.

c. Each group presents its design to the rest of the class. The class comments on the advantages and disadvantages of each design.
d. The teacher then describes the actual equipment that will be used in the laboratory to do the tests. The features of the equipment are compared with those of the students' designs
e. Each team is asked to plan an experiment to carry out the tests. They must identify the functional objective, the number of measurements to be taken, the reduction of data and the form in which the results will be presented.
f. Each team presents its proposed experiment to the teacher and the class. Consensus is reached as to the most suitable experiment to perform.
g. The chosen experiment is carried out. The results obtained are discussed in class, including a discussion on whether the initial objective of the practical session has been achieved, namely to obtain experimentally a mathematical correlation of the experimental data on heat transmission by forced convection.
h. Each team has to write up the results of the laboratory practical work.

Assessment of the student's practical work includes assessment of their designs, their presentation to the class and their report of the results.

### Case Study 2: Electrical Engineering

Our practical work in Electrical Engineering deals with the phenomenon of resonance in electrical circuits, its problems and applications.

Each group of three students has the following basic equipment for this laboratory practical work: various elements such as resistances, coils, condensers and small loudspeakers (with various characteristics as regards power, nominal voltage, etc.), sets of lamps of different kinds, three-phase balanced loads, an autotransformer, an insulation transformer, two power supplies, a variable frequency signal generator, two multimeters, two wattmeters, and energy counters, etc. An arrangement of the equipment used is presented in Figure 5.

In earlier sessions, students have learned to correctly measure the different electrical outputs. They know how to calculate errors, to choose the most suitable measurement apparatus for the job in hand, and have made prior calculations to correctly choose the right components for the predicted experimental conditions (e.g. calculations relative to power, potential difference, etc.). All this is necessary to avoid any damage to the equipment and to guarantee the safety of the students themselves.

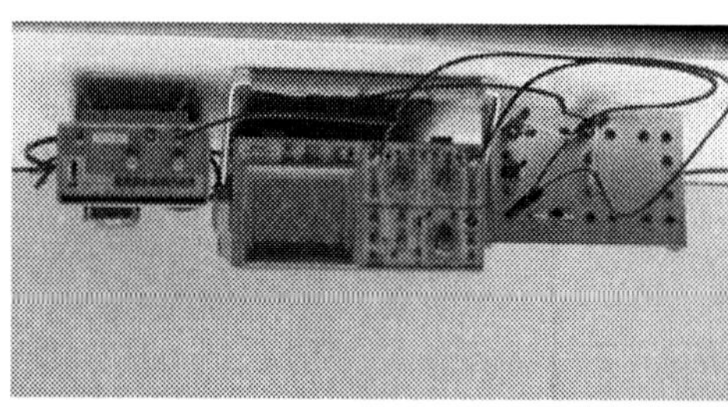

FIGURE 5
A VIEW OF THE EQUIPMENT USED TO STUDY RESONANCE IN ELECTRICAL CIRCUITS

The objective of the experiment is to enable students to gain a thorough understanding of the phenomenon of resonance, and to introduce them to the basic applications of this phenomenon.

In the teaching by transmission approach, the students will already have learned in theory classes about the phenomena they will study in the practical sessions: resonance in electrical circuits and its application to the filtering of signals. Prior to going into the laboratory, the students will have been given a set of instructions, prepared by the teacher, for the practical work. The text, the oral explanation by the teacher, and the assessment system are aimed at providing an understanding of the experimental techniques and a verification of scientific laws previously explained in the classroom. For students to make the most of the laboratory practical work, they must have a good understanding of the basic concepts that the practical session aims to reinforce. The experiment is totally guided by the teacher and all that is required of the students is that they use the apparatus correctly at each stage of the experiment. The possibility of developing any skills is rather small.

In the teaching by investigation approach, the students will likewise have a set of instructions for the practical work prepared by the teacher. This set contains the description of the objective of the practical work, the methodology used to achieve it, the assessment criteria, an introduction with antecedents to the scientific theme being studied, and a description of the specific experimental objectives. The approach will require the active participation of the students and of the teacher in the role of moderator. The most significant differences between this and the set of instructions described in the previous paragraph for the teaching by transmission approach are as follows:

a. The introduction is brief and focuses on a review of what the students already know. It is assumed that students already have some basic concepts, but the real introduction to the study of resonance and its possible problems and applications will be through the practical work. The coverage of the subject will be completed later in theory classes.
b. The experimental equipment is not described. Groups of three students are formed and they are asked to design an experiment with a fixed objective.
c. Each team presents its design to the remaining teams. Comments are offered to the other students and the teacher as to its advantages and disadvantages.
d. The teacher presents the most complete test, in accordance with the elements available and the best way to carry it out. Its characteristics are compared with those of the earlier designs.
e. The experiment is then performed. As a group, the results obtained are commented on and there is discussion as to whether the initial objective of the practical session has been achieved: to understand through experimenting the phenomenon of resonance and some of its possible applications in Electrical Engineering.
f. Each team has to write up a report of the practical work.

Assessment of the students' practical work includes assessment of their designs, the presentation to the group, and the report of the results.

## DISCUSSION

We have presented a practical class in Heat Transmission by Convection and another in Resonance in Electrical Circuits as pilot experiments of the new approach to learning. The laboratory practical in Heat Transmission by Convection is given in the first year of the degree in Electronic Engineering, as part of a course which studies the transmission of heat applied to electronic equipment. The practical in Resonance in Electrical Circuits is taught in the second year of Electronic Engineering, as part of Circuit Theory. The assessment of the effectiveness of these experiments is done in three ways: observation of the behaviour of students in the classroom; the scientific designs and the reports presented by the students; and through questionnaires. The main conclusions reached are the following:

a. The general attitude of the students during the session was positive.
b. In the teams, the role of each participant was not specified and a leader was not always chosen.
c. The teams found it difficult to organise their time adequately during the work sessions.
d. The designs proposed by students was an indicator of their creativity that otherwise would not have been discerned.
e. The students' communication skills were uneven and, in general, rather poor.
f. There were minimal to non-existent critical comments by students concerning their classmates during oral presentations.
g. The oral presentations given to the whole class in the presence of the teacher revealed that students used different approaches to a given problem.
h. Teachers had to learn to guide sessions through questions, and to limit the urge to simply explain things, as is done in a traditional classroom situation.
i. These sessions took longer than other types of practical work, although students gained a better learning experience.
j. The questionnaires did not show any correlation between higher learning and higher student satisfaction, although the indices of student satisfaction in laboratory work improved after the application of the new methodology.

## CONCLUSIONS

The training of engineers has been undergoing a great deal of change during the last several decades. The preponderance of scientific contents and techniques in this training has given way to the inclusion of personal attitudes and skills that will be very important later in the engineer's professional career. This implies a move from teacher-centred training to student-centred training.

We have presented two examples of laboratory practical work, the first one in Thermal Engineering and the other one in Electrical Engineering. Two approaches, the traditional teaching by transmission and the new teaching by investigation, are compared. The application of the approach of teaching by investigation has led to the following remarkable conclusions:

1. The percentage of students actively engaged with laboratory exercises has increased to 90% typically, which implies a high level of active learning.
2. This approach allows students to engage in activities directly related to an engineer's professional work, namely: design for an established purpose, design and implementation of experiments, communication and discussion of results, and writing reports.
3. In grading the class, teachers must give credit to the level of difficulty that students face in achieving the course objectives.
4. Since resources are limited and the sessions require more time to carry out, the teachers have to choose between running only a subset of the experiments using the new approach, or significantly reduce the subject matter to be covered in laboratory classes.

## ACKNOWLEDGEMENTS

The research team, and in particular the main researcher, E. Montero, wish to thank the Junta de Castilla y León (SPAIN) for supporting this project, through the Programa de

Recursos de Apoyo a la Enseñanza Universitaria de la Consejería de Educación y Cultura (Proyecto BU 16/02).

## References

1. Bar-Cohen, A., "Mechanical Engineering in the Information Age," Mechanical Engineering, ASME, New York, Vol. 117, No. 12, December 1995, pp. 66-71.

2. Buonopane, R.A., "Engineering Education for the 21st Century. Listen to the Industry," Chemical Engineering Education, Vol. 31, 1997, pp. 166-167.

3. Industrial Research and Development Advisory Committee of the European Commission, Quality and Relevance: the Challenge to European Education, Unlocking Europe's Human Potential, European Commission, 1994.

4. Yadarola, M.A., "Tecnología y formación socio-humanística del ingeniero para las necesidades de un mundo interdependiente," La Universidad del siglo XXI y su impacto social, Universidad de Las Palmas de Gran Canaria, Spain, 1996.

5. ASEE, The Green Report: Engineering Education for a Changing World, American Society for Engineering Education, Washington D.C., USA, 1994.

6. Grinter, L., *Report on Evaluation of Engineering Education,* American Society for Engineering Education, Washington D.C., USA, 1995.

7. ABET, Engineering Criteria 2000, Accreditation Board for Engineering and Technology, Baltimore, USA, 1997.

8. CAE, Engineering Education in Canadian Universities, Canadian Academy of Engineering, Ottawa, Canada, 1993.

9. Bates, I., Lloyd, B., Martinelli, F., and Vines, J., *Skills for the Future - Engineers and Scientists Achieving Enterprise Performance,* Association of Professional Engineers and Scientists of Australia, Melbourne, Australia, 1992.

10. Riley, M., Pickering, C., "Professional development skills for civil and environmental engineers," European Journal of Engineering Education, Vol. 20, No. 2, 1995, pp. 255-259.

11. Palmer, S.R., "Student Responses to Activities Designed to Develop Generic Professional Skills," Journal of Professional Issues in Engineering Education and Practice, Vol. 126, No. 4, 2000, pp. 180-185.

12. Edward, N. S., "Evaluation of a Constructivist Approach to Student Induction in Relation to Student's Learning Styles," European Journal of Engineering Education, Vol. 26, No. 4, 2001, pp. 429-440.

13. Hargreaves, D.J., "How Undergraduate Students Learn," European Journal of Engineering Education, Vol. 21, No. 4, 1996, pp. 425-434.

14. Felder, R.M., Silverman, L.K., "Learning and Teaching Styles in Engineering Education," Engineering Education, American Society for Engineering Education, Washington, USA, Vol. 78, No. 7, 1988, pp. 674-681.

15. Felder, R.M., Brent, R., "Learning by Doing," Chemical Engineering Education, Vol. 37, No. 4, 2003, pp. 282-283.

16. Hodson, D., "Re-Thinking Old Ways: Towards a More Critical Approach to Practical Work in School Science," Studies in Science Education, Vol. 22, 1993, pp. 85-142.

17. Magin, D., Kanapathipillai, S., "Engineering Student's Understanding of the Role of Experimentation," European Journal of Engineering Education, Vol. 25, No. 4, 2000, pp 351-358.

18. Khan, W.A., Al-Doussari, S.M.R., Al-Kahtani, A.H.M., "Establishment of Engineering Laboratories for Undergraduates and Postgraduates Studies," European Journal of Engineering Education, Vol. 27, No. 4, 2002, pp. 425-435.

19. Hodson, D., "Redefining and Reorienting Practical Work in School Science," School Science Review, Association for Science Education, London, United Kingdom, Vol. 73, 1992, p. 264.

20. Staskun, M.G. et al., "Do the Students Think We Are Running the General Chemistry Laboratories Better?" Journal of Chemical Education, American Chemical Society, Easton, PA, USA, Vol. 68, No. 2, 1989, pp. 152-154.

21. Moreira, M.A., "A Non-Traditional Approach to the Evaluation of Laboratory Instruction in General Physics Courses," European Journal in Science Education, Vol. 2, No. 4, 1980, pp. 441-448.

22. Tobin, K., "Research in Science Laboratory Activities in Pursuit of Better Questions and Answer to Improve Learning," School Science and Mathematics, School Science and Mathematics Association, Oregon State University, USA, Vol. 90, No. 5, 1990, pp 403-421.

23. Woolnough, B.E., Allsop, T., Practical Work in Science, Cambridge University Press, Cambridge, 1985.

24. Hodson, D., "Hacia un enfoque más crítico del trabajo de laboratorio," Enseñanza de las Ciencias, Vol. 12, No. 3, 1994, pp. 299-313.

25. Gil, D., "Tres paradigmas básicos en la enseñanza de las ciencias," Enseñanza de las Ciencias, Vol. 1, No. 1, 1983, pp. 3-26.

26. Gott, R., Foulds, K., "Structuring Investigations in the Science Curriculum," Physics Education, Vol 23, No. 6, 1988, p. 347.

27. Guisasola, J., Almudí, J.M., Ceberio, M., and Zubimendi, J.L., "A Teaching Strategy for Enhancement of Physics Learning in the First Year of Industrial Engineering," European Journal of Engineering Education, Vol. 27, No. 4, 2003, pp. 379-391.

28. Oehme, F., and Seitzer, D., "Learn by Doing: How to Include New Requirements of Research in Engineering Education," European Journal of Engineering Education, Vol. 25, No. 2, 2000, pp. 131-137.

29. Gil, D., and Valdés, P., "La orientación de las prácticas de laboratorio como investigación: un ejemplo ilustrativo," Enseñanza de las Ciencias, Vol. 14, Nº 2, 1996, pp. 155-163.

30. Aguilar, F., Alonso, C., and Montero, E., "El trabajo en equipo en el Laboratorio de Termodinámica," I Jornadas Nacionales de Ingeniería Termodinámica, Universidad de Extremadura, Badajoz, Spain, 1999.

31. Alonso, C., Zapater, C., Aguilar, F., Montero, E., and Meneses, J., "Nuevas Técnicas de Aprendizaje en el Laboratorio de Ingeniería Térmica," II Jornadas Nacionales de Ingeniería Termodinámica, Universidad Rovira i Virgili, Tarragona, Spain, 2001.

32. Aguilar, F., Alonso, C., Zapater, C., Montero, E., Díez, M., González, M.J., and Meneses, J., "Nuevas Técnicas de Aprendizaje en los Laboratorios de Ingeniería Térmica e Ingeniería Eléctrica," 2º Congreso Internacional de Docencia Universitaria e Innovación, Tarragona, Spain, 2002.

**Eduardo Montero** received his B.Sc. in Industrial Engineering, specialising in Energy Engineering, from the Polytechnic University of Madrid, Spain in 1984, and his Ph.D. from the University of Valladolid, Spain, in 1996. He has been a permanent Professor at the University of Burgos, Spain, since 1987 in the Department of Electromechanical Engineering. Previously, he worked at the Polytechnic University of Madrid. He leads the Innovation Teaching Group of Engineering Thermodynamics in its Department. His main areas of interest and research are engineering education and experimental determination of thermodynamic properties of fluids.

**Cristina Alonso** holds a Doctorate in Industrial Engineering and a B.Sc. in Physics and Electronic Engineering from the University of Valladolid. She has been a Lecturer in the Department of Electromechanical Engineering at the University of Burgos, Spain since 1996. She belongs to the Innovation Teaching Group of Engineering Thermodynamics of this Department and leads the Information and Communication Technologies activities of the group. Her main educational and research interests cover pedagogical innovation and experimental determination of thermodynamic properties of fluids.

**Maria Jesus Gonzalez** received her B.Sc. in Physics from the University of Valladolid, Spain in 1980. She has been a Lecturer of the Department of Electromechanical Engineering at the University of Burgos, Spain since 1988. She leads the Innovation Teaching Group of Electrical Engineering of this Department. Her main educational and research interests cover engineering education and computational methods in electrical engineering.

**Fernando Aguilar** received his B.Sc. in Industrial Engineering, specialising in Mechanical Engineering, from the Universidad Nacional de Educación a Distancia, Spain in 2000. He has been a Lecturer of the Department of Electromechanical Engineering at the University of Burgos, Spain since 1990. He belongs to the Innovation Teaching Group of Engineering Thermodynamics of this Department. His main areas of interest and research are engineering education and experimental determination of thermodynamic properties of fluids.

**Montserrat Diez** received his B.Sc. and Ph.D. in Physics from the University of Valladolid (Spain) in 1985 and 2001, respectively. He has been a Lecturer of the Department of Electromechanical Engineering at the University of Burgos (Spain) since 1992. He belongs to the Innovation Teaching Group of Electrical Engineering of this Department. His main areas of interest and research are engineering education and solar energy.

# Chapter 31

## Active Learning Environments for Automatic Control Courses

MAURICIO DUQUE, LUZ ADRIANA OSORIO, ALAIN GAUTHIER and FERNANDO JIMENEZ
*Universidad de los Andes, AA 4976, Bogotá, Colombia. E-mail: maduque@uniandes.edu.co*

*Over the last 6 years, different learning strategies have been tested in the Automatic Control course at Universidad de Los Andes, in Bogotá, Colombia. Problem-Based Learning, Teaching for Understanding, virtual peer evaluations of students' projects using the Internet, MATLAB simulations, and hands-on labs have been the main strategies. As a result of these experiences, the current Automatic Control course uses a variety of pedagogical approaches, adapting each of them toward one specific course goal, and covering different learning methods and thereby improving the overall results. Course objectives are structured around understanding goals, defined as 'student performance'. This concept resembles that of 'competence'. The proposed evaluation system conforms to the selected philosophical approach. In general, no lectures are given during this course.*

### INTRODUCTION

Throughout most of the twentieth century, engineering education was fundamentally based on pedagogical proposals centered on the following main principles:

- Knowledge is transmitted.
- Course objectives are basically defined in terms of information, statements, formulae, and procedures to be taught to students.
- A good professional/researcher makes a good teacher.

Although these statements may not explicitly appear in institutional documents, a close examination of what actually takes place in undergraduate programs and engineering courses reveals their constant presence. It was only towards the end of the century that curricular proposals showing departures from these three central tenets began to surface.

These changes are the product of evidence—more and more prevalent and conclusive, emerging from research in different fields—that shows that these three pillars of twentieth century education were fundamentally flawed, that they frequently fostered an inefficient, costly and superficial learning process [5,7,9,10].

Although teaching practices based on the three pillars responded to the needs of twentieth century society, present-day requirements seem much more difficult to fulfill. As a reference, see ABET Criteria 2000 [1], in particular the section on attitudes and abilities common to all engineering programs. The acquisition of these abilities and attitudes cannot be guaranteed within the framework of the three pillars.

The new curricular proposals are fundamentally based on new pillars, more in accordance with scientific research results related to the ways in which people learn [2]:

- Each individual builds knowledge starting from what he/she already knows. Learning takes place in the borderline between what is known and what is not known. It is fundamental that what is learnt makes sense for the person who is learning. Just-in-case learning is replaced by just-in-time learning.
- Abilities, attitudes, and competences are more relevant than information and procedures in educating professionals. More than learning, the objective then becomes understanding, i.e. being able to transfer knowledge into new problems, contexts and fields of knowledge.
- Active learning, understood as the transference of control in the learning process from the teacher to the learner, calls for a faculty with specific and important pedagogic competencies. These competencies are not achieved through the exercise of the profession or the research in the discipline to be taught.

In particular, Automatic Control System courses have been taught using classical teaching approaches based on lectures complemented by laboratory practices, which were oriented to a limited design of control systems and a verification of the theory using educational lab equipment. Normally, this type of course is supported by one textbook selected from the large offering of books available in the market. These books are frequently more structured around the discipline—a deductive approach—than around the process of learning the subject—a more inductive approach. The texts are therefore more appropriate as reference texts than as guiding texts for the learning process.

In the case of Control courses, there was a slow shift from courses based on the experimentation and empirical methods of the 1930s towards courses that followed a fundamentally deductive process with strong abstract mathematical content. Here, students would learn the central principles—in a specific framework—starting from algebraic and mathematical developments. The lab was used only as a less valued complement.

This shift in focus frequently gave rise to students who could answer theoretical exam questions about concepts, or exams where they were expected to actually carry out, on paper, a design that they often were able to validate in non-realistic simulations. But these students were unable to address real problems. They could not solve more open problems related to control. Moreover, they were incapable of transferring the acquired knowledge to other fields of knowledge. For example, very seldom would a student educated in this methodology be able to identify, in the configuration of an operational amplifier as an inverter amplifier, a feedback system that, depending on its feedback sign, may or may not be stable. Similarly, he would rarely be able to identify an unstable behavior in a real process.

This article presents an alternative learning proposal for undergraduate courses on automatic control systems, within an "Active Learning" modality that combines different didactic approaches. The next section outlines the conceptual design framework, and the final sections briefly describe the various versions of the courses taught in recent years.

## ACTIVE LEARNING IN AUTOMATIC CONTROL SYSTEMS TEACHING

As already mentioned, the different modalities that grant the students control over their own learning process can be classified as Active Learning. In "classical" teaching approaches, the professor holds much of such control: the professor decides what, when, and how learning occurs, and also evaluates what the students have learnt. Giving students control over their learning requires them to examine their learning process and to define their own strategies resolving difficulties. They are aware not only of what they are learning, but also of the learning process by which they learn it.

Not all environments in which students carry out many activities can be labeled active learning environments. Two seemingly similar scenarios can offer radically different learning experiences. The following two cases illustrate the differences.

**Case A**: The students arrive in a lab where they must test a proportional control system over a position control system. They have attended the theory class, where the professor has explained what a proportional regulator is, and showed how the transference function in close loop is calculated, how stability is theoretically evaluated, and how different system errors are calculated. In addition, the students have a laboratory guide where they find instructions on how to connect the lab equipment and the approximate steps they must follow in the course of the lab, as well as the results they must present (analytical calculations, simulated and real result curves, and conclusions). During the whole practice the students remain active, they follow the guide's instructions, collect data, and prepare the report.

**Case B**: The professor presents a process requiring a positioning system containing a control system meant to guarantee the achievement of certain specifications. To carry out their task, the students may use whatever elements they deem necessary. When finished, they must submit a final report demonstrating fulfillment of the expected requirements. The final report must also describe the procedure selected, the teamwork methodology chosen, and the results obtained. They must also submit a reflection on the new knowledge and concepts acquired. During the laboratory, students may formulate questions while analyzing the problem and identifying what they must learn in order to solve the problem. In response to those questions, the professor may either conduct a session to discuss the issues or appoint other students to prepare presentations the issues for the rest of the class. Throughout the laboratory, the students remain active, discussing, searching, assembling models, collecting data, and drafting the report.

An unprepared visitor would note very few differences between the two classes. In both cases, the professor would probably be ready to answer the students' questions while they work. But a more thoughtful examination shows some differences: in the case B, each team is working on a slightly different process, all designs do not reach the same results, and the sought-after solutions seem different.

There are indeed ample differences between the two cases. In case A, although the students are active, the professor remains in control of the learning process. In case B, each student controls a significant portion of his/her own learning process. The professor has not told the students what to study or which exercises from a textbook to carry out. The professor has simply presented a challenge for them to solve, and has asked them to think about what they are doing and the reasons for doing it. The evaluation will probably be done with the students examining their own results. Additionally, the students will not only be learning the content required in the discipline, but also they will be developing transversal abilities and attitudes (learning to learn, working as a team, solving open problems).

As was mentioned in the introductory section, the essentially deductive approach of the current offerings in system control courses keeps the students away from both practice and the real world. For many reasons, an inductive approach may prove more interesting in view of the consolidation of an active learning environment:

- Starting from a real problem (or one situated on the realistic limit), one that is open or poorly defined, the student may find the appropriate context for significant learning. This is one of the fundamental pillars in an effective learning environment.
- A problem that has been adequately presented allows the students to individually identify what they need to find and learn, thus assuming control over their learning process. Similarly, students begin to use their new knowledge immediately (just in time). The professor's presentations are pertinent to their laboratory activities, and the students know how to apply them [7].
- Working with real problems allows the students to develop concrete professional abilities.
- Evaluation becomes simpler, since it is related to real results that the students themselves may evaluate: the control system works adequately or it does not; specifications are either fulfilled or unfulfilled.
- Since they are being effectively used, the learned concepts should be better understood, thus facilitating transference to other contexts.

## A Short History of the Course

At the beginning of the 1990s, the Automatic Control course at UniAndes (Universidad de los Andes) followed a classic format: lectures about the topics, followed by the solution of some key exercises by the professor, eventually complemented by some exercises carried out by students in class or lab practice. Students were expected to read the material in advance and solve a number of exercises presented in the textbook. This type of course was not very popular among students, who found the topics to be tedious. Probably as a consequence of this, the general results were not the desired outcomes. Additionally, a few semesters later, the students showed a low level of knowledge in the topic, suggesting only superficial learning of the topics.

Towards 1992, some workshops—courses exclusively devoted to the solution of exercises and some more open problems by students—started to be formally introduced. These activities were always carried out after the professor had presented the theory. With these workshops, the students found courses more interesting, eventually achieving better results in exams in which they solved standard exercises

In 1996, the course objectives were redefined, using the Teaching for Understanding definition of understanding as a conceptual framework [3]. Such a framework evidenced the students' low understanding level, since although they achieved acceptable results in solving standard exercises in tests, they received very negative evaluations when solving more open problems in which they were required to perform a certain level of knowledge transference.

As a consequence of these poor results, a search for alternatives and new methodologies and didactics was initiated to improve the situation. In 1998, one of the course's sections introduced Problem-Based Learning (PBL) in one part of the course, achieving an important shift in the students' attitude: they would get involved, in a permanent fashion, in the course activities. Nevertheless, the lab sessions continued to follow a more or less classical format. Evaluation modalities based on students' portfolios were also tested. [6].

In 1999, an approach promoted by Professor Jerry Pine (Caltech, USA) was introduced. Pine had been working in the development of learning environments based on Hands-On practices, where the students learn by doing. This type of experience has also been developed in other institutions, such as the École de Mines de Nantes, where an introductory course for engineers follows this format. At Los Andes, the new lab design was developed in parallel with the classical version, allowing for an evaluation of a control group. The results were largely positive [4].

Similarly, a second word choice was carried out, this time with cooperation from Carnegie Mellon University, in which the impact of self-evaluation and peer evaluation on learning was analyzed. Again, a control group was involved in the final evaluation. In this instance, results also turned out to be highly positive [8].

By 2002, the first integrative version, which included the various elements and components already tested, was developed. The lab was totally integrated into the course, and the hands-on practices generally preceded the conceptualizing work. In comparison to the course taught in the 1990s, the structure had been completely reversed: first the problems and the lab, then formalization.

Designing a course with these characteristics requires a closer follow-up on the part of the professor. The course needs to be more personalized, which turns out to be more expensive and sometimes difficult implement. Handling the students' paper portfolios remains an important problem in terms of logistics. For this reason, the school decided to complement the course with information technology support, which may facilitate the relationship between the professor and the students.

In the following sections, this article presents the course design to be taught in the second term of 2003, which will include an impact evaluation.

## Conceptual Design

Table 1 presents the objectives of the automatic control system course divided in three main dimensions:

- Information, concepts and events, which may be evaluated either independently or in context, starting from the evaluation of the abilities that require them.
- Abilities, which may be developed and evaluated in the students' laboratory practices.
- Attitudes, whose development and evaluation are difficult.

| | | **Knowing / Doing / Notions** | **Understanding / Strategies / Concepts** |
|---|---|---|---|
| **Information, norms, concepts, laws, notions** | **Disciplinary** | Use of norms and standards, automatic control history, robust control, available technologies, Sensors and drivers | Transfer functions, state representation, **modeling**, stability, feedback, time response, frequency response, simulation, error, dynamic systems |
| | **Interdisciplinary** | Basic laws governing thermal, chemical, hydraulic, mechanical, biological, and electric systems. Systemic thought | Feedback in other disciplines |
| **Abilities, procedures, strategies** | **Disciplinary** | State space control, phase and gain margin, Nyquist, programming a regulator in a digital control system | Stability, bode and root locus analysis, system simulation, PID design, lea-lag and algebraic design, system modeling, error calculation and reduction, assembling a control system. |
| | **Interdisciplinary** | | MATLAB and LabVIEW use in control problems solution |
| | **Interpersonal** | Teamwork, presentation preparation, communication, circuit and system assembling, failure detection | Learning to learn |
| **Attitudes** | **Interdisciplinary** | Systemic thought | |
| | **Interpersonal** | Responsibility in project development | |

TABLE 1

AUTOMATIC CONTROL COURSE OBJECTIVES

Similarly, each of these dimensions has been subdivided in three categories:

- Disciplinary: issues directly related to automatic control systems
- Interdisciplinary work: issues related to other areas
- Interpersonal: that which responds to transversal abilities and attitudes

Finally, the desired learning level for each one of the objectives has been classified in two large levels: knowing and understanding. This is an important classification, since traditionally courses are overloaded with content, pretending to address all relevant topics. In practice, the professor may eventually manage to present all the topics, but what the students actually learn does not match the professor's expectations. The students may even fail to learn central concepts in favor of less important ones. The solution is to define central objectives and the ancillary objectives, and to make this information available to the students. Similarly, the objectives have been phrased in terms of understanding goals [3]. For example, an objective such as modeling (bold in Table 1) may be more precisely expressed in the following manner:

> **Modeling**: Presented with a simple process (almost invariant and linear dynamics, reduced order), the students will be able to find their model; to identify the proposed equations' coefficients; to validate and analyze its basic characteristics of response time, type of response, stability and frequency response. Based on its characteristics, the modeling may be carried out starting from either balance equations, or from the temporal or frequency response of the process.

This type of objective definition offers the additional advantage of suggesting the type of test that the students should be able to successfully pass. In turn, the students will have a clear understanding of precisely what is expected of them.

The active learning environment includes two working environments structured by the professor to support the course: the virtual environment and the classroom environment. Moreover, the students must carry out learning activities outside of both working spaces, either in groups or individually. The learning process is framed in the interaction of three central activities: problems and projects, hands-on laboratory practices, and real-time simulations, the latter being a support tool for the development of the former two. Although the students work mostly in groups, individual work is also required, especially in the advanced conceptualization stages.

For the realization of the whole course, a protocol was developed which lays out a guided inquiry process in which students must achieve proposed objectives through a series of problems and projects. This inquiry protocol is presented as a set of learning experiences. Each one of these learning experiences is a unit structured around one problem or one project, to be developed in one or several class sessions.

A typical two-session sequence in the virtual space of the course includes the initiation (presenting a challenge) and the end (final construction of meaning and conceptualization). Additionally, each sequence includes its own evaluation system elements as part of the simulation and ranking component of the virtual system. The component allows each student to introduce their proposed solutions in order to evaluate his/her performance, and also to obtain a ranking of his/her solution in relation to the solutions proposed by other students, based on previously defined criteria for each problem.

Moreover, the virtual environment facilitates portfolio follow-up activities. As a registry of the various tasks and papers, including all corrections and observations, the portfolio serves as evidence of the learning process.

Finally, the virtual environment's concept album, described later in more detail, lists for students with the fundamental concepts of the course, allowing them to identify which central concepts that have not been fully understood. Self-examination and the generation of relationships between concepts should support this space.

## Information Technologies – Virtual Environment

The course is supported by information technologies, with a virtual learning environment developed in WebCT. MATLAB is used for specialized simulation and control software, and LabVIEW is used for setting and testing some instruments. In particular, the virtual environment in WebCT is built around student-to-student and student-to-professor interactions. The environment was created to enrich lecture activity in a way such that it is integrated to the class' rhythm and activities. This environment offers opportunities for:

- Continuing the conceptualization initiated during the lecture session, both in groups and individually.
- Interacting around the course's learning sequences.
- Following the learning process of working teams.
- Following the individual learning process of each student registered in the course.

The virtual space offers components suitable for managing group interaction (sequences, concepts), individual interaction, interaction with the professor (portfolio), communication, resources, calendar, and general information about the course. The following section offers a description of some of the components of the proposed virtual space:

**Sequences**: As noted above, a sequence or learning experience corresponds to the work surrounding one study problem which will be solved by a team of students, and an interaction protocol for the construction of the solution. The sequences are arranged in the virtual space such that they parallel progress in the lecture sessions as the course advances. Each sequence contains interaction sub-spaces: challenges, resources, group construction, discussion of the topic, simulation, the ranking tool, and solutions. Sequences begin with the challenge or problematic situation, and continue in the following manner:

- Each group has access to the sequence's resources, which have been arranged by the professor (readings, links, and interactive control book).
- Each group undertakes the collective construction of the solution. In this space, each team must record its construction process: initial hypotheses, team strategies, and selected solution. The professor contributes to the groups' construction process by asking questions or offering contributions that stimulate group dynamics.
- The groups may use the related discussion space in order to solve conceptual questions emerging from the process. The professor participates in this space and facilitates interaction between the various groups.
- Each group must test its solution on the simulator. All sequences allow for teams to introduce their solution and receive feedback on its efficiency and performance.
- Based on the simulator results, the ranking system provides the groups with information about the efficiency of their solution (through comparison with other proposed solutions).
- Each group must finally state its solution in the solution space.

**Concept album**: In this space, the students may consult the list of concepts taught and tested in the course. The album includes a permanent conceptualization forum, permitting dialogue among course participants (students and professor) about their experiences with the sequences. The students may also find additional resources associated with the concepts and the learning sequences in which the concepts were used.

**Portfolios and Concepts**: This space is devoted to individual construction and conceptualization. It is also the space for the individual learning process registry and follow-up. Through portfolio activities, each student reflects on what has been learnt and undertakes the concept construction task. This space includes the following sub-spaces: reflection, concept album, conceptual self-evaluation, and conceptual map:

- After the experience, all students must work on the reflection and conceptualization tasks.
- They must register, in the concept album, the concepts that become clear in the course of the semester.
- By means of the conceptual self-evaluation, any student may test him/herself for conceptualization.

- As a means to evaluate both individual learning and conceptualization, the students must construct a conceptual map in which they establish connections between concepts, and write an essay describing the connections.

Additionally, the course's virtual environment includes sections such as **Resources**, **Calendar**, **Communications**, and **Course Information**.

## CONCLUSIONS AND FUTURE PERSPECTIVES

Through the last 10 years, results from the automatic control course in UniAndes have suggested strong evidence of improvement in learning efficiency when an active learning approach is chosen. In general students find this approach more interesting and challenging, working in a more continuous way during the course. Also, some evidence about deep learning has been found. Depending on the strategies selected, it is possible to manage courses of up to 60 students.

In general terms, in order to change their role in the classroom and effectively plan their activities, the faculty involved in this process require more hours of intensive and demanding design and evaluation activities before the term starts. However, the required workload during the actual course may even diminish in comparison to a course taught following more a conventional approach. In larger groups, this strategy might require graduate assistant support.

For the second term of 2003, the course will be taught using the complete set of developed tools, including follow-up and evaluation strategies. The results of this new format will be compared with those of previous versions. It is worth noting that in the past this course has been the subject of several institutional follow-up and observation studies.

In addition, this venture will enable an evaluation of the impact of guided inquiry on undergraduate level courses, which has been successfully used in innovative science education projects in several countries.

At present, the School of Engineering at Universidad de los Andes is committed to an ambitious curricular innovation project that not only includes a thorough revision of the desired profile of engineers—one that fulfills the new requirements of both society and industry at the same time that it complies with ABET2000 criteria—but also involves the development of efficient learning spaces in an Active Learning approach, supported by adequate new physical classroom and laboratory spaces.

As a consequence, this type of courses will, in the near future, rely on the physical spaces and technologies required for the development of pedagogical methodologies such as the one presented here.

Over time, a certain amount of data showing positive results has been found in the control systems course final examinations. The first National Engineering Examination[6] also adds interesting information in this direction. Nevertheless, further research is required in order to confirm the hypothesis that this type of course fosters both a deeper understanding and a more thorough development of the fundamental abilities that underlie the practice of engineering

[6] In October 2003, the Colombian government conducted the first National Engineering Test. All graduating engineering students from all universities took the test. Uniandes students obtained the best results in the area of control systems.

## Acknowledgements

This paper was possible thanks to the support of Universidad de los Andes. It has been nurtured by the exchange of ideas facilitated by ALE (Active Learning for Engineering Workshop). Professors Karl Raugh (École de Mines de Nantes, France) and Jerry Pine (Caltech, USA) have made immense contributions, offering an array of documents and materials as well as their invaluable comments and opinions.

## References

1. ABET, Criteria For Accrediting Engineering Programs, http://www.abet.org/criteria.html.
2. Committee on Developments in Science of Learning, *How People Learn,* National Academic Press, 2002.
3. Stone, M., *Teaching For Understanding: Linking Research with Practice,* Jossey-Bass Publishers, San Francisco, 1997.
4. Duque, M., "Learning By Doing," Workshop On Active Learning, Caracas Venezuela, January 2001.
5. Gardner, H., *Multiple Intelligences: The Theory In Practice,* Basic Books, N.Y., 1993
6. Duque, M., "Active Learning In Automatic Control Courses," SEFI2001, Copenhagen, September 12th to 14th, 2002.
7. Novak, J., *Learning How To Learn,* Cambridge University Press, 1984.
8. Duque, M., "On-Line Tool for Improving Feedback in the Learning Process," Internal report, Universidad de los Andes, 2002.
9. Felder, R., Rugarcia, A. "The future of engineering Education: I. a vision for a new century", Chemical Engineering Education, Vol. 3, No. 1, 2000, pp. 16-25.
10. Bransford, J. et al, *How People Learn: Brain, Mind, Experience, and School,* Commission on Behavioral and Social Sciences and Education, National Research Council, National Academy Press, 2003.

**Mauricio Duque** received his B.A. and M.S. in Electrical Engineering from Universidad de los Andes, Colombia, in 1981 and 1984, respectively. He received his D.E.A. and Ph.D. in Automatic Control and Signal Processing from the Institute Politechnique de Grenoble and Laboratoire d'Automatique de Grenoble, France, in 1985 and 1989, respectively. Since 1995, he has been Associate Professor of Electrical Engineering at University of los Andes, Colombia. Currently he is involved in two research groups concerning Control Systems and Engineering Education. He also acts as Academic Vice-dean of the School of Engineering at Universidad de los Andes, Bogotá, Colombia.

**Luz Adriana Osorio** received her B.A. in Systems Engineering from Universidad Autónoma of Manizales, Colombia, in 1993. She received her M.S. in Systems Engineering from UniAndes, Colombia, in 1998. Since 1998 she has been Professor and researcher in the Systems Engineering Department at UniAndes. She is currently leading UNIANDES-LIDIE, the R&D Lab on Applied Informatics in Education. She is also participating in the Virtual Learning Environments project.

**Alain Gauthier** received the B.A. and M.S. degrees in Electrical Engineering from the University of Grenoble, France, in 1973 and 1974, respectively. He received his Ph.D. degree in Automatics from the Institute Politechnique de Grenoble and Laboratoire d'Automatique de Grenoble, France in 1977. Since 1983, he has been a Professor of Electrical Engineering Department at UniAndes, Colombia. He is currently a member of research groups concerning Control Systems.

**Fernando Jimenez** received the B.A. degree in Electrical Engineering from UniAndes, Bogotá, Colombia, in 1979. He received his D.E.A. degree in Automatic Control from Ecole Nationale Superieure de l'Aeronautique et de l'Espace (Sup' aero, Toulouse, France), in 1983. He received his Ph.D. degree in Systems Engineering from Institute National des Sciences Appliqués (INSA-Toulouse) and UniAndes in 2000. He is currently participating in two research groups concerning Diagnostic, Supervision and Control Systems and Microsystems and Integration. Since 1994, he has been an Associated Professor at the Electrical Engineering Department at UniAndes, and he was accepted as an associated researcher at Laboratoire d'Analyse et d'Architecture de Systemes (LAAS-CNRS) in Toulouse, in 2003. Dr. Jimenez is a member of the IEEE Society.

# Chapter 32

## A Model for Synergistic Interaction between Industry and Universities with a Focus on Undergraduate Education

STEPHANIE FARRELL, ROBERT P. HESKETH, JAMES A. NEWELL, MARIANO J. SAVELSKI and C. STEWART SLATER
*Rowan University, Glassboro, NJ, USA. E-mail: farrell@rowan.edu*

*Rowan University has developed a program that fosters synergistic interaction between industry and academia. This program provides a mechanism for performing industrially-sponsored research or design projects in an academic environment. One of the unique features of this program is its integration into the undergraduate curriculum at our university. Students work in multi-disciplinary teams on semester-long or year-long projects that are supported by external sponsors. In addition to providing a mechanism to introduce emerging technologies relevant to regional industries, the clinics provide the students with exposure to industrial projects with real deadlines and deliverables, and an opportunity to develop their project management, teamwork and oral and written communication skills. This program offers the industrial sponsor a cost-effective approach to problem-solving with potential for a high return on investment, through technical assistance from advanced undergraduate engineering students supervised by faculty. This paper presents three case studies that` examine successful synergistic interaction between industry and academia through the Rowan Engineering Clinic Program. The case studies focus on the development of three successful and different relationships with companies in the region, and benefits to students, faculty and industry.*

### INTRODUCTION

Undergraduate engineering and technology students benefit from real-world experiences that are usually obtained through internship and co-op experiences. Through these work experiences, students have the opportunity to apply their technical skills to industrially-relevant problems, gain exposure to company culture, and build a foundation which helps provide motivation for future learning in an academic environment. While these real-world experiences are highly valuable to students, they are still separate from the engineering curriculum and academic experience. It would be desirable to integrate more real-world experiences into the undergraduate curriculum at universities; however, industry-academic relations have not been, in general, developed to their full potential. Rowan University has developed an Engineering Clinic Program that fosters collaboration between academia and industry and provides real-world project experiences to undergraduate students.

At Rowan University, all engineering students are required to participate in an eight-semester course sequence known as the Engineering Clinics [1]. During the first two years of Engineering Clinic, students learn basic technical engineering skills through hands-on, project-oriented work. Communication and teamwork skills are also emphasized in the early stages of the Clinic sequence. In the junior and senior years, these clinic courses involve multi-disciplinary student teams working on semester-long or year-long research and design projects sponsored by a company in our region. Every engineering student participates in these projects and benefits from project-based learning, exposure to emerging technologies [2], industrial contact, teamwork experience and technical communications.

Other universities have developed successful programs that integrate industrial experiences into the engineering curriculum. For instance, the Industrial Affiliates Program of the Department of Electrical and Computer Engineering at the University of Puerto Rico at Mayagüez combines features of a traditional, paid co-op experience with undergraduate industrial research [3,4]. Student participation is optional, and students are compensated financially for their work.

## Typical Project Life

The typical engineering clinic project starts well before the first day of the semester, and the preliminary work in defining a project and securing funding requires a substantial time investment by the faculty members. The initial contact between a professor and a scientist or engineer from a regional company often results from a connection made through a professional society meeting, recruiting event, student internship, or newspaper article about the university or company.

Representatives of the interested company are invited to the university for an informational visit. They are introduced to the unique nature of the Engineering Clinics and the particular advantages that the flexible nature of the clinic sequence offers their company. The representatives also receive a brief overview of the expertise and interests of the college faculty members, while these members learn about the engineering priorities of the company. After this visit, interested faculty members often visit the plant site.

The next stage is to match faculty interest with the operations of the company. Then further meetings are set up to brainstorm and sketch out project ideas. Professors research these ideas to develop and scope the difficulty level of the project to upper level engineering students. The professor must also engineer the project to have outcomes that can be achieved within one and two semesters that will satisfy the students and the sponsor. The duration of the project is determined during the discussions between industry and faculty, and is governed by the scope and level of difficulty as well as the company's timeline. Finally, a budget is prepared for the project and negotiations are undertaken with the company to finalize the agreement. The official agreement and budget administration for the Engineering Clinic is handled through the Development Office of the Rowan University Foundation. The average funding level for one Clinic project is about $30,000. A confidentiality agreement is established between the company and the university. Normally, the time between first contact and obtaining a defined and funded clinic project averages about one year.

Prior to the start of the semester, background work should be done so that undergraduate students will be able to "hit the ground running." One graduate (Masters) student is usually assigned to the project, and this student would initiate a literature search and become familiar with the technical aspects of the project before the semester begins. The learning curve is steep for both the graduate student and the faculty member, and the effort required for this aspect of the project should not be underestimated.

When the semester begins, all faculty members give a brief presentation on the Clinic projects they will be offering that semester. Undergraduate students are matched to projects based on their ranked project choices and on the needs of the project. The professor then provides a brief introduction to the project, along with an overview of its goals. The students are required to read introductory material for familiarization. Early in the semester, typically in the second week, the team meets with industry representatives who provide an overview of their industry as well as a description of the technical problem that is to be addressed. At this meeting, students begin to develop a rapport with the industry representatives. They begin to see what aspects of a project are important to industry, that industry has very short deadlines, and that they expect to see experimental results. They also see that these projects have a goal that will directly impact the operations of the plant and the engineers and scientists in the meeting room. Close interaction with industrial representatives is critical to success of the project, and regular meetings continue throughout the semester, typically on a bi-weekly basis.

Students begin immediately to review the technical literature, and then to develop a project plan. They then establish a budget and begin to purchase equipment and supplies necessary for their experiments. Students contact vendors directly for information, and all paperwork for purchases is handled between the students and the Department secretary, with authorization signatures required by the professor. Students learn that it is necessary to work diligently and aggressively through this phase of the project due to the inevitable lag time between choosing the equipment and receiving it.

The student team has weekly meetings with the faculty members involved in the project, as well as regular (biweekly or monthly) informal meetings with the industry representatives. The frequency of the meetings with the industry representatives depends on their availability and their desired level of involvement in the project. Our experience has been that frequent meetings are highly beneficial because accountability, progress and results are required. Formal presentations to the industry are given mid-semester and at the end of the semester. The final report submitted by the students is a detailed technical report. Typically the report requires substantial editing before it can be submitted to the company.

These projects also help the program address many of the "softer" skills required by ABET [5]. Students function in multi-disciplinary teams, design and conduct experiments, learn about safety and environmental issues, analyze and interpret data, communicate through oral and written reports, and use modern engineering tools.

# Case Studies

## Johnson Matthey Metals Purification Processes

Johnson Matthey, Inc. (JMI) is a global specialty chemicals company with a focus on precious metals, catalysts and fine chemicals. A precious metals "refinery" is operated at West Deptford, NJ, which is less than 10 minutes from our campus. This close proximity facilitates numerous interactions and projects that we have with Johnson Matthey. This company has provided significant support to our chemical engineering department and was a "charter member" of PRIDE, the acronym for the "Partners with Rowan in Developing Engineers" program. They have continuously employed Rowan chemical engineering students as interns, weekend shift workers, and permanent employees. Johnson Matthey was a pioneer of the Engineering Clinic Program, sponsoring two projects in the first year of the program and 1-2 projects for the five years that have since passed. These projects have focused on precious metals separation and recovery. The objective of all these projects has been to investigate novel techniques that have the potential to replace current "traditional" refinery process units.

At the refinery, precious metals such as Pt, Pd and Rh are purified from feed streams containing many unwanted metal species and other impurities. The feed streams are made up of spent catalysts from which precious metals are recovered and recycled to feed streams from mines. In the refinery, there are many dissolution, selective-precipitation, and filtration steps. Using innovative processes, like membranes, the plant capacity, product purity, and processing cost have the potential to be improved. In essence, students have an opportunity in the engineering clinic to conduct engineering projects that are equivalent in scope to those done by engineers in the plant.

One of the Johnson Matthey projects concerned the development of a membrane process in an intermediate process step that used electrodialysis separation in a precious metal chloride ion solution contaminated with unwanted acids and salts. Technical aspects of the project focused on process design, performance evaluation, economic evaluation, and scale-up. Other technologies investigated through different projects include liquid-liquid extraction, ceramic membrane microfiltration, ion exchange, and adsorption.

These projects have been beneficial to Johnson Matthey, students, and faculty. Based on the results of these projects, JMI has added several new process units to their refinery processes that result in increased capacity, higher product purity, and shorter "hold-up" time for precious metals. Students have learned new technology through industrial projects, have gained exposure to industrial culture, and in many cases have secured summer or full-time employment with Johnson Matthey. Faculty have developed valuable relationships with industrial partners, secured funding for research projects, and learned about new technologies.

## General Mills

General Mills is an international company producing a wide range of food products. The company's major business includes Big G cereals, the leader in the $7.5 billion U.S. ready-to-eat cereal category, with consumer favorites like *Cheerios, Wheaties* and *Lucky Charms.* Other business divisions focus on meals, baked goods, snacks, and yogurt. Our interaction with General Mills involves the Pillsbury division that produces dough-based products.

Our relationship with General Mills began with Pillsbury before it was bought by General Mills. Pillsbury has a manufacturing facility in nearby Swedesboro, NJ, and Pillsbury engineers were involved with the Rowan Chemical Engineering Capstone Design Course. As the relationship between Rowan and Pillsbury evolved, and Pillsbury saw other successful Clinic projects, we began to discuss possible Clinic projects with Pillsbury (now part of General Mills). In September 2002, General Mills boldly sponsored *three* Clinic Projects for improvement and optimization of their dough line processes. One project focused on the analysis of raw materials; the second aimed to optimize a process line; and the third project investigated wastewater minimization.

The structure and lifetime of the Clinic Project was very similar to that described above for the Johnson Matthey Projects. The nature of the work, however, required a significant amount of time on site at the manufacturing facility. Students were required to complete the General Mills safety-training program and to learn good manufacturing practice (GMP) requirements for food production before they were granted permission to work in the production area. They were given access cards so that they could enter the plant at any time without being required to sign in and to have an official host from the company.

The on-site work included careful observation of the current process, quantitative analysis of the individual process steps, and modification of the process to test proposed solutions. The wastewater minimization team investigated several novel technologies both in the laboratory at Rowan and on site at General Mills. They have designed a water recovery system that will be implemented at the plant in 2004, and is likely to be implemented at several other facilities around the world. The dough line optimization team performed a thorough analysis of all process steps and proposed a process modification that has already been implemented and will result in more than $1 million annual savings.

The students working on this project benefited tremendously from working directly with operators and technicians at the plant. Engineers from General Mills also reported to us that their operators had an increased sense of pride and ownership in their own work, because University involvement meant that their project was particularly important to the company. They were highly cooperative with the students who were working on the plant floor. The students were invited to give a short presentation to the entire company at their quarterly meeting last spring; this high visibility showed the students how significant their work was to the company.

Regular project meetings were limited to weekly meetings between students and faculty. General Mills engineers did not attend regular meetings at the University, as was done with Johnson Matthey. Frequent meetings with General Mills engineers were unnecessary because of the regular informal contact between students and sponsors on site at the plant. The students gave two formal presentations to the sponsor each semester. These presentations were given on site at General Mills, and were attended by engineers, technicians and operators, as well as managers from Finance Operations, Manufacturing, Maintenance, Quality and Regulatory Control, and Human Resources.

General Mills worked diligently with the students to produce different types of documentation, some for internal use and "sanitized" versions for external use. This helped overcome what is often a limitation of industry-sponsored projects: the inability to share/publish work externally. An undergraduate student presented some of the work at the 2003 AIChE Spring Meeting in New Orleans, LA [6]. The student's fall semester report was nominated for consideration as the best paper in the AIChE student paper competition – and received the First Place AIChE Zeisberg Award for the best report.

## Campbell Soup Company

Campbell Soup is the world's largest maker and marketer of soup and also a leading producer of juice beverages, sauces, biscuits and confectionery products. The Campbell Soup Company World Headquarters and Corporate Research and Development facilities are in nearby Camden, NJ. Campbell has sponsored Clinic projects for two consecutive years.

In a project sponsored during 2001-2002, a team of students researched cutting-edge technologies such as novel membrane processes and supercritical fluid technology for the processing of soups and juices. The multi-disciplinary team consisted of two undergraduate chemical engineering students, one civil engineering student, and one biology student. In addition, one master's student served as a project manager. Through this project students investigated advanced membrane separation techniques as well as enzymatic, thermal, and physical/mechanical treatment techniques applied to vegetable processing. Their responsibility included HAZOP analysis, project planning, budget formulation and management, literature and patent reviews, experimental design, and development of a proposal for a second phase of the clinic project. In addition to the engineering expertise the students acquired through this project, they gained familiarity with Food and Drug Administration regulations, Good Manufacturing Practices, and labeling requirements.

Engineers from Campbell demonstrated a high level of commitment to the project and to the student learning by attending monthly progress meetings. At these meetings, students gave oral presentations on their progress. This was followed by brainstorming and discussion sessions with the industrial representatives and faculty, which refocused and fine-tuned the project. This industrial interaction helped maintain a high level of motivation among the students, and maintained the focus and a fast pace of productivity. In addition to the progress meetings, the student team also conducted two "lunch-and-learn" seminars at Campbell's to share their research with engineers, scientists, and marketing representatives from the Company. The enthusiastic response of the audience at Campbell's reaffirmed the industrial relevance and impact of the team's clinic research project. Students were surprised by the high interest level exhibited by the marketing employees who believed that the project would significantly impact product marketing.

Campbell Soup Company is a strong supporter of our program, not only supporting the clinic project mentioned above, but also employing both full-time and internship students from our program. In the summer following the vegetable processing project, two Rowan undergraduate students accepted summer internships at Campbell's. The students had the rewarding experience of successfully implementing two of the technologies developed at Rowan into Campbell's processing facilities in California and New Jersey.

## SUMMARY

Rowan University has developed a program that fosters synergistic interaction between industry and academia that provides a mechanism for performing industrially-sponsored research or design projects in an academic environment. Undergraduate junior and senior engineering students work in multi-disciplinary teams on semester-long or year-long projects that are supported by external industrial sponsors. In addition to providing a mechanism to introduce emerging technologies relevant to regional industries, the clinics provide the students with exposure to industrial projects with real deadlines and deliverables, and an opportunity to develop their project management, teamwork, and oral and written communication skills.

This program offers the industrial sponsor a cost-effective approach to problem-solving with potential for a high return on investment, through technical assistance from advanced undergraduate engineering students supervised by faculty. Successful projects have led to the implementation of several new process modifications and process units that have resulted in increased capacity, higher product purity, decreased labor cost, and less process down time. In addition, the company has the opportunity to watch for potential interns and employees for future hire.

Students have learned new technology through industrial projects, have gained exposure to industrial culture, gain experience with deadlines, progress meetings, presentations and written deliverables, and in many cases have secured summer or full-time employment with the sponsoring company. Students have won external awards for their work and have presented their work at national conferences. Faculty members have developed valuable relationships with industrial partners, secured funding for research projects, learned about new technologies, and occasionally have published results externally.

The Clinic model provides a framework for academic-industrial interaction involving undergraduate students. To ensure success of the project, a significant effort is made to focus and define the goals so that they are achievable in a semester or year-long time frame. It is crucial to maintain the flexibility necessary to tailor the project to the company's needs and corporate structure, and the three case studies provide three different project structures that fit with the needs of the sponsor. It is necessary also for the University administration to work with the company on issues such as confidentiality agreements, intellectual property agreements, budgets and payment schedules. Rigid policies in any of these areas would be an impediment to the collaboration.

Industrial Affiliates have provided generous support to the Rowan Clinic Program, and in turn have seen real economic, technical and personnel benefits. Major budgetary changes, reorganization, or personnel changes at a company represent concerns that could result in a hiatus or possible termination of the collaborative relationship (much the same as funding cuts at government funding agencies). It has, however, been our very positive experience that most companies are willing to continue supporting the program after the completion of the first successful project.

**ACKNOWLEDGEMENTS**

The support of our Clinic Affiliates, Johnson Matthey, General Mills, and Campbell Soup Company, is very gratefully acknowledged. These companies have demonstrated a tremendous commitment to the Rowan Engineering Program and to undergraduate engineering education.

## REFERENCES

1. Newell, J.A., Marchese, A.J., Ramachandran, R.P., Sukumaran, B., and Harvey, R., "Multidisciplinary Design and Communication: a Pedagogical Vision," The International Journal of Engineering Education, Vol. 15, No. 5, 1999, pp. 376-382.
2. Newell, J.A., Farrell, S.H., Hesketh R.P., and Slater, C.S., "Introducing Emerging Technologies into the Curriculum through a Multidisciplinary Research Experience," Chem. Eng. Educ., Vol. 35, No. 4, 2001.
3. Ustáriz, C., and Toledo-Quiñones, M., "Industrial Affiliates Program at UPRM, an Undergraduate Research Program," Proceedings of the 2003 International Conference on Engineering Education (ICEE-2003), Valencia, Spain, July 2003, Ref. No. 6572.
4. Palomera-Garcia, R., "Combining Undergraduate Research with Industrial Practice, A Successful Approach to University-Industry Relationship," Proceedings of the 2003 International Conference on Engineering Education (ICEE-2003), Valencia, Spain, July 2003, Ref. No. 5179.
5. ABET, Criteria For Accrediting Engineering Programs, http://www.abet.org/criteria.html.
6. M. Savelski, S. Farrell, J. Kumar and B. Sirchio, "Simultaneous Process Water and Energy Optimization in Food Manufacturing," AIChE 2003 Spring Meeting, Paper 106d, New Orleans, LA, April 2003.

**Stephanie Farrell** is Associate Professor of Chemical Engineering at Rowan University. She received her B.S. in 1986 from the University of Pennsylvania, her MS in 1992 from Stevens Institute of Technology, and her Ph.D. in 1996 from New Jersey Institute of Technology. Prior to joining Rowan in September 1998, she was a faculty member in Chemical Engineering at Louisiana Tech University. Stephanie's research and teaching interests are in the area of drug delivery and bioengineering.

**Robert Hesketh** is Associate Professor of Chemical Engineering at Rowan University. He received his B.S. in 1982 from the University of Illinois and his Ph.D. from the University of Delaware in 1987. After his Ph.D. he conducted research at the University of Cambridge, England, before joining the faculty at the University of Tulsa. His research is in the areas of reaction engineering, novel separations and green engineering.

**James A. Newell** is Associate Professor of Chemical Engineering at Rowan University. Prior to joining Rowan in 1998, he spent three years as an Assistant Professor at the University of North Dakota. His technical expertise is in the area of high performance polymers.

**Mariano J Savelski** is Assistant Professor of Chemical Engineering at Rowan University. He received his B.S. in 1991 from the University of Buenos Aires, his ME in 1994 from the University of Tulsa and his Ph.D. in 1999 from the University of Oklahoma. His technical research is in the area of process design and optimization with over seven years of industrial experience. His prior academic experience includes two years as Assistant Professor in the Mathematics Department at the University of Buenos Aires.

**C. Stewart Slater** is Professor and Chair of the Department of Chemical Engineering at Rowan University. He received his B.S., M.S. and Ph.D. from Rutgers University. Prior to joining Rowan, he was Professor of Chemical Engineering at Manhattan College. Dr. Slater's research and teaching interests are in separation and purification technology, laboratory development, and investigating novel processes for fields such as bio/pharmaceutical/food engineering and specialty chemical manufacture.

# Chapter 33

# An Interactive MATLAB-Based Tutorial for Model Order Reduction Techniques

ALI EYDGAHI[1,2], JALAL HABIBI[2] and BEHZAD MOSHIRI[2,3]
[1]*Department of Engineering and Aviation Sciences, University of Maryland Eastern Shore, Princess Anne, MD 21853, USA.*
[2]*Department of Electrical and Computer Engineering, University of Tehran, Tehran, I.R. Iran.* [3]*Control & Intelligent Processing, Center of Excellency, ECE Department, Faculty of Engineering, University of Tehran, Tehran, I.R. Iran. E-mail: aeydgahi@umes.edu, j.habibi@ece.ut.ac.ir, moshiri@ut.ac.ir*

*This paper presents a MATLAB-based toolbox with a Graphical User Interface (GUI), which has been developed to assist educators, students, and professionals in the area of control systems with computation of reduced order models of a large system. The toolbox consists of twenty order reduction techniques and can be run in two different modes of Normal and Tutorial. In Tutorial mode, a teaching dialog appears, which has embedded tutorials on order reduction concepts. The methods that have been implemented in the toolbox include the Padé, Routh, Cauer, Continued-Fraction Expansion, and algorithms that provide mixtures of these techniques. The toolbox GUI consists of four frames named "Methods," "High Order System," "Output Options," and "Results." In the "Methods" frame, one or more of the reduction techniques can be selected. The high order system can be defined by its transfer function numerator and denominator or its state space representation in the "High Order System" frame. In this frame, a number of high order benchmark systems, in a pop up menu, are provided and can be selected by the user for testing purposes. The order of the reduced model and the types of the desired output plots can be selected in the "Output Options" frame on the "Option" menu. The impulse response, step response, Bode diagram, and root locus graph are available as an output. In the output plot, the software shows and compares the reduced models with nominal system response. The graphs of step and/or impulse responses are displayed in two different windows. The numerator and denominator of the computed reduced order models are shown in the "Results" frame. The toolbox can be used as a teaching aid to enhance the students' understanding and visualization of the various theoretical concepts in model order reduction techniques.*

# INTRODUCTION

The modeling of complex dynamic systems is one of the most important subjects in engineering. A model is often too complicated to be used in real problems, so approximation procedures based on physical considerations or mathematical approaches are used to achieve simpler models than the original one. The subject of model reduction is very important to engineers and scientists working in many fields of engineering, especially for those who work in the process control area. In the control engineering field, model reduction techniques are fundamental for the design of controllers where particular numerically heavy procedures are involved. This would provide the designer with low order controllers that have more modest hardware requirements. Efforts towards obtaining low-order models from high-degree systems are related to the aims of deriving stable reduced-order models from stable original ones and assuring that the reduced-order model matches some quantities of the original one. In past decades, many papers in the literature have efficiently addressed these two objectives [1-28].

Polynomial reduction methods are one of the important groups of the reduction techniques that are applied in the frequency domain. They are used to approximate low-order transfer functions where the model coefficients are chosen according to various criteria. The frequency domain model-reduction techniques that are mainly based on polynomial manipulations are Padé-type approximations, continued fraction expansion methods, model reduction using the Routh stability criterion, model reduction based on the Routh table criterion, model order reduction algorithms using stability equations, mixed methods of approximation, and energy-based methods.

Popular software packages such as MATLAB, Maple, Mathematica, and PSpice are essential tools for analysis and design of an engineering system. Out of these software packages, MATLAB appears to be the most popular among academic institutions. The software is a general-purpose high-performance numerical computation engine with over 500 callable kernel functions to support numerical and graphics applications across a wide range of disciplines. Its functionality is extended by a number of toolboxes. Its open architecture allows a third party to create his or her own toolbox where necessary. The benefits of using these packages have been confirmed by a number of new and revised textbooks that have incorporated new exercises and problems based on them. These packages have been found to greatly enhance readers' understanding of the material and can prepare them for design and development of more complex systems, much like those seen in industry.

Development of software tools for teaching control engineering is of great importance [29-30]. Some ideas about the design and implementation of root locus in educational software for control engineering are presented in [31]. WinFACT, a software tool for teaching digital control, is introduced in [32]. This software has been successfully used at Brazil and Bolivia universities in graduate and undergraduate courses. A number of computer aided control system design and analysis packages have been introduced in [33]. A set of educational Java applets employed as educational tools in process control have been introduced in [34]. These applets cover some basic topics in process control such as cascade and feed-forward controllers, multi-loop PID control and decoupling as well as tuning methods for PID controllers and model predictive control. To improve learning by using multimedia technology and computers, three learning environments have been developed at the University of Hanover [35]. These environments are TESS (Tutorial system for the design of programmable logic controllers), MILAS (Multimedia information system for a cordless drill driver), and VAL (Virtual automation laboratory). In [36], a graphical and somewhat automated tool for classical control design methods with iterations has been reported.

This paper presents a graphical MATLAB-based toolbox for performing the computation needed to obtain reduced order models of large systems. The software's GUI provides an easy way for students to see the effect of each method. The toolbox presented in this paper extends the capacities of control methods as they are presented in textbooks. The software has options for twenty order reduction methods such as Padé approximation, Cauer continued-fraction expansion, Routh stability criterion, differentiation method, and a number of different mixed methods of frequency techniques.

The package can be used in either Normal mode or Tutorial mode. In Tutorial mode, students can select tutorials about order reduction concepts and be presented with step-by-step lessons to learn the concepts. Furthermore, the Bode plot, root locus, and time response of the reduced order systems are available and are tied together in such a way that students can see how changes affect each response. An example is provided to demonstrate the use and effectiveness of this toolbox. Some MATLAB code for order reduction methods have been adopted from [28].

## MODEL ORDER REDUCTION METHODS

In this section, brief descriptions of the order reduction techniques that have been implemented in the toolbox are provided, along with their advantages and disadvantages.

***Padé technique [1]:*** This approach is based on matching a few coefficients of the power series expansion, about $s = 0$, of the reduced-order model with the corresponding coefficients of the original model. The power series expansion of a transfer function about $s = 0$ is an expansion of the form:

$$H(s) = c_0 + c_1 s + c_2 s^2 + \Lambda \tag{1}$$

This method is computationally simple and provides perfect agreement between the steady state of the output of the system and the model for polynomial inputs belonging to the class of form $\sum \alpha_i t^i$. A serious drawback of Padé-based approximants is that unstable reduced-order models can be generated from an original asymptotically stable high order system.

***Cauer first, second, and third form [2-4]:*** The continued fraction expansion method is one of the most attractive methods for the order reduction of transfer functions. It has many useful properties, including computational simplicity, the fitting of the time moments, and the preservation of the steady-state responses for polynomial inputs of the form $\sum \alpha_i t^i$.

There are three basic forms for continued fraction representation of a transfer function. Each form concentrates on a different frequency range for the approximation of the original transfer function. They are referred to as the three Cauer forms. The Cauer first form is represented as:

$$H(s)=\cfrac{1}{f_1 s+\cfrac{1}{f_2+\cfrac{1}{f_3 s+\cfrac{1}{f_4+\cfrac{1}{O}}}}} \tag{2}$$

A reduced order model can be obtained by truncating the last terms of expansion. It has been shown that the reduced order model obtained from the Cauer first form cannot preserve the steady state value of the original model. The restriction to the original transfer function is that it requires the order of the numerator to be one less than that of the denominator. Otherwise, the Cauer first form does not exist. The reduced order model from the Cauer first form is the matching of the original transfer function by the Markov parameters, which is the power series expansion over $1/s$.

The Cauer second form is given as:

$$H(s)=\cfrac{1}{h_1+\cfrac{1}{\cfrac{h_2}{s}+\cfrac{1}{h_3+\cfrac{1}{\cfrac{h_4}{s}+\cfrac{1}{O}}}}} \tag{3}$$

It has been shown that this kind of continued fraction is equivalent to the Maclaurin expansion about $s=0$. This form matches the steady state responses, but does not provide a good match for the initial transient response.

The Cauer third form, also referred to as the mixed Cauer expansion, combines the Cauer first and second forms. It is given by:

$$H(s)=\cfrac{1}{h_1+f_1 s+\cfrac{1}{\cfrac{h_2}{s}+f_2+\cfrac{1}{h_3+f_3 s+\cfrac{1}{\cfrac{h_4}{s}+f_4+\cfrac{1}{O}}}}} \tag{4}$$

The Cauer third form can only be applied when the order of the numerator of the original transfer function is one less than that of the denominator. The mixed form is equivalent to the Maclaurin series about $s=0$ and $s=\infty$. The reduced order model matches the first $m$ time moments and $m$ Markov parameters of the original model, where $m$ is the order of reduced order model. The one serious disadvantage of this method is that it may produce unstable (non-minimum phase) reduced models even when the original high-order system is stable (minimum phase).

***Order reduction based on stability equation [5]:*** This method is based on forming stability equations for the numerator and denominator polynomials of the high-order system, discarding the non-dominant poles and zeros, and obtaining the reduced-order model. This approach preserves the stability of the reduced-order model. The disadvantages of this approach are related to the fact that it is founded on heuristic considerations and, aside from the structural stability that is preserved in the reduced-order model, further correlations between the original and the reduced model are lost.

***Differentiation method [6]:*** In this approach, the reciprocals of the numerator and denominator polynomials of the high-order transfer function are differentiated suitably many times to yield the coefficients of the reduced order transfer function. This method is computationally simple and is equally applicable to unstable and nonminimum phase systems.

***Aggregation method [7]:*** The aggregation method is based on the concept of combining a defined set of state variables of the original system, involving chosen weighting factors. The set of state variables $(z)$ is often computed from the original system state vector $(x)$ using an aggregation matrix $(L)$ by $z = Lx$. The adequacy of finding the suitable aggregation state vector, and therefore the matrix $L$, depends upon the objective that the designer wants to reach, such as open-loop or closed-loop performance, optimisation of certain indexes, etc.

***Order reduction based on the Routh stability criterion [8]:*** A system with transfer function $M(s)$ can be expanded into the so-called $\alpha$ and $\beta$ parameters as:

$$M(s) = \beta_1 f_1(s) + \beta_2 f_1(s) f_2(s) + \Lambda + \beta_n f_1(s) f_2(s) \Lambda\ f_n(s) \qquad (5)$$

where $n$ is the degree of high-order system, $\beta_i$ are constants, and the functions $f_i(s)$ are defined as:

$$f_i(s) = \cfrac{1}{\alpha_i s + \cfrac{1}{\alpha_{i+1} + \cfrac{1}{\text{O} \cfrac{1}{\alpha_{n-1} s + \cfrac{1}{\alpha_n s}}}}} \qquad (6)$$

The first term is $1/(1+\alpha_1 s)$ instead of $1/\alpha_1 s$. Order reduction based on the Routh stability criterion requires computing the $\alpha_i$ and $\beta_i$ parameters, and then reconstructing the reduced-order models with certain algorithms after dropping out a few $\alpha_i$ and $\beta_i$ parameters. This method cannot be applied to non-asymptotically stable systems.

***Simplified Routh Approximation Method (SRAM) [9-11]:*** This method obtains the reduced-order models using only $\alpha$ parameters, where $\alpha$ parameters are defined above. The simulated version of SRAM in this toolbox also retains some time moments of the full system in the reduced model.

***Order reduction based on real Schur form decomposition [12]:*** Model order reduction, based on measures of controllability and observability, has been proposed by Moore [13] and extended and generalized by other researchers. In this approach, the system is transformed using a balancing transformation such that the input-to-state coupling and the state-to-output coupling are weighted equally. State components that are weakly coupled to both input and output are discarded while those that are strongly coupled to both input and output are retained and used as a low-order approximation for the high order system.

Calculating the balancing transformation in Moore's method tends to be ill-conditioned, especially when the original system is non-minimal or when it has very nearly uncontrollable or unobservable modes. Therefore, an alternative algorithm has been proposed based on finding the eigenspaces associated with large eigenvalues of the cross-gramian matrix using the real Schur-form decomposition. This method avoids calculating the balancing transformation.

***Stable Padé [14]:*** This approach combines the Routh-Hurwitz array method and the Padé approximation technique to obtain stable reduced-order models. It is computationally efficient and does not require computation of the poles of the original system for dominant pole retention. This is not the only method that solves the stability problem of Padé approximation.

***Condensed continued fraction expansion method [15]:*** Much attention has been diverted to overcoming the "stability" problem of order reduction techniques based on continued fraction expansion (CFE). Condensed CFE has been proposed to improve the chances of stability in the reduced model. One of the main causes of instability in the reduced-order models obtained by Cauer continued fractions is thought to be the dependence of the reduced denominator on the full numerator. This is especially true if the numerator has zeros in the right half-plane. The condensed CFE method proposes to use Cauer CFE only on $1/D(s)$, (where $D(s)$ is the denominator) instead of $G(s) = N(s)/D(s)$. The effect of $N(s)$ on the reduced denominator is thus removed, and the $k$-th order approximants of $1/D(s)$ can then be multiplied by $N(s)$, truncating the numerator after the term in $s^{k-1}$. This latter operation ensures that the first $k$ time moments are preserved in the reduced model. However it should be remembered that unstable poles might still occur, and experience has shown that lightly damped, highly oscillatory systems tend to cause the most problems.

***Moment matching technique [16]:*** The moment matching technique is in fact the matching of the time moments of the impulse response of the original model with those of the reduced order model. This reduction method is not restricted to rational transfer functions since other functions, once written in the Maclaurin form, can also be reduced using the moment matching method. This approach, though simple in theory and computation, appears to have serious drawbacks. For example, stable reduced-order models may be obtained from a high-order, unstable system.

***Truncation approach [17-18]:*** From the viewpoint of the condensed CFE order reduction approach, if $[p,q]$ is taken to be the Padé approximant about $s = 0$ whose numerator and denominator degrees are $p$ and $q$ respectively, then the following result holds for $k$-th order approximants of $1/D(s)$. If

$$\frac{1}{D(s)} = \frac{1}{a_0 + a_1 s + \Lambda + a_n s^n} \tag{7}$$

then the $[m,k]$ approximant of $1/D(s)$, where $0 \le m \le k-1$, is equivalent to the two-point Padé/continued-fraction approximant of $1/D(s)$ matching $(k+1+m)$ terms about $s = 0$ and $(k - 1 - m)$ terms about $s = \infty$. This follows from the fact that the first $(n - 1)$ terms of $1/D(s)$ about $s = \infty$ are all zero. In the special case of $m = 0$, let the approximant be given by

$$\frac{1}{e_0 + e_1 s + \Lambda + e_k s^k} \tag{8}$$

The series expansion of this must be identical to that of $1/D(s)$ for all terms in $s^k$. By equating the two series it is easily seen that $e_i = a_i$ for *i = 0, 1, 2, ..., k*. Forming the final reduced transfer function by multiplying $N(s)$ and truncating after the term in $s^{k-1}$ shows that the model is identical to that obtained by simple truncation of terms in the full denominator and numerator, respectively. Yeung in [18] has shown that for the truncation method, if the poles of the system are well damped (i.e. |real part| > |imaginary part|), then stability is guaranteed in the reduced model.

***Factor division (Padé with dominant mode retention) [19]:*** In this approach, the numerator of the reduced model is obtained in a direct way. It avoids finding system time moments and does not solve the Padé equations, but obtains the reduced models that still retain the initial time moments of the full system. This method retains the dominant modes of the system and has been claimed to be often quicker to use than the similar method of [20], at least when the poles of the high order system are known.

***Mixed method 1 [21]:*** Generally speaking, mixed methods are developed to ensure the stability of reduced models. They often use criteria that assure the stability of the reduced-order model if the original system has this property. The numerator is obtained by using the Padé method or the continued fraction expansion.

In the first mixed method, the denominator of the reduced order transfer function is formed by the dominant poles of high order system, while the numerator is obtained by fitting the first $r$ time moments of the reduced order system to the high order one.

***Mixed method 2 [22]:*** In this approach, the Routh stability criterion is used to compute the denominator of the reduced model. The numerator is obtained as in the first mixed method. This approach avoids the drawback of the first method, which is its need for computation of system poles.

***Mixed method 3 [23]:*** This method is based on determining the denominator using the Routh array method. The numerator is obtained by using the Padé approximant technique.

***Mixed method 4 [24]:*** This method leverages both the advantages of the stability equations method for obtaining a reduced order stable system and the Padé method for finding the numerator. In this way, it fits a number of the Markov parameters and time moments of the original transfer function.

***Mixed method 5 [25]:*** In this method, the differentiation procedure [6] is used to obtain the reduced order denominator, thus ensuring stability preservation, while the numerator is obtained by matching the first MacLaurin expansion coefficients of the original system with the model transfer functions using the Padé method.

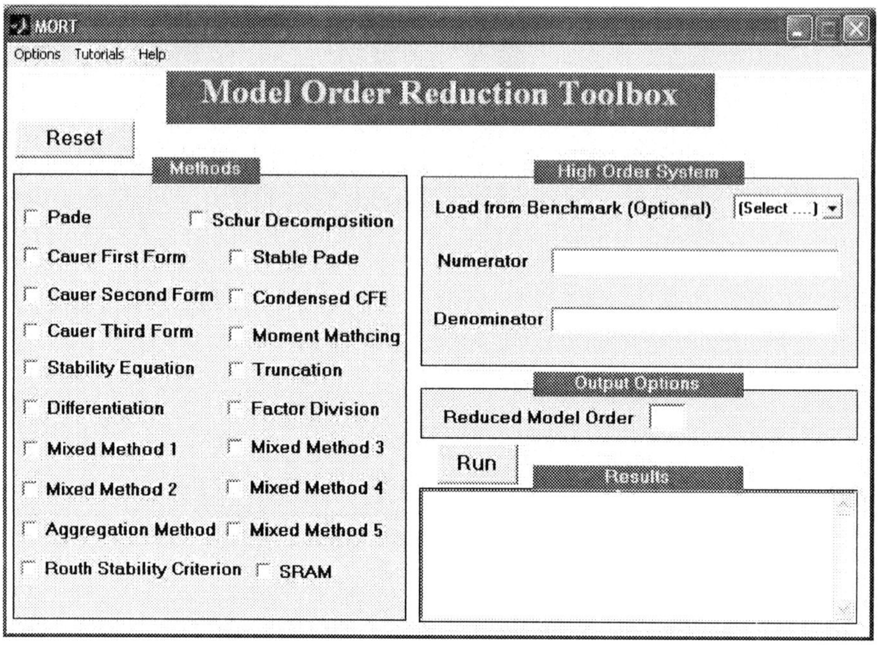


FIGURE 1

WORKSPACE OF THE GRAPHICAL USER INTERFACE OF THE TOOLBOX IN NORMAL MODE

## DESCRIPTION OF THE TOOLBOX

The toolbox has been developed for both professional users to compute order reduction methods and for students to learn the techniques of model order reduction. Therefore, the toolbox can be used in two modes: Normal and Tutorial. Upon execution of the toolbox, a GUI appears with four frames named '**Methods**,' '**High Order System**,' '**Output Options**,' and '**Results**.' Figure 1 shows the workspace of the GUI in the Normal mode of operation.

In the '**Methods**' frame, one or more of the reduction techniques can be selected. Information about the high order system can be input either as a transfer function or in a state space form. The transfer function is defined by its numerator and denominator in the '**High Order System**' frame context. In this frame, a number of high order benchmark systems are provided in a popup menu that can be selected by the user for testing purposes. The user can also select **High Order System Description** from the **Options** menu to input the high order system in its state space representation. If the user selects the state space description, the '**High Order System**' frame changes accordingly, as shown in Figure 2. The user can then define the high order system by four matrices of the state space representation.

In the '**Output Options**' frame, the order of the reduced model is selected. The types of the desired output plots can be selected from the **Options** menu as shown in Figure 3. Impulse response, step response, Bode diagram, and root locus are available as output plots. In the output plot, the software shows and compares the responses of the reduced models with nominal system response. If the root locus diagram was selected as an output

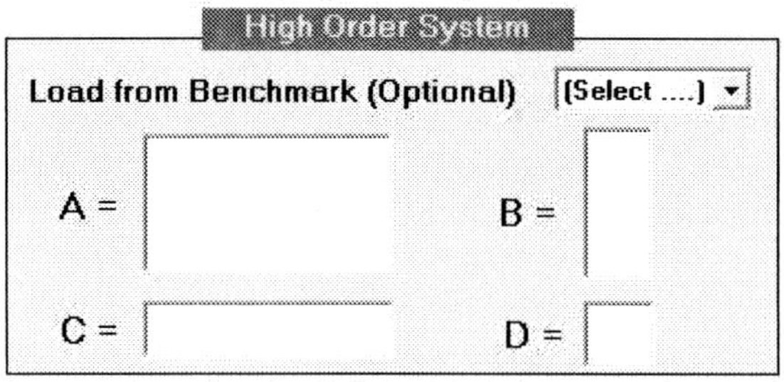


FIGURE 2
FRAME FOR STATE SPACE FORM OF HIGH ORDER SYSTEM

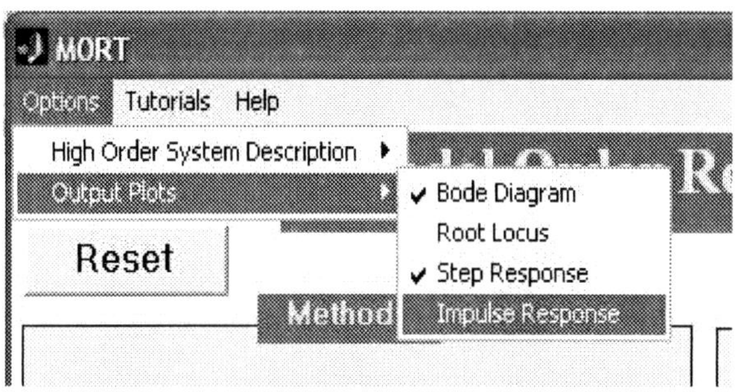


FIGURE 3
OUTPUT PLOTS OPTIONS

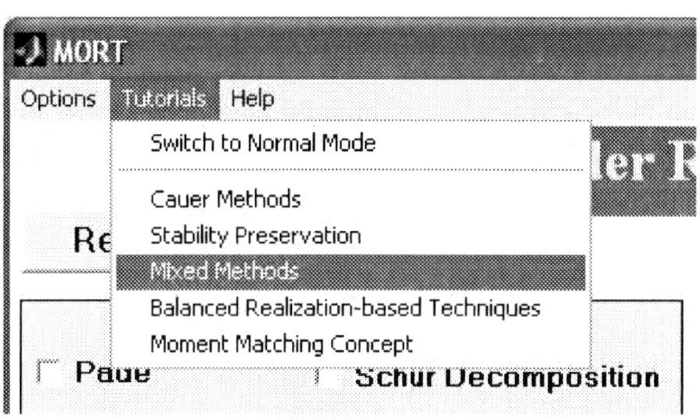


FIGURE 4
TUTORIAL SELECTION

plot, the distinct graphs with proper titles are plotted for each of selected methods of order reduction

When the user clicks the '**Run**' button, the software starts the computation of the reduced order model. The selected graphs appear in different windows. The numerator and denominator of the computed reduced order models will be shown in the '**Results**' frame. The user can easily compute another model or change coefficients for the current model through the GUI. A '**Reset**' button is also provided to clear all frames and start computations for a new system.

Some tutorials have been prepared and embedded in the toolbox. The user can select the desired topic from the **Tutorials** menu as shown in Figure 4. The user can switch from Normal mode to Tutorial mode by selecting '**Switch to Tutorial Mode**' from the **Tutorials** menu. In this mode, the toolbox is a self-learning tool and some lessons can be learned in a step-by-step manner. The appearance of the toolbox in this mode is shown in Figure 5. The selected tutorial appears in the '**Teaching Dialog**' frame and students can follow the steps in each lesson to learn the corresponding order reduction concepts.

Useful hints about order reduction methods can be found from the **Help** menu. The HTML-based help appears in the MATLAB help window as shown in Figure 6. This feature of the toolbox is implemented by using WHICH and WEB, two special MATLAB commands.

An example is provided to demonstrate the performance of the order reduction algorithms of the toolbox.

## EXAMPLE

In this section, we present an illustrative example to show the operation of the order reduction toolbox. Some methods inherently place some restrictions on the high order system. For example, Cauer first and third forms restrict the relative degree of the high order system to be equal to one. Thus, as mentioned in the previous section, some

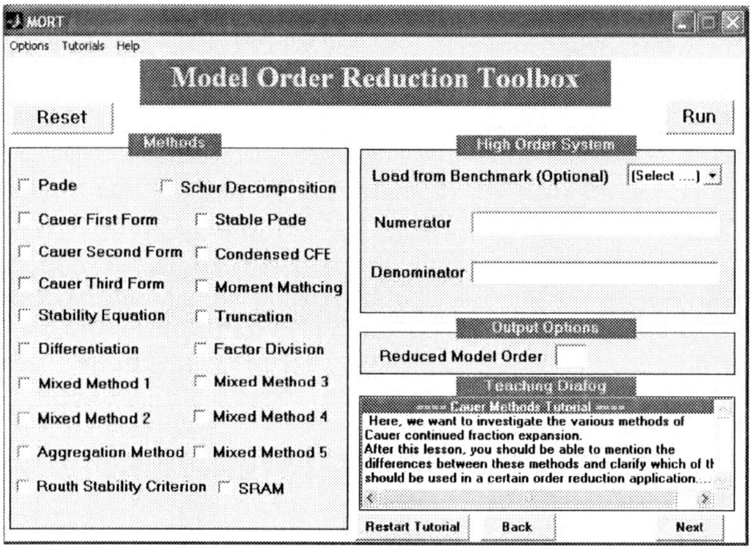


FIGURE 5

WORKSPACE OF THE GRAPHICAL USER INTERFACE OF THE TOOLBOX IN TUTORIAL MODE

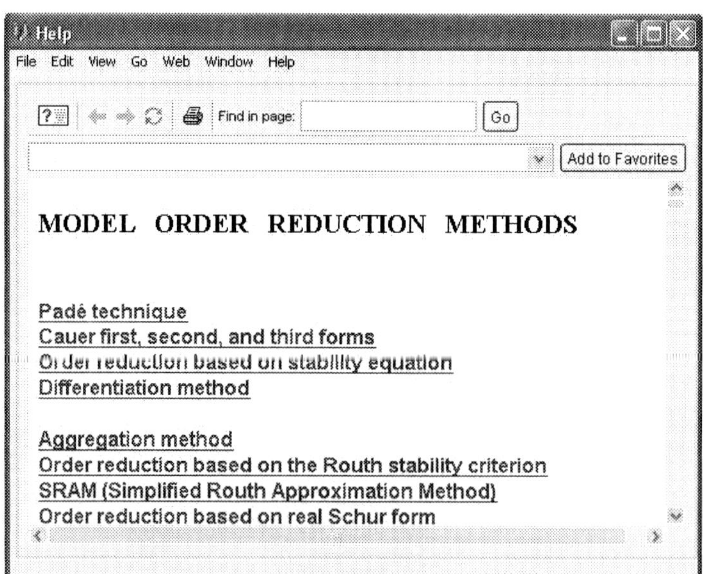


FIGURE 6

HTML-BASED HELP OF THE TOOLBOX

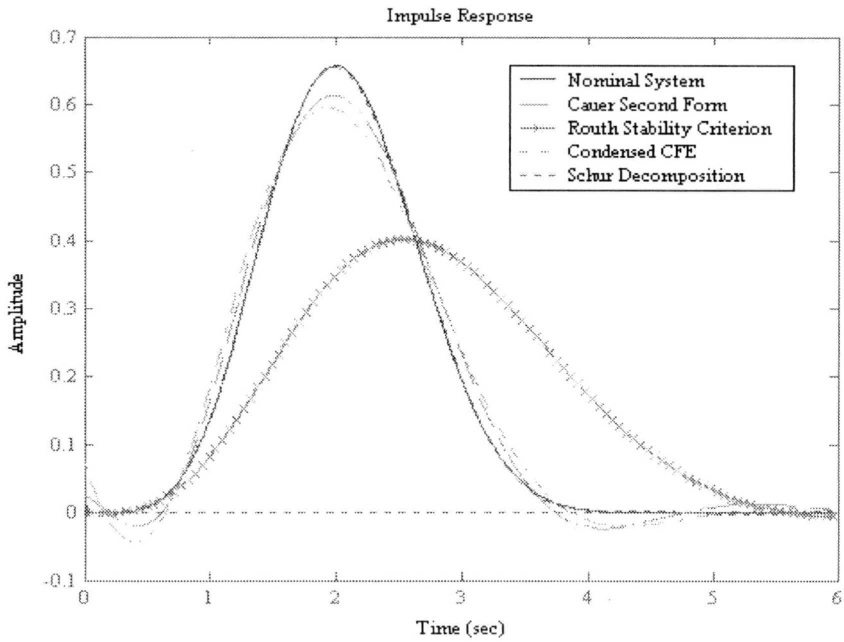


FIGURE 7
IMPULSE RESPONSE COMPARISON OF EXAMPLE

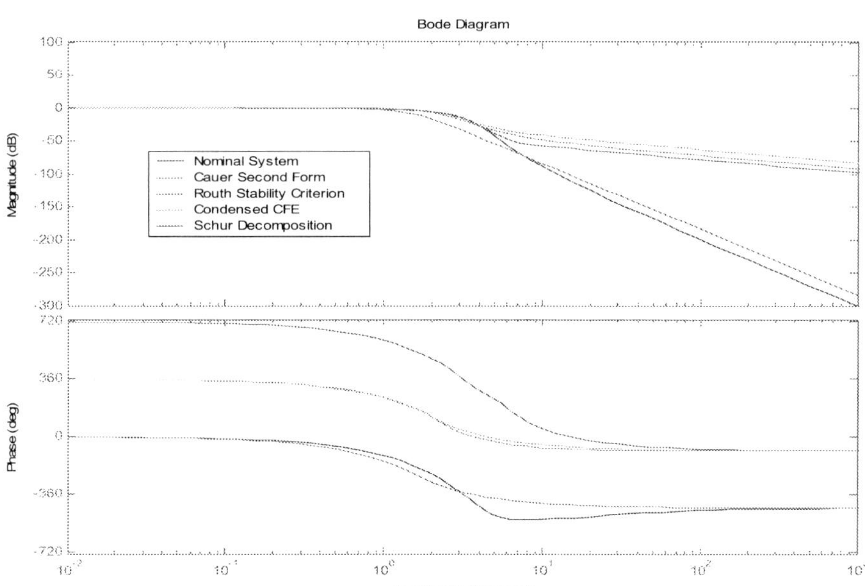


FIGURE 8
BODE DIAGRAM COMPARISON OF EXAMPLE

Assume the high order system from the benchmarks of the form:

$$H(s)=\frac{s^4+35s^3+291s^2+1093s+1700}{s^9+9s^8+66s^7+294s^6+1029s^5+2541s^4+4684s^3+5856s^2+4629s+1700} \tag{9}$$

Let us use Cauer second form, Routh stability criterion, Schur decomposition, and condensed CFE methods to find the reduced order models of order 5. Figures 7 and 8 show the impulse and Bode responses for the nominal system and reduced order models obtained using the selected methods.

It can be seen that the Schur decomposition method produces the most accurate model.

## CONCLUSIONS

Personal computers are a learning aid to students. The students' motivation and determination lead them to successfully and persistently work on the computer and learn about the subject in their own time. Simulations have been found to be useful in education as a way to gain practice before facing excessively complicated real problems. Simulations help to understand the dynamics of different systems and to find what is possible and acceptable. Computer aided engineering software is an important aspect of teaching, since it makes it possible to significantly increase personal efficiency in problem-solving. It enables us to deal with realistic problems, which makes teaching more interesting. The use of new methods motivates students and makes studies more flexible by allowing learning by doing. Software tools such as the one presented in this paper strengthen the link between theory and practice. Class project and homework assignments should be designed to challenge students to correlate observations from the simulation with the theory presented in class. By using the online help capabilities of the toolbox, students can progress on their own through the tutorials that accompany the software and the examples given in class. Control applications for illustrating and exposing students to different aspects of model order reduction are only limited by the imagination of the instructor and scope of the course.

From the students' view, the approximation process can be very confusing and laborious. Students would benefit greatly from an interactive tool with a graphical user interface that automates much of the computation and provides the appropriate results of the order reduction process. The toolbox presented in this paper has been used at Tehran University to enhance the way model reduction techniques are taught. The classical and theoretical methods are taught, and this interactive software is used to take the students to the next level beyond textbooks and lecture notes. The toolbox is designed for use at an introductory level for graduate and senior undergraduate students, and emphasizes the concepts of model order reduction. The software is self-explanatory and uses menus, dialog boxes with figures and context-sensitive online help. In this way, students are motivated to investigate different methods and concepts of model reduction. Students who have used this tool in the computer aided control system analysis and design class, which is a graduate level course at Tehran University, have given extremely positive feedback about it. The toolbox is available free of charge. Those interested in obtaining the package should send their request to the first author.

# REFERENCES


1. Bultheel, A. and Van Barel, M., "Padé Techniques for Model Reduction in Linear System Theory: A Survey," Journal of Computational and Applied Mathematics, Vol. 14, 1986, pp. 401-438.

2. Chen, C. F. and Shieh, L. S., "A Novel Approach to Linear Model Simplification," International Journal of Control, Vol. 8, 1968, pp. 561-570.

3. Shieh, L. S. and Goldman, M.J., "Continued Fraction Expansion and Inversion of the Cauer Third Form," IEEE Transactions on Circuits and Systems, Vol. 19, 1974, pp. 341-345.

4. Rathore, T. S., Singhi, B. M. and Kibe, A. V., "Continued Fraction Inversion and Expansion," IEEE Transactions on Automatic Control, Vol. 24, No. 2, 1979, pp. 349-350.

5. Chen, T. C., Chang, C. Y. and Han, K. W., "Reduction of Transfer Functions by the Stability-Equation Method," Journal of the Franklin Institute, Vol. 308, No.4, 1979, pp. 389-404.

6. Gutman, P., Mannerfelt, C. F., Molander, P., "Contributions to the Model Reduction Problem," IEEE Transactions on Automatic Control, Vol. 27, 1982, pp. 454-455.

7. Aoki, M., "Some Approximation Methods for Estimation and Control of Large Scale Systems," IEEE Transactions on Automatic Control, Vol. 23, 1978, pp. 173-182.

8. Hutton, M. F. and Friedland, B., "Routh Approximations for Reducing Order of Linear Time Invariant Systems," IEEE Transactions on Automatic Control, Vol. 20, 1975, pp. 329-337.

9. Sastry, G. V. K. R. and Krishnamurthy, V., "Relative Stability Using Simplified Routh Approximation Method," Journal of Institute of Elect. & Telecom Engineering, Vol. 33, 1987.

10. Sastry, G. V. K. R. and Krishnamurthy, V., "Biased Model Reduction by Simplified Routh Approximation Method," Electronic Letters, Vol. 23, No. 20, 1987, pp. 1045-1047.

11. Sastry, G. V. K. R. and Krishnamurthy, V., "State-Space Models Using Simplified Routh Approximation Method," Electronic Letters, Vol. 23, No. 24, 1987, pp. 1300-1301.

12. Aldhaheri, R.W., "Model Order Reduction via Real Schur-Form Decomposition," International Journal of Control, Vol. 53, No. 3, 1991, pp. 709-716.

13. Moore, B. C., "Principal Component Analysis in Linear Systems: Controllability, Observability and Model Reduction," IEEE Transactions on Automatic Control, Vol. 26, 1981, pp. 17-31.

14. Pal, J., "Stable Reduced-Order Padé Approximants Using the Routh-Hurwitz Array," Electronic Letters, Vol. 15, No. 8, 1979, pp. 225-226.

15. Lucas, T. N., "Model Reduction by Condensed Continued-Fraction Method," Electronic Letters, Vol. 21, No.16, 1985, pp. 680-681.

16. Lal, M. and Mitra, R., "Simplification of Large System Dynamics Using a Moment Evaluation Algorithm," IEEE Transactions on Automatic Control, Vol. 19, 1974, pp. 602-603.

17. Shamash, Y., "Truncation Method of Reduction: A Viable Alternative," Electronic Letters, Vol. 17, 1981, pp. 97-99.

18. Yeung, K. S., "Stability of Reduced Model Obtained by Truncation," Electronic Letters, Vol. 17, 1981, pp. 374-375.

19. Lucas, T. N., "Factor Division: A Useful Algorithm in Model Reduction," IEE Proceedings, Vol. 130, Pt. D, No. 6, 1983, pp.362-364.

20. Shamash, Y., "Linear System Reduction by Padé Approximation to Allow Retention of Dominant Modes," International Journal of Control, Vol. 21, 1975, pp. 257-272.

21. Davison, E. J., Chidambara, M. R., "On a Method for Simplifying Linear Dynamic Systems," IEEE Transactions on Automatic Control, Vol. 12, 1967, pp. 119-121.

22. Shamash, Y., "Model Reduction Using the Routh Stability Criterion and the Padé Approximation Technique," International Journal of Control, Vol. 21, No. 3, 1975, pp. 475-484.

23. Pal, J. and Ray, L.N., "Stable Padé Approximant in Multivariable Systems Using a Mixed Method," Proceedings of IEEE, Vol. 68, No. 1, 1980, pp.176-178.

24. Chen, T. C., Chang, C. Y. and Han, K. W., "Model Reduction Using the Stability-Equation Method and the Padé Approximation Method," Journal of the Franklin Institute, Vol. 309, No.6, 1980, pp. 473-490.

25. Lepschy, A., Viaro, U., "A Note on Model Reduction Problem," IEEE Transactions on Automatic Control, Vol. 28, 1983, pp. 525-527.

26. Eydgahi, A.M., Singh, H.; "Family of Reduced-Order Models for Linear Multivariable Systems," Electronics Letters, Vol. 23, No. 22, 1987, pp. 1178-1180.

27. Eydgahi, A.M., "A Mixed Method for the Simplification of High Order Systems," Proc. American Control Conference, Atlanta, GA, 1988, pp. 1064-1067.

28. Fortuna, L., Nunnari, G. and Gallo, A., Model Order Reduction Techniques with Application in Electrical Engineering, Springer-Verlag, New York, 1992.

29. Prendergast, D.P., Eydgahi, A.M., "EDCON: An Educational Control Systems Analysis and Design Program," IEEE Transactions on Education, Vol. 36, No. 1, 1993, pp. 42-44.

30. Wahid, M.S., Eydgahi, A.M., "Computer Simulation for Teaching System Stability Concepts," International Journal of Applied Engineering Education, Vol. 7, No. 5, 1992, pp. 392-397.

31. Eydgahi, A.M. and Ghavamzadeh, M.; "Complementary Root Locus Revisited," IEEE Transactions on Education, Vol. 44, No. 2, 2001, pp.137-143.

32. Bol, E. W. et al. "WinFACT - A Tool for Teaching Digital Control," Proc. 38th Conference on Decision and Control, Phoenix, AZ, 1999, pp. 353-358.

33. Jamshidi, M. and Herget, C. J., Computer-Aided Control Systems Engineering, North-Holland, New York, 1985.

34. Yang, D. R. and Lee, J. H., "Java Applet Modules for Undergraduate Process Control Education," Proc. American Control Conference, Denver, CO, 2003, pp. 694-699.

35. Wagner, B., "From Computer-Based Teaching to Virtual Laboratories in Automatic Control," Proc. 29th IEEE Frontiers in Education Conference, San Juan, Puerto Rico, 1999, pp. 6-11.

36. Garcia, R. C. and Heck B. S., "Enhancing Classical Controls Education via Interactive GUI Design," IEEE Control Systems Magazine, Vol. 19, 1999, pp. 77-82.

**Ali Eydgahi** is a Professor of Electrical Engineering at the University of Maryland Eastern Shore. He received his Ph.D. and M.S. in Electrical and Computer Engineering from Wayne State University. Since 1985 and prior to joining the University of Maryland Eastern Shore, he was with the Rensselaer Polytechnic Institute, State University of New York, University of Tehran, Wayne State University, and Oakland University. Dr. Eydgahi was the recipient of the Dow Outstanding Young Faculty Award from the American Society for Engineering Education in 1990, and the Silver Medal for outstanding contribution from the International Conference on Automation in 1995. He has served as a regional and chapter chairman of IEEE and SME and as a session chair and a member of scientific and international committees for many international conferences. He has published more than ninety papers in refereed international and national journals and conference proceedings.

**Jalal Habibi** received his B.Sc. and M.Sc. in Electronics and Control Engineering from Sharif University of Technology in 1996 and 1999, respectively. Currently, he is a Ph.D. student in Control Engineering in the Electrical and Computer Engineering Department of the Faculty of Engineering at the University of Tehran. His main research interests are predictive control and hybrid systems.

**Behzad Moshiri** was born in Tehran, Iran, on May 16, 1959. He received his B.Sc. degree in Mechanical Engineering from Iran University of Science & Technology (IUST) in 1984. He pursued higher education studies to obtain the M.Sc. and Ph.D. degrees in Control System Engineering from the University of Manchester Institute of Science & Technology (UMIST), UK in 1987 and 1991, respectively. He joined the Department of Electrical & Computer Engineering at the University of Tehran, Iran in 1992, where he is currently an associate professor and senior member in Control & Intelligent Processing, Center of Excellence (CIPCE). He has also been an adjunct professor at the University of the Ryukyus, Okinawa, Japan, since January 2004. Dr. Moshiri is currently the vice-president of the Iranian Society of Instrument and Control Engineers (ISICE). He has published more than eighty papers in refereed international and national journals and conference proceedings. His major fields of research are advanced process control, intelligent control system, sensor-data fusion and mechatronics.

# Chapter 34

# A Virtual Verification and Execution of Grafcet Using VRML

MOHAMED MHAMDI[1], RAFIK BRAHAM[2], MOHAMED MOALLA[3] and HAMADOU SALIAH-HASSANE[4]
*[1]Research Unit PRINCE, ISET Sousse, Tunisia. [2]Research Unit PRINCE, ISITC H. Sousse, Tunisia. [3]Laboratory LIP2, Faculty of Sciences of Tunis, Tunisia. [4]Research Center LICEF, TÉLUQ, Quebec, Canada. E-mail: Mohamed.Mhamdi@isetso.rnu.tn*

*A given system design description must be structured into both an operative and sequence control part. From the system point of view, GRAFCET is the tool of choice to implement control structures according to the French Standard NF C03-190/European Standard CEI IEC 848 programming language standards, and is a well-known and recognized graphic tool in the educational and industrial world for describing the specification of a sequence control system. In a teaching context, the operative part is often represented by didactic components, therefore a student must: (a) elaborate the Grafcet according to a system viewpoint; (b) implement the Grafcet in whole or in part on a programmable logic controller (PLC); (c) validate the behavior of the operative part according to the implementation mentioned in (b).*

*In this paper we describe a system that allows one to accomplish these three tasks without the need of a real PLC or an operative part. A visual simulation is possible by using a Virtual Reality Modeling Language (VRML) viewer to validate the conformity of the Grafcet with the expected behavior. The virtual didactic models are stored in a library of reusable components. The reusability of components is based on the following features: (i) the 3D description of the components is made by using the standard VRML language, which makes them portable on many types of platforms; (ii) the storage of the components in a spatiotemporal database based on a relational model gives them a modular structure that facilitates reusability; (iii) the component is a software component that can be used in a virtual environment (simulation) or in a real environment (cosimulation). The system can also be combined with different types of sequence control systems in order to create new types of automatic operations.*

## INTRODUCTION

The purpose of our research is to make the teaching of sequential automatics more interesting, more active, more individual and yet based on many media. To meet these objectives, the traditional approach based on theoretical courses and written exercises must be complemented by an approach based (partially or completely) on experimentation.

The principal question that must be treated in this paper is how to integrate an experimental approach into an automatics course. In fact, it could be the pedagogical effect of integrating some simulation exercises or practical sessions into the process. We have chosen the use of simulation since it can be done anywhere, any time.

Also, simulation makes it possible to replace experimental installations with cheap, computer-run software models. Students can in this way freely explore these models instead of real systems.

On the one hand [1], the simulation tools can be presented to students in two forms, namely, either by using a multi-function software program (for example MATLAB), or programs specialized in a single topic. These programs are intended for specific uses, like illustrating elements in difficult theories or acquiring given knowledge in a given field. We call this last type of teachware a tutorial.

On the other hand, e-learning and technologies on which it is based are progressing rapidly; corporate use of e-learning is growing exponentially. Currently, there is an ever-increasing demand for next generation e-learning: simulation-based e-learning [2], that is, a blend of e-learning and simulation. This is a simulation-based online training that immerses the student in the proper context while learning.

Consequently, we were ready to invest in setting up a tutorial that simulates in 3D the sequential automatic operations based on the use of GRAFCET, and which can be used in a simulation-based e-learning context. This requirement implies that this teaching tool is a shareable resource used by several partner-students working in parallel or in an autonomous way. Note that Grafcet is currently the tool of choice to implement control structures according to the French Standard NF C03-190/European Standard CEI IEC 848 programming language standard; it can facilitate teaching basic control systems and provide thorough accompanying documentation for pneumatic, hydraulic and electrical projects.

With the defining specification, we were able to formalize our objectives for the realization of a 3D simulator for vocational teaching. The system can be considered as a tool for learning Grafcet and automatics in a soft way, without worrying about the evolution of the manufacturer's languages, or about the availability of the real operative parts.

The elaborated specification dissociating C.P. (control part) and O.P. (operative part) necessitated dealing with three types of problems:

1. Firstly, there are problems connected with the C.P. simulation:

Grafcet's expressive capabilities led industrialists to develop and incorporate many Grafcet languages into their material. But none of these implementations have the same semantic behavior. Above all, these languages interpret Grafcet's basic rules and integrate badly the evolution model's assumption of zero-time. This has multiple consequences. Firstly, users can confuse the industrial language and model; secondly,

Grafcets are not compatible between industrialists; and finally, one can program systems specified by Grafcet in different ways by using different Grafcet languages [3].

To circumvent some of these problems, we have conceived and implemented within our system a "Grafcet Engine". This engine can be considered as a simple and straightforward Grafcet and automatics training tool, without worrying about the evolution and heterogeneity of industrial languages. This type of problem is dealt with in one of the following sections.

2. Secondly, there are problems connected with the O.P. simulation:

- No standard language really exists to describe models for system components with dynamic behaviors such as engines, arm manipulators, etc.
- No one proposes reference models to represent system components and their complex, dynamic behaviors.
- Discrete event simulators and robot simulators are costly. Moreover, simulation environments of many products are not open and are limited to specific computer and operating system platforms.

In order to overcome these problems, three things are required: a standard language, a reference model and an inexpensive environment that is independent of all platform types. To meet these needs, we find that a simulation system based on a Virtual Reality Modeling Language (VRML) viewer is a good choice. In fact, VRML can be used as a component modeling language (operative part) and the simulation can be run via a VRML viewer. We detail the adopted solutions to these problems in the following sections.

3. Thirdly, there are problems connected with the use of the simulation environment:

In a teaching context, the operative part is often represented by didactic components, therefore a student must: (a) elaborate the Grafcet according to a system viewpoint; (b) implement the Grafcet in whole or in part on a programmable logic controller (PLC); (c) validate the behavior of the operative part according to the implementation mentioned in (b).

While completing these three tasks, the student encounters difficulties stemming from various sources:

- Difficulties due to the automatic system, such as identifying system input/output information, coordinating operative tasks and defining the relative movements of kinematically linked component subsets.
- Difficulties due to the operative part, such as identifying actuators, pre-actuators and sensors.
- Difficulties due to the sequence control component, such as making logical equations associated with the transition, or identifying parallelism and alternative sequences.
- Difficulties linked with the dialogue between the operator and the sequence control component, such as identifying the various dialogue components and their functions.
- Difficulties arising when implementing the automated system, such as connecting a pre-actuator to an actuator, or a sensor to a PLC Input/Output module and implementing the Grafcet in whole or in part on a PLC.

To solve these problems [4], the best idea is to be able to create an integrated environment that is entirely configurable, thus making it possible to create and develop programs in a simulation mode. Our system includes an implementation of a "Grafcet Engine" and a virtual 3D simulation of operative parts. The next section discusses a general architecture designed to solve some of these difficulties.

## General System Architecture

Based on an integrated approach, Figure 1 shows the general architecture of the Grafcet and automatics training system we have developed. The components simulated by our system are those mainly found in teaching laboratories and are generally didactic components.

The system offers students a tool for visualizing the behavior of their sequence control systems specified in Grafcet. The system can create a VRML representation of the O.P and render it to the student.

The system architecture follows a three-tier scheme:

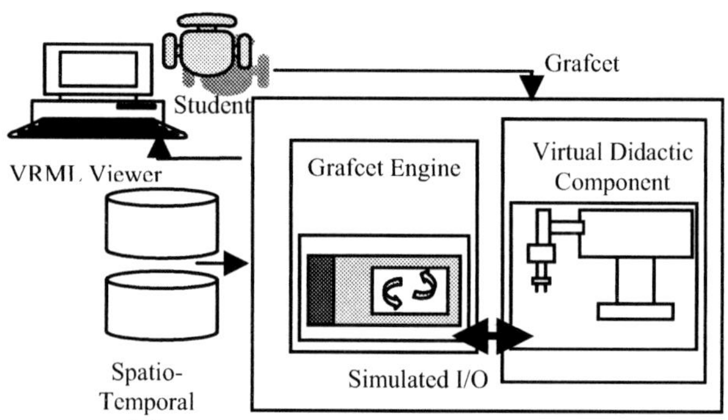


FIGURE 1
General system architecture

- The database contains information on the various mechanical components with their virtual representations in VRML. This database is considered a learning object repository. The learning objects are useful not only in a single context, but can be used in different ways, for different purposes, and in different contexts.
- The application server is responsible for execution of the student Grafcet using:
  - A Grafcet Engine implemented in VRML script.
  - A Grafcet Engine implemented in reactive script (a more complicated implementation which allows better simulation).
- The application client offers a highly interactive interface and communicates with the application server to provide the user with a simulation.

As shown in Figure 1, the developed system is based on a client/server (C/S) architecture. The server side consists of a Grafcet Engine simulating the Grafcet introduced by the student. The application server receives requests to connect to the engine, the Grafcet to be simulated, commands to be executed by the engine and simulated input signals in order to modify Grafcet transitions.

A client is an Internet application that is comprised of a Web browser, a VRML viewer, an EAI (external authoring interface) and a Java applet that controls communication with the server.

Inside the client applet, the VRML sends student/component events that are translated into commands that will be sent to the server. The EAI enables Java applets to manipulate the VRML world dynamically. The applet translates the commands injected by the server into the received control frames, into a manipulation of the VRML visualization of the didactic component.

Problems arising from such an architecture are:

- How to model and implement the virtual didactic components in VRML, keeping in mind that each has a specific geometry and behavior;
- How to store the VRML implementation of components within a reusable and public component library (database);
- How to represent the virtual model as a software component that can be used in a virtual environment and can also be combined with different types of sequence control systems in order to create new types of automatic operations;
- How to model and implement a Grafcet Engine in a VRML environment.

The following sections present the solutions adopted for each type of problem.

This environment can be considered a simulation-based e-learning application in which learners intervene with incidents in order to determine the course of events and then observe the consequences of their actions, something we call "learn-by-example simulation."

## VRML AND GRAFCET

### What are VRML and Grafcet?

The Virtual Reality Modeling Language (VRML) 2.0 was originally designed as the standard specification language of 3D geometry and its dynamic behaviors. It can be executed on any platform or operating system. There are many VRML viewers and authoring tools available commercially, and they are much less expensive than commercial discrete-event simulators. The advantages of VRML as the modeling language of components of the sequence control system are:

1. The VRML specification includes the behavioral mechanism of the scene based on the *event cascad*e. The execution engine is also installed in all VRML viewers.
2. The language format is specified as the ISO/IEC international standard (1997).
3. The state transitions of each model can be also described in the script nodes by combining the external JAVA code with the VRML.

And now we'll take a look at how Grafcet came to be. In 1975, forty French-speaking researchers and industrial managers, involved in complex, discrete system control, would meet every two months to compare and investigate models and methods for building sequential system control. Based on their own experience, they came up with seventeen techniques to be used. Some used empirical questionnaires. Other used a technological module model. Still others used a purely theoretical model derived from State Machines or Petri Nets. They decided to build a custom model that would be easier and more suitable for complex system, particularly manufacturing systems. After two years of hard and laborious meetings, talks, chats, and studies, they proposed a model called GRAFCET. This name came from "graph", because the model had a graphic basis, and "AFCET" from the scientific association that supported the work, Association Française de Cybernétique Économique et Technique. The basic concepts of this discrete system model were, and still remain today, quite clear and simple: the "step", the "action", the "transition" and the "condition associated to transition".

## Implementation of the Grafcet Engine

Up until now, the classic method used in implementing sequence control systems specified with Grafcet is to use a PLC. In this section, our aim is to replace the PLC by a programmable component (for the simulation). We believe that the PLC sequential core as the input/output modules can be replaced by a specific programmable component designed in Grafcet, while maintaining the capability that characterizes the PLC input/output. This is done through what we have called a "Grafcet Engine".

For modeling the Grafcet Engine, one can adopt one of two approaches:

- Modeling based on local Grafcet evolution rules. This approach consists in expressing the rules in concurrent equations. The translation is simple and systematic.
- Modeling based on the translation of Grafcet to another model. This approach consists in translating Grafcet into another model characterized by the capability of the mathematical aspect, or formal tools. More specifically, some work based on this approach is already performed [5], namely finite state automata, Petri nets, transition systems and signal language.

In this section, we will focus on the first approach, simply because we are only interested in the simulation aspect, where implementing local Grafcet evolution rules is sufficient.

Here we present the five Grafcet evolution rules:

- Rule 1: At the outset, the initial steps are active.
- Rule 2: A transition can be fired if its condition is true and if all the preceding steps are active.
- Rule 3: When a transition is fired, all the preceding steps are simultaneously deactivated and all the following steps are simultaneously activated.
- Rule 4: If several transitions can be fired, they are done so simultaneously.
- Rule 5: If a step is simultaneously activated and deactivated, it remains active.

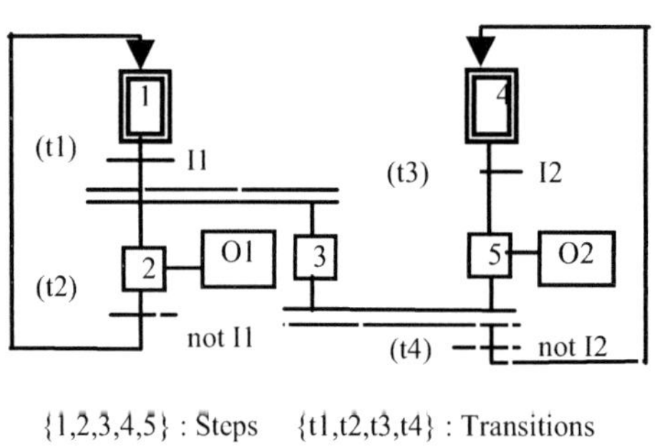


FIGURE 2
PRODUCER-CONSUMER G7

To implement the engine in a VRML environment, it should be noted that VRML 97 allows programming using different high-level languages. In practice, the standard defines only the interface between the viewer (Cosmo Player, for example) and the program that is written in another language. Thus, in theory, any language (C/C++/Java, etc.) can be implemented by a viewer [6,7].

Based on this, we present an engine implementation based on the VRMLScript language which is supported by most VRML viewers, and we also present a second implementation based on an extension of VRML via the concept of "Reactive Script". In both sections that follow, we will illustrate both implementations, based on the example shown in Figure 2. It represents a producer-consumer model. The "I1" condition implies the production of an element and the "I2" condition implies its consumption.

### Implementing the Engine in VRML Script

This approach consists of writing local Grafcet evolution rules as concurrent VRML equations. The fourth evolution rule says that all simultaneously clearable transitions must be immediately and simultaneously cleared. Thus, for each transition ti, we can write a VRML equation Ti to represent the firing of the transition. As has been specified by the second evolution rule, it depends on the transition condition (Ri) in conjunction with the preceding steps Pre(ti): Ti = Pre(ti) and Ri.

The third evolution rule specifies the consequences of a transition clearing according to preceding steps (deactivated) and succeeding steps (activated). Thus, we can formulate the activation equation Xi for each step Xi as follows: Xi = not succ(xi) and (Xi or Pre(xi)).

```
DEF     EngineG7 Script {
eventIn    SFTime   Cyclee
#Input events (simualte the PLC inputs)
eventIn    SFBool    I1
eventIn    SFBool    I2
eventIn    SFBool    RESET
#Output events (simulate PLC outputs)
eventOut SFBool O1
eventOut SFBool O2
#Input field to latch inpt events
field SFBool FI1 FALSE
field SFBool FI2 FALSE
field SFBool Reset FALSE
#Step fields to calculate the step equation
field SFBool X1 FALSE
field SFBool X2 FALSE
#Transition fields to calculate the
# transition equations
field SFBool T1 FALSE
field SFBool T2 FALSE
url "javascript:
        var i1;
        var i2;
        var reset;
        function reset (v){RESET=v}
        function I1(v){FI1=v}
        function I2(v){FI2=v}
        function Cycle(){
        // The latching of input events
        i1=FI1
        i2=FI2
        reset=Reset
        //processing transition clearability
        T1= i1 && X1
        T2=!i1 && X2
        T3=i2 && X4
T4=!i2 && X3 && X5
// processing activation of steps
// X1: initial step
X1=reset || !T1 && (X1 || T2)
X2=!T2 && (X2 || T1)
X3=!T4 && (X3 or T1)
//X4: initial step
X4=reset || !T3 && (X4 || T4)
X5=!T4 && (X5 || T3)
// Sequential process
if reset
        O1=FALSE
        O2=FALSE
else
        O1= X2
        O2=X5
        }" }
```

FIGURE 3

IMPLEMENTATION OF THE GRAFCET ENGINE USING VRMLSCRIPT

Obviously, these equations suppose the existence of a stable state; otherwise, they will have a non-deterministic behavior. In order to ensure that actions are stable during a constant time and that they are deactivated during a reset time, synchronization is done according to a sequential output process.

VRML is based on a discrete time model, through a special node called "time sensors." Nodes of this type allow us to implement the execution of the Grafcet Engine. Once active, a "Time Sensor" generates a temporal event at the beginning of each cycle that activates the execution of the script node gathering transition and step activation equations. The "Time Sensor" frequency must be sufficiently small to guarantee access to a stable state. The "Time Sensor" period must be higher or equal to the maximum reaction time. Figure 3 shows a summary of VRML scripts that represents the implementation of the Grafcet Engine relative to the example in Figure 2.

### Implementing the Engine in Reactive Script

The main problem in the previous solution is that we have supposed that a stable situation exists. But synchronous languages like Esterel propose an approach for resolving the problem of boolean stability. Since the synchronous hypothesis is at the basis of Grafcet, it would be more suitable to implement it in a synchronous language like Esterel [8]. Based on steps and transitions, Grafcet formalism integrates the main characteristics of Esterel: the hierarchy by the embedding of macro-steps, parallelism by the divergence in "AND", determinism by the limited number of accessible states, concurrence by the description of an independent Grafcet in the same macro-step, and communication by the instantaneous broadcasting of signals. A Grafcet can then be translated automatically into an Esterel program.

The proposed translation considers each step of the Grafcet as a reactive system. Each step can react instantaneously to a state modification generated by other steps. Thus, the third evolution rule of Grafcet applied to steps 3, 4, and 5 of the example can be written in Esterel as: *present RTL3 and RTL5 and not I2 then*

```
emit X4_SET;
emit X3_RESET;
emit X5_RESET;
end present
```

where Xi_SET and Xi_Reset indicate respectively set and rest orders of activation regarding Step xi, RTLi indicates the "ready to leave" signal that is emitted by Step Xi if it is active.

In this approach, the Grafcet Engine is considered as a reactive system that must always be able to provide an immediate response when solicited by the environment. The Grafcet evolution is then a series of reactions. Relative to an environment scale time, each reaction is considered as instantaneous. The synchronous assumption defines a scale of discrete logical time, constituted of instants corresponding to each reaction of the system. The events having started the reaction are considered as simultaneous events. Moreover, the reactions are done at null time; the emitted signals during a reaction are then simultaneous to the signals that have caused the reaction (the production of the reaction at the same logical time). A reaction is therefore instantaneous, which avoids concurrent partial reactions, the source of indeterminism. Let's now see how we can integrate Esterel into a VRML model [9-10]. The "reactive script" concept is based on the use of SL language, a type of Esterel language. The implementation is done using one of either C Reactive, Tcl/Tk or Java languages. These scripts are very well adapted to developing web-based

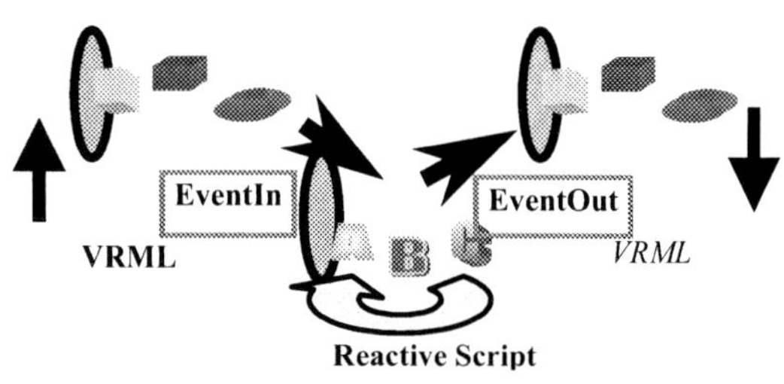


FIGURE 4
REACTIVE SCRIPT

synchronous reactive applications. The VRML Script of the preceding implementation will now be replaced by a reactive script.

As shown in Figure 4, VRML must generate temporal events to the reactive script; in other words, a "Time Sensor" node sends cyclic events to generate temporal points to the script component. Esterel treatment must be done between two events generated by a "Time Sensor" node. The main supposition here is that the time is external and controlled by VRML.

The concurrent execution relative to the Grafcet is executed at the beginning of each temporal point (generated by VRML) until the end of the treatment. At the beginning of each cycle, the script receives from the scene a set of events that represents the PLC inputs. At the end of the cycle, it generates other returned events that represent the PLC outputs.

## VMRL AND THE DIDACTIC COMPONENT

### Modeling the Didactic Component

The need for a reference model for representing a didactic component is even more critical when the component is more complex and constituted of a variety of modules.

The didactic component represents the operative part that regroups all the tools that are at the base of the physical process. Component inputs are made up of output orders sent by the sequence control part (Grafcet Engine) and process parameters at the entry of the system. Outputs correspond to both the information returned to the control part, as well as the executed actions that give an added value [11].

In modeling the component and in order to have a valid virtual pattern, we have used advanced concepts in the simulation and visualization fields [12]. The functional models of simulation and visualization tend to regroup the visual data and their treatments within objects that encapsulate them. Generally such types of models can be divided into three components:

- The visual model (in our case, the geometrical part of the component) that represents the simulated phenomenon.
- The graphic user interface (GUI), used to control user input values, the visual representation and the simulation execution.
- The simulation control (the dynamic part of the component) that controls the simulation behavior in response to the user or other events (mainly output events of the Grafcet Engine).

We can divide simulation models into two types. The first type, called the "off-line" simulation, represents the simulation in which the system obtains the needed key values for simulation from a mathematical model implemented outside of the simulation model. So the calculation of these simulation values is not directly implemented into the simulation module. Figure 5(b) shows an "off-line" simulation functional model. The "off-line" simulation is no more convenient for modeling our didactic components because the "off-line" simulation data are treated by a very sophisticated algorithm before being simulated. Once results are ready, they are sent to the simulation control module. This leads us to say that the real time aspect needed to react to user and I/O events exchanged with the Grafcet Engine is not possible. In other words, after modifying one of the simulation parameters, we cannot obtain a real time modification result.

The second type, called "online" simulation, represents simulations that are integrated into the simulation control module, the mathematical model of simulated phenomena. The user can modify simulation parameters in order to immediately affect the simulation behavior. This type of simulation can immediately show the modification of input values.

The simulation resolution engine calculates, without stopping the visualization model behavior, using actual input values and the local time within the simulation. This simulation model is the more appropriate model to simulate our didactic components that are characterized by a behavior based on a real-time animation and reaction to events. Figure 5(a) shows an "online" simulation functional model.

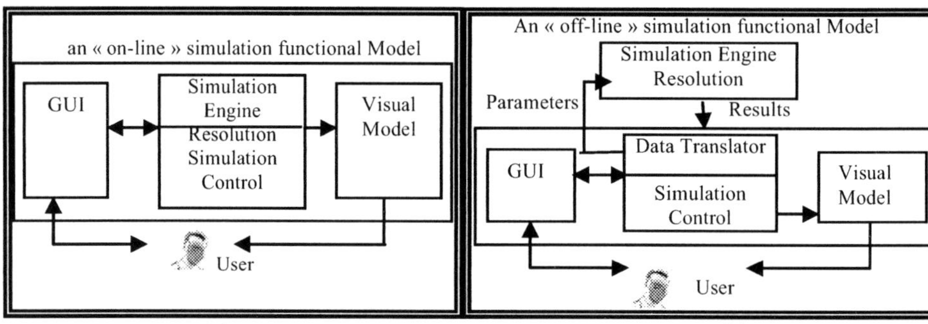


(a) (b)

FIGURE 5
THE SIMULATION FUNCTIONAL MODELS

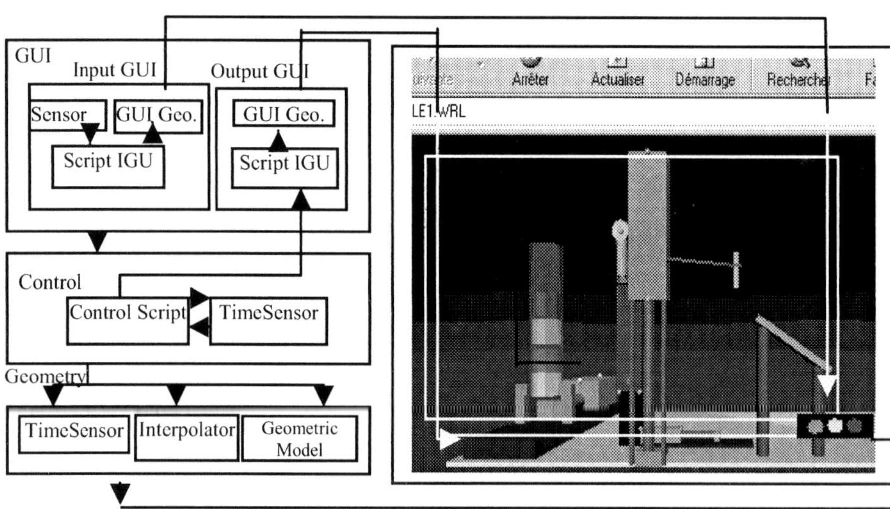


FIGURE 6
THE DIDACTIC COMPONENT MODEL

## Implementing the Didactic Component in VRML

In order to make the process of modeling and VRML implementation of the various types of components more systematic, more uniform and simpler, we propose a reference model for implementing each module that constitutes the virtual didactic component.

Referring to the "online" simulation, the representation of the reference model can be implemented in VRML according to the graphic model shown in Figure 6.

## Implementing the Visible Part of the Component

Visually, the component is recognized by:

1. The user graphic interface (GUI) that is made up of two parts: the input interface (user inputs) and the output interface (presents the visual data to the user). The GUI can contain some blocks known as "Sensors" (having only an input function), a visible geometry and some control scripts. Input and output events generated by the GUI are then routed to the control component.
2. The "Geometry" part that is used to visualize the geometric aspect of the didactic component. In regard to representing and grouping 3D objects, the VRML architecture is based on the same concepts as classic graphic scenes. The scene can be constructed using the basic nodes (boxes, spheres, cones or polygons) or user-defined nodes that are generally PROTO type nodes.

### Implementing the Dynamic Part of the Component

Time is of primary importance in representing the component. This can be explained by the fact that the dynamic aspect represents the operational evolution of the component. This may not be the appropriate time to describe in detail how time contributes to our system, but it may be more interesting to describe how this time must be used [4,13]. By creating links between the variation of one of the module characteristics and the time, we provide a description of both, the periods of time and the evolution of the component state during these time periods.

VRML is based on a discrete model of time, implemented through a special node

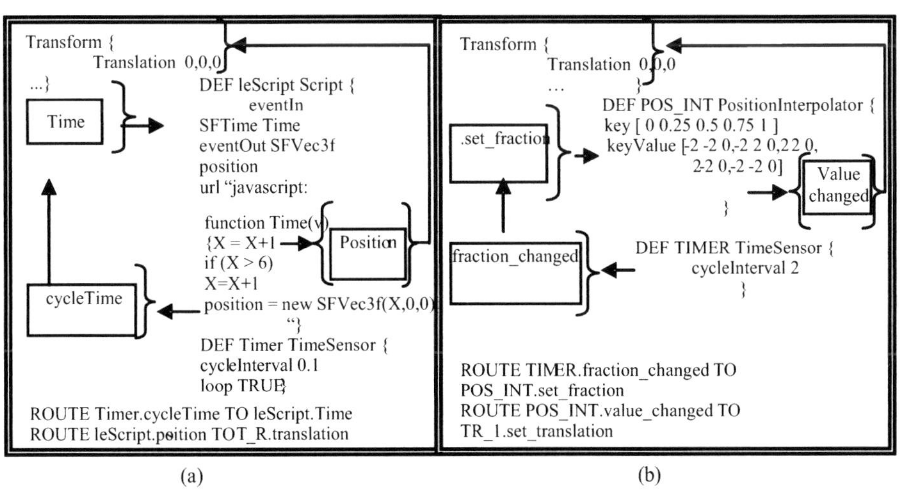


FIGURE 7
VRML ANIMATIONS

called "time sensors" that control animation. To animate an object, it must receive periodic events in order to modify one of its characteristics (position, color, etc.).

VRML introduces primitives for animation and interaction based on a simple event model, that is, each object in the virtual environment can receive input events and send output events, these events are communicated between the objects through defined "Routes" before and completely independent from the graphic scene structure.

In the proposed model, the "control" component controls the 3D scene animation that reflects component behavior. So the module behavior, defined by its reactions to the requested operational supplies (from PLC or IGU) according to its state, is described in a complementary view called "control". This behavior is implemented in a textual form using languages like VRMLScript or JavaScript. In order to respect VRML implementation concepts, the use of a language is not absolutely needed and the behavior can be implemented using another technique called "KeyFraming" [14]. This consists of specifying position values of the component and allowing the viewer engine to calculate the mechanical evolution of the component between these positions during an interval of time.

We illustrate the two possible animations in Figure 7(b). The KeyFraming technique is based on the use of an interpolator type node that, for an initial value, an end value and a given fraction of cycle, calculates the corresponding interpolation. The linear interpolation is the only possible interpolation through these nodes. Figure 7(a) shows the implementation of the same animation in a programmed way by using a script node.

## STORING THE SPATIOTEMPORAL CHARACTERISTICS OF THE DIDACTIC

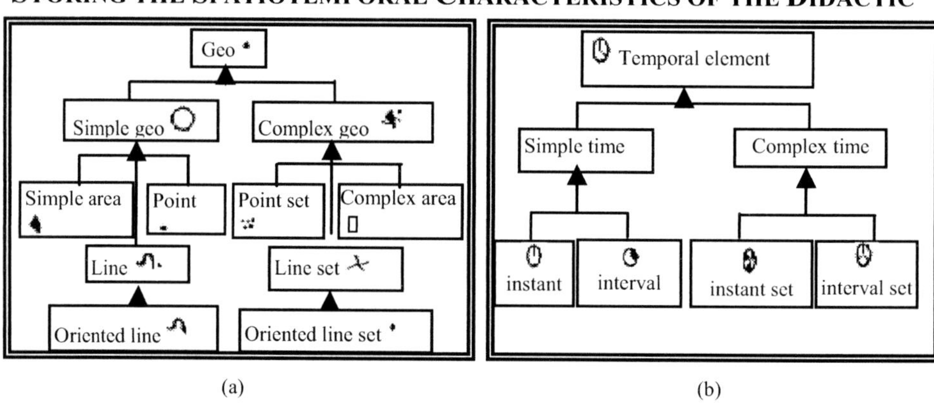


FIGURE 8
MADS PICTOGRAMS

## COMPONENT

We have presented the didactic component's spatial (visual aspect) and temporal characteristics (dynamic aspect), which we call spatiotemporal characteristics [15]. In this section, we present our proposal for storing these characteristics within a format that must achieve the following goals:

- Recording the component's spatiotemporal characteristics within an open format.
- Facilitating the reusability of the components and the translation from and to VRML format.
- Allowing a modularity description of the component by clustering 3D objects.
- Facilitating the checking and modifying of the objects encoded in VRML by using a relational language.
- Facilitating the checking and modifying of the objects encoded in VRML by using a relational language.

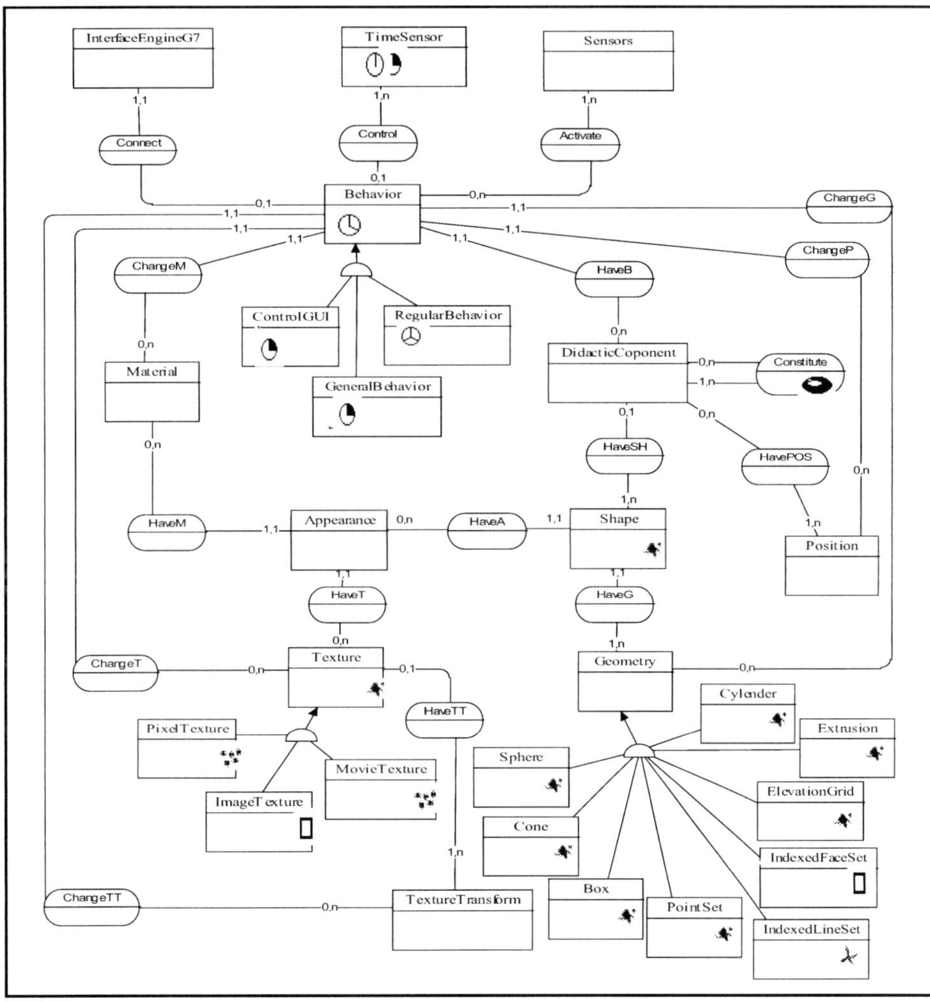


FIGURE 9

SPATIOTEMPORAL DATABASE MODEL

It is difficult to find a standard for representing spatiotemporal data, and even more so in the case of spatiotemporal data having several geometries. However, we can speak of two approaches for modeling this type of data:

- PERCEPTORY, developed by the geomatic research center at Laval University.
- MADS, developed at Lausanne University in partnership with the computing laboratory of the Brussels Private University and the database laboratory of the Federal Polytechnic School in Lausanne, an approach that has been used and accomplished in the MurMur project [16].

In our work, we have been inspired by MADS formalism, in order to extend the entity-association model used to conceive our spatiotemporal database. Like the PERCEPTORY approach, MADS formalism uses pictograms to represent the spatial and temporal characteristics of an entity, an attribute or an association between two attributes. Thus, in regard to geometry, it is possible to have cases such as those shown in Figure 8(a). The hierarchy emphasizes the inheritance which can propagate. So, a geometry can be redefined by refinement, for example, a "Simple Geo" can then be redefined then into "Line" or "Point", but not in "Line Set". As for the temporal aspect, Figure 8(b) shows temporal pictograms. For the same entity, it is possible to mix various pictograms, as shown in Figure 9. The left side is reserved for temporal characteristics and the right side for spatial characteristics. The spatiotemporal specification can be also attributed to associations as shown in Table 1. Currently, data modeling is done manually, as shown in Figure 9. This type of modeling provided us with a better approach in regard to the nature of the VRML data and their interaction. Once the model is developed, one must implement it within a management database system. To create our first prototype, we used Oracle8i [18].

| ***Spatial Type*** | **Pictograms** | **Definition** |
|---|---|---|
| Disjunction | 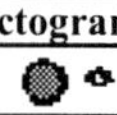 | The linked entities have some disjointed geometry |
| Adjacency |  | The linked entities have some adjacent geometries (without common surface) |
| Crossing |  | The linked entities have a common part of which the dimension is less than the dimension of the two entities |
| Overlapping |  | The linked entities have a common part of which the dimension is equal to the dimension of the two entities |
| Inclusion |  | The geometry of an entity is completely inclusive in the geometry of the other entity |
| Equality |  | The geometry of the two linked entities is equal |

TABLE 1
MADS TOPOLOGICAL RELATIONS

Once the base is structured and completed, one will need to execute queries based on this data. In order to do so, a query editor will be also developed and will of course be linked to a viewer, thus allowing for an interactive selection of didactic components and the execution of complex student queries.

To record a component within this base, one must record three parts having the following parameters:

- The visual component, that is Shape, Geometry, Appearance, Cylinder, Extrusion, ElevationGrid, IndexedFaceSet, IndexedLineSet, PointSet, Box, Cone, Sphere, Texture, TextureTransform, MovieTexture, ImageTexture, PixelTexture, Appearance, Material, Sensors, Position;

- The dynamic component, that is Behavior, ControleGUI, GeneralBehavior, RegularBehavior, TimeSensor ;
- The connection component: the related entities to this part are (InterfaceEngineG7).

The main idea is to structure the virtual didactic component so as to be able to model a highly complex component. As shown in the conceptual data model (Figure 9), the relation that exists between the didactic component modules is one of composition presented by the association "Constitute." This relation is retained by our model in order to accomplish the modularity and hierarchy aspects.

The obtained result from the application of this approach contains a single root module (presented by one record within the "Didactic Component" entity), that is the totality of the didactic component. The module is broken up into a set of child modules that replaces it, each module encapsulating its own decomposition. Thus, the obtained model is strongly structured and modular.

## CONCLUSION

Figures 10 and 6 show the prototype we have implemented based on the three main ideas developed in this article: a) designing and implementing a Grafcet Engine under VRML; b) modeling and implementing the virtual didactic component using VRML; c) storing the spatiotemporal characteristics of the virtual didactic component in a relational database. Based on the virtual laboratories concept [19,20], that implies an informational rebuilding of the real laboratory and its state through animations and 3D scenes, we are currently attempting to extend our system so as to allow the teaching of Grafcet and automatics in a connected mode. In other words, we seek to achieve tele-experimentation on remote PLCs and didactic components through a virtual environment that uses the modeling and implementation mentioned in (a) and (b).

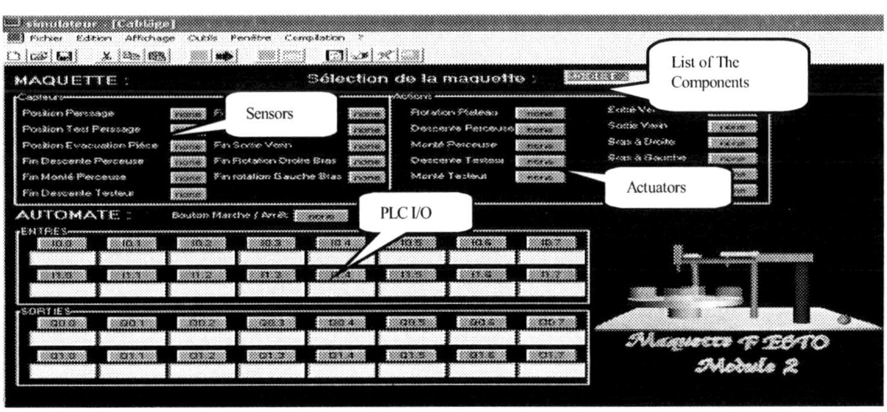


FIGURE 10

THE CONNECTION INTERFACE BETWEEN THE PLC AND THE DIDACTIC COMPONENTS

## ACKNOWLEDGEMENTS

This work is partially supported by the Learning Object Repository Network project (LORNET) under a National Science and Engineering Research Council of Canada (NSERC) grant.

## REFERENCES

1. Gillet, D., "Towards Flexible Learning in Engineering Education," Chapter 11, Aung, W. et al. (eds.), *Innovations 2003: World Innovations in Engineering Education and Research*, Int. Network for Eng. Ed. and Res. (iNEER), Arlington, VA, 2003, pp. 95-102.

2. Bosman, K., "Simulation-Based e-Learning," web.syr.edu/~kjbosman, 2002.

3. Mallet, F., Gaffé, D., and Boéri, F., "Concurrent Control Systems: from Grafcet to VHDL," Proceedings of the EuroMicro Conference, 2000.

4. Kanai, S., and Kishinami, T., "A Virtual Verification Environment for the Sequence Control System Using Vrml and Java," Proceedings of The 1999 ASME International Design Engineering Technical Conferences and Computers and Information in Engineering Conference, September 1999, Las Vegas, Nevada.

5. Moalla, M., "Réseaux de petri interprétés et grafcet," Rairo Automatique, 1985.

6. VRML 97 specification, http://www.vrml.org/Specifications/VRML97/index.html

7. X3D, "Extensible3D(X3D)Specification," http://www.web3d.org/news/x3d/index.html, 1999.

8. Richard, N., "Description de comportements d'agents autonomes évoluant dans des mondes virtuels habités," Thèse de doctorat, 2001.

9. Boussinot, F., "Reactive Scripts," Proceedings of the Third International Workshop on Real-Time Computing Systems Application, October-November, 1996.

10. Codognet, P., "Animating Autonomous Agents in Shared Virtual Worlds," Proceedings of the Sixth International Conference on Distributed Multimedia Systems, 1999.

11. L.I.R.M.M, "Conception et Validation de Systèmes Automatisés de production: un état de l'art," www.lirmm.fr/ftp, 1990.

12. Cibulka, K., and Zara, J., "Using VRML for Creating Interactive Demonstrations of Physical Models," Proceedings of Central European Seminar on Computer Graphics Conference, 1998.

13. Rohrmeier, M., "Web Based Robot Simulation Using Vrml," Proceedings of the 2000 Winter Simulation Conference, 2000, pp. 1525-1528.

14. Vrml Groupe Francophone, "Cours Vrml en ligne," http://kdo.chez.tiscali.fr/vrml, 1998.

15. Vakaloudis, A., and Theodoulidis, B., "Formalising Interactive Behaviour in 3D-Spatiotemporal Worlds," Proceedings of the Advanced Visual Interfaces International Working Conference, 2000.

16. B. Roland, "Apports réciproques d'un projet de recherche en base de données et d'une unité de recherche en risques naturels," Proceedings of the Systèmes d'Information pour les Risques Naturels Conference, 2002.

17. Muller, P., and Gaertner, N., "La notation UML," Chapter 3, *La modélisation objet avec UML*, Eyrolles, 2000, pp. 91-261.

18. Chapuis, R., "Procédure sockées en Java," *Les bases de données Oracle 8i*, 2001.

19. Saliah, H.H., "Design of a Generic, Interactive, Virtual and Remote Electrical Engineering Laboratory," Proceedings of 29th ASEE/IEEE Frontiers in Education Conference, 1999.

20. Saliah, H.H., Villardier, L., and Kedowide, C., "Resource Management Strategies for Remote Virtual Laboratory Experimentation," Proceedings of $30^{th}$ ASEE/IEEE Frontiers in Education Conference, 2000.

**Mohamed Mhamdi** received a B.Sc. degree in computing from the High Institute of Telecommunication and Post, Tunis, in 1993. Following a period in industry as a systems programmer, he received the "Agrégation" degree in 1996. In 1997, he received a M.Sc. degree in Industrial Informatics from the National School of Computer Sciences. In the same year he received his Technologue degree after which he moved into the field of educational technology in the High Institute of Technology at Sousse, Tunisia, where he has worked for the past six years. His areas of main interest are computer science education research, particularly pedagogical and technological innovations.

**Rafik Braham** received the M.S. and the Ph.D. degrees from Georgia Tech, Atlanta, USA, in 1990. He is currently professor of computer science and director of the High Institute of Informatics and Telecommunication at Sousse, Tunisia. His work focuses on technological innovation in computer science education.

**Mohamed Moalla** is a computer science professor. For nine years, he held the post of director of the Department of Computer Sciences in the Faculty of Sciences of Tunis. He is currently director of Laboratory LIP2 at The University of Tunis El Manar. He spent 25 years in research at Petri Nets.

**Hamadou Saliah-Hassane** obtained a diploma in electronics from Cheik Anta Diop University (Senegal) in 1976, and the Bachelor's degree in Engineering and the Master of Applied Science degree from École Polytechnique in Montreal in 1981 and 1983, respectively. He received the Ph. D. degree in Computer Aided Analysis and Design from the Electrical and Computer Engineering Department at McGill University in Montreal in 2000. Since 1999, he has been a member of the faculty at Télé-Université in Montreal, Canada. His areas of interest are related to telelearning, knowledge-based systems, computer aided analysis and design and laboratory innovation.

# Chapter 35

# Building Good Engineering Vocabulary: Adopting Technical Terminology in Non-English Speaking Countries

HUA-LI JIAN[1] and FRODE EIKA SANDNES[2]
*[1]National Cheng Kung University, Tainan City 70101, Taiwan, ROC.*
*[2]Oslo University College, N-0130 Oslo, Norway. E-mail: hljian@mail.ncku.edu.tw, frodes@iu.hio.no*

*Most engineering disciplines, such as computer engineering, are in constant and rapid change. Technology and paradigm shifts take place every few years and every major change brings new vocabulary and terms. Clearly, these changes also affect the engineering education establishments, the educators, and the students. The changes do not impose any serious language problems for the educational institutions of English speaking countries, because new inventions and advancements are generally published in English-language journals, magazines, conference proceedings and eventually textbooks. English is the de-facto international engineering language; however, many non-English speaking countries, including most European, Asian, African and Latin American countries, struggle with these changes from a linguistic point-of-view. It generally takes time for textbooks written in the local language to appear. More obscure engineering subjects may never enjoy a textbook written in the local language at all. The low-budget textbooks that do appear frequently use ad-hoc and poor translations of the new technical terms associated with the new technology, or they simply use the English terms or some phonetically close adaptations. As a result, English textbooks are frequently used. Consequently, students may fall into the habit of using an ad-hoc and anglicised terminology. For teachers authoring their own textbooks, lecture notes and teaching materials, the term-translating job can be overwhelming as it takes time and requires inspiration to find fitting and meaningful terms. In this paper we discuss how students can serve as an invaluable resource in the process of inventing good local variations of technical terms - terms that enhance the correct associations founded in the local culture and language. Experiences from Norwegian engineering education are described.*

## From Research Laboratory to Syllabus

Subject areas such as electrical engineering and computer science are constantly advancing. New inventions and technological breakthroughs are accompanied by new paradigms, practices, techniques, technology and terminology. The curriculum in higher education is constantly revised to reflect the most recent technological advances. Technological and scientific findings are usually first published as technical reports on a research group's web site and subsequently appear in the proceedings of international conferences. It generally takes about six months for a research result to first appear in print. Conference proceedings usually reach a narrow audience. To reach a wider audience, the inventors and researchers publish their results in renowned international journals with a high impact factor. This whole process can take up to a few years. "Important" results also make their way into textbooks, and these textbooks serve as the main source of inspiration for most educators. Once included in a textbook, a topic is likely to be included into the syllabus of an undergraduate course. The language of these forums is usually English. English has become the de-facto international language. New paradigms, techniques and technologies are accompanied by new terminology, usually introduced by the authors and subsequently altered by other researchers. Over time, terminologies usually converge onto a small set of generally accepted "standard" terms [1,2].

## From English to Foreign Language

Shifts in terminology are unproblematic in English speaking countries – at least from a linguistic point of view. However, the majority of the countries in the world are non-English speaking. Also, the English abilities of the locals vary in these countries. Teaching, especially at the undergraduate level, is often conducted in the local language. Local languages need local terminology, which is problematic.

Educators have influence on the terminology when orally delivering the materials and when producing teaching material such as lecture notes and textbooks. Students often adopt the terminology presented to them by the teacher. Simultaneously, teachers have a moral and ethical obligation to introduce good terminology. But, how would one describe something that has no name? The obvious answer is to consult the textbooks. Unfortunately, many smaller non-English-speaking countries only have textbooks covering the mainstream subjects where the publishers are guaranteed to profit from sales. In addition, it takes a long time for new subjects to reach print in a local language. There are thus two solutions: (a) adopt the English terminology, or (b) invent new terminology. General teaching fatigue and lack of inspiration and time leads to teachers often opting for alternative (a). Further, most engineering teachers have no theoretical background and training in translation, as translation is a profession in its own right. This paper discusses strategies for discovering terminology in the local language. The target language for this study is Norwegian; nonetheless, the technique extends to other language environments, such as Chinese, that struggle with similar problems [3,4,5,6].

## Why not English?

It is an undisputed fact that any non-English engineering degree program should have some element of English in the curriculum, as engineers often have to work in an international setting [7]. However, there are pedagogical, cultural and practical reasons for not always using English terminology. Students' English skills do not match their linguistic abilities in their native language. It is unnecessary to complicate and obscure an already difficult degree course using an unfamiliar language. One should expect that the students learn more and faster when taught in their native language.

The solution can be simple. This can be illustrated with an example: in mobile and wireless communication, there is often talk of the Bluetooth technology. One often hears students talk of and read media articles about Bluetooth. However, the Bluetooth technology is named after the Scandinavian King Harald Blåtann who united Norway and Denmark from A.D. 911 to A.D. 985. Why not use the term "blåtann"? Another similar example is the use of the term "stack" to describe a common data-structure. An old Norwegian term, "stakk," has a similar meaning and pronunciation, and yet computer professionals rarely use it. As Temmerman [8] suggests, effective but forgotten terms can be successfully revitalised or rediscovered for new contexts.

| **Term** | **Official translation (explanation)** | **Other viable suggestions (explanation)** |
|---|---|---|
| Check -box | Avkryssningsrute (check window), avmerkingsrute (tick window) [10] | Avkryssningsboks (checkbox), ja-nei svar (yes-no reply) |
| Cookie | Informasjonskapsel (information capsule), kjeks (biscuit) [10] | Merkelapp (label), småkake (shortbread), kinesisk kake (Chinese fortune cookie), kuruke (cow dung), visittkort (business card), tilstandskapsel (state capsule) |
| Pop-up menu | Flytende meny (floating menu) [10] | Rullgardinmeny (pulldown curtain menu), meny (menu), sprettopp meny (pop-up menu), dukkopp meny (suddenly appear menu), popmenyer (pop menu), rullemeny (roll menu) |
| Dead links | Lenkeråte (link decay), døde lenker (dead links) [10] | Død peker (dead link), brutte lenker (broken links), uvirksomme forbindelser (inactive connections), nullreferanser (null references) |
| Hub | Nettnav (network hub) [10] | Nav (hub), Høbb (phonetic adaptation), distribusjonsboks (distribution box), forgreningspunkt (branching point), Knutepunkt (center), nettfordeler (network distributor), Stjernebuffer (star buffer), dings (thingy), linjedeler (line divider), samlingspunkt (collection point) |
| FAQ | OSS – ofte stilte spørsmål (FAQ), SOS – spørsmål ofte stilt (QFA) [10] | Vanlige spørsmål (ordinary questions), sentrale spørsmål (key questions), aktuelle spørsmål (current questions), OBS – ofte besvarte spørsmål (frequently answered questions) |
| Mode -rated list | Redigert postliste (moderated maillist) [10] | Kontrollert liste (controlled list), moderert liste (moderated list), justert liste (modified list), overvåket liste (monitored list), endret liste (changed list) |
| Form | Skjema (form) [10] | Formular (form) |
| Offline | Frakoblet modus (disconnected mode) [11] | Passiv (passive), ikke tilkoblet (not connected), Avkoblet (switched off), utilkoblet (switched off) |
| Session | Økt (session) [11] | Runde (round), sesjon (session), epoke (epoch), seanse (rendezvous) |
| Spam | Søppelpost (garbage mail) [10] | Spam (spam), dritt (poo), møkkaspredning (fertilization), uønsket informasjon (unwanted information), sludder (rubbish), vås (rubbish), forstyrrelse (disturbance), uønsket mail (unwanted mail), skinke (ham), flesk (pork), kødd e-post (nag e-mail) |

TABLE 1

ENGLISH COMPUTING TERMINOLOGY, OFFICIAL NORWEGIAN TRANSLATIONS AND SUGGESTIONS THAT EMERGED DURING THE INVESTIGATION

Misguided translations come in other flavours too. Take for instance the web-technology term "cookie." The term "cookie" has no connection to what it is actually referring to and Callaway [9] claims that this term was chosen arbitrarily. However, the official Norwegian term is "informasjonskapsel" (meaning "information capsule") [10], which is more informative than the original term. Still, many practitioners continue to use the term "cookie" both in spoken and written Norwegian. To illustrate that the official computing terms are not generally known, a list of common English computing terms (Table 1) was distributed via e-mail to a number of subjects. These included faculty members from the computer science, electrical engineering and mathematics departments at Oslo University College, as well as the computing service officers and a class of undergraduate final-year computer science students. The subjects were asked to provide the official Norwegian equivalent for each of the English terms without consulting any reference works. Further, the subjects were asked to use their imagination and make suggestions if they were unfamiliar with the official Norwegian term.

| Expression | Teachers | Students |
|---|---|---|
| Checkbox | 50,0 % | 60,0 % |
| Cookie | 30,8 % | 50,0 % |
| Pop-up menu | 0,0 % | 10,0 % |
| Dead links | 26,9 % | 60,0 % |
| Hub | 19,2 % | 0,0 % |
| FAQ | 50,0 % | 90,0 % |
| Moderated list | 0,0 % | 0,0 % |
| Form | 57,7 % | 70,0 % |
| Offline | 34,6 % | 20,0 % |
| Session | 11,5 % | 30,0 % |
| Spam | 34,6 % | 40,0 % |

TABLE 2
STATISTICS SUMMARISING THE RESPONSES OF THE VOCABULARY SURVEY

Table 1 shows the English terms, their official Norwegian translations (acquired from the Norwegian Language Council [10] and official Microsoft documentation [11]) and the non-official terminology suggested by the subjects. The results of the experiment are summarised in Table 2, showing the percentage of subjects that knew the official translations. The first column lists the English expressions, the second column lists the correct responses among the teachers and staff and the third column lists the correct responses from the students. A total of 26 out of 46 teachers and technical staff from the departments of computing, electrical engineering, mathematics and computer services responded to the survey. A total of 10 students from a class of 34 students responded as well. Interestingly, none of the respondents recognised the official term for "moderated list." Only a minority knew the Norwegian equivalents for the common computing terms "pop-up menu," "session" and "hub." A larger proportion of the subjects were familiar with the Norwegian equivalents for the terms "form," "checkbox" and "FAQ" (approximately every fifth person). The differences across different groups were small. However, students performed better than the teachers. Further, two of the respondents were negative to the suggestion that English terminology be used unconditionally. We can conclude from this investigation that the official equivalents for common computing terms are not generally known and used by students and academic staff in this educational institution. This study also revealed that the respondents provide a diverse set of terms. Such diversity can be a problem, in the global sense, as it can be hard to achieve standardisation of terminology (see Saraireh [12]).

Efforts to use local terminology are not always appreciated by teachers and students. The coexistence of English and other languages is constantly debated around the world and continues to be a multi-faceted controversial issue [13]. The purpose of this article is not to fuel this debate, but rather to suggest constructive approaches for inventing meaningful and useful local terminology.

## PROFESSIONAL TRANSLATION

Translation is a profession in its own right [14-16]. Terminology is a sub-topic of translation involving the study and the collection of terms for LSP (Language for Specific Purposes). It can be divided into three main topics: theory of terminology, terminology mining and terminology management. The theory of terminology addresses the origin of terminology, social aspects, evolution, control and standardisation. Terminology mining involves identifying the terminology from text corpora (see Chung, [17]) and terminology management involves the cataloguing and classification of terminology – especially by the means of computers and databases. Researchers are even developing techniques for extracting terminology from text corpora automatically (see for instance the study by Justeson and Katz [18]). Further, Nakagawa [19] describes a technique for automatically building dictionaries for special purposes by extracting Japanese-English term pairs based on parallel Japanese-English corpora. The purpose is to assist professional translators, technical writers and information specialists. These subjects and research areas place little emphasis on the creation of terminology, although some techniques exist such as concept analysis (see Meyer [20]) where the concept of a term is linked to the context in the target language. Further, few engineering and science scholars have formal translation training, but are still actively involved in technical writing and translation processes. Moreover, most textbook projects and teaching-material development projects lack sufficient budgets to allow for translation consultants.

## TERMINOLOGY MINING

Authors of engineering textbooks must pay special attention to the vocabulary used. There are a number of methods that are used interchangeably. Reference works are obvious and essential tools for authors. These include technical dictionaries, encyclopaedias [21] and computer-assisted translation tools [22]. A problem with these tools is that only the very established terminology is included, while terms covering emerging terminologies are absent. These sources of information are therefore of limited value for authors addressing current technology. Another source of information is governmental language institutes. In Norway, the Norwegian Language Council [23] is responsible for overlooking the conservation and evolution of the Norwegian language, and subject-specific wordlists are occasionally posted on their website. However, only very mainstream terms make their way into these wordlists, and not all entries are useful. For example, the English term "tag," referring to a mark-up tag as can be found in HTML documents, is acceptable according to the Language Council. "Tag" is not a Norwegian term while the term "markør" is a valid alternative.

A valuable source of information is timely published materials such as subject specific newspapers, magazines, and reports. In particular, Microsoft has done an excellent job in translating their documentation into foreign languages [11]. Their product documentation is translated by translation experts. During the authoring of a web-technology textbook in Norwegian [24], the reference documentation for Internet Explorer provided an invaluable source of information, with many well-translated terms. Sadly, many of the technical terms used in this documentation are not recognised by the computer experts.

When this fails, one is forced to invent and introduce new terms. Sometimes this is easy, but often it proves to be difficult. Co-workers and students can provide valuable input.

## Evaluating Terminology Quality

According to the theory of terminology, there are a number of social, practical and even completely random reasons terminology converges in particular directions (see, for example, the theoretic overview by Castellví [1] and the specific case by Bowker [2]). First, a term must be descriptive. The term should describe the object or process in the best way possible, be it a noun, verb or an adjective. If possible, it should be associative – when hearing or reading the term the listener or reader should make the "correct" associations. Further, the term should be unambiguous and distinctive thereby reducing the chance of mistaking it for a different term. Also, the term must be practical in terms of length and pronunciation. A long term can be hard to read, and if it is difficult to pronounce then the "users" of the terminology will probably adopt a shorter term instead. The practicability of a term is also related to the context in which the term will be used. A term should be memorable. A memorable term is more likely to survive and become the accepted standard. Finally, humorous terms can be quite successful as demonstrated by many of the terms found in modern computer science.

## Using Students to Invent Terminology

In an effort to assemble suitable terminology for course textbooks, students have actively been involved in the terminology mining process. Much can be gained from listening to and conversing with students informally. Some students are creative and will intuitively and naturally construct and use their own terms. The teacher should always be alert to new terms and make notes of them as they are discovered. Teachers can also try out new terms with the students and get immediate feedback. Our experience is that students are quick to complain when confronted with unsuitable terminology.

Good terminology can be discovered by inspecting coursework submitted by students. For example, at Oslo University College, students have to do an extensive final-year group project developing some software system. Nearly every year we discover creative and impressive translations invented by students. For example, in computing, the English term "garbage collector" refers to a mechanism of automatically recycling memory that is no longer in use so that it can be reused for new purposes. The direct translation of the term "garbage collector" into Norwegian is "søppletømmer," which may sound vulgar and is therefore not popular. Recently, however, a group of students referred to the "garbage collector" as the "vaskemaskin" (meaning "washing machine") – quite a fitting and descriptive term that we now have adopted. The "washing machine" washes the memory "clean" of old "stuff."

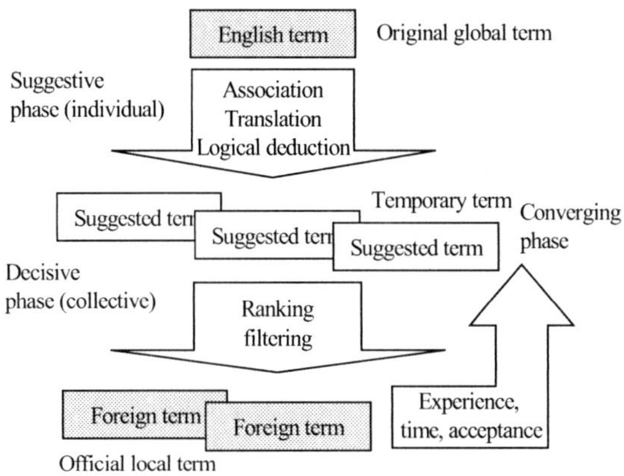


FIGURE 1
TECHNICAL TERMINOLOGY THEORY

Both student interviews and inspection of written work are forms of passive terminology mining. Terminology mining can also be performed actively both in writing and in class. We call our approach the "protocol contract," where the students and the teacher agree upon the valid terminology to be used (see Figure 1). This agreement is a contract stipulating the protocol or the terms that are used during oral and written communication. The fact that students are involved in the "negotiations" makes the students more interested in the subject and willing to accept the new terms.

There are several ways to establish the protocol contract. One approach is as follows: The teacher writes a list of English terms on the blackboard or on a handout during a lecture. The students are organised into groups of 2-4 individuals and each group has to find translations and come to some consensus. After each group has settled for a set of terms, the teacher summarises the groups' suggestions in plenary on the blackboard, and students vote for the most fitting terms.

The students already must be familiar with the concepts behind the terminology, i.e. have a working understanding of the context and the meaning. Therefore, the protocol contract negotiations should typically take place after the introduction to the subject, but before reaching advanced levels building upon the fundamentals, so that the terminology can be used actively and become part of the practice.

Terminology introduced by students can be reused and extended in subsequent academic years, but students should be allowed to renegotiate the protocol. The negotiations can also be performed in writing or even using a collaborative web-based dictionary.

## AN EXAMPLE

Table 3 summarises the results of one attempt at using the "protocol contract" technique in practice. During a lecture in the course "mobile information systems," a final-year optional computer science course, seven expressions were written on the blackboard. The expressions in the first column of Table 3 are all central to the syllabus of the course. The terms are all relatively new and there are no official Norwegian translations. The students were organised into groups of three and four depending on their seating, and they were asked to discuss and agree upon a set of alternatives. Ten minutes later, the teacher summarised the results in plenary, and the students voted for the best alternatives. Column two in Table 3 shows the suggestions; the terminology chosen by the student majority is underlined. The in-class atmosphere was friendly, concentrated and the students seemed to enjoy the activity. Certain expressions were apparently more difficult than others. For instance, "wearable computing" is originally an awkward and misleading term, and thus received numerous suggestions. The term "Bluetooth" posed no difficulties; nevertheless, some students were surprised to learn about its origin.

| Term | Student's in-class suggestions |
|---|---|
| Wearable computing | Bekledningsbehandling (in clothes processing), bærbar byte (wearable byte), bekledningsprosessering (wearable processing), kroppsprosessering (body processing), **påtagbar maskinvare** (wearable hardware). |
| Voice portal | **stemmeportal** (voice portal), lydportal (sound portal), taleportal (speech portal) |
| Location-based service | **stedbasert tjeneste** (place-based service), nærværstjeneste (presence service), lokasjonsbaserte tjenester (location-based service), tilstedetjeneste (presence service) |
| MIDlet | MIDlett (MID-light - light as in lightweight software – phonetically similar), **MIGlett** (MIG is the Norwegian translation of the acronym MID – Mobile Information Device) |
| Bluetooth | **Blåtann** (bluetooth), Bluetooth |
| Microbilling | Minifakturering (tiny-billing), småbetaling (tiny-payment), **minimynt** (tiny-coins) |
| Tilt-technology | **vippe-teknologi** (tilt-technology), vri teknologi (turn-technology), hellingsteknologi (slope-technology) |

TABLE 3
STUDENTS' TERMINOLOGY DISCOVERED IN-CLASS

## CONCLUSIONS

This paper has addressed some of the difficulties associated with technological paradigm shifts from a terminology-translation point of view. Our experience is that the students are a valuable resource for discovering suitable terminology for use in language for specific purposes. The fact that the students create their own terminology collectively means that they are also more likely to accept, use and include the new terms into their own vocabulary.

## ACKNOWLEDGEMENTS

The authors are grateful to the students and faculty at Oslo University College, Norway for participating in the experiments, and the six anonymous reviewers that have helped us improve the quality of this manuscript.

## REFERENCES

1. Castellví, M.T.C., "Theories of Terminology: Their Description, Prescription and Explanation," Terminology, Vol. 9, No.2, 2003, pp. 163–199.

2. Bowker, L., "You Say "Flatbed Colour Scanner," I Say "Colour Flatbed Scanner": A Descriptive Study of the Influence of Multidimensionality on Term Formation and Use with Special Reference to the Subject Field of Optical Scanning Technology," Terminology, Vol. 4, No. 2, 1997, pp. 275-302.

3. Scott, S., "A Journey into Chinese-English Environmental Translation," Translation Journal, Vol. 8, No. 1, 2004.

4. Chuansheng, H.E. and Xiao, Y-N., "Brand Name Translation in China: An Overview of Practice and Theory," Babel, Vol. 49, No. 2, 2003, pp. 131–148.

5. Hu, Q-P. and Xu, J., "Semantic Transliteration: A Good Tradition in Translating Foreign Words into Chinese," Babel, Vol. 49, No. 4, 2003.

6. Cao, D., "Finding the Elusive Equivalents in Chinese/English Legal Translation," Babel, Vol. 48, No. 4, 2002, pp. 330–341.

7. Jian, H-L and Sandnes, F.E., "Written Group Assignments – Experiences with Collaborative Web Spaces," Proceedings of the 2002 ICEE International Conference on Engineering Education (ICEE-2002), Manchester, United Kingdom, August, 2002, Paper 0132.

8. Temmerman, R., "The Process of Revitalisation of Old Words: 'Splicing,' A Case Study in the Extension of Reference," Terminology, Vol. 2, No. 1, 1995, pp. 107-128.

9. Callaway, D.R., *Inisde Servlets – Server Side Programming for the Java Platform,* 2nd Edition Addison Wesley, 2001.

10. Short Norwegian Computer Science wordlist, Norwegian Language Council, http://www.sprakrad.no/dataord.htm

11. Microsoft Corporation, Internet Explorer ver 6.0 User Manual, Online Documentation, 2001.

12. Saraireh, M.A., "Inconsistency in Technical Terminology: A Problem for Standardization in Arabic," Babel, Vol. 47, No. 1, 2001, pp. 10–21.

13. Elbro, C., "Snabel a," Mål og Mæle, Vol. 20, No. 2, 1997, 20-22.

14. Baker, M. (editor), *The Routledge Encyclopedia of Translation Studies,* Routledge, 1998.

15. Pearson, J., *Terms in Context,* John Benjamins Publishing Co, 1998.

16. Sager, J., *A Practical Course in Terminology Processing,* John Benjamins Publishing Co, 1990.

17. Chung, T.M., "A Corpus Comparison Approach for Terminology Extraction," Terminology, Vol. 9, No. 2, 2003, pp. 221–246.

18. Justeson, J.S. and Katz, S.M., "Technical Terminology: Some Linguistic Properties and an Algorithm for Identification in Text," Natural Language Engineering, Vol.1, No. 1, 1995, pp. 9-27.

19. Nakagawa, H., "Disambiguation of Single Noun Translations Extracted from Bilingual Comparable Corpora," Terminology, Vol. 7, No. 1, 2001, pp. 63–83.

20. Meyer, I., "Refining the Terminographer's Concept-Analysis Methods: How Can Phraseology Help?" Terminology, Vol. 3, No. 1, 1996, pp. 1-26.

21. Parker, S.P. and Licker, M.D., *McGraw-Hill Dictionary of Scientific and Technical Terms*, 6th Edition, McGraw-Hill, 2002.

22. Kenny, D., "CAT Tools in an Academic Environment: What Are They Good For?" Target, Vol. 11, No. 1, 1999, 65-102.

23. Norwegian Language Council, http://www.sprakrad.no/

24. Sandnes, F.E., *Modern Application Development: Thin Clients and Fat Servers,* Tapir Academic Press, Trondheim, 2002.

**Hua-Li Jian** received a B.A. in foreign languages and literature in Taiwan in 1991, a M.A. in TESOL from the University of Newcastle upon Tyne, U.K. in 1993, a M.Sc. in Computer Science from the University of Newcastle upon Tyne, U.K. in 1995 and a Ph.D. in linguistic science from the University of Reading, U.K. in 1999. Dr. Jian is currently with the Department of Foreign Languages and Literature at National Cheng Kung University, Tainan, Taiwan. Dr. Jian's research interests include English language learning, computer assisted pedagogy, Western and Oriental phonetics and phonology.

**Frode Eika Sandnes** received a B.Sc. in computer science from the University of Newcastle upon Tyne, U.K. in 1993, an honorary M.Sc from the Norwegian University of Science and Technology, Norway in 1993, and a Ph.D. in computer science from the University of Reading, U.K. in 1997. Dr. Sandnes has worked as a software engineer in the British space industry developing low earth orbit satellites. Dr Sandnes is currently an associate professor in the department of Computer Science at Oslo University College and also serves as the international coordinator for the engineering faculty. Dr Sandnes research interests include parallel processing, error-correction, mobile user interfaces and university level pedagogy.

# Chapter 36

# The Magic of Dynamic Programming

M. ASHRAF IQBAL and ATIF ALVI
*Department of Computer Science, Lahore University of Management Sciences, D.H.A., Lahore-54792, Pakistan. E-mail: aiqbal@lums.edu.pk, alvi@lums.edu.pk*

*Dynamic Programming is an artistic intellectual activity depending in part on the specific structure of the sequential decision problem [1]. Around the world, students of mathematics, engineering and science struggle every day to learn the difficult art of dynamic programming, an important algorithm design technique. We contend that the primary reason for this struggle is the inadequate treatment of this topic in popular algorithm textbooks. In this paper, we provide vital pedagogical aids, the inclusion of which, in the teaching mix, would make the learning curve a lot gentler for students. These aids enable the students to build powerful graphical templates and help them find the missing links required to understand the underlying concepts. We propose that teachers should highlight multiple recursive formulations of the same problem followed by drawing of their recursion trees and DAGs (directed acyclic graphs). It then becomes natural for students to visualize the problem and to devise a computationally cheap dynamic programming solution. We use typical optimization problems from textbooks and the research literature to demonstrate our technique. Overall, we strive to demystify dynamic programming by taking advantage of the power of visual processing.*

## INTRODUCTION

How people learn and how a teacher might facilitate learning are two of the fundamental questions that have occupied educational theorists for a long time and will continue to do so in the future. It has been argued that instructors need to develop different ways of teaching to suit the different learning styles of students, as determined, for example, through methods such as the Meyer/BriggsType Indicator, commonly known as MBTI [2]. As a result, a number of diverse pedagogical approaches have been proposed as supplements to formal classroom lectures [3]. It is also important to mention here the teaching reform movement, which started in the eighties when a number of mathematicians organized a meeting at Tulane University to lament the sorry state of mathematics education in the US [4]. The movement, with the help of government funding, has enabled researchers to buy additional equipment, to arrange teacher-training workshops and to hold conferences. Some of the hallmarks of the new methodology proposed by this movement are as follows:

1. Students should discover mathematical facts themselves; they can help each other discover these facts in small groups while the instructor is acting as a guide.
2. Students can use technology to help them visualize mathematical concepts and to aid in the construction of these concepts in their own minds.

The motivation for (a) perhaps came from the Socratic dialog, which is usually initiated by a student who needs help with a problem. In this methodology, the teacher does not deliver a formal lecture but asks questions, helping students find connections in their current knowledge. In recent times (1920-1970), Professor R. L. Moore incorporated the principle of student discovery in his famous style known as the Moore Method [5]. What was so different about his style of teaching was that he did not lecture. He sat in the back of the class, remaining quiet for the most part, but occasionally asking a question and thus enabling his students to discover answers in their own way. A somewhat similar approach has been described in [6], which argues that while we are teaching students standard algorithms, we should not formally teach anything. Instead, we should encourage, or rather provoke, students to create algorithms themselves using some very fundamental concepts. The objective is that students should experience the tension and excitement of discovery even during the initial phases of understanding.

Understanding is known to be the collection of maps and templates which we have at our disposal to fit facts into as we learn. These templates are the places where facts and ideas can be connected to gain understanding and appreciation. Thus one of the primary roles of a teacher is to help students build or find the relevant maps and templates that might help them learn and solve problems [7]. It is precisely here that visualization tools can immensely help students construct concepts and design graphic templates. According to [8]: "Visualization...is more apparition than actual, drawing attention inward to a more primal state where thoughts are not subject to the limitations of physicality." This close encounter with the nature of a problem, having seen it with the mind's eye, enables students to understand a problem and possibly arrive at an efficient solution with the help of visual processing [9,10].

In this paper we contend that some very popular textbooks on algorithms have not done full justice to the presentation of dynamic programming and that is why we have serious problems teaching it [11-15]. Our teaching experience indicates that it is quite simple to show the actual working of this technique on a specific example but it is extremely difficult to make the students conceptualize the problem in general. It seems as if there is some magic behind this approach or that there is something missing. Until and unless we find that missing link, it will not be possible to effectively teach this otherwise very simple but powerful approach. Our experience has been substantiated by [12]: "After you understand it, dynamic programming is probably the easiest algorithm design technique to apply in practice…until you understand it, however, dynamic programming seems like magic." This paper, in its quest to simplify dynamic programming, presents various visualization techniques that unravel the nature of dynamic programming more fully than what is possible by mere mathematical formulation. It encourages students to build powerful graphical templates, which help them find the missing links required to understand the underlying concept.

Transforming a given complex problem into an easier analog and using appropriate symbolism are two mechanisms that move a problem from an unfamiliar domain to a friendlier and more familiar setting. References [10,16] cite the use of analogies and symbols to make the mathematics more intelligible for students. It is in fact a standard practice in engineering to make domain transformations while finding solutions; e.g. Fourier and Laplace transforms. This paper also includes an earlier transformational approach used by researchers for dynamic programming [17,18] and improves upon it. We illustrate through two typical dynamic programming problems how our approach is different from the conventional way of teaching. We also take a brief look at some researchers' use of dynamic programming and compare it with our methodical approach.

## MISSING LINKS OF DYNAMIC PROGRAMMING

We propose that an instructor should go through the following sequence of steps while teaching dynamic programming:

- Make *multiple* recursive formulations of the optimization problem.
- Draw the *recursion tree* corresponding to the top-down recursive formulation of the problem.
- Convert the recursion tree into a *directed acyclic graph (DAG)* by merging similar vertices into a single vertex.
- Compute the optimal solution by moving bottom-up in the DAG.

The italicized phrases highlight what is missing from the traditional approach. It is important to have multiple recursive formulations for the same problem in order to make a comparison between various approaches. It is very much possible that one recursive formulation is more efficient in solving one problem while another is more effective in another situation. The students should have multiple choices instead of relying on just one formulation. The DAG is also a very useful tool. It not only helps us design the bottom-up dynamic programming algorithm but is also meaningful otherwise: the directed edges represent the dependencies in the problem and its solution. In fact, the number of edges in the DAG will determine the time complexity of the dynamic programming solution (remember that the number of edges in the recursion tree is an exact measure of the time complexity of the simple recursive implementation of the problem).

We shall now apply our technique to some typical dynamic programming problems.

## THE MATRIX-CHAIN MULTIPLICATION PROBLEM

If we have a chain of $A_1A_2...A_n$ of $n$ matrices to be multiplied, where for $i = 1, 2, ..., n$, matrix $A_i$ has dimensions $p_{i-1} \times p_i$, we can minimize the number of scalar multiplications if we can discover an optimal parenthesization of the chain of matrices. This is a classic dynamic-programming-amenable problem as it exhibits an *optimal substructure*: optimal solutions to its subproblems can be combined to give an optimal solution to the whole problem. An optimal parenthesization of $A_iA_{i+1}...A_j$ that splits the product between $A_k$ and $A_{k+1}$, for some $i \le k < j$, contains within it optimal parenthesizations of $A_i A_{i+1} \dots A_k$ and $A_{k+1} A_{k+2} \dots A_j$. We should set up a recursion that gives the cost of the optimal solution in terms of optimal solutions to subproblems. For a chain of $n$ matrices, our subproblems are the problems of determining the minimum cost of a parenthesization of $A_i A_{i+1} \dots A_j$ for $1 \le i \le j \le n$. Let $M[i..j]$ be the minimum

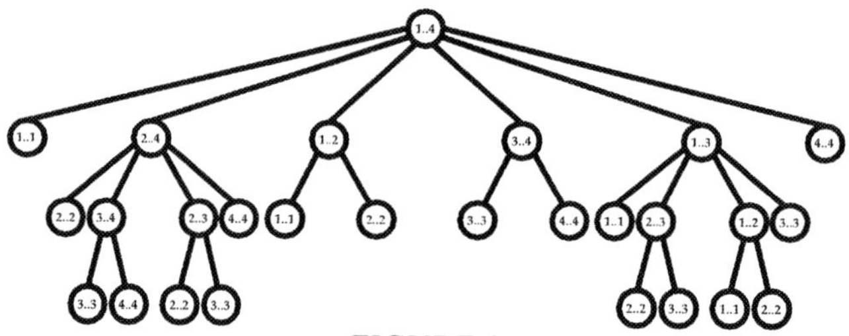


FIGURE 1
RECURSION TREE FOR THE MATRIX-CHAIN MULTIPLICATION PROBLEM

number of scalar multiplications needed to compute the product $A_i A_{i+1} \dots A_j$. The optimal solution to the whole problem would thus cost $M[1..n]$.

It is obvious that $M[i..i] = 0$ for $i = 1, 2, \dots, n$. For $i < j$, let us assume that an optimal parenthesization splits the product $A_i A_{i+1} \dots A_j$ between $A_k$ and $A_{k+1}$, where $i \le k < j$. Keeping in mind that each matrix $A_i$ has dimensions $p_{i-1} \times p_i$, we see that computing the matrix product of the two subproblems entails $p_{i-1} p_k p_j$ scalar multiplications. Thus, we can set up the following recurrence:

$$M[i..j] = \min_{i \le k < j} \left( M[i..k] + M[k+1..j] + p_{i-1} p_k p_j \right) \qquad (1)$$

## CONVENTIONAL APPROACH

If recurrence (1) is implemented directly, we end up doing an exponential amount of work. Instead, a dynamic programming solution that computes in a bottom-up manner, storing intermediate results and using them whenever required, does only a polynomial amount of work. This is exactly how [11] attempts to solve the problem. Cormen et al. [11] also claim that a recursive implementation of the recurrence takes exponential time, but this claim is substantiated only in a later section. These authors [11] then compute the optimal cost in a bottom-up dynamic programming way. The pseudocode of the dynamic programming algorithm, given on page 336 of [11], operates on a triangular table whose sudden appearance puzzles students. The issue of arriving logically at this table becomes all the more important when one finds that some problems require a rectangular table while others need a three dimensional table for a dynamic programming solution.

## OUR APPROACH

Our teaching experience shows that it is best to ask students to draw a recursion tree for a reasonable size matrix-chain immediately after arriving at (1). An example is shown in Figure 1 for a chain of size four. As soon as they draw a recursion tree, they can actually see that it grows exponentially at each layer downwards from the root. They are therefore convinced that a direct recursive algorithm in this case would be expensive, in fact exponential. The number of edges in the tree is exactly $3^{n-1}-1$. The recursion tree not only captures the essence of the problem in all its complexity, but also holds the key to its solution: the students can see that many nodes (subproblems) are repeated at various

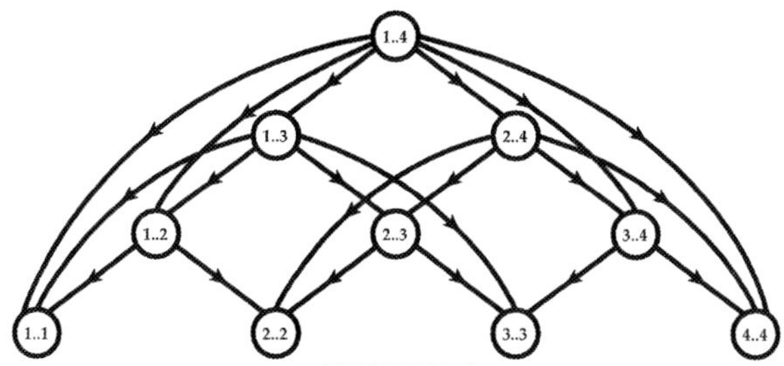


FIGURE 2

DAG FOR THE MATRIX-CHAIN MULTIPLICATION PROBLEM

layers. Hence, both the exponential nature of the problem and the existence of only a polynomial number of subproblems are evident even to an average student.

As the next step, the class should be asked to transform the recursion tree into a DAG. The DAG (Figure 2) removes all unnecessary repetitions and gives a clear picture of the reduced amount of work that would have to be done using a dynamic programming approach. The drawings of the recursion tree (with exponential growth) and the DAG (with constrained growth) are indeed powerful vizualization tools for understanding the magic behind dynamic programming. The DAG essentially tells us about the various dependencies and naturally leads to the bottom-up dynamic programming algorithm. The triangular table of [11] is in fact a simplified version of the DAG where edges are removed and it is implied that a vertex on top will depend upon all the vertices below it. It is important to note that a triangular matrix does not convey all the information contained in the DAG.

As the number of distinct subproblems is exactly equal to $n(n+1)/2$, which is only $O(n^2)$, we can use the bottom-up technique of dynamic programming to avoid the numerous recomputations of the top-down recursive formulation. This cuts down the complexity to the number of edges in the DAG of Figure 2, which grows as $\sum_{n=1}^{n} n(n-1)$. This sums to $n(n^2-1)/3$, yielding an only $O(n^3)$ solution. For example, the number of edges in Figure 2 is exactly 20.

## THE PARTITION PROBLEM

We next look at another interesting and seemingly unrelated problem. Given a fixed sequence $\{s_1, s_2, ..., s_n\}$ of $n$ nonnegative numbers, we want to divide the set into $p$ partitions such that the maximum sum over all the partitions is minimum.

We can find an optimal solution by exhaustive recursive search. If we let $M_p[i..j]$ denote the optimal solution to dividing a sequence of numbers with indices from $i$ to $j$ for some $i \le j \le n$ into $p$ partitions, we can form the following recurrence:

$$M_p[i..j] = \min_{i \le k < j} \max\left(M_{p-q}[i..k], M_q[k+1..j]\right) \tag{2}$$

Below, we discuss two interesting cases of the above recurrence.

**Case 1:** We can set $p-q=q$, which means $q=p/2$. The recurrence in (2) is reproduced below with this condition:

$$M_p[i..j] = \min_{i \le k < j} \max\left(M_{p/2}[i..k], M_{p/2}[k+1..j]\right) \tag{3}$$

Equation (3) represents the so-called *binary* solution. We put a divider in the middle of the sequence and try to optimally partition the subsequences formed to the left and right into $p/2$ partitions each. This process can be repeated recursively until the base case is reached. The recursion tree is shown in Figure 3 for a sequence of five numbers and four partitions. It is interesting to note that the recursion in (3) as well as the recursion tree for the partition problem are exactly the same as those for the matrix-chain multiplication problem ((1) and Figure 1), except for one difference: the recursion tree for the matrix multiplication problem has a depth proportional to $n$, while the depth of the

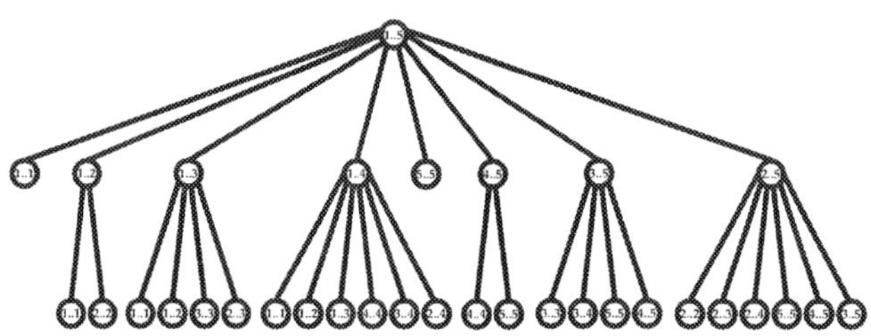

FIGURE 3

RECURSION TREE FOR THE BINARY SOLUTION OF THE PARTITION PROBLEM (CASE 1)

partition problem's tree is proportional to $log_2$ (*number of partitions*). Surprisingly, the DAG for the partition problem (Figure 4) is very different from the DAG for the matrix multiplication problem (Figure 2). The number of vertices in the entire DAG of Figure 2 is proportional to $n^2$, while the number of vertices in any one of the layers (except the top two) of the DAG in Figure 4 is proportional to $n^2$.

**Case 2:** If we set $q=1$, which means $p-q=p-1$, we get the following recurrence:

$$M_p[i..j] = \mathit{min}_{i \le k < j} \max\left(M_{p-1}[i..k], M_1[k+1..j]\right) \tag{4}$$

It is easy to see that $M_1[k+1..j]$ is the sum of the numbers with indices from $k+1$ to $j$. This leads to the so-called *incremental* partitioning strategy as we form the $p^{th}$ partition by placing the last divider right after the $k^{th}$ element for some $i \le k < j$. The maximum sum over all the partitions would then be the larger of the sums to the left and right of this last divider. The sum to the right of this last divider,

$M[k+1..j]$, is simply $\sum_{a=k+1}^{j} s_a$. The recursion tree for (4) is shown in Figure 5 for a sequence of six numbers and four partitions. Its DAG is shown in Figure 6.

## CONVENTIONAL APPROACH

Skiena [12] includes the partition problem on pp. 56-59. Skiena's treatment of this problem has the same defects as those of [11] for the matrix multiplication problem. No recursion tree or DAG is drawn and there is no mention of the binary way of solving it. In addition, the dynamic programming rectangular tables that are shown (p.59) neither give any clue about the various dependencies that exist among their entries, nor are they any help in understanding the complexity of the algorithm. The common form of (1) to (4) is also not
found as such in textbooks. There, the students are introduced to each problem in isolation and the common thread that runs through all is never explored.

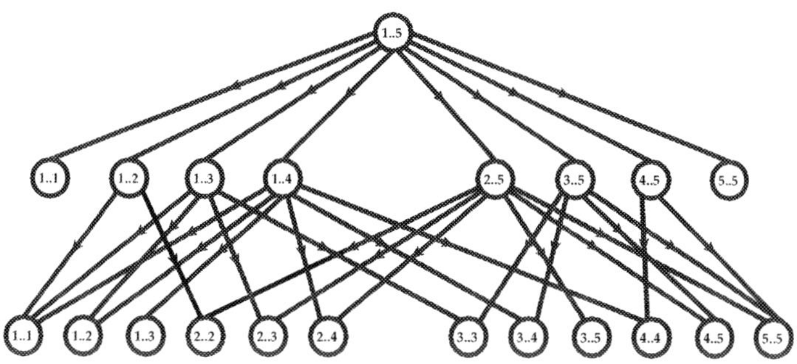


FIGURE 4
DAG FOR THE BINARY SOLUTION OF THE PARTITION PROBLEM (CASE 1)

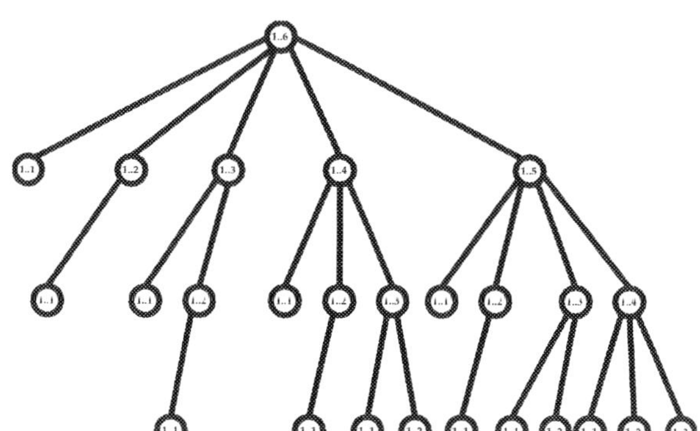

FIGURE 5
RECURSION TREE FOR THE INCREMENTAL SOLUTION OF THE PARTITION PROBLEM (CASE 2)

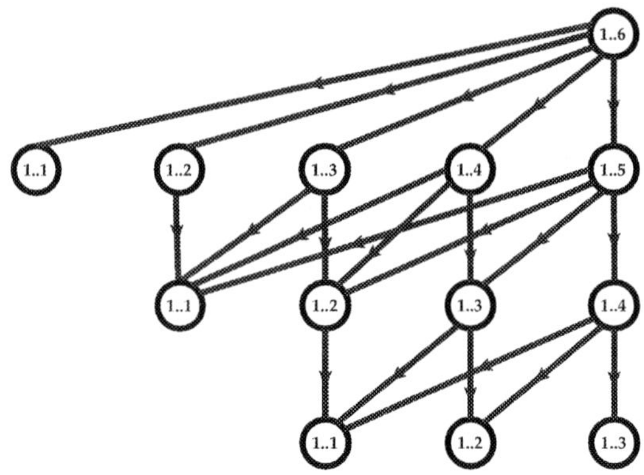


FIGURE 6
DIRECTED ACYCLIC GRAPH FOR THE INCREMENTAL SOLUTION OF THE PARTITION PROBLEM (CASE 2)

## OUR APPROACH

We believe in showing the students at least two recursive formulations for some of the dynamic programming problems. With this experience, the students are able to pick the better recursive approach.

The directed acyclic graph of Figure 6 is essentially a rectangular table similar to the one given in [12]. In fact, the DAG also paints a vivid picture of the dependencies. For example, in order to calculate $M_4[1..6]$ we need to calculate all the vertices connected to it. In the classroom, we also stress (1) to (4), encouraging students to write them on their own after they have understood the problems. Their facility in this indicates their level of understanding.

The recursion tree and DAG are visual tools that correctly depict the complexity of the problem: the number of edges in them is a measure of the complexity of the respective solution. For the binary solution (Case 1) of the partition problem, the number of vertices in the DAG in a typical layer in Figure 4 is $O(n^2)$, which makes the number of edges $O(n^3)$. The overall complexity thus turns out to be $O(n^3 \log p)$. For the incremental solution (Case 2), the number of edges per layer in Figure 6 is $O(n^2)$ and thus the overall complexity is $O(n^2 p)$. For comparison, the complexity of the recursive solution can be determined by counting the edges in the recursion tree of Figure 5, which grow as $1+2+4+7+11+16+...$, i.e. $O(n^3)$. The interesting thing to note here is that if we increase the number of partitions, then the number of edges becomes the summation $1+2+4+8+16+32+...2^n$, which is $O(2^n)$. Thus, the number of partitions to be formed drastically affects the complexity of the brute-force recursive solution.

## LAYERED GRAPH APPROACH OF RESEARCHERS

Bokhari [17] and Nicol et al. [18] present a layered graph solution to the partitioning problem, which is inherently a dynamic programming solution. The only difference is that the DAG approach is derived in a formal manner after making a recursive formulation, whereas the layered graph approach seems like an elegant trick as it directly arrives at the layered graph. Moreover, it fails to model the binary version of recursive formulations and works only for the incremental one. Hence, the layered graph is just a special case of dynamic programming and does not always guarantee an efficient solution. Without following the steps that we prescribe, the researchers were forced to rely on their ingenuity to arrive at a solution, which, although beautiful, was neither straightforward nor easily reproducible.

## CONCLUSIONS

In this paper, we have proposed an enhanced methodology for teaching dynamic programming. We felt the need to do so after finding the treatment of popular textbooks on algorirthms inadequate. These textbooks leave the average student in awe of this paradigm that seems to produce efficient solutions magically. We have identified certain ingredients missing from the mix of teaching methods whose inclusion takes the magic out of dynamic programming. These ingredients include the following: the recursion tree, the directed acyclic graph, the different recursive formulations of the same problem, and discovering the similarities and differences of the various problems chosen to demonstrate dynamic programming. Our experience has shown that after the students are convinced of the optimal substructure property of the problem, it is best to let them experiment with its various recursive formulations. They should draw graphs (i.e. recursion trees) of various recursive formulations to appreciate how much redundant work is being done in the various branches. In this visual processing, a DAG is then the next logical step, making the students realize how much work can be saved by moving bottom-up in the DAG. By using these graphical templates, the quest for a general dynamic programming solution becomes a methodical task that, along the way, also provides visual insight into the nature of the problem and its solution. Iqbal and Alvi [19] present a comprehensive and detailed treatment of the topics discussed in this paper. Therein we also discuss the all-pairs shortest path problem and show that its binary recursive formulation is more efficient than the corresponding incremental formulation.

## REFERENCES


1. Shapiro, J. F., *Mathematical Programming,* John Wiley, Chichester, UK, 1979.
2. Krantz, S. G., *How to Teach Mathematics,* American Mathematical Society, 1999.
3. Froyd, J., "Model for Inquiry and Reflection about Learning and Teaching," Chapter 17, Aung, W., et al. (ed.), *Engineering Education and Research – 2001: A Chronicle of Worldwide Innovations,* Int. Network for Eng. Ed. and Res. (iNEER), Arlington, VA, 2002, pp. 169-178.
4. National Commission on Excellence in Education, A Nation at Risk: The Imperatives of Educational Reform, U.S. Government Printing Office, 1983.
5. Halmos, P., Moise, E., and Piranian, G., "The Problem of Learning to Teach," The American Mathematical Monthly, Vol. 82, No. 5, May 1977.

6. Iqbal, M.A., and Tahir, S., "Should We Teach Algorithms?" IJECE, Vol. 2, No. 2, Summer-Fall 2003.

7. Davis, W.J., "Remember the Students," Appendix, Krantz, S. G., *How to Teach Mathematics,* American Mathematical Society, 1999, pp. 183-195.

8. Patton, A.H., and Bannerot, R.B., "A Theoretical Basis for Integrating Rational and Intuitive Processing in Engineering Curricula Through Art," Chapter 5, Aung, W., et al., *Innovations 2003: World Innovations in Engineering Education and Research,* Int. Network for Eng. Ed. And Res. (iNEER), Arlington, VA, 2003, pp. 37-47.

9. West, T., *In the Minds Eye,* Prometheus Press, New York, 1991.

10. Polya, G., *How to Solve it, A New Aspect of Mathematical Method,* Princeton University Press, NJ, USA, 1957.

11. Cormen, T., Leiserson, C., Rivest, R., and Stein, C., *Introduction to Algorithms,* MIT Press, Cambridge, MA, USA, 2001.

12. Skiena, S.S., *The Algorithm Design Manual,* Springer-Verlag, New York, 1997.

13. Michalewicz Z., Fogel, D. B., *How to Solve It: Modern Heuristics,* Springer-Verlag, 2000.

14. Horowitz, E., Sahni, S., and Rajasekaran, S., *Computer Algorithms,* Computer Science Press, 1997.

15. Sara Baase, S., and Gelder, A.V., *Computer Algorithms,* Addison Wesley, 2000.

16. Boas, R., "Can We Make Mathematics Intelligible," The American Mathematical Monthly, Vol. 88, No. 10, December 1981.

17. Bokhari, S. H., "Partitioning Problems in Parallel, Pipelined, and Distributed Computing," IEEE Transactions on Computers, Vol. 37, No. 1, January 1988.

18. Nicol, D.M., and O'Hallaron, D.R., "Improved Algorithms for Mapping Parallel and Pipelined Computations," IEEE Transactions on Computers, Vol. 40, March 1991.

19. Iqbal, M. A., Alvi, A., "The Magic of Dynamic Programming," Technical Report, Department of Computer Science, Lahore University of Management Sciences, 2003.

**M. Ashraf Iqbal** received his B.Sc., M.Sc. and Ph.D. degrees in Electrical and Computer Engineering from the University of Engineering and Technology (UET), Lahore, Pakistan, in 1975, 1983 and 1991, respectively. He was a Fulbright and DAAD scholar in 1992 and 1996. He worked in the Electrical Engineering department of UET from 1975 to 2000. He is currently a Professor in the Computer Science Department of the Lahore University of Management Sciences (LUMS). His current research interests are innovations in teaching and discovery-based learning. He is also an Urdu poet.

**Atif Alvi** did his B.Sc. in Electrical Engineering from UET and his M.S. in Computer Science from LUMS in 1991 and 2002, respectively. Currently, he is a Teaching Fellow in the Computer Science department at LUMS.

# Chapter 37

# The Development of the Engineering Scholars' Program at FAU

SAM HSU, SHARON SCHLOSSBERG and KARL STEVENS
*College of Engineering, Florida Atlantic University, Boca Raton, FL 33431, USA. E-mail: sam@cse.fau.edu, sschloss@fau.edu, stevens@fau.edu*

*Known as the Engineering Scholars' Program, a summer enrichment program at Florida Atlantic University that focuses on disciplines in engineering and computer science for high-achieving gifted high school students, has served more than 500 students from 23 different high schools in its service area since its first offering in 1997. This program is virtually free for its participants, making the program accessible to all students who are academically qualified within the defined geographic region. Experienced faculty, caring staff, and motivated students, together with a sound plan and implementation, have made this program a great success. The program enjoys a high student satisfaction and ever-increasing enrollment. This paper discusses the initial design, subsequent development, present situation, and future plan of the program.*

## INTRODUCTION

There are numerous universities and colleges in the U.S. offering summer enrichment programs for gifted and talented high school students to provide summer academic advancement activities, and to some extent, functioning as a recruitment tool as well. Some of the summer programs focus on disciplines in math, science and engineering, as can be seen in [1-11]. Some other programs are in a wide range of disciplines, as typified in [12-16]. While most summer programs are sponsored as independent efforts of individual colleges/universities, some academic institutions join together, pooling resources to offer worthwhile educational programs in an entrepreneurial manner. References [4], [17] and [18] show examples of this practice. Most of the summer enrichment programs are costly for participants, ranging from hundreds to thousands of dollars for a short period of summer time. Programs in [2], [4], and [7-17] are in this category. Only a few, as exemplified in [3], [5] and [6], are free for participants.

As a leading academic institution in south Florida and in keeping with trends, Florida Atlantic University (FAU) has a summer enrichment program focused on disciplines in engineering and computer science for high school students in its service area. The program, known as the Engineering Scholars' Program (ESP) [19], was established in the College of Engineering (COE) in 1997. Features of the program include: (a) expert engineering faculty who are experienced in applying math and science to real-world problems, and experienced in educating and working with high-achieving high school students; (b) a professional staff experienced in working with secondary school students and assisting them in making reasoned personal, educational, and career choices; and (c) FAU/COE state-of-the-art laboratories and computer systems comparable to those in business and industry. Furthermore, ESP is offered via a dual enrollment program that covers tuition costs for all participants [20]. In essence, it is a free program for those who have satisfied all academic evaluation and geographic criteria for acceptance. This is the realization of the deep belief of the program directors that the program should be accessible to all who meet the criteria, regardless of their family finances. ESP has been a very successful program since its inception, with high student satisfaction and ever-increasing enrollment.

The next section of the paper focuses on the development of the ESP program, including student selection, tuition, instructor/course selection, course structure, and program funding. The following section lists the features and improvements that have been added over the past several years to make enhance the program. These changes include course placement, course mentors, program orientation and coordination, program assessment, marketing, etc. The section "ESP Today" gives an overall view of the current status of the program, followed by a brief description about current ongoing efforts on a tracking plan, and then the conclusion.

## DEVELOPMENT OF THE ESP

The ESP program was first offered in 1997 in response to a request from Palm Beach County Schools to provide a program that would meet the advanced academic needs of students in their Math, Science and Engineering (MSE) Magnet Program. Responsibility for developing the summer program was given to the Associate Dean for Academic Affairs and to the Director of the Division of Engineering Student Services. Upon receiving the assignment, the two newly assigned program directors quickly laid out several major elements in their initial design of the program, namely: student selection, faculty/course selection, course structure, and program funding. To make the program competitive, it was decided that the program must focus on course selection and cost. More details on initial concerns and issues are described below.

- **Student selection**: The first students targeted were the high-achieving and gifted students from Suncoast High School's MSE Magnet Program. This program requires students to have a university level course experience between their sophomore and junior year, and again between their junior and senior year. A formal request from Palm Beach County on behalf of the MSE program at Suncoast was presented to FAU by county officials. ESP was created in response to this request.
- **Tuition**: The program was required to be accessible to all who were academically qualified. Free tuition was the idea that the program directors had in the beginning.

Based on previous experience in pre-college programming and dealing with students from all backgrounds, the program directors were aware of an articulation agreement between the university and the school systems of Palm Beach and Broward Counties. This agreement would enable qualified high school juniors and seniors to take university courses as dual-enrolled students to earn high school graduation credit and college credit simultaneously [20]. Furthermore, this program would entitle participating students to be exempt from the payment of registration and laboratory fees. Taking advantage of this existing articulation agreement resolved the tuition issue.

- **Instructor/course selection**: Instructor and course selections for the program were another concern. While faculty were contacted to evaluate their level of interest, one suggestion was to choose twelve-month faculty as program instructors. Although an excellent cost saving measure for the college, the program directors vetoed the idea as a selection criterion. They believed strongly in the importance of teacher methodology and approachability, particularly with such young students. The directors felt that ultimate program success rested heavily in the type of faculty selected. Several professors from different departments who expressed interest were then asked to write a brief description of a course they might offer. The course would be at the 3000 level, with content suitable for college juniors. Years of experience with pre-college activities had convinced the program directors that the instructors were as important to program success as the actual course offerings, and yet providing a cross section of college courses was important too. As a result, four courses by faculty from different departments with experience in pre-college outreach were selected. These courses were: Automation and Robotics, Electronic Design with Operational Amplifiers, Introduction to Inventive Problem-Solving, and Introduction to Web Authoring and Programming.
- **Course structure**: Since this program was defined as a summer enrichment program, and considering that summer vacation is a hectic time for high school students, it was decided that the program should begin the Monday following the last day of school and run for two weeks (Monday thru Friday from 9:00am to 4:00pm). Each school day would closely resemble two weeks of classes during a regular semester. That is, an ESP course would need to be a fast-paced 3000 level course, comparable to what a college junior might take.
- **Program funding**: Major expenses in this program included tuition for students, salaries for faculty, and expendable laboratory supplies. Since tuition costs were already addressed by the articulation agreement, salaries and laboratory experiment supplies remained the issues. These costs would be absorbed primarily by COE, with some help from the university.

The initial implementation of ESP in 1997 was an instant success. We received compliments from students, parents, and school system representatives, as well as suggestions for improvement. The positive feedback and general interest encouraged the program directors to continue offering ESP.

## ESP in Progress

Over the years, ESP has changed substantially. These changes are the direct results of feedback from all constituencies and have markedly improved the program, resulting in enrollment growth each year.

### Student Selection

In 1997, students recommended by Suncoast High School were all accepted to the program. Thanks to outstanding program success, many requests to expand the opportunity for 1998 came from other high schools in Palm Beach and Broward Counties. These two large school districts represent over 50 high schools. Due to the ever-increasing student population and the limited resources available, a formal application process has been established over the years, reflecting changing times. Now, there are three pages for students and parents to complete, which include students' e-mail addresses, cell phone numbers and more specific course-related information, and two pages to be completed by the guidance department. Students are asked to write a short essay, submit letters of recommendation and a school transcript. They must have a minimum GPA of 3.0 on a 4.0 scale and be staffed gifted or high-achieving in their home high school. Students participating in rigorous academic programs like the International Baccalaureate Program or Engineering Magnets are also considered for acceptance. A student's level of maturity is a very important component for acceptance because of the academic demands placed on participants. Guidance counselor feedback, teacher recommendations and students' essays help administrators to evaluate maturity. Time constraints and travel issues do not allow for personal interviews, which would be the preferred method for evaluating the level of student maturity.

Over the years, about one third of the applicants have not been accepted to the program due to academic reasons. Of those accepted, about 40% are female, and about 35% are underrepresented minorities.

### Course Placement

A formal procedure for course placement has been established over the years. Students are asked to make course selections in order of their preference and every effort is made to give students their first or second choice. However, issues may arise because students tend to underestimate the importance of the required prerequisites necessary to succeed in each course (prerequisites are clearly stated in the syllabus for each course included with the application packet). In cases where student choices do not match their academic background a personal phone call is made and the choices and explanations are discussed.

### Course Mentors

Suncoast High School had a teacher/chaperone on campus to help their students when ESP was first offered in 1997, and it has continued to provide this each year. When the program was expanded to include more schools in 1998, the directors were concerned about student assimilation. It was unrealistic to expect schools with one or two students in attendance to provide a teacher/chaperone for their students. The challenge was to bring these individual students together into a cohesive unit. From this concern, the idea of class mentors was born. Each course would have a mentor who would help bring the students together, shepherd them, interpret college life for them, and at the same time would act as the eyes and ears of the program directors. These mentors would be engineering, science or education graduate students, or high school teachers in math, science or engineering. Mentors, it has turned out, are instrumental in providing program directors timely feedback contributing to the successful implementation of the program.

## Program Orientation and Coordination

It was learned from the first offering of ESP that a synchronized and coordinated approach among participants would make the program run more smoothly. Plans were laid out to help faculty/staff and students interact in a well-managed manner. After several years of implementation and updates, the following has become a standard procedure.

A few weeks prior to the program start a general faculty, a mentor and administrator meeting is held. Concerns and changes are discussed and if possible student rosters are distributed. The first day of class begins with a general student/parent orientation meeting where the directors explain how the program is run, what students' responsibilities are and general campus rules. Parking and lunch questions have been explained through various mailings, but they are reviewed at this time. Students are also made aware of office locations that may be important to them. Faculty and mentors are introduced, and then students are taken to their classes by their instructors. Mentors are asked to meet with a designated program director on a daily basis. They can stop by her office for five minutes during lunch or at any other convenient time during the day. These debriefing meetings have played a critically important role in effective program management and student success.

## Program Funding

The program expense for 1997 was covered by the college and the university. In late 1997, a RFP (Request for Proposal) from the Florida Department of Education came to the directors' attention and seemed to fit the ESP model perfectly. It was called the Governor's Summer Program, and it provided funding for sponsoring meaningful summer educational activities for gifted and high-achieving high school students. Maximum funding from the state was capped at $25,000, but that additional funding helped expand the program. Funding was also obtained from Region V Area Center for Educational Enhancement[7] to support course mentors.

## Program Assessment

Starting 1998, students at the conclusion of the program have been asked to complete two class evaluation forms, one from the State, and the other from the college. The state form is brief and very succinct. The college form is more detailed and provides much more constructive program feedback. These forms are carefully reviewed and summarized and shared with faculty and program administrators. Student concerns and suggestions are taken quite seriously and have resulted in significant program changes over time. After the conclusion of the program, early in the fall, there is a faculty and staff debriefing

---

[7] Region V Area Center for Educational Enhancement is a multi-year research project, funded jointly by National Science Foundation and Florida Department of Education, for improving reading achievement through science contents for elementary and middle school students in the FAU service area.

which provides the opportunity for faculty suggestions and input. Four months after course completion a follow-up survey is sent to participants to help evaluate the lasting impact of program participation. These results are carefully reviewed, summarized, and shared with faculty, college administrators, university administrators, and high school principals.

## Marketing

In 1997, ESP was provided to serve only one high school (Suncoast) with a very select student body and a teacher chaperone on campus. Program marketing was not an issue. However, along with the expansion of the program in 1998 came the concern of community awareness and marketing. Through a variety of pre-college activities in both counties, the Director of Engineering Students Services (serving as a program director of ESP) had many contacts in the K-12 arena which enabled her to attend meetings for math and science department chairs, guidance directors, magnet coordinators, etc. With these opportunities word began to spread. In addition, flyers and brochures were created and mass mailings were sent to Southeastern Consortium for Minorities in Engineering (SECME) Coordinators, high school JETS (Junior Engineering Technical Society) team coaches and others in the school system. Presentations were made at various K-12-related meetings and articles were placed in local newsletters and papers.

## Program Directors Responsibilities

Along with the development of ESP over the years, the two program directors have full responsibilities for the following areas:

- Budgeting
- Curriculum development
- Faculty selection
- Marketing
- Mentor selection
- Parent/student/school communication
- Program assessment and evaluation
- Program debriefing
- Student recruiting and selection
- Scholarship selection

## Enrollment Growth

There were four courses with about 70 students enrolled when ESP was first started in Summer 1997. Due to the program's success and popularity, it expanded to five courses with over 100 students enrolled in Summer 2002, a growth of 30% in the number of students registered. Starting with just one school, Suncoast High School in 1977, ESP has so far served over 500 students from the following 23 schools:

- Atlantic High School
- Blanche Ely High School
- Boca Raton High School
- Boyd Anderson High School
- Coral Spring Charter High School
- Deerfield Beach High School

- Dillard High School
- Ely High School
- Flanagan High School
- Hollywood Hill High School
- JP Taravella High School
- Nova High School
- Olympic Heights High School
- Pembrook Pines Charter High School
- Plantation High School
- Sheridan Technical High School
- Spanish River High School
- St. Thomas Aquinas High School
- Suncoast High School
- University School
- Wellington High School
- Western High School
- W. T. Dwyer High School

## ESP Today

After six years in existence, the value of ESP has become known community-wide. Parents, teachers, and students begin calling in early fall, prior to any program announcement, requesting information about ESP and applications for ESP for the following summer. Siblings and other family members of previous participants are among the first to apply each year. Teachers in local high schools are recommending program participation for their eligible students. To date, ESP has reached the following status.

### Program Goal

The goal of the ESP program is to provide high-achieving high school students an educational opportunity, not available at regular high school settings, to develop their potential to place math and science concepts into the context of real-world problems. The goal is carried out via the offering of dual enrollment credit courses in engineering and computer science at the university junior level with the active involvement of experienced faculty and staff, and supported by state-of-the-art engineering and computer science laboratories.

### Student Population

The ESP program is open to rising juniors and seniors in both Broward and Palm Beach counties who are enrolled in science/engineering magnet programs, International Baccalaureate magnet programs, or who have been staffed into the gifted program in their home high school. Students must have a cumulated GPA of 3.0 on a 4.0 scale and have scored in the $85^{th}$ percentile or higher on standardized tests. A letter of recommendation from an instructor (science, mathematics, computer) or a guidance counselor is also required. In addition, student groups not traditionally served by programs for gifted are encouraged to apply.

## Student Products

The students demonstrate their learning primarily through the course projects they complete. Student teams demonstrate and report on their projects replicating practices in business and industry (these demo sessions are open to the public). Typical of university level work, students further demonstrate their learning through class participation, homework, quizzes, examinations, and laboratory reports. Each student receives a university grade from the course instructor based on his or her overall course performance. The grade received also counts toward the participant's high school graduation credit. Each course has a variety of small projects and laboratory exercises, culminating in a major project experience.

## Instructors

Currently, there are six instructors from three major departments in College of Engineering involved in ESP (the three departments are Computer Science and Engineering, Electrical Engineering, and Mechanical Engineering). They have been selected because of their expertise within their individual disciplines, and their experience in educating and working with high-achieving high school students.

## ESP Courses

What follows are brief descriptions of the five courses offered in the most recent year.

- **Automation and Robotics**: This course introduces students to problems and solutions encountered during the application of automation and industrial robots. Class material will cover: design of pneumatic systems, robotics fundamentals, and computerized numerical control (CNC) machine tools. Lab activities include: construction of a small programmable robot, programming of a robotic assembly line, and programming of a CNC machine tool.
- **Electronic Design with Operational Amplifies**: This course is to provide a hands-on circuit design using standard electronic components and electronic lab instruments. The theoretical material and the design methods are covered through interactive PSPICE simulation software sessions. The course culminates in a wireless optical communication project competition.
- **Introduction to Inventive Problem-Solving in Engineering**: This course introduces students to inventive thinking. After an introduction to creativity and creativity styles, students are introduced to topics such as "paradigm shift" and "mental blocks" and how to avoid them. Using some hands-on out-of-the-box activities, students learn to identify "the real problem", and generate/evaluate solutions using systematic and non-systematic methods.
- **Introduction to Systematic Problem-Solving**: The course is designed to expose the students to practical strategies of problem-solving and design through various activities, case studies and projects. Students are guided through the systematic methodology of problem-solving and design: from problem exploration and definition to solution identification, decision and implementation. Emphasis is placed on distinguishing the various types of problems and applications of several proven techniques as well as relevant tools (e.g., PS/DS, Duncker diagram, brainstorming, cause mapping, K.T. tables) in solving real life problems and interesting design tasks. Note: This course was added in 2002 because of the overwhelming popularity

of the above Introduction to Inventive Problem-Solving in Engineering course. Although not exactly the same, this course is similar to the above-mentioned course.

- **Introduction to Web Authoring/Programming**: The course is designed to teach students about the basic concepts of the Internet, including the major services it provides. Two main components of the course are web authoring and web programming. Students are expected to create their own web pages using tools available on the Internet. Some basic web programming using languages such as Java and JavaScript are also a focus of the course. Note: The level of this course has been upgraded significantly, as the student population has become increasingly more sophisticated in computer knowledge and usage.

## ESP Instructional Outline

Each course in the ESP program is a 60-hour course, comparable to a regular 3-credit hour university course in a semester. All ESP courses have met daily for a period of two weeks, purposely held in mid-June each year, right after the end of the high school academic year. The program opens with an orientation session for students, parents, teachers, program faculty, and staff. Inclusion of parents and teachers in the opening event helps disseminate information about ESP and helps students relate their program experiences to others.

Instructors arrange the daily schedules for their individual courses, including lecture and lab sessions, time to work on projects, "self-exploration" periods, study periods, and student presentations. Experienced mentors are assigned to each course. They are in class with students daily to serve as learning facilitators, advisors for student projects and campus guides. They are invaluable in helping students negotiate the rigors of university level work and to successfully complete their courses.

## Program Assessment

ESP is assessed using four important evaluation methods to ensure the correct and successful implementation of the program: course evaluations, program debriefing, follow-up survey, and daily mentor feedback. Course evaluations and follow-up survey are designed to seek feedback from the students' perspective to assess the effectiveness, relevance, and success of the program, and to plan for the future. The program debriefing sessions serve the ESP faculty and staff to review successes and problem areas also for future program improvement. The program has evolved because of student feedback. Another assessment method is daily mentor feedback. It is used for quality control to ensure the success of the current program. Mentors are also evaluated by students. Their evaluations are incorporated into course evaluation. It is worth mentioning that analyses of the course evaluations and follow-up surveys over the last few years clearly indicate that ESP enjoys an over 80% student satisfaction rating in general.

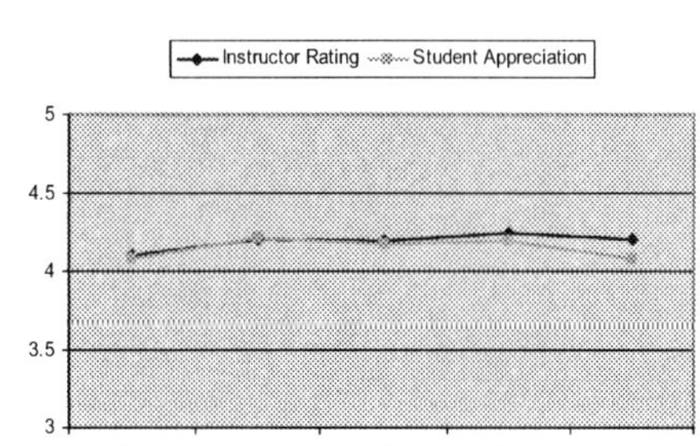


FIGURE 1
COURSE EVALUATION RESULTS
(RATING SCALE: 1-5, 5 BEING THE HIGHEST)

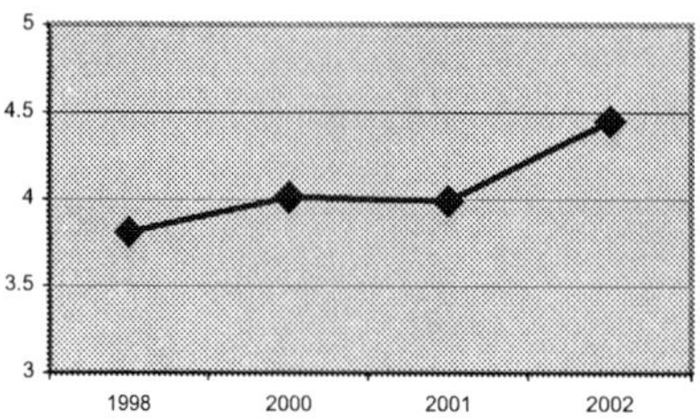


FIGURE 2
FOLLOW-UP STUDENT SURVEY RESULTS
(RATING SCALE: 1-5, 5 BEING THE HIGHEST)

Figure 1 shows the mean results of all ESP courses in 1998-2003 (2000 data is missing) with respect to instructor rating and overall student appreciation of the program. These two aspects are considered most relevant to the effectiveness and success of ESP. The data are extracted from the college evaluation form since it is more comprehensive than the state form. The rating scale goes from 1 to 5, with 5 being the highest. As can be seen from the figure, students have rated the program with an over 80% satisfaction level consistently in terms of both the instructor rating and student appreciation. A close correlation between these two factors is clearly observed in the figure. Figure 2 shows the follow-up survey results for the years of 1998-2002 (1999 data is missing; 2003 data is not yet available). The curve shows the mean results of all ESP courses with respect to the students' perceptions of engineering and science knowledge learned in the ESP courses, and their applications of the knowledge acquired to the real-world problems. The figure clearly shows that students have evaluated the program with an average overall 80% satisfaction rating.

### Registration

All students receive ESP tuition and textbooks free. In addition, the program also offers 10 need-based scholarships of $100 each to financially disadvantaged students. This money is intended to defray the costs of transportation and lunch. All accepted students receive a letter announcing the scholarships and an application if they choose to apply.

### Program Funding and Support

For the past few years, the State of Florida has been funding the ESP program through its Governor's Summer Program, earmarked for gifted and high-achieving students, with a fixed maximum amount of $25,000 annually [21-26]. The university and the college have largely funded the rest of the expense, usually totaling about another $25,000. Region V Area Center for Educational Enhancement has been providing funds for course mentors. In addition, they also help publicize ESP throughout Broward and Palm Beach Counties. Other support for ESP, although non-monetary, includes strong endorsement from the school boards of both the Broward and Palm Beach Counties, and free use of course textbooks.

## ONGOING EFFORTS

While the high satisfaction level of ESP students with respect to this program has demonstrated clearly the sound design and successful implementation of the program, an important aspect of measuring the impact of this program is to track how many participants have selected engineering as their majors after they have graduated from high school. We have laid out some ground work for this undertaking, and the actual implementation is scheduled to set off early next year. To be specific, a follow-up survey focusing on previous ESP students' chosen colleges and majors is being developed. A two-pronged approach is considered: to contact the previous ESP participants directly based on the contact information we have for them, and to go through the appropriate high school resources, such as guidance offices and counselors. Subsequent analysis will then be performed after the data is obtained, and results will be reported.

## CONCLUSION

ESP has been in existence for six years after its first offering in 1997. To date, it has served over 500 students from 23 different schools, with a wait-list for participation. This program now has a select student population, student-oriented program assessment, well-defined procedures for course placement, program orientation and coordination, and good marketing. Currently, over 80% of the student participants report a high level of satisfaction with the program. Experienced faculty, caring staff, and motivated students, together with a sound plan and implementation, have made this program a great success. As a consequence, the program directors are facing more challenges in coping with an ever-increasing demand.

## ACKNOWLEDGEMENTS

ESP requires effective team work on the part of both faculty and staff to ensure its success. We would like to thank all who have participated in the program, in particular, Drs. Ming Huang, Oren Masory, Roger Messenger, Daniel Raviv, Zvi Roth, Sandra Tsurutome, Ali Zilouchian, and the many mentors. Without them, ESP would not have been so successful.

## REFERENCES

1. Summer Institute, Arizona State University, http://www.fulton.asu.edu/fulton/cor/summer_institute.php
2. The Program in Mathematics for Young Scientists (PROMYS), Boston University, http://www.promys.org/
3. Young Engineering and Science Scholars (YESS) Program, California Institute of Technology, http://www.msa.caltech.edu/outreach.html
4. COSMOS, California State Summer School for Mathematics and Science, http://www.ucop.edu/cosmos/
5. Young Scholars Program, University of Chicago, http://hep.uchicago.edu/outreach/programs.html
6. Jim Holland Summer Enrichment Program (SEP), Indiana University, http://www.bio.indiana.edu/events/HollandSEP/
7. Michigan Science and Math Scholars Summer Program, University of Michigan, http://www.math.lsa.umich.edu/mmss/
8. Ross Mathematics Program, Ohio State University, http://www.math.ohio-state.edu/ross/

9. Penn Summer Science Academy, University of Pennsylvania, http://www.upenn.edu/summer/youth/pssa/pssaindex.html

10. Stanford University Mathematics Camp (SUMaC), Stanford University, http://math.stanford.edu/sumac/

11. SWT Honors Summer Math Camp, Southwest Texas State University, http://mathworks.txstate.edu/student/HSMC/description.htm

12. Summer College for High School Students, Cornell University, http://www.sce.cornell.edu/sc/index.php

13. Summer Enrichment Program, University of Northern Colorado, http://www.unco.edu/sep/

14. Summer PreCollege Program for High School Students, Pratt Institute, http://www.pratt.edu/precollege/pres/index.html

15. GERI Summer Camps for Gifted and Talented Students, Purdue University, http://www.geri.soe.purdue.edu/youth/summer/default.htm

16. Summer Enrichment Program, University of Virginia, http://curry.edschool.virginia.edu/centers/enrich/index/home.html

17. Academic Summer Camps for Kids and Teens, Education Unlimited, http://www.educationunlimited.com/Camps/2003-SummerCatalog.pdf

18. Ventures Scholars Member Institution Enrichment Programs, Ventures Scholars, Inc., http://www.venturescholar.org/enrichment_programs.html

19. Engineering Scholars' Program, Florida Atlanta University, http://www.dess.eng.fau.edu/esp.htm

20. High School Dual Enrollment for Florida High School Students, http://www.fau.edu/academic/freshman/hsdualenroll.htm

21. Governor's Summer Program: Engineering Scholars' Program, Florida Department of Education, P# 501-92150-80051, 1998.

22. Governor's Summer Program: Engineering Scholars' Program, Florida Department of Education, P# 501-92150-90051, 1999.

23. Governor's Summer Program: Engineering Scholars' Program, Florida Department of Education, P# 501-92150-00051, 2000.

24. Governor's Summer Program: Engineering Scholars' Program, Florida Department of Education, P# 501-92150-1QC01, 2001.

25. Governor's Summer Program: Engineering Scholars' Program, Florida Department of Education, P# 501-96570-2D001, 2002.

26. Governor's Summer Program: Engineering Scholars' Program, Florida Department of Education, P# 501-92150-3D001, 2003.

**Sam Hsu** received his B.S. degree from the Chinese Military Academic at Taiwan in 1968. He received his MAS degree in CS, and Ph.D. degree in CE from Florida Atlantic University in 1984 and 1993, respectively. He is currently an Associate Professor, and Director of the Telecommunication and Computer Networking Laboratory at the Department of Computer Science and Engineering, College of Engineering, Florida Atlantic University. His research interests include web technologies, computer networking, and web-based distance learning. Integrating technology into education is his major research focus at the present time. He is also interested in projects that have practical and educational applications.

**Sharon Schlossberg** received the B.A. degree in Elementary Education and Sociology from the City University of New York (Queens College) in 1962, and an MS degree in Special Education and Education of the Gifted from Long Island University (C.W. Post Center) in 1979. She received her M.Ed. in Guidance and Counseling and an Ed.D. in Curriculum and Instruction and Higher Education from Florida Atlantic University in 1984 and 1987, respectively. As of July 1, 2002, she has been serving as the first Assistant Dean for Student Affairs in the College of Engineering. In addition to existing responsibilities as Director of the Division of Engineering Student Services, she assists the Dean with management and development of the college, serves on the Executive Committee, and oversees student development issues including orientation, recruitment and retention.

**Karl Stevens** received his M.S. and Ph.D. degrees in Theoretical and Applied Mechanics from the University of Illinois at Urbana-Champaign and a B.S. degree in Mechanical Engineering from Kansas State University. His prior positions include Associate Dean of Engineering for Academic Affairs, Professor and Chair of Mechanical Engineering, Professor and Chair of Ocean Engineering at Florida Atlantic University, Professor of Engineering Mechanics at Ohio State University, and staff member at Sandia Corporation, Bell Telephone Labs, and the U.S. Army Ballistic Research Labs. He is currently Dean of Engineering at Florida Atlantic University. His current interests focus upon educational administration, development of academic and distance learning programs, development of cooperative educational programs with business, industry, community colleges and public school systems, and teaching and research in engineering mechanics.

# Chapter 38

# PanAmerican Collaboration on Micro- and Nano-Technology Education

IDALIA RAMOS[1], ANA NEILDE RODRIGUES[2], MARIA L. P. DA SILVA[3], ROGERIO FURLAN[1], JORGE J. SANTIAGO-AVILES[4] and LUIS SOLA-LAGUNA[5]

*[1]University of Puerto Rico at Humacao, Dept. of Physics and Electronics, Humacao, PR, 00791. [2]LSI-PSI-EPUSP, São Paulo, FATEC-SP, São Paulo, Brazil. [3]Dept. of Electronic Systems, Polytechnic School, University of São Paulo, Brazil. [4]University of Pennsylvania, Dept. of Electrical Engineering, Philadelphia, PA 19104, USA. [5]DuPont Electronic Materials, Manati, Puerto Rico, 00674. E-mail: iramos@cuhwww.upr.clu.edu*

*Members of the faculty at the University of Puerto Rico at Humacao (UPRH) have developed collaborative ties with the University of Pennsylvania (PENN), the University of São Paulo (USP), and Dupont Electronics. The focus of this international collaboration is research and education on micro- and nano-materials and structures. Personnel of the different institutions interact through the Internet, videoconferences and on-site visits by faculty, and graduate and undergraduate students to the participating institutions. The collaboration has served to promote the participation of women in research and educational activities. As the industry partner, Dupont Electronic Materials' manufacturing facilities in Puerto Rico provide materials, the use of instrumentation as well as technical expertise.*

## INTRODUCTION

Three educational institutions are involved in a unique collaborative project dealing with research and education in the field of micro and nano-materials and structures: University of Puerto Rico at Humacao, University of Pennsylvania, and the University of São Paulo in Brazil.

The University of Puerto Rico at Humacao (UPRH) is a public undergraduate institution located on the southeastern side of the island of Puerto Rico. UPRH has approximately 4,300 students and strong programs in applied science and technology. The Physics and Electronics and Computational Mathematics programs have more students engaged in these academic areas than any other university on the island. Sixty-eight percent of the students at UPRH are women. Women constitute about 50% of the majors in Computational Mathematics and 27% in Physics and Electronics. The programs are based on a four-year curriculum with general and major courses. Undergraduate research is considered a principal activity for all students, and students are initiated in research as early as their freshman year. Faculty research is designed to involve undergraduates. Funds for these efforts are provided by U.S. National Science Foundation, NASA, and other agencies. The University of São Paulo has over 26,000 students enrolled in graduate studies and 39,000 in undergraduate programs. A typical undergraduate program lasts five years. In Brazil, the state invests heavily in funds for research and scholarships for students. The University of Pennsylvania (PENN) is a private Ivy League institution with over 20,000 students in graduate, undergraduate and research programs.

The institutions initiated collaborative projects over ten years ago. The focus of the current collaboration is research and education on nanomaterials and structures. Communication between the participants is conducted primarily through the Internet and videoconferences, but faculty and student exchange programs allow the research collaboration teams to interact personally. As the industrial partner, Dupont Electronic Materials' manufacturing facilities in Puerto Rico provides materials and access to instrumentation for nanomaterials fabrication and research, as well as technical expertise. The company also promotes a healthy and mutually beneficial relationship with universities where industrial research (or research/development with a strong industrial flavor) is being emphasized in the participating institutions.

One of the project's primary goals is to promote diversity by promoting the participation of women at all levels of the collaboration in both research and educational activities. At the undergraduate level, 50% of the students participating in research activities are women. The visiting faculty program has brought women researchers from Brazil to PENN and UPRH where they interact and serve as role models for women students. Similarly, women faculty from Brazil have participated and learned about efforts at UPRH towards the advancement of women faculty in science and technology. These experiences have helped them assess the status of women faculty in Brazil and the need for and suitability of similar programs.

This paper will describe the main aspects of the collaboration, the current outcomes and its future direction.

## Program Participants

The participants in this international collaboration include three faculty mentors (one woman), two post-doctoral researchers (two women), one graduate student and six undergraduates (three women). The distribution per country is as follows: Puerto Rico: two professors, four undergraduates; Pennsylvania: One professor, one graduate, one undergraduate; Brazil; one professor, two Post Docs, one undergraduate. As the industry collaborator, Dupont Electronics provides, in addition to materials, access to instrumentation, technical expertise, and the involvement of a senior scientist, Dr. Sola-Laguna, who brings his experience and unique industry perspective to the effort.

## COMMUNICATION BETWEEN COLLABORATORS

Day-to-day communication is carried out primarily through the Internet. A videoconference link was established between PENN and UPRH four years ago under NSF sponsorship. Through videoconferencing, a series of workshops, seminar presentations and teleconferences have brought faculty, graduate and undergraduate students from both institutions together. With funds from NSF, UPRH and PENN faculty and students have visited back and forth between the two collaborating institutions during the summer, making intensive use of laboratories and research facilities. Visits to and from Brazil are more limited, but during the last three years one member of PENN's faculty spent a semester in Brazil, one UPRH faculty member spent a summer month in Brazil and two female Brazilian researchers have been to UPRH.

## IMPACT ON WOMEN'S RECRUITMENT AND RETENTION

UPRH has been very successful in granting B.S. degrees to women in applied sciences and mathematics (over 60% overall [1]). This is also true for other universities on the island. However, women do not pursue graduate studies at the same rate as men. As a consequence, the number of women in the faculty is lower as shown in Figure 1, and those in faculty positions do not advance in academic positions at the same rate as their male colleagues [2].

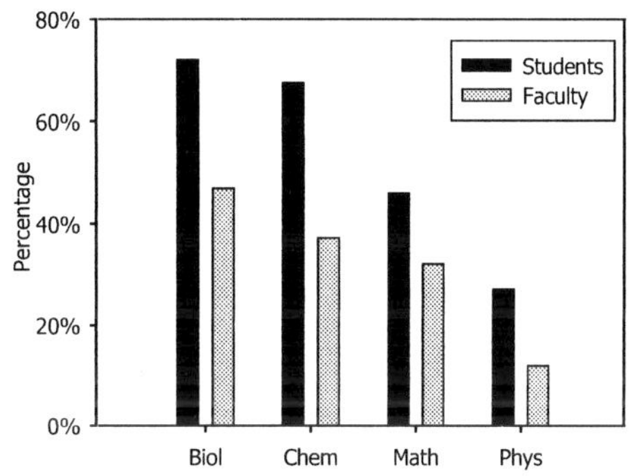


FIGURE 1
WOMEN IN SCIENCE AND TECHNOLOGY AT UPRH

UPRH is involved in an effort to increase the participation and advancement of women in our science programs. Our experience demonstrates that participation in undergraduate research is one of the best ways to encourage female students to succeed in their studies and to pursue graduate degrees [3,4]. Figure 2 shows the changes in the participation of women in undergraduate research during the last 10 years in the Physics and Electronics Department. The increase in women's participation is more significant if we consider that only 27% of the students in this department are female. As evidence of our commitment to diversity, at least 50% of the undergraduate research students in our collaborative program are women.

The collaboration has provided us the opportunity to exchange women faculty. The Department of Physics and Electronics at UPRH has only one female faculty member, but during the last year, two visiting women researchers from Brazil have benefited the department by doing top-caliber research, providing role models for our students, and making the institution aware of the need to recruit more women. At the same time, due to interaction with UPRH, PENN and Brazil have become more aware of the need of developing similar efforts to increase the participation of women.

## Research Activities on Nanomaterials and Structures

The focus of the current collaboration is research and education in micro- and nano-materials and structures. The specific topics are the generation of micro patterns in ceramic substrates [5-7], and generation of nanomaterials by electrostatic means [8-11], including the design, modeling, simulation and fabrication of nano-structures, their physico-chemical characterization, as well as utilization in sensing and actuating devices.

The research projects have been designed to allow students at all levels to apply basic concepts and to acquire hands-on experience. The experiences range from very simple experiments to the use of advanced instrumentation and analysis techniques.

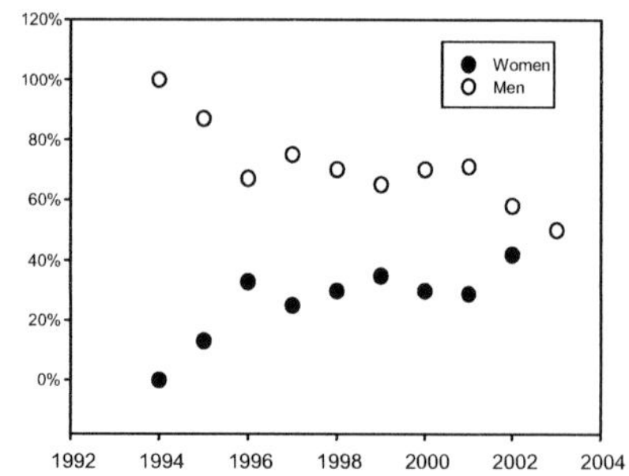


FIGURE 2
PERCENTAGE OF PHYSICS AND ELECTRONICS STUDENTS IN UNDERGRADUATE RESEARCH BY GENDER

## Educational and Outreach Activities

The international collaboration in a high technology field like nanotechnology provides an excellent opportunity to teach new skills to students at different levels. The students involved in this effort receive training on research techniques and scientific presentations, and are taught to be sensitive to cultural and gender issues. All undergraduates are required to dedicate at least 10 hours every week to the effort, including group seminars and videoconferences. The results of their work are presented in an annual meeting at the home institution and one additional meeting in their country of origin. In addition, some of the students with the best projects are rewarded with the opportunity to participate in a conference in the United States, like the National Conference on Undergraduate Research (NCUR).

The Summer Program at PENN is designed for faculty and students to continue the work they do during the academic year and to learn new techniques and use equipment not available at UPRH. The faculty and student exchange provides an opportunity for participants to learn science and also to familiarize themselves with other cultures and languages.

## Future Direction

In order to expand our education and outreach activities, presentations on the subject of micro- and nano-technology will be given at UPR-Humacao during our annual hands-on workshops and open houses for elementary, middle, and high school teachers and students. The subject will also be incorporated into our regular high school visits by faculty to recruit students for our science programs. A special section for elementary, junior high and high school students will be incorporated to our current website[8]. It will include activities and information on materials sciences with the aim of attracting motivated students.

[8] http://www.uprh.edu/~cire

## CONCLUSIONS

A successful collaboration in research and education between Puerto Rico, US and Brazil involves a myriad of techniques and participants. Undergraduates and graduate students have acquired new knowledge, research skills and motivation to pursue further studies and research activities in micro- and nano- technology. As a result of this interaction, faculty and students have an enhanced sensitivity to cultural and gender issues. UPRH is involved in an effort to increase the participation of women in our science programs and significant results have already been achieved. Future activities will be aimed at also impacting elementary, middle and high school students and teachers.

## ACKNOWLEDGEMENT

This work is supported by NSF-DMR-9872689 and NSF-SBE-0123654 in the US and PR and by FAPEST and CNPq in Brazil.

## REFERENCES

1. Planning Office, University of Puerto Rico at Humacao, Statistical Data - Students and Faculty, 1990-2004.

2. Ramos, I., Sanchez, Y., Hernandez, B., and Rivera, L., "Hacia una transformación institucional para el avance de las mujeres en la UPR-Humacao (Towards an Institutional Transformation for the Advancement of Women at UPR-Humacao," Proceedings of 4th National Colloquium about Women, April 11, 2003, pp. 168-171.

3. Negrón, D., and Ramos, I., "Increasing the Quality of Physics and Technology Programs at UPR-Humacao by Promoting Gender Diversity," Proceedings of IEEE Frontiers in Education Conference, 1999, pp. 13b5-7.

4. Muller, R.J., and Guerra-Vela, C., "A Successful Applied Physics Program at the University of Puerto Rico at Humacao," Am. J. Phys., Vol. 71, No. 5, May 2003, pp. 469-473.

5. Wagner Simoes, E., Ramos, I., Garcia, I., Furlan, R., Santiago-Aviles, J.J., and Tadeu Pereira, M., "Numerical Modeling of a Process for Definition of Features in Low Temperature Co-Fired Ceramics," Electrochemical Society Proceedings, Vol. 2002-8, Porto Alegre (Brazil), SBMicro 2002, September 9-14, 2002, pp. 244-252.

6. Gongora-Rubio, M.R., Wagner Simoes, E., Fontes, B.A., Furlan, M., Ramos, I., Pereira, M.T., and Santiago-Aviles, J.J., "LTCC Hybrid Technology for Microfluidic Systems in Analytical Chemistry," III Congreso Iberoamericano de Sensores y Biosensores (IBERSENSOR2002), Lima, Peru, Nov. 6 – 8, 2002, pp. 72 – 76.

7. Tressino de Carvalho, A., Garcia, L., Pereira da Silva, M.L., Ramos, I., Furlan, R. and Santiago-Aviles, J.J., "Modeling and Characterization of LTCC Tape Etching Using Acetone as Solvent," Proceedings of Tenth Annual International Conference on Composites Engineering (ICCE/10), New Orleans, July 20-26, 2003, pp. 83-84.

8. Wang, Y., Santiago-Aviles, J.J., Furlan, R., and Ramos, I., "Pyrolisis Temperature and Time Dependence of Electrical Conductivity Evolution for Electrostatically Generated Carbon Nanofibers," IEEE Transactions on Nanotechnology, Vol. 2, No. 1, 2003, pp. 39 – 43.

9. Wang, Y., Furlan, R., Ramos, I., and Santiago-Aviles, J.J., "Synthesis and Characterization of Micro/Nanoscopic Pb(Zr0.52Ti0.48)O3 Fibers by Electrospinning," Published Online in Applied Physics A, August 2003.

10. Pinto, N.J., Rodrigues da Silva, A.N., Fachini, E., Carrión, P., Furlan, R., and Ramos, I., "Electroless Deposition of Thin Metallic Films on Polymer Fibers Prepared via Electrospinning," Polymer Preprints 2003, Vol. 44, No. 2, pp. 138-139.

11. Wang Y., Ramos I., Rogerio Furlan, and Santiago-Avilés J.J., "Electronic Transport Properties of Incipient Graphitic Domains Formation in PAN Derived Carbon Nano Fibers," accepted for publication in IEEE Transactions on Nanotechnology, November 2003.

**Idalia Ramos** is an Associate Professor in the Department of Physics and Electronics at the University of Puerto Rico, Humacao. Her research interests are in the areas of materials and implementation of micro and nano fibers and devices.

**Ana Neilde Rodrigues** is a researcher at the Laboratory of Integrated Systems of University of São Paulo. Her primary areas of interest are micro and nanofibers and also PECVD deposition of thin dielectric films.

**Maria L. P. da Silva** has been a professor in the Faculty of Technology of São Paulo since 1992 and a researcher in U. of São Paulo, Brazil, since 1984. Her research areas include Clean Technologies and Silicon Chemistry, especially their application to environmental issues.

**Rogerio Furlan** is an Associate Professor of the Physics and Electronics Department of the University of Puerto Rico at Humacao. His primary areas of interest are micro and nano fibers and microfluidic devices.

**Jorge J. Santiago-Avilés** is currently an Associate Professor of electrical and system engineering at the University of Pennsylvania, Philadelphia. Currently, his research interests are focused on the field of nanostructures and nanomaterials.

**Luis M. Solá-Laguna** serves as Technical Manager in DuPont Electronics Microcircuit Industries Ltd., Manati, Puerto Rico. He is responsible for Technical and Engineering support to manufacturing plant operations and Characterization and Quality Control laboratories for the Americas region.

# Chapter 39

# Improving Engineering Education for Ethnic Minorities in Norway: A Study of the Situation at Oslo University College

ANN-MARI TORVATN and FRODE EIKA SANDNES
*Faculty of Engineering, Oslo University College, P.O. Box 4, St. Olavs plass, N-0130 Oslo, Norway. E-mail: ann-mart@iu.hio.no, frodes@iu.hio.no*

*Up until the 1970s Norway was an ethnically homogenous society. Few students with an immigrant background pursued higher education. Currently, Oslo University College hosts a large proportion of students with an immigration background, but this group of students is not evenly distributed across the different disciplines and departments. A majority of these students decide to study engineering-related fields, especially the computer science programme. It is estimated that in 2002 approximately 40% of the students that enrolled in computer science had an immigrant background. The policy and tradition at Oslo University College has been to handle all students uniformly and equally; however, students with immigrant backgrounds have not performed as well as their peers. Further, a significant number of these students drop out at the start of the first year, and those who stay on frequently spend more time finishing their engineering degrees. A committee was established to identify the particular problems this group of students face in Oslo, as a precursor to improving service provided to these students. Four students with an immigrant background were used during the investigation. This paper addresses the qualitative and quantitative findings of this committee based on student interviews and questionnaires.*

## Oslo and Its Ethnic Minorities

Norway was until quite recently an ethnically homogenous society. The first foreign workers arrived in Norway in the 1970s. They were mostly men, and the original intention was for them to stay for only a limited number of years and then return.

Gradually, the foreign workers were joined by their wives and children, as many chose to make Norway their new home. Refugees and asylum seekers followed, and once asylum applications were approved, other family members came. There is now a relatively large proportion of young people in Norway, especially in Oslo, whose parents or even grandparents immigrated to Norway and have lived in Norway for several decades. Norway has become a heterogeneous multi-cultural society with significant student diversity. All the national educational establishments face new challenges. These challenges are on many accounts similar to those described in the global arena [1,2,3,4].

Currently, Oslo University College hosts a relatively large number of WIB (With Immigrant Background) students [5]. This group of students is not evenly distributed across the different disciplines and departments. A majority enrolls in engineering and related disciplines, especially computer science. It is estimated that in 2002 approximately 40% of the students who enrolled in the computer science programme were WIB students.

The policy and tradition at Oslo University College has been to handle all students uniformly and equally; however, WIB students do not perform as well as NIB (Non Immigrant Background) students. Further, a significant number of these students quit at the beginning of their first year, while those who stay on frequently spend more time completing their engineering degrees. According to a study by Mehlen [6], WIB students obtained on average one grade lower than the NIB students in an elementary statistics course. Further, during 2002, 63% of the NIB students passed the examination in an introductory programming course, while only 43% of the WIB students passed the same examination. Similar observations are reported by Borland and Pearce [7]. A new law introduced with the intention to improve the effectiveness of higher education makes the funding of universities and colleges dependent on the number of credits taught and the number of students who have successfully obtained their degrees. The education authorities have for a number of years endeavoured to make an efficient higher education system accessible to everyone – now the objective is for more students to succeed and pass through the system.

At the same time, the recruitment to higher education is low. Courses in mathematics, science and engineering are decreasing in popularity. To survive financially, engineering faculties are effectively forced to "get more out of the students."

How can we get more of the WIB students to succeed? Perhaps the "equality principle" is not very effective? Maybe WIB students have needs or problems that we have not yet identified? In addition to expectations and specific learning needs, WIB students may also be accustomed to different learning styles than those we are practicing, and this mismatch can result in problems (see Ballard [8], Ballard and Clanchy [9,10], Ramburuth [2], Kember and Gow [11] and Niles [12]). Another problem can be rooted in misconceptions amongst teachers and NIB students [13,14] about students from different cultures.

In this study we primarily focus on student expectations.

## THE APPROACH – IDENTIFYING STUDENTS' EXPECTATIONS AND EXPERIENCES

A committee was established to gain insight into the particular problems WIB students face at our institution. The committee consisted of two lecturers from the Computer Science Department in the Faculty of Engineering and a faculty member from SEFIA

(Centre for Multicultural and International Studies (see [5])). The work was jointly funded by the Engineering Faculty, SEFIA and the Central Administration of Oslo University College.

The committee set out to acquire quantitative and qualitative data by means of questionnaires - reaching a large number of students - and in-depth one-to-one interviews with a few selected students. We decided to focus on the freshmen as it is in their first year of study that the largest number of WIB students quit. We decided to limit our study to the Department of Computer Science due to its large proportion of WIB students compared to other departments.

NIB students were also included in the study to serve as a control group. Experienced WIB students (successful WIB students in their final year) were also consulted during the study. A focus group of four final-year students was paid to provide qualitative data through interviews. They also generated qualitative questions for the questionnaire.

Two questionnaires were distributed to all first-year computer science students. The first one, the autumn-semester questionnaire, was distributed during the first week of August, 2002, with the purpose of identifying students' earlier experiences regarding language problems and teamwork. A total of 137 students responded to this survey. The second one, the spring-semester questionnaire, was issued in the middle of the second semester (March 2003). The latter questionnaire was more comprehensive, and was designed based on interviews with the four members of the focus group. Exactly 100 students responded to this survey.

## Quantitative Findings – The Autumn-Semester Questionnaire

The initial questionnaire was given to the students during their first week of study. A large portion of the questionnaire addresses language-related issues. Language issues are often central to studies addressing student diversity in lower, intermediate and higher - education [7,15-17]. The students were asked to declare their native language and answer five simple multiple-choice questions about their experiences from primary and secondary school. They had the option of agreeing with, partly agreeing with or disagreeing with the following statements:

- It was easy to comprehend the teacher.
- It was easy to understand the textbooks.
- It was difficult to identify the essence of the texts.

The final questions addressed the students′ experiences and expectations collaborating with students fluent in other mother tongues. The rationale behind the questionnaire was as follows: The most visible difference between the NIB students and the WIB students is that the WIB students are not as drilled in using the Norwegian language since they are using their native language to some degree. Many students have distinct foreign accents and some difficulties comprehending Norwegian speech and expressing themselves orally. Others have no difficulties at all. Still some students have hidden language problems – some researchers claim that these students have an innate weaker ability for abstracting concepts in Norwegian - and they have problems understanding more subtle distinctions and unexpressed suppositions in the Norwegian language [5,17].

As the main emphasis throughout the engineering programme is on project work, we found it important to gain insights into how students perceive teamwork and collaboration with fellow students mastering different native languages. Of the 137 students who completed the questionnaire, 85 were NIB students and 52 were WIB students. The forms were designed to be easy and quick to complete.

First, the questionnaires revealed that a total of 22 foreign languages are spoken among the students in the first year. It is possible that this number is even higher, as about 30 students did not answer this question.

Most of the NIB students expressed the opinion that it was easy to understand the teacher, but not all of them agreed completely. Five students out of the 85 only partly agreed.

The corresponding statistics for the WIB students are as follows: 40 out of 52 agreed completely, and 12 agreed only partly. Hence, less than 6% of the NIB students and 23% of the WIB students occasionally experienced problems comprehending the teacher.

Only 58 NIB students found the textbooks easy to understand. Twenty-four (24) of them partly agreed that textbooks were easy to understand - which also means that they sometimes found the textbooks difficult - and two disagreed completely. Corresponding figures for the WIB students are as follows: 21 out of 52 agreed that it was easy, 28 only agreed partly, and 3 disagreed completely. Thus, 30% of the NIB students had some difficulties reading the textbooks, while nearly 60% of the WIB students had the same experience.

When asked about the difficulty of extracting the essence from a text, 6 out of the 85 NIB students agreed, 49 agreed partially, and 29 disagreed. This means that 55 out of 85 (64%) occasionally have problems abstracting. Correspondingly, 40 out of 52 (77%) WIB students reported having similar problems.

## Conclusions – The Autumn-Semester Questionnaire

There are measurable differences in language proficiency when comparing NIB and WIB students. Still, both NIB and WIB students struggle with language-related problems, but the difficulties are more pronounced among students without Norwegian as a native language.

It is interesting that both groups find the teacher easier to understand than the textbook. One would expect that dialectal variations and poor diction could play a role especially for the WIB students. The difference in difficulties is greatest among the groups with respect to textbook reading. Thirty percent (30%) of the NIB students have problems while the corresponding figure is 60% for WIB students.

Teachers must be aware of the implications of having students with weaker language skills; it is more difficult for them both to understand the teachers and the textbooks. Consequently, the topics are perceived by these students as difficult and will take them longer to master. Further, these students may also be frustrated more quickly than their NIB peers. Being uncertain about their own understanding of the subject may prevent or inhibit them from participating in discussions.

Insufficient language competence may be the root of yet other social problems. The NIBs often communicate implicitly in a cultural context where vital information is implied. This can easily lead to misunderstandings. Misunderstanding about the agreed time and place often leads to bad feelings between students in the same team.

## THE SPRING-SEMESTER QUESTIONNAIRE

Interviews with members of the focus group revealed important dimensions of students' needs. The focus group indicated that students do not solely engage in educational activities. Other factors are often of greater importance. To become an engineer is not the main objective for many of the freshmen. Students often have no clear perception of what an engineer is and what engineers do in their profession. They do not think of engineering as just an educational or occupational option. Social life is of higher importance, especially finding companions with whom to study, participating in extracurricular activities and socialising. The first priority for students is to establish a social network as they commence their engineering degrees.

These insights affected the design of the second questionnaire. In addition to questions addressing language proficiency problems – i.e. how they managed to follow lectures and read textbooks, the students were asked about their previous computer experience and their social life during their time at our institution.

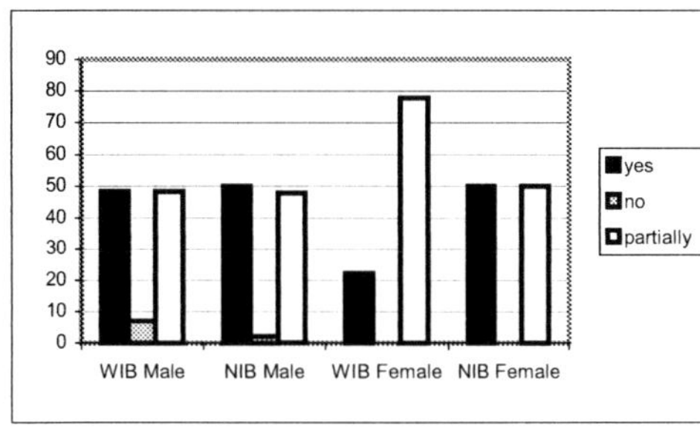


FIGURE 1
PERCENTAGE OF STUDENTS WHO FOUND IT "EASY TO FOLLOW LECTURES"

One hundred (100) students completed the second and more comprehensive form. There were 62 NIB students (48 men and 14 women), and 38 WIB students (29 men and 9 women).

Our understanding of the students′ situation was broadened considerably after we analysed this second, more comprehensive questionnaire, which the freshmen filled out in March of 2003.

## LANGUAGE AND PRIOR KNOWLEDGE

We asked if it was easy, partly easy or not easy to follow the lectures (see Figure 1), and if it was easy, partly easy or not easy to understand the textbooks (see Figure 2).

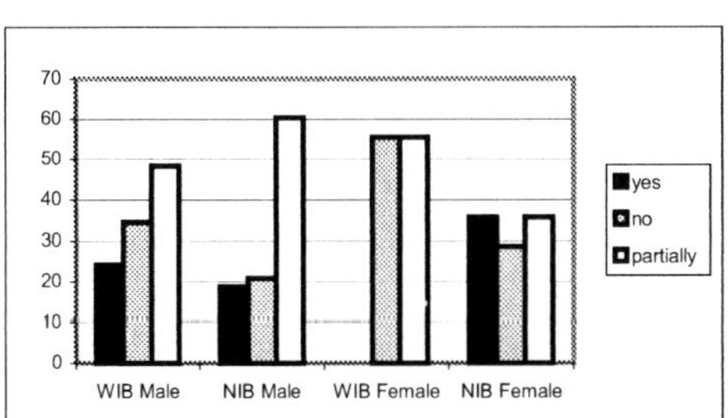


FIGURE 2
PERCENTAGE OF STUDENTS WHO FOUND IT "EASY TO UNDERSTAND THE TEXTBOOKS"

There were only small differences between the NIB and WIB students' ability to comprehend lectures. About half of the students indicated that they had no difficulties following the lectures, and about half the students replied that they had some difficulties. Only a few students found it difficult. However, the responses from the WIB women deviated from the trend as only two of nine had no difficulties following lectures – the others found it partly difficult (see Figure 1).

As for the textbooks, all the WIB women found them either difficult or partly difficult to understand, and none of them found the textbooks easy. In contrast, 36% of the NIB women had no problems with the textbooks. Corresponding figures for NIB men were 19% and 24% for the WIB men (see Figure 2).

There is a noticeable connection between prior experience or knowledge in the field and the ability to understand textbooks (see Figure 3). All the students who today find it easy to read textbooks had either "much" or at least "a little" knowledge when they started. Sixty (60) out of the 79 students who had some difficulty understanding textbooks indicated that they had "little" or "no" prior knowledge. Four of the WIB female students had "a little" experience with information technology at the beginning, while 5 of them had no prior experience. Only 4 out of the 33 students who reported having "much" prior experience later found the textbooks difficult, while 16 of them found the textbooks "partly" difficult.

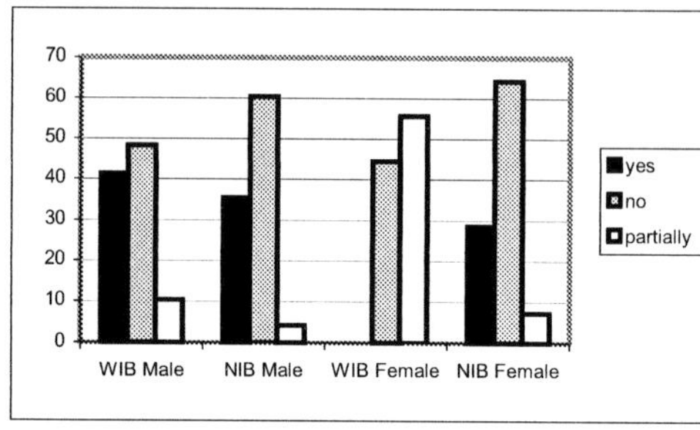


FIGURE 3
PERCENTAGE OF STUDENTS WITH "PRIOR EXPERIENCE"

Currently, there are no computing-related admittance requirements to become a computer science student. Unfortunately, this policy signals that no such knowledge is needed. Perhaps that can deceive some students?

Prior knowledge in itself is not enough to succeed. Several students find it hard to understand the textbooks, despite their prior knowledge. Forty-one percent (41%) of the WIB male students had "much" knowledge when they started, but only 24% find the textbooks easy. Thirty-five percent (35%) of the NIB men claimed to have "much" prior knowledge, but only 19% of them find it easy to understand the textbooks. An exception is the NIB women, who manage with a limited amount of prior experience: 31 had "much prior computing experience" at the beginning and 34% had no difficulties with the textbooks. The difference between the sexes can probably be ascribed to the fact that women often have more realistic and modest ideas about themselves and their own competence. None of the WIB women had "much" prior computer experience, and as mentioned previously, none of them find the textbooks easy.

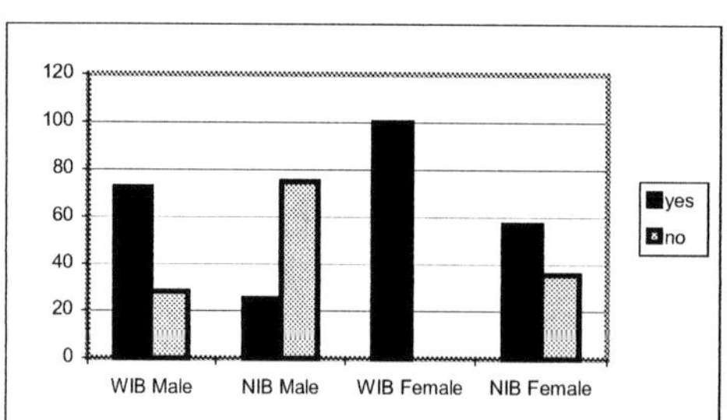


FIGURE 4
PERCENTAGE OF STUDENTS WHO WOULD "AGREE TO PARTICIPATE IN AN OPTIONAL FOUNDATION COURSE"

The differences in language proficiency also affect these results. Male WIB students have more difficulties reading textbooks than male NIBs, despite a higher percentage of them claiming to have prior computing experience.

We asked the students if they would have attended a voluntary summer foundation or bridge course in elementary computing if given the option (see Figure 4). Half of the students replied "yes." This half constituted 25% of the NIB men, 72% of the WIB men, 57% of the NIB women and 100% of the WIB women.

## TEAMS AND TEAMWORK

All freshmen are initially organised into groups of 4-5 members for the purpose of learning cooperation, teamwork, communication and publishing skills. These groups are formed by the teacher, with a policy to mix students with respect to the WIB/NIB category and gender. Students were asked about their experiences working in these teams.

Happily, nearly 80% were both content with the composition of their team and with the cooperation within the team. Students are different, and the chemistry between the group members cannot always be optimal, so the result could not possibly be perfect. As much as 94% indicated that they had influence on the decision-making in the team and that their voices were heard. All the women indicated that they had a say, and 96% of the WIB men replied likewise. NIB men were least content, as only 86% felt that they had been heard. We interpret this as an indicator that we have an open and inclusive environment where most students feel accepted.

However, about a third of the students found it difficult to find friends at the college. Seventy-one percent (71%) of the male NIB students find it easy to establish contact with fellow students, while only 43% of the NIB female students think likewise (see Figure 5). Further, only a minority of students engages in activities organised at the college (movie club, diving club, etc.). Their social networking is therefore mostly done during the hours of study.

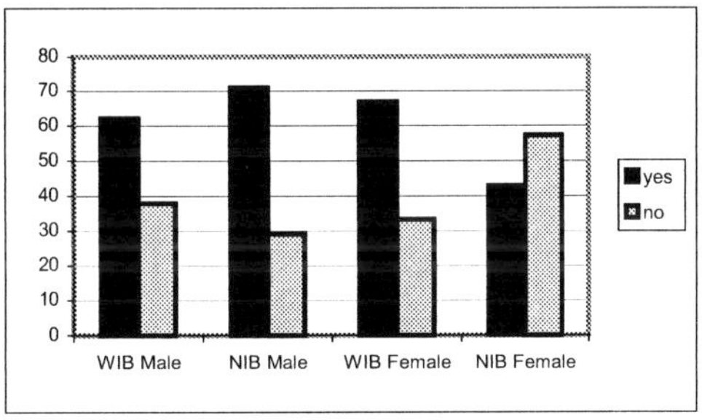


FIGURE 5
PERCENTAGE OF STUDENTS WHO FOUND IT "EASY TO MAKE FRIENDS"

It is important for the students′ well-being to get acquainted and make friends, and some of the students can be modest, shy and quiet. If the assigned team does not provide the necessary friendship, it can be difficult to find friends because there are no distinct groups with whom to associate. The large lecture theatres and the lack of communal areas is a drawback.

Students are by default assigned to different standard "classes" – a group of about 30 individuals. A few years ago lectures were given for individual classes, but now all lectures are given in plenary. Only tutorials are conducted on a class basis. Perhaps it would be helpful to make the class affiliation more visible for the students by giving the class a distinct function and helping the students to make the class their own environment.

The questionnaire had one open question. We asked what the computer science department could do to improve the learning environment. Some of the answers do not belong to the issues discussed here, such as better food and better air conditioning, but two of the themes that occurred were as follows:

- Textbooks in Norwegian (about 10%)
- Lectures in smaller groups, stronger class affiliation (about 10%)

## CONCLUSIONS

The insights obtained through this study can be used to improve the students' learning environment and help increase their success rate. First, the study elucidated the differences in language abilities when comparing NIB and WIB students. This problem is difficult to eradicate; however, simple remedies can help reduce the language barriers. Generally, teachers need to express themselves in clear, simple and unambiguous language. Further, prior computing experience plays a role in competence building, and action should be taken to reduce the gap between students with and without prior experience. We must tell the students beforehand that it is advantageous to possess prior knowledge. Students could, for instance, be offered a soft introduction to the degree programme through a voluntary summer course. Finally, action must be taken to stimulate social networking among students. The students' well-being and feeling of belonging will improve, which will in turn make it more likely that they will complete their degrees successfully on time.

## FUTURE RESEARCH

Future research activities involve assessing the effects of deploying the recommended actions outlined in this paper. Further, we need to explicitly investigate the teaching styles of the teachers at engineering colleges in Norway, the learning styles of the largest WIB groups enrolled in engineering courses in Norway, and the interrelationships among them.

## ACKNOWLEDGEMENTS

The authors are particularly grateful for the enthusiastic efforts of Michael Cheffena Gebresilass, Yen Pham, Thuong Van Dang and Shahid Idrees; and the helpful consultations with Per-Inge Båtnes, Center for Multicultural and International Studies, and Agnes Lorange Hayman, Olav Talberg and Mari Mehlen of the Engineering Faculty. The authors are also grateful for the comments and suggestions of the anonymous reviewers for this manuscript.

## REFERENCES

1. Otten, M., "Intercultural Learning and Diversity in Higher Education," Journal of Studies in International Education, Vol. 7, No. 1, 2003, pp. 12 – 16.

2. Ramburuth, P., "Cross-Cultural Learning Behaviour in Higher Education: Perceptions versus Practice" in *Global Learning*, Singh, M. (editor), Altona, Vic, Common- Ground, 2002.

3. Sillitoe, J., Webb, J. and Zhang, C. M., "Valuing Student Diversity in International Education in Value and Vision – Virtual and Reality," Proceedings of the 14th Annual Conference of the ISANA International Education Association, Queensland, 2003.

4. Harris, R., "Overseas Students in the United Kingdom University System," Higher Education, Vol. 29, 1997, pp. 77-92.

5. Greek, M. "På Vei Mot et Flerkulturelt Studiested? Minoritetsstudenter i et Majoritetsmiljø," HIO-Report 9, 2002.

6. Mehlen, M., "Science Exam Results and the Multicultural Student," Oslo University College, unpublished study, 2003.

7. Borland, H. and Pearce, A., "Identifying Key Dimensions of Language and Cultural Disadvantage at University," Australian Review of Applied Linguistics, Vol. 25, No. 2, 2002, pp. 101 – 127.

8. Ballard, B., "Academic Adjustment: The Other Side of the Export Dollar," Higher Education Research and Development, Vol. 6, No. 2, 1987, pp. 109-119.

9. Ballard, B. And Clanchy, J., *Study Abroad - A Manual for Asian Students,* Longman Press, Malaysia, 1993.

10. Ballard, B. And Clanchy, J., *Teaching Students from Overseas - A Brief Guide for Lecturers and Supervisors,* Longman Cheshire, Melbourne, 1991.

11. Kember, D. and Gow, L., "Cultural Specificity of Approaches to Study," British Journal Educational Psychology, Vol. 60, 1990, pp. 356-363.

12. Niles, S., "Cultural Differences in Learning Motivation and Learning Strategies: A Comparison of Overseas and Australian Students at an Australian University," International Journal of Intercultural Relations, Vol. 19, No. 3, 1995, pp. 369-385.

13. Chalmers, D. and Volet, S., "Misconceptions about Students from South-East Asia Studying in Australia," Higher Education Research and Development, Vol. 16, No. 1, 1997, pp. 1-11.

14. Kember, D. and Gow, L., "A Challenge to the Anecdotal Stereotype of the Asian Student," Studies in Higher Education, Vol. 16, No. 2, 1991, pp. 117-128.

15. Collier, V. P. and Thomas, W. P., "How Quickly can Immigrants Become Proficient in School English?," Journal of Educational Issues of Language Minority Students, Vol. 5, 1989, pp. 26-38.

16. Dolson, D., "The Effects of Spanish Home Language Use on the Scholastic Performance of Hispanic Pupils," Journal of Multilingual and Multicultural Development, Vol. 6, 1985, pp. 135-155.

17. Engen, T. O. and Kulbrandstad, L. A, Tospråklighet og Minoritetsundervisning, Ad Notam, 1998, pp. 143-159.

**Ann-Mari Torvatn** has a Cand. phil. in Norwegian language and literature from the University of Oslo. Torvatn has written several textbooks on communications for engineers. Torvatn is currently an associate professor in the Department of Computer Science at Oslo University College. Research interests include linguistics and human computer interaction. Torvatn is leading a research project addressing the coherence of factual texts.

**Frode Eika Sandnes** received a B.Sc. in computer science from the University of Newcastle upon Tyne, U.K. in 1993, an honorary M.Sc. from the Norwegian University of Science and Technology, Norway in 1993, and a Ph.D. in computer science from the University of Reading, U.K. in 1997. Dr Sandnes is currently an associate professor in the Department of Computer Science at Oslo University College and also serves as the international coordinator to the Engineering Faculty.

# Chapter 40

# Online Process-Oriented Mathematics Learning for Engineering Courses

BARBARA M. GRÜTER, JÖRN LOVISCACH, THOMAS RISSE, HEIDE-ROSE VATTERROTT and ULRIKE WILKENS
*Hochschule Bremen, University of Applied Sciences, Center for Computer Science and Media Technologies, Flughafenallee 10, 28199 Bremen, Germany. E-mail: grueter@informatik.hs-bremen.de*

*The results of the latest PISA study in Germany are mirrored in typical Mathematics lectures: students lack motivation, the failure rate is high, and the students display insufficient capabilities in applying mathematics to real-world problems. To improve mathematics education for students of engineering, we have designed, implemented and deployed "teleVISE," a web-based service providing mathematical problems and tutor-supported work for students. The service is utilized by several departments of Hochschule Bremen (University of Applied Sciences) to enhance presence education. teleVISE is intended to integrate calculation practice as well as abstract mathematics into the full process of mathematical work, i.e. forming and using mathematical models. It follows a process-oriented approach in the didactics and the design of the online service: modeling of real-world problems in mathematics, computing solutions, and applying and verifying the computed results. teleVISE consists of two parts: The first is a growing database of mathematical problems indexed using metadata so that both students and lecturers can easily search for appropriate exercises; the second is an online collaboration and tutoring environment. An electronic graphical notepad integrated into teleVISE aids the communication of students, tutors, and teachers. Notebook computers with wireless connections and graphics tablets allow for flexibility in time and space. Our report covers the didactics, the technology, the evaluation and the lessons learned during the deployment of teleVISE.*

## INTRODUCTION

There is a wide gap between what actually is and what, according to the curricula, should be achieved in undergraduate mathematics courses. Thus, in early 2002 we set out to look for a better way of teaching, one that is supported by web and wireless technology. This was the start of teleVISE (Tutorial Enhancement of Learning Environments: Virtual Exercises and Student Expertise), a web-based tutoring system for mathematics, strongly oriented towards mathematical modeling and the process of mathematical reasoning.

teleVISE is used by four departments at our university (Electrical Engineering and Computer Science, Economics, Mechanical Engineering, and Civil Engineering), in seven study programs ranging from the International Degree Course in Digital Media to the International Degree Course in Aviation Systems Engineering and Management. Since the summer semester of 2003, approximately 150 students and 10 tutors (advanced students) have been using the system. From August 2002 to December 2003 the work was supported with approximately EUR 450.000 from the German Ministry of Education and Research (BMBF). During this period, more than ten people worked full- or part-time on the development of the system and its content.

The main user group consisted of about 70 students in the first year and about 115 students in the second year. The students were participants of one of two different computer science study programs, namely the Digital Media Study Program and the Computer Engineering Study Program. Two of the authors (Jörn Loviscach and Thomas Risse) teach the corresponding mathematics courses, one for each of these two programs. Each course starts in the winter term (Part I) and continues in the summer term (Part II). For the students in these courses, teleVISE use is compulsory for admission to the final written examination. As an alternative, the students might submit paper work. No grades are given for their work, but they have to put sufficient work into at least 80 percent of the problems. The tutors judge whether or not the students are meeting this requirement. The students also receive specially prepared problems that are strongly oriented toward exploration and mathematical modeling. In the lectures of the two courses involved, we present not only the standard mathematical topics, but also discuss modeling processes, the students' attempts at problem-solving, and their strategies and experiences. This is accomplished primarily by presenting the electronically archived solutions or attempts at solutions that students have submitted through teleVISE.

To work with teleVISE, students can use the computer labs (if not occupied) or work at home. The main intended mode of use, however, is via notebook computers with a wireless connection. Using the WLAN installed at our university, teleVISE can be accessed from anywhere in the building, perhaps from a place more comfortable than a classroom. To sketch diagrams or to quickly write mathematical equations, the students can borrow or purchase inexpensive graphics tablets.

teleVISE does not replace a textbook nor is it intended for interactive exercises like those in Wallis [1]. Our software does not perform mathematical procedures in contrast to systems such as math-kit [2], a construction tool based on computer algebra software. Given that our objective is mathematical modeling rather than computation, we do not try to generate automatic answers like WeBWorK [3] does. Due to the complexity of the domain, we also did not try to implement an intelligent tutoring system such as the ones generated by WEAR [4]. The SmartTutor system [5], developed at Brooklyn College, is aimed at a setting resembling ours: a computer science course with a drop-out/failure rate above 50%. It offers content organized through concept maps as well as e-mail contact to peer tutors. With regard to the integration of modeling experience into the academic program, our didactical concept is akin to concepts of the Greenfield Coalition [6].

## MATHEMATICS, THE TROUBLESOME SUBJECT

Many students perceive mathematics courses in computer science degree programs as a burden. Throughout all mathematics courses in engineering and computer science programs in Germany, failure rates in written examinations typically exceed 50%. At our university and elsewhere, a large percentage of students only pass the third (and last) repetition examination. Topics such as differential equations are part of the formal curriculum and are supposedly covered in detail; however, in practice, advanced students cannot even draw simple statistical diagrams. Learning happens largely by rote, and connections between different mathematical topics are seldom drawn. The problems easily solved by students consist of simple substitution of values into mathematical equations. Conversely, the most challenging problems are those formulated without any numbers or equations and not allowing an answer consisting of a single number.

We attribute these findings to the following: mathematics education in secondary schools, students' perceptions and misconceptions about computer science curricula, and our style of teaching and learning, which traditionally is focused on computation instead of mathematical modeling.

Large international studies such as TIMSS and PISA (see below) have researched the results of school education. For instance, TIMSS [7] found that less than one-fifth of the students in the final year of secondary school could correctly sketch a graph showing the relationship between the height of a person and his/her age from birth to 30 years. Also, only one student out of five could master the following task [7]: "A TV reporter showed this graph and said: 'There's been a huge increase in the number of robberies this year.' Do you consider the reporter's statement to be a reasonable interpretation of the graph? Briefly explain." The graph shows a number of 508 robberies last year and 518 this year; the difference is exaggerated, however, by using a vertical axis with the range 505 to 520.

Given the superficial and rote learning, it is no surprise that even elementary mathematical knowledge vaporizes soon after school. In a study of 24 adults, Karsenty [8] concludes that, "the fact is that most of the subjects were unable to draw a correct sketch of a simple linear function. Of the seven subjects who succeeded in this task, five were graduates of the highest level track." Compare this to the vision of mathematical literacy of PISA [9]: "[...] to analyse, reason and communicate ideas effectively by posing, formulating and solving mathematical problems. Mathematical literacy skills include thinking skills [...]; argumentation skills [...]; modelling skills [...]; problem posing and solving skills; representation skills [...]; symbolic skills; technical skills [...]; communication skills; and skills in using mathematical tools and aids."

We believe that this kind of mathematical literacy is one of the major objectives in the education of not only engineers, but also of citizens in general. However, the discrepancy of where we start in undergraduate mathematics and where the curricula propose to arrive could not be greater.

To bridge the gap, more is needed than classic engineering mathematics or the definition-theorem-proof style typically taught by mathematics departments. Lakoff and Núñez [10] infer the following from their theory of embodied mathematics: "The point is not to be able to prove that $e^{\pi i} = -1$ but, rather to be able to prove it knowing what $e^{\pi i} = -$

1 means, and knowing why $e^{\pi i} = -1$ on the basis of what $e^{\pi i} = -1$ means, not just on the basis of formal proof."

Lakoff and Núñez observe further that the ideas of mathematics do not exist independently of mankind. In their view, the "Romance of Mathematics" is an illusion. This illusion is characterized by such assumptions as, "Mathematics is an objective feature of the universe; mathematical objects are real; mathematical truth is universal, absolute, and certain. What human beings believe about mathematics therefore has no effect of what mathematics really is." The "Romance of Mathematics" has consequences for education: "It intimidates people. It makes mathematics seem beyond the reach of even excellent students with other primary interests and skills."

## teleVISE as a Medium of Modeling: the Didactical Concept

We want to discuss most aspects of mathematical literacy—and we want to do so on the assumption that there is no single universal truth. In order that students learn to act with mathematical understanding, we supplement the mathematics lectures by process-oriented learning.

Our main focus is the process of mathematical modeling, which for PISA is described [9] as, "structuring the field or situation to be modelled; mathematising (that is translating from 'reality' to 'mathematics'); de-mathematising (that is interpreting mathematical models in terms of 'reality'); tackling the model (working within the mathematics domain); validating the model; reflecting, analysing, offering critique of models and model results; communicating about the model and its results (including the limitations of such results); monitoring and control of the modelling process." This concept encompasses the problems, concepts, strategies, approaches and, in particular, the modeling experiences of the students, the teacher and the groups.

Our didactical concept is also loosely connected to the developmental psychology tradition of Jean Piaget (1896–1980) and Lew S. Vygotsky (1896–1934). This approach, possibly unfamiliar to many mathematicians, but not to experts in the didactics of mathematics, takes into account non-mathematical sources of mathematical thinking. The trainer, accompanying the learner from the posing of the problem to the emergence of a solution, has to understand and to respond to the particularities of the individual problem-solving activity. According to this view, modeling is one step in each kind of human activity. Modeling is that process of thinking and/or acting by which an actor empirically sets his or her own realm of possibilities in which the next step of activity will happen. This might occur by an intention, by the setting of a goal, by the touching of something or by the desire for something. The activity will be structured by that first step. Pursuing the goal, the actor will experience differences between the goal and the results of his or her own thoughts and/or actions. In many cases the actor may remedy the differences by actions that are possible within the given realm. But the rule in the case of learning processes and in the case of creative activities like mathematical modeling is that the difference between goal and result is caused by systematic problems. These cannot be solved within the given realm of possibilities. The actor has to transcend the border of his or her realm of possibilities and, consequently, has to change him- or herself in that dimension.

Modeling in this case is a developmental process, the emergence of novelty or innovation for the actor. Thus, modeling is the transformation of the old structure of an activity to another, more powerful, novel structure. Compare the discussion of Jean

Piaget and Jerry Fodor [11]. Jerry Fodor pointed out the logical problem that arises if one views cognitive development as a succession of more and more powerful structures: The stronger, later structures cannot be deduced logically from the prior, weaker structures. Therefore, Fodor believes that cognitive development is a sequence of specifications of a general, inborn structure each provoked by certain triggers. However, this point of view is also fraught with problems: Why does the special trigger appear in the right place and at the right time? Thus, within both Piaget's and Fodor's approach there remains a severe problem as to how the gap between the cognitive structure of the learner and the empirical problem he has to solve and the gap between the preceding and the subsequent cognitive structure of the learner are closed. Approaches in the tradition of Vygotsky enable us to bridge the gap between lower and higher mental functions theoretically and methodically and to understand how novel knowledge is developed by the learner within collaborative activities involving learners and experts.

The subjectively given possibilities of action are misleading. Another enhanced realm should emerge. But how will one achieve it? This problem is not trivial for the learner or even for the mathematical genius. To transcend the border of one's own given realm of possibilities is a private, daring venture. The actor may fail and may risk the destruction of his or her current structures, which secured the success of the activities in the past. In this transitory situation the learner's attempts meet with the responses of the experts; thus the learner gains security on his or her own pathway.

Understanding, realizing, recording, supporting, and evaluating those processes are in our view the main issues of mathematical didactics. To follow the developmental view of modeling means to take the individual processes of mathematical modeling as a foundation, a point of reference for the teaching and learning of mathematics.

The problems we offer in teleVISE are formulated in the spirit of TIMSS and PISA, covering the topics of undergraduate mathematics. To illustrate the nature of the problems and the interaction between students, tutors and teachers via teleVISE, we describe one example: In one mathematical problem we sketched how the remainder after division by nine can be used to easily verify if a sum has been computed correctly. The students had to answer if and possibly how this could be carried over to subtraction, multiplication, and division. They were asked how one can compute the remainder efficiently and which percentage of false results will be caught. In this and similar problems we could observe the benefits of offering fast response from the tutor via teleVISE and giving the students time to think about the answers. The solutions clearly improve with each turn of feedback, one prominent reason being that the tutors ask the students to clarify weak spots of their submission: Did you plug in some test values for your proposed method of checking the division? Can you be more specific concerning the percentage of "nearly all" errors? In addition, several rounds of feedback helped to acquaint the students with giving mathematical arguments in problems such as, "Prove that the modulus of the product of two arbitrary complex numbers is equal to the product of their moduli." For instance, the tutors guided the students to sensible decisions concerning when to use or not to use graphical sketches, numerical examples, and the like. During the projects tutors were supervised in their attempts to support the students offline and online.

In addition, the format of the lectures has changed: The teachers present examples of students' solutions and discuss their approaches in detail. In the tradition of Pólya [12], we discuss meta-cognitive questions such as those proposed by Kramarsky [13] concerning comprehension (e.g., "What is the problem/task all about?"), connection (e.g., "How is this problem/task different from/similar to what you have already solved?"), strategy (e.g., "How can I organize the information to solve the problem/task"), and reflection (e.g., "Does it make sense?").

The integration of modeling experiences into the routine of learning and teaching is based on teleVISE as a mediating link. teleVISE mediates between learner and expert. The students work on the problems on their own or in small groups and can ask tutors for help via the Internet (see next section). The students submit solutions or at least their attempts at solutions, and in most cases the tutor will respond with individual feedback. Once a week, students and tutors meet face-to-face.

teleVISE is used as a mediating link between lectures, such that the ordering would be lecture, online exercise, lecture. In the first lecture the teacher presents concepts of mathematics as usual. In the online exercise the students work on mathematical problems alone or in groups with tutorial support on demand. In the next lecture the teacher will refer to some of the electronically archived individual solutions or attempts and use them as prototypes. Projected onto the wall, they will be a subject of common modeling in the class, a platform in which professor and students can discuss the experiences the students have made in working on the problems.

## TECHNOLOGY AND WORKFLOW

teleVISE is a web-based client-server application. The server is a LAMP (Linux, Apache, MySQL, PHP) system managing a database of mathematical problems and communication about those problems. On the client side, i.e., on students', tutors' and lecturers' computers, we use a web browser with Sun's Java™ 2 plug-in (see Figures 1 and 2). The problems can be stored in any data format appropriate for the web; at this time, we only use HTML pages with embedded GIF images. The lecturers prepare problems and prototypical answers as HTML, LaTeX or Word documents. To facilitate re-use, the problems are tagged with metadata according to an appropriate subset of IMS and to the mathematical subject classification of the AMS. We added a few metadata elements of our own such as the subject area of the application (computer graphics, biology, etc.) and the type of problem (computation, modeling, or both). At the time of writing, approximately 700 mathematical problems are available in the system. This number includes 220 specifically developed problems that are oriented towards mathematical modeling. Several hundred problems from archived collections have been imported into the

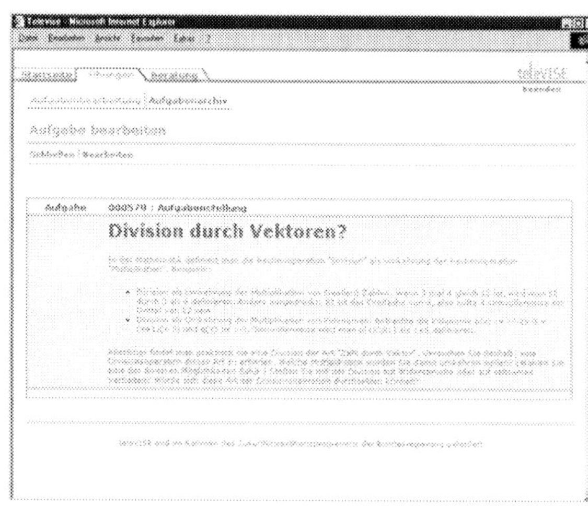


FIGURE 1

A PROBLEM AS DISPLAYED TO THE STUDENT

system. To this end, we used existing open source solutions to develop auxiliary software such as a converter for files given in TeX format. This can drastically reduce the manual work needed to import files.

From the viewpoint of the student, teleVISE offers the following features: a list of compulsory problems together with their state of work; a list of archived problems; and free choice of problems from the database using a metadata-based search. The student can work on solutions with a graphical editor (the notepad), submit solutions or consult the tutor using the graphical editor, and discuss with peers (one-to-one or via a forum) using the graphical editor.

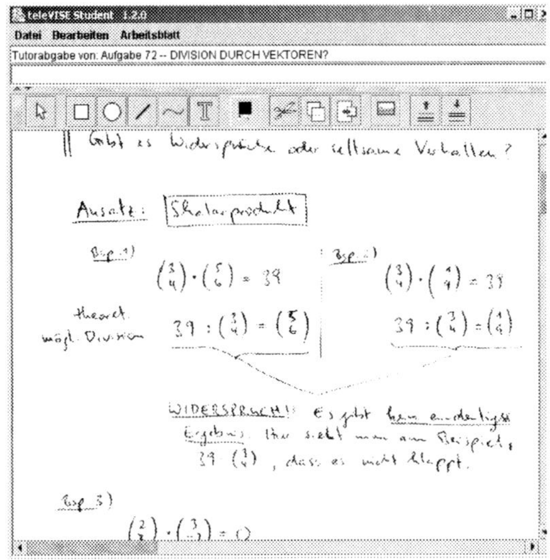


FIGURE 2

A STUDENT'S ANSWER TO A PROBLEM

These features unfold within the workflow as follows: For every student, the server stores a list of problems together with their state—new, in progress, sent to tutor, marked by the tutor as needing reworking, accepted by tutor as sufficient, and whether it has been archived. After logging in, a student examines the list and selects a problem. This is displayed as a web page, on top of which opens a Java™ applet containing a notepad for text and graphics. The student can create new pages or duplicate old ones in the notebook, type onto them or draw graphics using the usual tools (straight or freehand line, rectangle, etc.), or import a bitmap image file. Students who favor paper and pencil work usually employ the latter feature. After the Java™ notepad has appeared on the screen, the student may work offline—a money-saving feature for students working at home.

The Java™ notepad is intended both to collect ideas and to serve as a communication tool. A page of the notepad can be submitted as a solution to the tutor. It may also be sent as a request for help to the tutor, to other students or to a discussion forum. The notepad keeps a history of all transactions. By offering multiple ways of communication we hope to stimulate discussions.

The tutors receive the students' messages and their submitted pages with a similar Java™ notepad. They can add comments to the submitted pages as graphical elements and return the annotated pages. On their own notepad they see a history of each student and his or her questions, answers, and proposed solutions for each problem. This allows them to observe the mathematical process and give the students appropriate help.

## EVALUATION

At the beginning of the 2002/03 winter term, tutors were hired and trained. During the term, they worked with the students both offline and online using e-mail until teleVISE was set up. Thanks to experienced developers, we had no trouble getting the whole system to work correctly and on time starting in April 2003. Since then, teleVISE has been running in the affiliated study courses and during part II in the summer of 2003. A second student generation started part I of the course in the winter 2003/04 term.

According to the process-oriented approach, the whole project course, from design to deployment of teleVISE, was accompanied until February 2004 by a range of usability engineering activities: interviews of the students and tutors at the end of each part of a course; usability tests of the system in January 2003 and in the summer 2003; participant observation during lectures and the meetings of students and tutors; analysis of system use; process analysis of the interaction during lectures and of thinking-aloud problem-solving activities by students; and comparison of grades of classes. All of these activities primarily served the purpose of detecting application problems in time, optimizing the service, counteracting erroneous trends, and inventing novel solutions, if necessary. Here we present only some numbers and qualitative observations on the effect of using teleVISE.

The percentage of students possessing a notebook increased from winter 2002/03 to winter 2003/04 in the digital media group from 21% to 52% and in the computer engineering group from 23% to 80%. Two explanations offer themselves for the difference between the groups: The students of the computer engineering program have a stronger affinity to technology. In addition, many of them can access financial means earned in jobs prior to studying.

From the rudimentary version in winter 2002/03 to the first complete version in summer 2003, the percentage of students who participated in teleVISE as opposed to submitting paper work increased drastically, from 35% to 51% in the digital media group, and from zero to 33% in the computer engineering group. The percentage of students participating in teleVISE increased further in the transition to the second generation of students who started the course in winter 2003/04: the figures went up to 73% and 82% in the digital media group and the computer engineering group, respectively. From summer 2003 to winter 2003/04, the average percentage of submissions through teleVISE increased from 31% to 60%.

Sixty-four percent (64%) of the students characterized the online service as helpful in general. When asked for specification, these students gave several reasons, including reduction of examination nerves (3%); development of interest in mathematics (5%); mathematics became visible and tangible (21%); teleVISE was helpful in preparation for the examination (25%); acquisition of a routine in working on mathematical problems (33%); better understanding of approaches to mathematical problems (41%); refreshing of foundations (41%); regular exercises (46%); and better understanding of the lecture (46%).

When asked about teleVISE in group interviews, tutors made statements such as the following: "Yes, teams have formed because of the tutorial, ...[and] teams [have] prepared the tutorial collaboratively"; "A strong team spirit emerged between them, where one of them always knew what happened if the other one was missing or had difficulties with a problem"; "If they didn't understand the problems, they explained them to each other. And they presented the ideas they worked out together"; "I believe the benefit has been the development of collective learning above all"; and "They may ask 'stupid' questions, work together and assist one another."

An oft-heard explanation of the problems of mathematics education is the lack of student engagement, initiative, and motivation. However, eight of twelve tutors stated that all of their students displayed motivation and a willingness to perform. The other four tutors considered this to be true for at least half to a third of their student group.

According to the analysis of the thinking-aloud protocols, the feeling of insecurity forms an eminent problem for students dealing with mathematical problems. When different options are of equal value, the students succumb to their own uncertainty and hence fail because of their inability to deal productively with doubts. The trail of different ideas gives rise to growing confusion and results in a mental blockade. If, in this situation, an apparent expert comes to the students' rescue, many students show a knee-jerk reaction and transfer their own responsibility to the expert. However, problems without self-evident solutions belong to the everyday life of any professional computer scientist. Mastery in attacking complex problems is indispensable. In this context it is helpful that the tutors are peers, not experts. One of the tutors comments, "The most concentrated and engaged discussions emerge if a tutor gets stuck working on a problem. Some students develop ambition to get ahead of the tutor in solving the problem." The tutor is competent, but not omniscient as teachers seem to be.

teleVISE offers electronic communication between students and tutors in the course of problem-solving. Such "callbacks" indicate interaction during online support. A tutor commented, "After his or her first submission, the difficulties of the individual student became visible. Only then could you be responsive to them and to support them well, directing them with questions through their coping with difficulties. I believe this call-back system to be a great achievement compared with the submission of solutions, which come back simply corrected." This feature of teleVISE has not been designed. It emerged in the use of teleVISE and exploits the specific potential of mobile devices. The number of callbacks increased from summer term 2003 to winter term 2003/04 from 230 to 562. This figure still seems low; however, it represents a doubling.

Meanwhile, one tutorial group of about 15 students is willing to pay for their tutor if the university cannot afford this teaching style any longer. This statement is more than astonishing given that university education is largely provided free of charge in Germany and that many students are strongly opposed to tuition fees.

Assesing the deployment of teleVISE in the lecture, students' responses were as follows: "not good" (7%), "confusing" (7%), "helpful and good in general" or similar statements (27%), and "helpful to solve mathematical problems" (40%).

Finally, we compared the grades achieved in the study course Digital Media (summer 2003) with the grades of the study course Digital Media in the preceding year (Figure 3). Note that in the German system, a "1" denotes an excellent achievement, comparable to an "A" mark, while a grade of "5" is equivalent to a "Fail." The failure rate decreased from 2002 to 2003 from 55% to 25%. The percentage of students with a better grade than 3 increased from 2002 to 2003 from 11% to 50%. However, we attribute these findings more to the

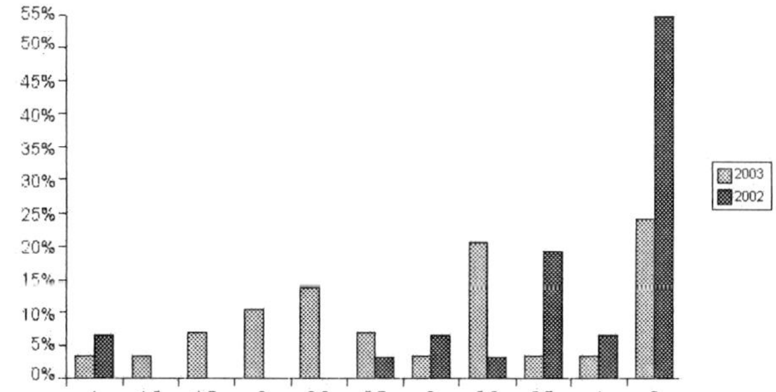


FIGURE 3

DIGITAL MEDIA GROUP: GRADES OF THE STUDENT GENERATIONS 2002 AND 2003 students than to teleVISE. The same students showed similar performances not only in mathematics, but also in the parallel introductory science course. This may be ascribed to a change in inscription policy: Every applicant received a personal letter that explained the mathematical challenges of the study program. Before that, this information only appeared in the department's brochures and on the web. It would have been interesting to observe the effect that teleVISE had in the other study program, Computer Engineering, where it has also been applied. However, the conditions changed so severely that a comparison does not make sense. Due to external factors half of the group switched to a different parallel lecture during the summer term.

In conclusion, teleVISE seems to be on the right track, but problems still remain. We detail these in the following section.

## LESSONS LEARNED

The main lessons we learned are as follows: (1) The surplus value of a (mobile) learning application like teleVISE is a dynamic phenomenon; and (2) The design strategies concerning the system, the service, the embedding into user activities, and finally the sustainability have to account for this.

Though we started the project in autumn 2002 with a focus on online collaboration, we realized in February 2003 that the students explicitly preferred paper solutions and would not accept the teleVISE system. Furthermore, we realized that we would have difficulty guaranteeing the sustainability of the online service because of the costs of the tutorial system.

Both reasons prompted us to redirect the focus of the project. From summer 2003 until February 2004 we emphasized the use of teleVISE within the lectures instead of using it exclusively for online collaboration. Of course we also realized a number of limitations there. Teaching and learning routines are always enabled and limited by the personal style of teachers and students. But, nevertheless, we succeeded in introducing teleVISE as a medium to demonstrate mathematical modeling within lectures.

Though at first unnoticed by us, another feature of teleVISE emerged during the second generation of students' use of teleVISE in the winter 2003/04 term: the online call back. The call back allows for a tutor response to the individual problem-solving process, a feature which is genuine for this medium and cannot be reproduced by a paper-based solution. Additionally, the students of the second generation accepted the ideas underlying teleVISE and the teleVISE system itself as a matter of course.

Nevertheless, the problems to finance the tutorial support have not yet been solved. We are looking for solutions within the economic scheme of the university. The conditions to find a solution are better now than they were a year ago since the students accept the system. The faculty dean and the university management are positively attuned to teleVISE. Together we hope to find a lasting solution.

## ACKNOWLEDGEMENT

teleVISE has been funded in the framework of the "Zukunftsinvestitionsprogramm" (Future Investment Program) of the German Federal Government.

# REFERENCES


1. Mavrikis, M.P., and Maccioccia, A., "Developing Wallis: A Web-based System to Enhance Mathematics Teaching," Proceedings of the 2nd International Conference on the Teaching of Mathematics, July 1-6, 2002, Paper No. 358, Hersonissos, Crete, Greece, 2002.

2. Unger, L., Oevel, G., and Mertsching, B., "Web-based Teaching and Learning with math-kit," Proceedings of the 2nd International Conference on the Teaching of Mathematics, July 1-6, 2002, Paper No. 438, Hersonissos, Crete, Greece, 2002.

3. Gage, M.E., Pizer, A., and Roth, V., "Generating, Delivering, and Checking Math Homework via the Internet," Proceedings of the 2nd International Conference on the Teaching of Mathematics, July 1-6, 2002, Paper No. 189, Hersonissos, Crete, Greece, 2002.

4. Moundridou, M., "Analysis and Design of a Web-based Authoring Tool Generating Intelligent Tutoring Systems," Computers & Education 40, 2003, pp. 157–181.

5. Kopec, D., Whitlock, P., and Kogen, M., "Enhancing Post-Secondary Instruction with Peer Tutors, SmartBooks(TM) and SmartTutors," Chapter 10, Aung, W., et al. (eds.), *Innovations 2003. World Innovations in Engineering Education and Research,* iNEER, Arlington, VA, 2003, pp. 85–94.

6. Falkenburg, D., and Schuch-Miller, D., "Strategies for the Development of Web-Based Engineering Case Studies," Proceedings of 2003 International Conference on Engineering Education (ICEE-2003), Paper No. 4478, Valencia, Spain, 2003.

7. TIMSS International Study Center, TIMSS Mathematics Items: Released Item Set for the Final Year of Secondary School, Boston College, 1997.

8. Karsenty, R., "What do Adults Remember from their High-school Mathematics? The Case of Linear Functions," Educational Studies in Mathematics, Vol. 51, 2002, pp. 117–144.

9. OECD, Measuring Student Knowledge and Skills: The 2000 PISA Assessment of Reading, Mathematical and Scientific Literacy, DEELSA/PISA/BPC (2000), 10, 2000.

10. Lakoff, G., and Núñez, R., *Where Mathematics Comes From,* Basic Books, New York, 2000.

11. Fodor, J., "Fixation of Belief and Concept Acquisition," in Piattelli-Palmarini, M., (ed.), Language and Learning. The Debate between Jean Piaget and Noam Chomsky, Routledge & Kegan, London 1980, pp. 143-149.

12. Pólya, G., *How to Solve It,* Princeton University Press, 1945, reissued 1971.

13. Kramarsky, B., "The Effect of Metacognitive Instruction on Solving Mathematical Authentic Tasks," Educational Studies in Mathematics, Vol. 49, 2002, pp. 225–250.


**Barbara Grüter** received her Diploma degree and her doctorate degree in Psychology from the Free University Berlin in 1975 and 1989, respectively. Since 1975 she has been working as a consultant, initially in the welfare and education area and later on in the software- and media-industries. From 1985 to 2000 she worked in the first instance as a scientific assistant, then as an assistant professor in the Department for Educational Science and Psychology at the Free University Berlin. From 1996 to 1997 she worked as a Visiting Scholar in the Department of Psychology at the University of North Carolina at Chapel Hill,

NC, USA. Since 2001 she has been Professor of Media-management. She teaches in this field and in Usability-engineering. Her research interests include human-computer-interaction, media-psychology and developmental problems in both the basic and the applied areas.

**Jörn Loviscach** received his Diploma degree and his doctorate degree in Physics from Bielefeld University in 1990 and 1993, respectively, with a specialization towards Mathematical Physics. Subsequently, he worked as an editor and senior editor at various German computer magazines. For three years he held the position of Deputy Editor-in-Chief at "c't Magazin für Computertechnik," Hannover. Since 2000 he has been Professor of Computer Graphics, Animation, and Simulation at Hochschule Bremen, University of Applied Sciences. He teaches in this field and in Mathematics; his major research interests are hardware-accelerated algorithms for graphics.

**Thomas Risse** received his Diploma degree in Mathematics in 1977 and his doctorate degree in Physics in 1982 from the University of Tübingen. From 1985 to 1986 he worked as 'World Trade Visiting Scientist' at IBMs T.J. Watson Research Center in Yorktown Heights, NY, USA. Since 1994 he has been Professor of Computer Science at Hochschule Bremen, University of Applied Sciences. His research interests include computer architecture and implementation of algorithms in image and signal processing.

**Heide-Rose Vatterrott** received her Diploma degree in Mathematics in 1978 and her doctorate degree in Engineering (Computer Science) in 1986 from the University of Rostock. Since 2002 she has been Professor of Applied Computer Science at Hochschule Bremen, University of Applied Sciences. She teaches programming, including the development of distributed web applications, and human-computer-interaction. Her research interests include mobile computing as well as usability engineering and scientific visualization.

**Ulrike Wilkens** received her doctorate degree in Engineering (Computer Science) from the University of Bremen in 2000. Since 1996, she has worked as a scientist in the Department of Computer Graphics and Interactive Systems. Her research interests concern the relationships among computer science, education and the arts. Since 2001 she has been head of the Multimedia Competence Center at the University of Applied Sciences Bremen.

# INDEX

## F



## G



## H



## I



## J



## K

## L



## M



## N



## O



## P

## R



## S



## T



## U



## V



## W